Lecture Notes in Computational Science and Engineering

140

Editors:

Timothy J. Barth
Michael Griebel
David E. Keyes
Risto M. Nieminen
Dirk Roose
Tamar Schlick

More information about this series at http://www.springer.com/series/3527

Oliver Sander

DUNE — The Distributed and Unified Numerics Environment

 Springer

Oliver Sander (iD)
Institut für Numerische Mathematik
TU Dresden
Dresden, Germany

Additional material to this book can be downloaded from http://extras.springer.com.

ISSN 1439-7358   ISSN 2197-7100 (electronic)
Lecture Notes in Computational Science and Engineering
ISBN 978-3-030-59704-7   ISBN 978-3-030-59702-3 (eBook)
https://doi.org/10.1007/978-3-030-59702-3

Mathematics Subject Classification (2020): 65-04, 65N30, 65N08, 65Y99

This Springer imprint is published by the registered company Springer Nature Switzerland AG
The registered company address is: Gewerbestrasse 11, 6330 Cham, Switzerland

Preface

The idea that a book about DUNE would be a good thing to have has been on my mind for a long time. Many people have expressed interest in DUNE, right from the start. Many have started to use it, but it was a steep learning curve for a lot of them, because DUNE is a complex piece of software, and not every part of it is well documented.

There were two things that kept me from considering to write the book myself. First, it was obvious that it would be a large book, and I was intimidated by the amount of work that this would mean. Secondly, DUNE is the joint work of a large group of people, and I did not know how to properly and sufficiently give them credit.

For a while, I thought that I had a plan that would elegantly solve both worries at once. When that plan failed, I had already produced too much of a book to simply throw it away again. Therefore I simply kept going, and the result is what you hold in hands. I hope it will be useful to you.

Many people helped me write this book. Some proof-read parts of it and suggested improvements, others explained technical details to me that I hadn't known before. Several people donated text fragments and images that existed previously, and some fixed bugs in DUNE at my request. For all this I would like to thank Peter Bastian, Markus Blatt, Ansgar Burchardt, Andreas Dedner, Martin Drohmann, Christian Engwer, Jö Fahlke, Bernd Flemisch, Sebastian Götschel, Carsten Gräser, Dominic Kempf, Robert Klöfkorn, Sven Marnach, Dmitry Mazilkin, Steffen Müthing, Martin Nolte, Benedikt Oswald, Simon Praetorius, Susanne Schmidt, Justus Schock, Michaël Sghaïer, Henrik Stolzmann, Edscott Wilson Garcia, and Jonathan Youctt. Special thanks go to Christoph Grüninger for reading the entire manuscript twice. Finally, I would also like to thank Jobst Hoffmann for his help with the LaTeX listings package, and Leonie Kunz and Martin Peters from Springer Verlag for their support.

Dresden,   *Oliver Sander*
July 2020

Contents

Contents

1. Introduction

This is a book about the DUNE software system. DUNE, the *Distributed and Unified Numerics Environment*, is a collection of C++ libraries for different aspects of finite element (FE) and finite volume (FV) methods. It is distributed under an open source license,[1] which allows to obtain and see the source code, and modify it to suit one's needs. DUNE libraries are called *modules*. They form an interdependent network, but must be downloaded and installed separately. Figure 1.1 shows the set of modules covered in this book, although there are many more.

The DUNE project was started in 2003 by Peter Bastian as an effort to implement the lessons he had learned when co-developing the successful UG3 simulation code [13]. Early on, other people started to contribute, and the development model quickly became as *distributed* as the project name implies. Today, DUNE is used by many people mainly in academic research, and is developed in the classical bazaar-style free software fashion. The main point of contact is the project website www.dune-project.org, which is where project development is coordinated. The site provides the usual features, such as software downloads, a GITLAB instance that keeps the source code and bug trackers, mailing lists, and a detailed class documentation. The website is where users and developers alike should go to get further information.

1.1. The Case for Standardization

DUNE was written with the idea in mind that standardization of certain parts of finite element and finite volume codes should be possible.

The last decades have seen a proliferation of computer codes for finite element and finite volume methods. At the time of writing, WIKIPEDIA lists roughly 45 codes, but many more are mentioned in the page history.[2] The list of programs ranges from small research codes developed by a small number of people in a particular research unit, to comprehensive powerful software developed by large teams with a large number of third-party users. Many of these codes were developed in industry for commercial purposes, but there are also a few projects from academic teams such as DEAL.II[3] [8], FEniCS[4] [120], and FEAP[5] [152].

---

[1] GPLv2, with "runtime exception", see www.dune-project.org/about/license

[2] https://en.wikipedia.org/wiki/List_of_finite_element_software_packages, last checked on February 9., 2020

[3] www.dealii.org

[4] fenicsproject.org

[5] www.ce.berkeley.edu/projects/feap

© Springer Nature Switzerland AG 2020

O. Sander, *DUNE — The Distributed and Unified Numerics Environment*, Lecture Notes in Computational Science and Engineering 140, https://doi.org/10.1007/978-3-030-59702-3_1

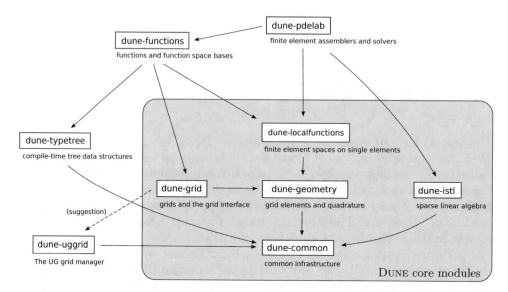

Figure 1.1.: DUNE is organized as a collection of modules, and this figure shows the modules that are discussed or used in this book. Arrows denote module dependencies. The modules in the gray box are called the *core modules*, and they form DUNE in the strict sense of the word.

Since all these programs implement similar algorithms, they all need similar infrastructure. A finite element code always needs grids, shape functions, and linear algebra. Most codes also need discrete function spaces, and multi-process codes will need ways to communicate grid-based data from one process to another. Many of these topics are "done" in the sense that there are few people actively researching these topics, compared to the people that simply use them. There is not much research in grid data structures anymore. People keep inventing new finite elements, but it is pretty clear what a finite element space is. Therefore, users should be able to use discrete functions for their research without having to think about the implementation details. This does not imply that these topics are necessarily easy. Features like adaptive red–green refinement on a distributed grid are very difficult to implement correctly, and therefore only a few codes actually support them.

Even though the requirements are very similar, most codes still provide their own implementations of many of these features. While some codes do use dedicated external libraries for linear algebra, virtually all codes implement their own grid data structures, shape functions, and discrete function spaces. The disadvantage of this approach is two-fold. On one hand, it binds resources. Precious developer time is needed to implement, debug, and maintain these data structures. On the other hand, chances are that the result will not be as good as possible. Features may be missing, or the implementation may be of inferior quality, because the main developer interest is elsewhere.

One of the central ideas of DUNE is that standardization can help here. First of all, standardization is possible, because, as explained above, various aspects of finite element codes are in principle "done" and agreed upon. Standardization saves development time, because less time is spent reinventing wheels that have already been invented in other codes. At the same time, the standardized components will be more powerful and of better quality, because they are used by more people and can therefore be expected to receive more developer attention: one really good component replaces several competing not-quite-as good ones.

Besides these technical advantages, standardization also facilitates knowledge transfer, because standardized components form common knowledge that is shared by a larger group of people. That means that conversations between programmers of different simulation codes become easier. Also, people can move easier from developing one particular simulation code to another, because they have less new things to learn in the new job. Of course, sharing code between different simulation projects becomes easier as well.

A good example for the benefits of low-level standardization is the MPI standard for message passing on distributed computers. [6] Before the advent of MPI, many different implementations provided low-level communication and message passing. Many of these implementations were supplied by the hardware vendors, they were closed-source, and optimized for a particular hardware. Codes that were expected to run on a range of different hardware platforms had to cope with these implementations, different programming interfaces and idiosyncrasies. For example, until 2014, the above-mentioned UG3 code contained eight different backends for different communication layers. This changed when people realized that all these communication libraries implemented similar things, and that the problem space was well-enough understood to hide all necessary functionality behind one common interface. The result was the MPI standard, which slowly replaced all competing approaches. And while there may be things to dislike about MPI, few people will dispute the claim that its introduction has made distributed scientific computing simpler and more productive. The same is the goal of DUNE.

1.2. Goal of the Book

This book was written to teach people how to use DUNE for their own simulation codes. It tries to target two groups of people: First of all, there are those who have not used DUNE yet, and want to give it a try. For those, parts of the book are written in the style of a tutorial. The introductory Chapter 3 shows how to solve first finite element and finite volume problems, and throughout the book there are chapters that explain complete examples that demonstrate particular aspects of DUNE. These chapters all have titles that start with "Example:". People that are new to DUNE, or who want to use it to solve a given PDE problem quickly, should start by reading Chapter 3, look at the example chapters of interest, and read Chapters 10 and 11 on

---

[6]www.mpi-forum.org

the `dune-functions` and `dune-pdelab` modules. They should then be able to decide whether DUNE is the right tool for their needs.

On the other hand, the book also targets readers who already have some familiarity with DUNE, and who want to learn more. For those, the book contains chapters roughly dedicated to individual modules. These chapters provide background information on the concepts used in these modules, and try to present the interfaces somewhat systematically. In those parts, the book has a bit more of a "reference" flavor. Be warned, however, that "systematically" does not mean "complete". There is more to DUNE than what is covered in this book; for a more complete view consult the online documentation at the project website www.dune-project.org. Some recent developments in the DUNE ecosystem are described in [17].[7]

DUNE is intended to serve as the infrastructure for numerical simulations of the highest complexity and the largest scale. However, in this book there are only fairly simple examples: the Poisson equation, the Stokes problem, the p-Laplace and others, and all of them on two-dimensional domains only. The reason is that the book should inform rather than impress. Each example shows a particular aspect of DUNE, and unnecessarily complex example programs would just distract from the main points. To see impressive applications of the DUNE software we recommend to search the research literature.

Somewhat related, attentive readers will notice that high performance computing (HPC) is never mentioned, even though it is one of the deliberate aims of DUNE development. Chapter 6 describes how to work with distributed grids, but no other HPC technique is mentioned. Other aspects of HPC have also received great attention in DUNE recently [18, 19], but the results are not mature enough yet to describe them in a book.

1.3. Structure of the Book

The book is divided into three parts. The first comprises Chapters 2 to 4, and discusses introductory material. Chapter 2 reviews the finite element and finite volume methods, which are the motivation for the design of DUNE. The exposition is brief and superficial, and anyone interested in these methods should only take the chapter as a starting point, before diving into dedicated textbooks and research literature. Still, it gives a few details related to implementations of these methods not typically found in textbooks.

Chapter 3 is a tutorial on how to get started with DUNE. This is where to start reading for people that have never used DUNE before. It briefly shows how to install DUNE and create a DUNE module. Then, it walks through the complete implementations of a finite element code for the Poisson equation, and a finite volume method for a transport equation.[8]

---

[7]One interesting example is the comprehensive set of Python bindings that has recently been added to DUNE [50]. They are omitted in this book for lack of space.

[8]On first reading, that code may seem overly verbose. It is an advantage of DUNE, however, to

Chapter 4 explains the overall structure of DUNE, and some of the underlying design ideas. It does not contain much information directly helpful for day-to-day coding, and can be safely skipped on first reading. Come back to it later at any time to get a deeper understanding of the DUNE system.

The following part of the book, which is made up of Chapters 5–9, discusses individual DUNE modules. Just as DUNE modules are only loosely coupled among themselves, most of these chapters can be read independently from each other. Chapters 5 and 6 cover the DUNE grid interface. This is the historical and conceptual center of DUNE. The chapters explain what a DUNE grid is, what the programming interface looks like, and how to use it. It also presents some of the grid implementations currently available. The code discussed in these chapters is contained in the DUNE modules dune-grid and dune-geometry. Chapter 5 covers the grid interface for single-process situations. Chapter 6 shows how these ideas are extended for grids distributed across more than one process.

The next chapter presents the DUNE linear algebra implementation, contained in the module dune-istl.[9] While there are several good linear algebra implementations available, dune-istl has a few interesting features particularly geared towards finite element and finite volume computations. For example, it allows to nest arbitrary vector and matrix types to give the compiler as much information about the blocking and sparsity structure of sparse linear algebra problems as possible. It also contains the implementation of a robust, highly scalable algebraic multigrid preconditioner.

Chapter 8 describes the dune-localfunctions module, which collects finite element bases on single elements. The text describes the interfaces that dune-localfunctions offers for such finite elements, and explains a few of the elements currently available. The dune-localfunctions module does not depend on the dune-grid module, so if desired its finite elements can be used with non-DUNE grids.

Chapter 9 covers the subject of numerical quadrature. In DUNE, there is no separate module for this; rather, the quadrature rules are provided in the dune-geometry module. The chapter reviews a few aspects of quadrature rule construction, and explains how the different rules are provided by DUNE.

The previous modules, which are sometimes called the DUNE *core modules*, provide mainly low-level functionality. The last two chapters, which make up the third part of the book, provide functionality on a higher level of abstraction.

Chapter 10 describes dune-functions, a module that provides unified access to functions and discrete function spaces. The functions interface follows the approach taken by the C++ standard library with the std::function class, but extends it to differentiable functions and functions defined on finite element grids. Spaces of finite element functions are represented by specifying bases of such spaces. The dune-functions module provides various standard bases, and a powerful mechanism to combine simpler bases into more complicated ones.

---

allow to program at various levels of abstraction. The codes in Chapter 3 strike a middle ground. Examples of how to solve similar problems with much less code (and less control over the details) are given in Chapter 11.

[9]The acronym "istl" is an abbreviation of "Iterative Solver Template Library".

None of the previous code, not even `dune-functions`, contains assemblers for partial differential equations. One implementation of such assemblers is presented in Chapter 11 on `dune-pdelab`.[10] The `dune-pdelab` module builds on top of `dune-functions`, and provides assemblers and solvers for many standard partial differential equations. These work on sequential and distributed machines, and can be combined with local grid adaptivity. Solving PDEs with `dune-pdelab` only requires a few lines of C++ code, if one of the existing local assemblers can be used. Writing new assemblers is also discussed in the chapter.

The book ends with a two-part appendix. Appendix A explains the DUNE build system. While each DUNE module itself uses a standard `cmake` build system, the outer logic that ties together the different modules is nonstandard and warrants some explanation. This information becomes important once the DUNE user needs to go beyond writing simple programs, e.g., when depending on locally installed third-party software, or when writing in-code documentation or unit tests.

Appendix B, finally, lists all complete example program source codes that are discussed in the book. C++ code can be fairly long, and complete source codes are usually not shown in the chapter where they are discussed. Still, in particular for beginners, it is important to be able to see the complete code with all the details, and therefore they are all listed in Appendix B.

1.4. Source Code in this Book

This book contains a lot of source code; pretty much all of it in C++. On one hand, short code snippets are given to demonstrate individual features. On the other hand, complete example applications are also discussed. All code has been tested to work with the DUNE release 2.7; if it does not then there is a misprint in the book, and we would like to hear about it. Most code is also likely to run with newer and slightly older versions of DUNE, possibly after a small bit of tweaking. The chapters that start with "Example:" discuss complete example applications. All these applications are constructed such that the entire code is contained in a single file, and hence only that single file and the DUNE libraries from Figure 1.1 are needed to build and run the example. The easiest way to build a particular example is to set up a new

---

[10]Here, an interesting effect occurs: The more we move up towards higher levels of abstraction, the more difficult it is to reach a consensus about how reusable software components should look like. While there is basic agreement about the design of the DUNE core modules (arguments about details notwithstanding), and `dune-functions` is gaining a lot of support, there is no such agreement on the level of PDE assembler modules. As a result (besides the number of finite element software systems that use the DUNE core modules and implement their own assemblers) in the DUNE ecosystem there are several modules roughly providing the same functionality, but following different design ideas. The big two are `dune-fem` (`www.dune-project.org/modules/dune-fem`) [48] and `dune-pdelab` (`www.dune-project.org/modules/dune-pdelab`) [20], after which comes `dune-fufem`, (`www.dune-project.org/modules/dune-fufem`) trailing by quite a bit. Covering only `dune-pdelab` in this book does not imply any objective judgment of the relative merits of the three modules. Discussing all of them would make the book even longer than it already is, and the author is simply more familiar with `dune-pdelab` than with `dune-fem`.

DUNE module as explained in Chapter 3, to copy the source file into the module's src directory, and to add it to the CMakeLists.txt file.

The chapters that discuss complete example files do not show the complete source. C++ files are very wordy, and printing the entire file as part of its discussion would make the text rather difficult to read. Nevertheless, it is also necessary to be able to look at the complete file. There are two ways to get the complete source files. First, all of them are printed in Appendix B. Also, they are embedded in electronic versions of this book. Right-clicking on the pin icons in the margin provides the complete source codes for download.

The following points should be kept in mind when reading source code in this book.

- All code in the DUNE modules treated in this book is contained in the namespace Dune. However, we will usually omit the namespace prefix, in order to save space. The example programs work nevertheless, because they all have the line

```
using namespace Dune;
```

somewhere near the top of the file.

- The code in the modules dune-functions and dune-pdelab is additionally contained in the nested namespaces Dune::Functions and Dune::PDELab, respectively. While we omit the prefix for the outer namespace Dune, we do write the inner one, to give people a hint on where a particular object is coming from. For example, to declare a first-order Lagrangian finite element basis from the dune-functions module we write

```
Functions::LagrangeBasis<GridView,1> firstOrderLagrangeBasis(gridView);
```

but this only compiles if

```
using namespace Dune;
```

has been set previously.

- Similarly, we will omit **typename** and **template** specifiers that are sometimes needed to disambiguate between nested values and nested types. That is, we will usually write

```
template<class GridView>
struct Foo
{
  using Element = GridView::Codim<0>::Entity;
  [...]
}
```

instead of

```
template<class GridView>
struct Foo
{
  using Element = typename GridView::template Codim<0>::Entity;
  [...]
}
```

Only the second version is valid C++, but the first one is easier to read, and takes up less space.

- When showing class interfaces we usually only show the relevant parts. Some classes will have additional methods and export a few additional types. Consult the class documentation at www.dune-project.org for all the details. If the DOXYGEN program[11] is installed, it is also possible to create a module's class documentation by calling make doc, and pointing a browser at build-cmake/doc/doxygen/html/index.html.

- All example programs hard-wire the problem parameters within the C++ code. This is of course very inconvenient, and only done to keep the examples as simple as possible. The dune-common module contains code to maintain trees of run-time parameters and to read them from text files, but any other way to get parameters into C++ code works just as well.

Finally, keep in mind that DUNE is a software under active development. To be able to innovate rapidly and keep the code simple, there are only a few promises regarding stability of the API (see Chapter 4.6 for what the DUNE team promises). This implies that some information in this book may become outdated over time. Again, for up-to-date information on the interface consult the automatically created class documentation at www.dune-project.org.

---

[11]www.doxygen.org

Part I.

Preliminaries

2. Mathematical Concepts

DUNE is designed to form the foundation for implementations of finite element and finite volume methods. To give prospective users a quick idea these two methods are briefly revisited here. We stick to the basics, but try to point out some of the features that have influenced the design of DUNE. Readers with a deeper interest in the methods should consult one of the numerous books on the subject.

2.1. The Finite Element Method

The finite element method was introduced in the 1950s in the engineering community to solve solid mechanics problems. The text book [44] contains a historical overview. While the method is used nowadays for all types of partial differential equations (PDEs), it is most easily explained for elliptic ones.

Let $\Omega \subset \mathbb{R}^d$ be a domain, i.e., an open bounded set, and $f : \Omega \to \mathbb{R}$ a given function of sufficient smoothness. Further, let $\mathcal{A} : \Omega \times \mathbb{R} \times \mathbb{R}^d \to \mathbb{R}^d$ be a first-order elliptic differential operator, and $R : \Omega \times \mathbb{R} \to \mathbb{R}$ another given function. We call R the reaction term, and we look for a function $u : \Omega \to \mathbb{R}$ solving the equation

$$- \operatorname{div} \mathcal{A}(x, u, \nabla u) + R(x, u) = f \qquad \text{in } \Omega. \tag{2.1}$$

In order not to have this exposition become too technical, we assume directly that the operators $\mathcal{A}$ and R are independent of x. Additionally, we take them to be linear, that is, there is a vector $\mathbf{v} \in \mathbb{R}^d$, a matrix $C \in \mathbb{R}^{d \times d}$, and a number $r \in \mathbb{R}$ such that

$$\mathcal{A}(x, u, \nabla u) = C\nabla u + \mathbf{v}u$$
$$R(x, u) = ru.$$

Then, the model equation is

$$- \operatorname{div}(C\nabla u + \mathbf{v}u) + ru = f \qquad \text{in } \Omega. \tag{2.2}$$

The equation is elliptic if the matrix C has certain properties, which are not of importance here. The general nonlinear case is discussed in Chapter 11.2.

Equation (2.2) comprises many relevant partial differential equations. One is the *Poisson equation*

$$-\Delta u := - \operatorname{div} \nabla u = f,$$

where C is the identity, and $\mathbf{v}$ and r are zero. Others are the linear *reaction–diffusion equation*

$$-\Delta u + ru = f,$$

© Springer Nature Switzerland AG 2020

O. Sander, *DUNE — The Distributed and Unified Numerics Environment*, Lecture Notes in Computational Science and Engineering 140, https://doi.org/10.1007/978-3-030-59702-3_2

2. *Mathematical Concepts*

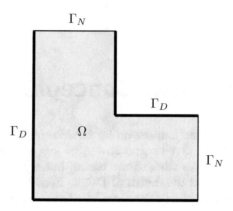

Figure 2.1.: A simple domain Ω with Dirichlet boundary Γ_D (thick lines) and Neumann boundary Γ_N

where r is nonzero, and the *convection–diffusion equation*

$$-\Delta u + \mathbf{v}^T \nabla u = f,$$

where $\mathcal{A}(x, u, \nabla u) = \nabla u + \mathbf{v}u$. These will appear in later chapters of this book. The equations of linear elasticity are a direct generalization to vector-valued functions [37].

To obtain a well-posed problem, boundary conditions have to be applied. We assume that the boundary $\partial\Omega$ of Ω is composed of two disjoint parts Γ_D and Γ_N (Figure 2.1), and we use $\mathbf{n}$ to denote the unit outer normal vector to $\partial\Omega$. On each part of the boundary we describe a different type of conditions. Classical examples are Dirichlet boundary conditions

$$u = \mathbf{g} \qquad \text{on } \Gamma_D,$$

and Neumann boundary conditions

$$\langle (C\nabla u + \mathbf{v}u), \mathbf{n} \rangle = \mathbf{j} \qquad \text{on } \Gamma_N.$$

The combination of a partial differential equation with a set of boundary conditions is called a *boundary value problem*. Depending on the precise type of equation, different types and combinations of boundary conditions are allowed for a well-posed boundary value problem. For a general introduction to well-posedness of partial differential equations see, e.g., [64].

2.1.1. Weak Formulation

For definiteness we consider now the elliptic boundary value problem

$$-\operatorname{div}(C\nabla u + \mathbf{v}u) + ru = f \qquad \text{in } \Omega \tag{2.3}$$

$$u = \mathbf{g} \qquad \text{on } \Gamma_D, \tag{2.4}$$

$$\langle (C\nabla u + \mathbf{v}u), \mathbf{n} \rangle = \mathbf{j} \qquad \text{on } \Gamma_N. \tag{2.5}$$

This way of writing the model problem is called the *strong formulation*. It assumes implicitly that its solution u is twice differentiable in the classical sense. However, in many situations such a solution does not exist. It is therefore more natural to write the problem differently, in what is called the *weak formulation*. While this formulation is somewhat more technical, it makes less assumptions about the solution function u.

We expect the reader to be familiar with the basic notions of Sobolev space theory [37, 162]. Denote by $H^1(\Omega)$ the space of all scalar L^2 functions on Ω with weak derivative in $L^2(\Omega)$. As the space for solutions we introduce

$$H^1_D(\Omega) := \left\{ v \in H^1(\Omega) \ : \ v|_{\Gamma_D} = \mathbf{g} \right\},$$

where $v|_{\Gamma_D} = \mathbf{g}$ is to be understood in the sense of traces. As a space of test functions we introduce

$$H^1_{\Gamma,0}(\Omega) := \left\{ v \in H^1(\Omega) \ : \ v|_{\Gamma_D} = 0 \right\}.$$

Multiplying (2.3) with a test function v from $H^1_{\Gamma,0}(\Omega)$, integrating over Ω and using Green's identity we obtain

$$\int_\Omega \langle (C\nabla u + \mathbf{v}u), \nabla v \rangle \, dx + \int_\Omega ruv \, dx = \int_{\partial\Omega} \langle (C\nabla u + \mathbf{v}u), \mathbf{n} \rangle v \, ds + \int_\Omega fv \, dx.$$

By construction of the test function we have $v = 0$ on Γ_D. Hence the boundary integral reduces to an integral over Γ_N. There, we can insert the Neumann boundary condition (2.5) and obtain

$$\int_\Omega \langle (C\nabla u + \mathbf{v}u), \nabla v \rangle \, dx + \int_\Omega ruv \, dx = \int_{\Gamma_N} \mathbf{j}v \, ds + \int_\Omega fv \, dx,$$

which must hold for all $v \in H^1_{\Gamma,0}$. This is called the *weak formulation* (or *variational formulation*) of the boundary value problem [37, 42, 63]. Observe that it only assumes u to be *once* weakly differentiable. This is in contrast to the strong form (2.3), where u has to be twice classically differentiable for the equation to make sense.

The weak formulation is typically written in a shorter way. Given the bilinear form

$$a(u,v) := \int_\Omega \langle (C\nabla u + \mathbf{v}u), \nabla v \rangle + ruv \, dx$$

and the linear form

$$l(v) := \int_{\Gamma_N} \mathbf{j}v \, ds + \int_\Omega fv \, dx,$$

find $u \in H_D^1(\Omega)$ such that

$$a(u, v) = l(v) \qquad \forall v \in H_{\Gamma,0}^1(\Omega). \tag{2.6}$$

This is simplified further by eliminating the affine space H_D^1: let u_g be any function in H_D^1, i.e., a function in $H^1(\Omega)$ that fulfills the Dirichlet boundary conditions. Then the weak formulation (2.6) is equivalent to finding $\tilde{u} \in H_{\Gamma,0}^1$ such that

$$a(\tilde{u}, v) = l(v) - a(u_g, v) \qquad \forall v \in H_{\Gamma,0}^1(\Omega).$$

Indeed, $u := \tilde{u} + u_g$ will solve the original problem. It is therefore sufficient to look for solutions in the vector space $H_{\Gamma,0}^1$. Writing u instead of $\tilde{u}$ and $l(v)$ instead of $l(v) - a(u_g, v)$, we therefore consider the following problem from now on: Find $u \in H_{\Gamma,0}^1$ such that

$$a(u, v) = l(v) \qquad \forall v \in H_{\Gamma,0}^1(\Omega). \tag{2.7}$$

This problem has a unique solution under certain assumptions on the bilinear form $a(\cdot, \cdot)$. See monographs like [37, 63, 64] for existence and well-posedness results.

2.1.2. Discretization by Finite Element Methods

The weak formulation forms the starting point of the finite element method. To make the problem tractable by a numerical algorithm we replace the space $H_{\Gamma,0}^1$ by a finite-dimensional subspace. This approximation is called *Galerkin discretization*.

Let V_h be a finite-dimensional subspace of H^1. The subscript h is traditionally used to denote objects related to such spaces. Once a grid for Ω is introduced, h is interpreted as a measure of the element size. The discrete variational formulation corresponding to (2.7) is to find a $u_h \in V_h \cap H_{\Gamma,0}^1$ such that

$$a(u_h, v_h) = l(v_h) \qquad \text{for all } v_h \in V_h \cap H_{\Gamma,0}^1, \tag{2.8}$$

with the forms $a(\cdot, \cdot)$ and $l(\cdot)$ as before.[1] We expect a solution u_h of (2.8) to be an approximation of the solution u of (2.7), with the approximation quality depending on the space V_h.

Let $\{\phi_i\}_{i=0}^{n-1}$ be a basis of $V_h \cap H_{\Gamma,0}^1$. Then (2.8) is equivalent to finding a $u_h \in V_h \cap H_{\Gamma,0}^1$ with

$$a(u_h, \phi_i) = l(\phi_i) \qquad \text{for all } 0 \leq i < n.$$

This in turn is equivalent to the linear system of equations

$$A\bar{u} = b, \tag{2.9}$$

---

[1] There also exists an important theory of *non-conforming* finite element methods, where the approximation space is not a subspace of the solution space (here: H^1) [37]. We disregard this possibility for the sake of simplicity.

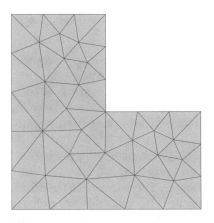 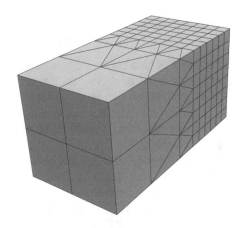

Figure 2.2.: A two-dimensional simplex grid, and a three-dimensional grid with several element types

where

$$A \in \mathbb{R}^{n \times n}, \qquad A_{ij} := a(\phi_i, \phi_j) = \int_\Omega \langle (C\nabla\phi_i + \mathbf{v}\phi_i), \nabla\phi_j \rangle \, dx + \int_\Omega r\phi_i\phi_j \, dx,$$

$$b \in \mathbb{R}^n, \qquad b_i := l(\phi_i) = \int_{\Gamma_N} \mathrm{j}\phi_i \, ds + \int_\Omega f\phi_i \, dx,$$

and $\bar{u} \in \mathbb{R}^n$ are the coefficients of u_h with respect to the basis $\{\phi_i\}$. The matrix A is called *stiffness matrix*. This expressions comes from the early uses of the finite element method for problems in solid mechanics. In such a setting, A describes the stiffness of the simulated objects, and the vector b describes the external loads.

Without further assumptions the matrix A is dense, i.e., practically every entry of A is nonzero. Working with such matrices is very expensive, because time and space requirements grow quadratically with the dimension of V_h. Finite elements solve this problem by using approximation spaces and corresponding bases that lead to sparse stiffness matrices. In other words, finite elements are a particular way to choose the subspace V_h.

For simplicity, suppose that the boundary of Ω is a (possibly non-convex) polytope. The finite element method constructs the approximation spaces V_h, by introducing a *grid*, i.e., a partition of Ω into simple convex subsets called *elements*.

Definition 2.1 *A set $\mathcal{T}$ of closed convex polytopes T is called a* conforming grid *of Ω, if the following holds:*

1. *(Partitioning) The union of the polytopes is the closure of the domain*

$$\overline{\Omega} = \bigcup_{T \in \mathcal{T}} T.$$

2. *(Conformity) The intersection of two polytopes from $\mathcal{T}$ is either a common face of lower dimension, or empty.*

For the rest of this book, a face of a d-dimensional polytope can have any dimension less than or equal to d. We use the word *facet* for a face of dimension $d - 1$.

In the simple-most case, all elements of a grid are simplices. An example of such a grid is shown in Figure 2.2. However, grids with quadrilateral or hexahedral elements are also very common, and three-dimensional grids may contain prisms and pyramids (see also Figure 2.2).

While the definition for a grid given here is typical for the ones given textbooks, it is actually too restrictive for many applications, and practitioners have used more general constructs for computational grids. For example, approaches like the Virtual Element Method [21] and Mimetic Finite Differences [119] use arbitrary polygons as elements (Figure 2.3(a)), which may even be non-convex. Also, the requirement that Ω should have a piecewise flat boundary is often not met in practice. Therefore, approximating such a domain with polytope elements leads to errors. Such errors can be bounded (see, e.g., [42]); nevertheless, grids have been proposed that use images of polytopes under polynomial functions as elements [67, 69, 166] (Figure 2.3(b)). Also, many PDEs of interest today are not posed on open sets in Euclidean spaces, but rather on lower-dimensional sets in such spaces. These lower-dimensional sets may have manifold structure [60], but they can also exhibit network topologies [38, 125] (Figure 2.3(c)).

Finally, various methods like the Discontinuous Galerkin (Chapter 11.2.6) or finite volume (Chapter 2.2) methods waive the conformity requirement of Definition 2.1. Then, intersections of two elements may be true subsets of common lower-dimensional faces [134] (Figures 2.3(d) and 2.3(e)). This can simplify the grid construction considerably, without making the numerical methods more complicated. Given a precise definition for a grid that also encompasses all these cases is difficult. One attempt can be found in [16].

On the other hand, various grids also have *more* structure than what is required by Definition 2.1. As an important example, in the grids in Figures 2.3(e) and 2.3(f), all element facets are perpendicular to one of the coordinate axes. This can be exploited by implementations for extra-efficient element normal and volume computations.

The finite element method now constructs discrete spaces V_h as spaces of functions that are piecewise polynomial with respect to the grid. The most simple ones are the Lagrange elements on simplex grids.

Definition 2.2 *Let the grid $\mathcal{T}$ consist of simplex elements only, and let Π_p be the set of all polynomials in d variables of order not higher than p. Then*

$$V_h^{(p)} := \{v \in C(\overline{\Omega}) \ : \ v|_T \in \Pi_p \quad \forall T \in \mathcal{T}\}$$

is called the space of p-th order Lagrange elements with respect to the grid $\mathcal{T}$.

This is a finite-dimensional vector space, and it is indeed a subspace of H^1 [37]. Figure 2.4 shows example functions from $V_h^{(1)}$ and $V_h^{(2)}$.

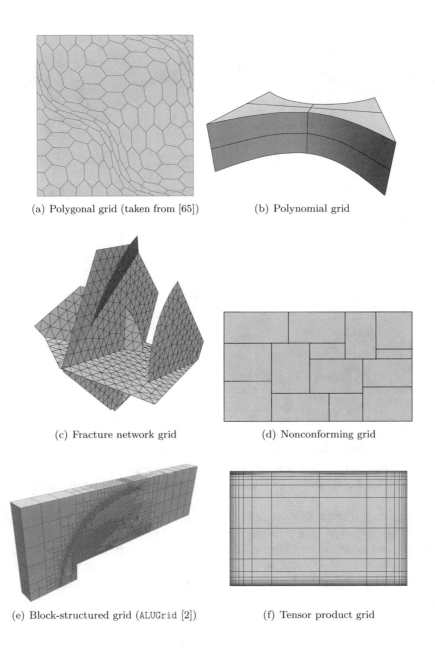

(a) Polygonal grid (taken from [65])

(b) Polynomial grid

(c) Fracture network grid

(d) Nonconforming grid

(e) Block-structured grid (ALUGrid [2])

(f) Tensor product grid

Figure 2.3.: Grids that are more general or more structured than what is captured by Definition 2.1 of a finite element grid

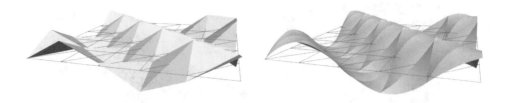

Figure 2.4.: First- and second-order Lagrangian finite element functions

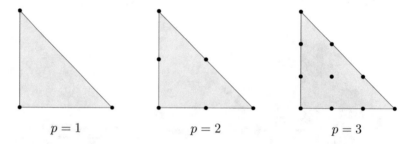

$$p = 1 \qquad\qquad p = 2 \qquad\qquad p = 3$$

Figure 2.5.: Lagrange nodes on a single element

A large variety of alternative finite element spaces appears in the literature. The best choice depends on the equation, the grid, and the desired accuracy. The spaces differ mainly in the type of polynomial space for each element, and the interelement continuity requirements. A few are explained in Chapter 8, which deals with how these spaces are implemented in DUNE. For an even longer list have a look at [106].

The algebraic system (2.9) requires the choice of a basis of V_h. The canonical basis for the Lagrange spaces is the *nodal basis*. On each element define a set of unisolvent Lagrange nodes (Figure 2.5), and identify corresponding nodes on the common boundaries of adjacent elements. Let $\mathcal{V}$ be the set of Lagrange nodes for the entire grid. Then the nodal basis is the set of functions $\phi_i \in V_h^{(p)}$ with

$$\phi_i(a_j) = \delta_{ij} \qquad \text{for all } a_j \in \mathcal{V},$$

δ_{ij} denoting the Kronecker symbol. Three such functions for spaces of first and second order are shown in Figure 2.6.

For implementing finite element methods it is important to note that nodal basis functions of Lagrange spaces can be associated to element faces in the grid by the position of the corresponding Lagrange node. Depending on whether the Lagrange

(a) Nodal basis function of $V_h^{(1)}$ (b) Vertex and edge nodal basis functions of $V_h^{(2)}$

Figure 2.6.: Elements of the nodal bases of $V_h^{(1)}$ and $V_h^{(2)}$

node of a basis function is in the interior of an element, or on a facet or a vertex, etc., the basis function itself is associated to the element, facet, or vertex. This idea prevails for many other spaces. It forms an important device in the implementation of finite elements that helps organizing the different basis functions. The concept resurfaces in Chapter 8.2.

The most important property of the nodal basis functions is the locality of their supports. Indeed, if a given basis function ϕ_i belongs to a Lagrange node in the interior of an element, then it is zero on all other elements. Likewise, if it belongs to a Lagrange node that is shared by a number of elements, then ϕ_i will be nonzero only on these elements. As a consequence, matrix entries

$$A_{ij} := a(\phi_i, \phi_j) = \int_\Omega \langle (C\nabla\phi_i + \mathbf{v}\phi_i), \nabla\phi_j \rangle \, dx + \int_\Omega r\phi_i\phi_j \, dx$$

are nonzero only if the basis functions ϕ_i and ϕ_j belong to the same or adjacent elements. The matrix A is *sparse*, and the linear system (2.9) turns into a sparse linear system[2]

$$Au = b. \tag{2.10}$$

This is an important property: Only with a sparse matrix the workload stays manageable even for high-resolution finite element grids.

The final step of the finite element method is to solve the linear system (2.10) for the coefficients $\bar{u} \in \mathbb{R}^n$ of the solution function u_h with respect to the finite element basis $\{\phi_i\}_{i=0}^{n-1}$. Even though A is sparse, this can be an arduous task. Depending on the grid and the finite element space, the matrix A can get extremely large. A great deal of effort has been put into the development of fast solvers and preconditioners for such linear systems, and such solvers form an integral part of any finite element software. Very large problems typically do not fit into the memory of a single compute node; therefore, unless matrix-free methods like [114] are used, the matrix will have to be

---

[2]When the partial differential equation is not linear, then the finite element method results in a sparse nonlinear system of algebraic equations for the unknown coefficients $\bar{u}$. The most common way to solve these is the Newton method, which considers a sequence of linearized problems. Then again we have to solve sparse linear systems of equations.

distributed across several compute nodes, and assembled in parallel. Consequently, the solution algorithms will have to be designed and implemented for parallel architectures as well.

2.1.3. Computing the Stiffness Matrix

The previous presentation of the finite element method shows that data structures for grids, function space bases, and sparse linear algebra must be central components of any finite element implementation.[3] These three concepts interact particularly during the setup of the stiffness matrix A and the load vector b of the algebraic system (2.10). This process, traditionally called *assembly*, is therefore one of the main driving forces behind the design of data structures. Many design choices of DUNE can only be understood with some knowledge of how finite element assembly works, and we therefore take a closer look.

Let n be the dimension of the space V_h. Then the stiffness matrix A is a sparse $n \times n$ matrix with entries

$$A_{ij} := a(\phi_i, \phi_j) = \int_\Omega \langle (C\nabla\phi_i + \mathbf{v}), \nabla\phi_j \rangle + r\phi_i\phi_j \, dx. \tag{2.11}$$

Correspondingly, the load vector b has the entries

$$b_i := l(\phi_i) = \int_{\Gamma_N} j\phi_i \, ds + \int_\Omega f\phi_i \, dx, \qquad i = 0, \ldots, n-1.$$

In these simple examples, the indices i, j are natural numbers. However, it can also make sense to consider matrices with a block structure, and number rows and columns using multi-indices (Chapter 10.2.4).

Element Stiffness Matrices and Load Vectors

In the model problem, each load vector entry b_i is the sum of an integral over the domain and an integral over the domain boundary. Matrix entries consist only of integrals over the domain, but more advanced discretizations can contain other terms as well. All terms can be computed by summing over the grid elements

$$A_{ij} = \sum_{T \in \mathcal{T}} \int_T \langle (C\nabla\phi_i + \mathbf{v}), \nabla\phi_j \rangle + r\phi_i\phi_j \, dx \tag{2.12}$$

$$b_i = \sum_{T \in \mathcal{T}} \left[\int_{T \cap \Gamma_N} j\phi_i \, ds + \int_T f\phi_i \, dx \right]. \tag{2.13}$$

However, naively using these expressions will not lead to an efficient algorithm. First of all, as the basis functions ϕ_i have small support, most element integrals in (2.12)

---

[3] As with anything in this chapter, this is not universally true. There exist finite element methods that do not need to store a matrix [114], and sometimes grids and bases are so simple as to not warrant dedicated data structures.

and (2.13) will be zero anyway. It is more efficient to turn around the loops. For each element T introduce the set of basis functions $\{\phi_p\}_{p=0}^{n_T-1}$ whose support intersects T in a non-trivial way. Then, the algebraic problem can be assembled with the following algorithm:

1   $A \longleftarrow 0$
2   $b \longleftarrow 0$
3   **foreach** $T \in \mathcal{T}$ **do**
4   // Compute element stiffness matrix
5   **foreach** *basis function* ϕ_p *with support in* T **do**
6   **foreach** *basis function* ϕ_q *with support in* T **do**
7

$$A_{i(p)j(q)} \longleftarrow A_{i(p)j(q)} + \int_T \langle (C\nabla\phi_p + \mathbf{v}), \nabla\phi_q \rangle + r\phi_p\phi_q \, dx$$

8   **end**
9   **end**
10   // Compute element load vector
11   **foreach** *basis function* ϕ_p *with support in* T **do**
12

$$b_{i(p)} \longleftarrow b_{i(p)} + \int_{T \cap \Gamma_N} \mathbf{j}\phi_p \, ds + \int_T f\phi_p \, dx$$

13   **end**
14   **end**

Written in this way, the work needed for the assembly of the entire algebraic problem is linear in the number of grid elements.

In order to implement this algorithm the software infrastructure has to allow certain things. First of all, it must be possible to iterate over all grid elements in some order. This is already an important point: While random access to the grid elements may seem convenient, it is hardly ever truly required for a finite element method. This fact is directly reflected in the design of the DUNE grid interface (Chapter 5.2).

Secondly, we need to be able to compute the *element stiffness matrices*

$$(A_T)_{pq} = \int_T \langle (C\nabla\phi_p + \mathbf{v}), \nabla\phi_q \rangle + r\phi_p\phi_q \, dx \qquad \forall T \in \mathcal{T}, \tag{2.14}$$

and the *element load vectors*

$$(b_T)_p = \int_{T \cap \Gamma_N} \mathbf{j}\phi_p \, ds + \int_T f\phi_p \, dx \qquad \forall T \in \mathcal{T}, \tag{2.15}$$

which are small dense matrices and vectors of size $n_T \times n_T$ and n_T, respectively. This issue will be investigated in the following section.

Finally, the set of basis functions $\{\phi_p\}_{p=0}^{n_T-1}$ whose support intersect the element T are given with local numbering $p = 0, \ldots, n_T - 1$. For each such local number p, the global number $i(p)$ needs to be computed.

Transformation to the Reference Element

The inner-most steps of the previous algorithm are the integrals over a single grid element

$$(A_T)_{pq} := \int_T \langle (C\nabla\phi_p + \mathbf{v}), \nabla\phi_q \rangle + r\phi_p\phi_q \, dx \qquad (2.16)$$

and

$$(b_T)_p := \int_{T \cap \Gamma_N} j\phi_p \, ds + \int_T f\phi_p \, dx. \qquad (2.17)$$

To simplify the use of numerical quadrature these integrals are typically transformed to a standard element. For each combinatorial type of element, we introduce a *reference element*. The most important ones are the *reference simplex*

$$T_{\text{ref}} := \left\{ \xi \in \mathbb{R}^d \mid \xi_1, \dots, \xi_d \geq 0, \ \sum_{i=1}^d \xi_i \leq 1 \right\},$$

and the *reference cube*

$$T_{\text{ref}} := [0, 1]^d.$$

By an abuse of notation we denote all reference elements by the same symbol T_{ref}. To connect the grid element to its reference element we introduce a sufficiently smooth map

$$F_T : T_{\text{ref}} \to T, \qquad (2.18)$$

and assume that it is invertible with a sufficiently smooth inverse. Basis functions on T can then be pulled back to the reference element by introducing

$$\hat{\phi}_{T,p} := \phi_p \circ F_T.$$

These new functions are called *shape functions*.

Computing the gradients $\nabla\phi_p$ of basis functions ϕ_p requires the chain rule. Deriving $\hat{\phi}_{T,p}$ with respect to coordinates ξ on the reference element produces

$$\nabla_\xi \hat{\phi}_{T,p}(\xi) = \nabla_x \phi_{T,p}(F_T(\xi)) \cdot \nabla_\xi F_T(\xi). \qquad (2.19)$$

Here, ∇_ξ is the derivative with respect to coordinates on T_{ref}, and ∇_x is the derivative with respect to coordinates on T. The matrix $\nabla_\xi F_T$ is the Jacobi matrix of F_T, with entries

$$(\nabla F_T)_{ab} = \frac{\partial (F_T)_a}{\partial \xi_b}.$$

Provided the Jacobi matrix ∇F_T can be inverted at ξ, Equation (2.19) can be solved for the desired quantity

$$\nabla_x \phi_{T,p}(x) = \nabla_\xi \hat{\phi}_{T,p}(\xi) \cdot (\nabla_\xi F_T(\xi))^{-1}. \qquad (2.20)$$

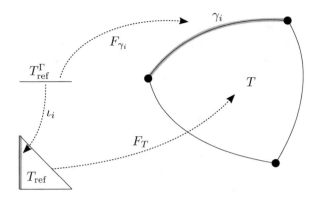

Figure 2.7.: Transformation of an element T and one of its facets γ_i to corresponding reference elements

For scalar-valued basis functions, both ∇_x and ∇_ξ are row vectors. The DUNE grid interface, however, prefers to treat them as column vectors. Therefore, DUNE assemblers typically implement the expression

$$(\nabla_x \phi_{T,p})^T(x) = (\nabla_\xi F_T(\xi))^{-T} \cdot (\nabla_\xi \hat{\phi}_{T,p}(\xi))^T.$$

rather than (2.20).

Using these expressions and the integral transformation formula we then obtain

$$(A_T)_{pq} = \int_{T_{\text{ref}}} \Big\langle \Big[C(\nabla_\xi F_T(\xi))^{-T} \cdot (\nabla_\xi \hat{\phi}_{T,p}(\xi))^T + \mathbf{v}\Big], (\nabla_\xi F_T(\xi))^{-T} \cdot (\nabla_\xi \hat{\phi}_{T,q}(\xi))^T \Big\rangle$$
$$+ r\hat{\phi}_p(\xi)\hat{\phi}_q(\xi)|\det \nabla_\xi F_T(\xi)| \, d\xi. \quad (2.21)$$

To allow the easy evaluation of expressions (2.16)–(2.17) or (2.21), finite element software has to provide implementations of the basis functions $\{\phi_i\}_{i=0}^{n-1}$ or the shape functions $\{\hat{\phi}_p\}_{p=0}^{n_T-1}$. These implementations need to be highly efficient, because basis and shape functions are evaluated many times during problem assembly. For second-order PDEs, the values and first derivatives are required, but higher-order PDEs need higher-order derivatives of the shape functions as well. Finally, the grid needs to provide the maps F_T together with their derivatives.

A priori, the shape functions $\{\hat{\phi}_p\}$ for an element T depend on the geometry of that element. For many important finite element spaces, however, this dependence vanishes. For these cases, software systems usually hold implementations for one shape function set per reference element. In DUNE, the dune-localfunctions module provides such sets of functions. Chapter 8 describes the module, and also gives some more information on finite element spaces.

The element load vector b_T is handled similarly: While the treatment of the volume load term $\int_T f\phi_p \, dx$ is straightforward, the Neumann boundary term $\int_{T \cap \Gamma_N} j\phi_p \, ds$

requires extra consideration, because it involves an integral over a part of the domain boundary. Assume for simplicity that the relevant element boundary $T \cap \Gamma_N$ consists of a set of element facets $\gamma_0, \ldots, \gamma_{m-1}$. The integral over the element boundary can then be written as a sum of facet integrals

$$\int_{T \cap \Gamma_N} \mathrm{j} \phi_p \, ds = \sum_{i=0}^{m-1} \int_{\gamma_i} \mathrm{j} \phi_p \, ds.$$

To each facet corresponds a $d-1$-dimensional reference element, which, by abusing notation again, we call T_{ref}^Γ for all $\gamma_0, \ldots, \gamma_{m-1}$. The corresponding maps are $F_{\gamma_i} : T_{\mathrm{ref}}^\Gamma \to \gamma_i$ (Figure 2.7). Using the integral transformation formula again, we rewrite the boundary integral as a sum of integrals over $d-1$-dimensional reference elements

$$\int_{T \cap \Gamma_N} \mathrm{j} \phi_p \, ds = \sum_{i=0}^{m-1} \int_{T_{\mathrm{ref}}^\Gamma} \mathrm{j}(F_{\gamma_i}(\xi)) \hat{\phi}_p(\iota_i(\xi)) |\det \nabla F_{\gamma_i}(\xi)| \, ds. \tag{2.22}$$

The map $\iota_i : T_{\mathrm{ref}}^\Gamma \to T_{\mathrm{ref}}$ is the embedding of the facet reference element T_{ref}^Γ into the element reference element T_{ref}, given by

$$\iota_i(\xi) := F_T^{-1}(F_{\gamma_i}(\xi)).$$

Note that this is a different map for each of the facets of T! With its help, there is no need to introduce shape functions for element facets. Instead, the element shape functions $\hat{\phi}_p$ are integrated over individual element facets.

Numerical Quadrature

In the model problem, the integrals of the previous section contain only polynomial expressions, and can therefore be computed symbolically. However, most applications require more generality, and therefore the integrals are usually computed using numerical quadrature. In other words, (2.21), i.e.,

$$(A_T)_{pq} = \int_{T_{\mathrm{ref}}} \left\langle \left[C(\nabla F_T(\xi))^{-T} (\nabla_\xi \hat{\phi}_{T,p}(\xi))^T + \mathbf{v} \right], (\nabla_\xi F_T(\xi))^{-T} (\nabla \hat{\phi}_{T,q}(\xi))^T \right\rangle$$
$$+ r \hat{\phi}_p(\xi) \hat{\phi}_q(\xi) |\det \nabla_\xi F_T(\xi)| \, d\xi$$

is approximated by

$$(A_T)_{pq} \approx \sum_{k=0}^{n_q-1} \omega_k \left\langle \left[C(\nabla F_T(\xi^k))^{-T} (\nabla_\xi \hat{\phi}_{T,p}(\xi^k))^T + \mathbf{v} \right], (\nabla_\xi F_T(\xi^k))^{-T} (\nabla \hat{\phi}_{T,q}(\xi^k))^T \right\rangle$$
$$+ r \hat{\phi}_p(\xi^k) \hat{\phi}_q(\xi^k) |\det \nabla_\xi F_T(\xi^k)|,$$

where $\omega_0, \ldots, \omega_{n_q-1} \in \mathbb{R}$ and $\xi^0, \ldots, \xi^{n_q-1} \in T_{\mathrm{ref}}$ are the weights and points of a quadrature rule on the reference element. The expression for the element load vector

b_T is done similarly—the main difference being that the facet integrals in (2.22) require a separate quadrature rule. All finite element implementations provide a collection of quadrature rules for the common reference elements.

In many frequent cases the approximation by numerical quadrature can be exact. For example, for first-order Lagrange elements on a simplex grid, the expressions $(\nabla F^{-1})^T$, $\nabla \hat{\phi}_i$, and $\nabla \hat{\phi}_j$ are all constant (as derivatives of affine functions). Consequently, if C, $\mathbf{v}$, and r are constant as well, the entire integrand is constant, and a one-point quadrature rule is sufficient.

Quadrature rules of higher order are needed for polynomial basis function of higher order, and when the grid contains, e.g., cube elements (because then F is generally not affine anymore). If C is not a polynomial it is in general impossible to integrate (2.21) exactly. The influence of the error introduced by numerical quadrature is discussed in [42, Sec. 4.1].

Finally, quadrature loops can be very time-consuming, in particular if shape functions of high polynomial order are used. Works like [121] and [129] investigate reformulations and automatic transformations of the nested sets of loops that make up quadrature. Dedicated compilers and code generators exist for this type of loops.

2.1.4. Dealing with Dirichlet Boundary Conditions

The previous section has shown two ways of dealing with the Dirichlet boundary condition

$$u = \mathbf{g} \qquad \text{on } \Gamma_D.$$

The first weak formulation (2.6) constrained the solution space to only those functions that comply with the boundary condition

$$u \in H^1_{\Gamma,D}: \quad a(u,v) = l(v) \qquad \forall v \in H^1_{\Gamma,0}. \tag{2.23}$$

To avoid the affine space $H^1_{\Gamma,D}$, the problem was then transformed to a pure vector space setting

$$\tilde{u} \in H^1_{\Gamma,0}: \quad a(\tilde{u},v) = l(v) - a(u_{\mathbf{g}},v) \qquad \forall v \in H^1_{\Gamma,0}. \tag{2.24}$$

The corresponding finite element problems were posed in intersections of these spaces with finite element spaces V_h, and it was tacitly assumed that these spaces are nonempty. Finally, the algebraic problem (2.9) resulted from picking a basis $\{\phi_i\}$ of the space $V_h \cap H^1_{\Gamma,0}$.

While this approach is elegant in theory, it is hardly ever used in practice. In such a setting, elements on the domain boundary have less shape functions than elements in the interior. While it is by no means impossible to program such element-dependent shape function sets, it is more convenient to deal with Dirichlet boundary conditions on the algebraic level.[4] Therefore, for the rest of this book we assume that $\{\phi_i\}$ is a basis for the full space V_h, and that matrix A and load vector b are assembled for this full space.

---

[4]We completely ignore the important approaches to incorporate Dirichlet boundary conditions in a weak sense, such as the Nitsche method [127] or penalty methods [11].

Writing Constraints into the Stiffness Matrix

For the model problem and Lagrange finite element spaces, the weak formulation (2.23) leads to algebraic constraints of the form

$$u_i = \mathbf{g}_i := \mathbf{g}(v_i) \tag{2.25}$$

for all degrees of freedom i whose Lagrange node a_i is on the domain boundary. The most common approach to incorporate these constraints consists of writing them directly into the linear system (2.10). For this, the i-th equation of (2.10) is simply replaced by the equation $u_i = g_i$. In terms of matrices and vectors this means filling the i-th row of the matrix A with the corresponding row of the identity matrix. The i-th entry of the load vector b is replaced by g_i:

$$\begin{pmatrix} \end{pmatrix} \begin{pmatrix} u_0 \\ \vdots \\ u_{n-1} \end{pmatrix} = \begin{pmatrix} \end{pmatrix}$$

$$\downarrow$$

$$\begin{pmatrix} \\ 0 \cdots 0 \; 1 \; 0 \cdots \\ \end{pmatrix} \begin{pmatrix} u_0 \\ \vdots \\ u_{n-1} \end{pmatrix} = \begin{pmatrix} \\ g_i \\ \end{pmatrix}.$$

This method is efficient, easy to implement, and does not require additional memory. As a drawback, the modified matrix is not symmetric, even if the original matrix was.

Static Condensation of Dirichlet Degrees of Freedom

Symmetry can be recovered by a second elimination step. Simply use the newly introduced 1 in the i-th row to eliminate all other entries in the i-th column. For each $j \neq i$, this amounts to replacing entry b_j by $\tilde{b}_j := b_j - A_{ji}g_i$:

$$\begin{pmatrix} \\ 0 \cdots 0 \; 1 \; 0 \cdots \\ \end{pmatrix} \begin{pmatrix} u_0 \\ \vdots \\ u_{n-1} \end{pmatrix} = \begin{pmatrix} \\ g_i \\ \end{pmatrix}$$

$$\downarrow$$

$$\begin{pmatrix} & & & 0 & & \\ & & & 0 & & \\ & & & 0 & & \\ 0 & 0 & 0 & 1 & 0 \\ & & & 0 & & \end{pmatrix} \begin{pmatrix} u_0 \\ \\ \vdots \\ \\ u_{n-1} \end{pmatrix} = \begin{pmatrix} -A_{0,i}\mathbf{g}_i \\ -A_{1,i}\mathbf{g}_i \\ \vdots \\ \mathbf{g}_i \\ -A_{n-1,i}\mathbf{g}_i \end{pmatrix}.$$

After such a second modification, solvers for symmetric problems can be used again, if the original problems was symmetric.

From the symmetrically-eliminated problem of the previous paragraph it is only a short step to eliminating the Dirichlet degrees of freedom completely from the linear system. This elimination is known as *static condensation*; it is the algebraic equivalent of (2.24).

This approach appears to be very attractive: it is mathematically elegant, and even reduces the size of the linear system to be solved. Nevertheless, the approach is rarely used in practice. To eliminate rows and columns from a matrix is difficult in many common sparse matrix data structures. Consequently, the price would either be a more complicated matrix data structure, or copying the relevant lines into a completely new matrix (which would at least temporarily require a lot of additional memory).

Modifying the Solvers

A disadvantage of the previous approaches is that they destroy the original stiffness matrix. This can be a problem: For example, it may still be needed to compute energy norms or coarse-grid approximations in a multi-level solver. A third approach therefore leaves the matrix and load vector untouched, but modifies the solver algorithms to always skip degrees of freedom marked as "Dirichlet". The modified solvers keep a boolean value for each entry of the iterate vector $\bar{u}$, and change an entry u_i only if the corresponding bit is not set. This approach, which is used in the DUNE extension module dune-solvers,[5] simplifies the implementation of and the reasoning about multilevel algorithms and algorithms for nonlinear equations. In particular, not modifying the stiffness matrix for Dirichlet values can simplify algorithms that do several transformations of the stiffness matrix anyway (as for example multi-body contact solvers [163] or certain nonsmooth descent methods like TNNMG [83]).

The main disadvantage of the approach is that all solver algorithms have to be modified. Off-the-shelf implementations of standard linear solvers do not accept lists of degrees of freedom to be skipped. On the other hand, general optimization codes for nonlinear problems like IPOPT[6] [158] allow to prescribe general linear side constraints. In such a flexible setting, Dirichlet constraints like (2.25) are merely an easy special case.

A second disadvantage is run-time efficiency. Checking a boolean flag every time before touching a vector component can lower execution speed, especially on processor

---

[5] www.dune-project.org/modules/dune-solvers
[6] https://github.com/coin-or/Ipopt

architectures with a deep pipeline [92]. This matters most for linear equations, where only few arithmetic operations are performed per access to any coefficient u_i, and is of lesser importance when the problems are nonlinear.

2.2. The Finite Volume Method

The finite volume method is the second large class of methods for partial differential equations that DUNE aims to support. Unlike the finite element method, which is based on the weak formulation of the PDE, the finite volume method uses an integral formulation. This makes the method particularly suited for equations with a conservation form, such as the linear transport equation that we consider in Section 2.2.1. Elliptic and parabolic equations are more challenging, but can also be dealt with, as we briefly show in Section 2.2.2. Good introductions to finite volume methods are [12, 66].

The nature of the finite volume method motivates several design choices in the DUNE grid interface. Simple cell-centered finite volume methods work naturally on non-conforming grids, and with element types other than merely simplices and cubes. To accommodate this, the DUNE grid interface puts very few restrictions on the element shape and relations between neighboring elements. On the other hand, finite volume methods typically need more information about the mutual relations between neighboring grid elements than many finite element methods. This need is reflected in the design of the *intersection* concept of the DUNE grid interface (Chapter 5.4).

2.2.1. Conservation Laws

The finite volume method is most easily explained for a simple scalar conservation law. Let Ω be a domain in $\mathbb{R}^d$, and $[0, t_{\mathrm{end}}]$ a time interval. We are looking for the time evolution of a scalar concentration $c : \Omega \times [0, t_{\mathrm{end}}] \to \mathbb{R}$ under the influence of a given stationary velocity field $\mathbf{v} : \Omega \to \mathbb{R}^d$, starting from a given initial concentration $c_0(x)$ for all $x \in \Omega$. In the simple-most case, the evolution is described by the linear transport equation

$$\frac{\partial c}{\partial t} + \mathrm{div}(c\mathbf{v}) = 0, \tag{2.26}$$

which is to hold for all points $x \in \Omega$ and all times $t \in [0, t_{\mathrm{end}}]$.

As previously, we use the letter $\mathbf{n}$ to denote the outer unit normal of the domain Ω. The velocity field separates the boundary $\partial\Omega$ of Ω into two parts. We call *inflow boundary* Γ_{in} that part of the boundary where the velocity $\mathbf{v}$ points towards the inside of Ω, i.e.,

$$\Gamma_{\mathrm{in}} := \{x \in \partial\Omega \mid \langle \mathbf{v}, \mathbf{n} \rangle < 0\},$$

and *outflow boundary* Γ_{out} the entire rest. Boundary conditions can only be posed on the inflow boundary Γ_{in}. The standard choices are Dirichlet conditions, i.e., prescribed concentration c, or Neumann boundary conditions, i.e., prescribed flux $\mathbf{j} := c\mathbf{v}$. On the outflow boundary Γ_{out}, no boundary condition can be posed at all, as

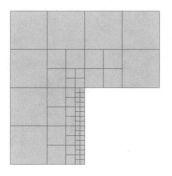 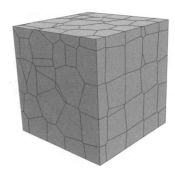

Figure 2.8.: Finite volume grids can contain hanging nodes, and elements that are general convex polytopes. The grid on the right was taken from [65].

the solution there is completely determined by the differential equation. The implicit condition $\mathbf{j} = c\mathbf{v}$ is sometimes called *outflow boundary condition*.

To derive the integral formulation of (2.26), we partition the domain Ω into a set $\mathcal{T}$ of closed convex polytopes T_i, such that

1. (Partitioning) the union of the polytopes is the closure of the domain

$$\overline{\Omega} = \bigcup_{T \in \mathcal{T}} T, \qquad (2.27)$$

2. (Nonintersection) for any two polytopes from $\mathcal{T}$, the intersection of their interiors is empty

$$\operatorname{int} T_i \cap \operatorname{int} T_j = \emptyset \qquad \forall T_i, T_j \in \mathcal{T}, \ i \neq j. \qquad (2.28)$$

Such a partition is again called a *grid*, but note how the definition is less restrictive than the corresponding Definition 2.1 for the finite element method. In the context of finite volume methods, the polytopes T are called *control volumes*. Two example finite volume grids are shown in Figure 2.8.

Given the partition $\mathcal{T}$, we apply the Gauss theorem on each control volume $T_i \in \mathcal{T}$ and obtain a set of coupled problems

$$\frac{\partial}{\partial t} \int_{T_i} c \, dx + \int_{\partial T_i} \langle c\mathbf{v}, \mathbf{n} \rangle \, ds = 0, \qquad (2.29)$$

one for each T_i. In an abuse of notation we use $\mathbf{n}$ here also for the unit outer normals of the control volumes.

To discretize in time, we subdivide the time interval into discrete steps

$$0 = t_0 < t_1 < t_2 < \ldots < t_m = t_{\text{end}}.$$

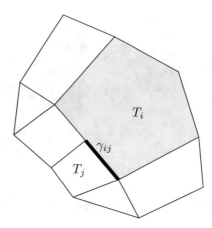

Figure 2.9.: A control volume and some of its neighbors. The intersection γ_{ij} between the elements T_i and T_j is drawn in bold.

A simple explicit Euler discretization yields

$$\int_{T_i} c(x, t_{k+1})\, dx - \int_{T_i} c(x, t_k)\, dx + \Delta t_k \int_{\partial T_i} \langle c\mathbf{v}, \mathbf{n} \rangle\, ds = 0 \quad \forall T_i \in \mathcal{T}, \qquad (2.30)$$

where $\Delta t_k := t_{k+1} - t_k$ is the k-th time step size. Then we approximate c by a cell-wise constant function $\bar{c}$, where

$$\bar{c}_i^k := \frac{1}{|T_i|} \int_{T_i} c(x, t_k)\, dx$$

is the average value in element T_i at time t_k, and where we have used $|T_i|$ to denote the volume of T_i. This allows to reduce (2.30) to

$$\bar{c}_i^{k+1} |T_i| - \bar{c}_i^k |T_i| + \Delta t_k \int_{\partial T_i} \langle c\mathbf{v}, \mathbf{n} \rangle\, ds = 0. \qquad (2.31)$$

With this construction, the degrees of freedom are associated to the grid elements (which are sometimes called *grid cells*). For this reason, the method is frequently called a *cell-centered* finite volume method.

The problem now is how to set the unknown flux $\langle c\mathbf{v}, \mathbf{n} \rangle$ across the control volume boundary ∂T_i into relation with the algebraic unknowns $\bar{c}_i^k$. For a given element T_i, we subdivide the boundary ∂T_i into facets γ_{ij}. These are either intersections with other elements $\partial T_i \cap \partial T_j$, or intersections with the boundary $\partial T_i \cap \partial \Omega$ (Figure 2.9). For each such intersection, let $\mathbf{v}_{ij}$ be the velocity on the midpoint of the facet γ_{ij}, and $\mathbf{n}_{ij}$ the unit outer normal of the facet γ_{ij} with respect to T_i there. The normal flux $\langle c\mathbf{v}, \mathbf{n} \rangle$ is approximated by a constant on each facet, which depends on the flow direction. We set

$$\int_{\gamma_{ij}} \langle c\mathbf{v}, \mathbf{n} \rangle \, ds = |\gamma_{ij}| \cdot \begin{cases} \bar{c}_j^k \langle \mathbf{v}_{ij}, \mathbf{n}_{ij} \rangle & \text{for } \langle \mathbf{v}_{ij}, \mathbf{n}_{ij} \rangle < 0 \\ \bar{c}_i^k \langle \mathbf{v}_{ij}, \mathbf{n}_{ij} \rangle & \text{otherwise.} \end{cases}$$

Such an approximation is usually called a *flux function*. Neumann boundary conditions on the inflow boundary are implemented by replacing the term $\bar{c}_j^k \mathbf{v}_{ij}$ by a suitable approximation of the given boundary flux $\mathbf{j}$ over γ_{ij}. Dirichlet boundary conditions $c = g$ are implemented by turning them into flux conditions via $\mathbf{j} = g\mathbf{v}$.

The flux function can be rewritten as

$$\int_{\gamma_{ij}} \langle c\mathbf{v}, \mathbf{n} \rangle \, ds = |\gamma_{ij}| \cdot \bar{c}_i^k \max(0, \langle \mathbf{v}_{ij}, \mathbf{n}_{ij} \rangle) - |\gamma_{ij}| \cdot \bar{c}_j^k \max(0, -\langle \mathbf{v}_{ij}, \mathbf{n}_{ij} \rangle).$$

Inserting this into (2.31) and solving for $\bar{c}_i^{k+1}$ we obtain

$$\bar{c}_i^{k+1} = \bar{c}_i^k \left(1 - \Delta t_k \sum_{\gamma_{ij}} \frac{|\gamma_{ij}|}{|T_i|} \max(0, \langle \mathbf{v}_{ij}, \mathbf{n}_{ij} \rangle) \right) + \Delta t_k \sum_{\gamma_{ij}} \bar{c}_j^k \frac{|\gamma_{ij}|}{|T_i|} \max(0, -\langle \mathbf{v}_{ij}, \mathbf{n}_{ij} \rangle)$$

for all elements T_i. This is an explicit scheme—no system of equations needs to be solved. Stability for sufficiently small time steps can be shown using properties of the update matrix.

The scheme is mass conserving: When computing the total mass by summing the element masses $\bar{c}_i |T_i|$ over all elements, each term for an interior intersection appears twice, and once with each sign. The interior intersection terms therefore cancel, and the total mass change is only caused by the boundary flux.

Observe how the element geometry does not enter the scheme in full detail: What is used are only element and facet volumes, and facet centers. On the other hand, in an explicit scheme the computation of these seemingly simple quantities can take a surprisingly large percentage of the overall run-time. Grid data structures for cell-centered finite volume grids should therefore be able to compute them as efficiently as possible.

2.2.2. Second-Order Elliptic Equations

The finite volume method can also be applied to second-order elliptic equations. There, the flux contains derivatives of the unknown function, which poses additional challenges.

As a model problem consider again the Poisson equation. Let $\Omega \subset \mathbb{R}^d$ be a domain and $f : \Omega \to \mathbb{R}$ a volume source term. We look for a scalar function $u : \Omega \to \mathbb{R}$ that solves

$$-\Delta u = f \qquad \text{on } \Omega,$$

subject to suitable Dirichlet and Neumann boundary conditions.

To derive a discrete integral formulation of the Poisson equation, again partition the domain Ω into a finite set of convex polytopes $\mathcal{T}$ that fulfills the two properties (2.27) and (2.28). Integrating the equation over each control volume $T \in \mathcal{T}$ and applying the Gauss theorem gives

$$-\int_{\partial T} \langle \nabla u, \mathbf{n} \rangle \, ds = \int_T f \, dx. \tag{2.32}$$

The boundary integral is very similar to (2.29), but the element boundary flux is now

$$\mathbf{j} = \nabla u,$$

instead of $\mathbf{j} = c\mathbf{v}$. The central question here is how to compute the gradient ∇u on an intersection γ_{ij} between two adjacent control volumes T_i and T_j. One way to do this is to suppose that for each control volume T there is a point $x_T \in T$ such that if two control volumes T_i and T_j share a common facet, then the segment $[x_{T_i}, x_{T_j}]$ is orthogonal to that facet. We then associate the degrees of freedom to these points x_T, and interpolate linearly between points of adjacent elements. With this approximation, the normal flux $\langle \nabla u, \mathbf{n} \rangle$ at the intersection of the common facet γ_{ij} and the segment $[x_{T_i}, x_{T_j}]$ is given by

$$\langle \nabla u, \mathbf{n} \rangle = \nabla u \frac{x_{T_i} - x_{T_j}}{\|x_{T_i} - x_{T_j}\|} \approx \frac{u_{T_i} - u_{T_j}}{\|x_{T_i} - x_{T_j}\|}.$$

To incorporate Dirichlet and Neumann boundary conditions into the discrete problem assume first that the Dirichlet and Neumann boundaries are resolved by the grid. For each boundary facet γ there is then either a prescribed boundary value $\mathbf{g}_\gamma$ or a prescribed boundary normal flux $\mathbf{j}_\gamma$. For a Neumann boundary facet γ_N, the normal flux $\langle \nabla u, \mathbf{n} \rangle$ in (2.32) over that facet can be replaced directly by $\mathbf{j}_\gamma$. For a Dirichlet facet γ_D of control volume T, the normal derivative $\langle \nabla u, \mathbf{n} \rangle$ is approximated as a one-sided difference quotient

$$\langle \nabla u, \mathbf{n} \rangle \approx \frac{\mathbf{g}_\gamma - u_T}{h/2},$$

where $h/2$ is an approximate distance from x_T to γ_D.

Altogether, this results in a linear system of equations

$$-\sum_{\gamma_{ij}} |\gamma_{ij}| \cdot \frac{u_i - u_j}{\|x_{T_i} - x_{T_j}\|} = \int_{T_i} f \, dx \qquad i = 1, \dots, |\mathcal{T}|$$

for the unknowns u_i, with the modifications listed above for the boundary values.

In this scheme, the unknowns are associated to the element centers. These form the vertices of the so-called *dual-grid* of $\mathcal{T}$ [12], and the method is therefore called a *vertex-centered* finite volume method.

The required orthogonality property holds, for example, if the control volume grid is Cartesian. An alternative approach is to choose Voronoi cells as control volumes [66, Example 3.2]. The dual grid is then the Delaunay triangulation of the control volume centers x_{T_i}, see Figure 2.10.

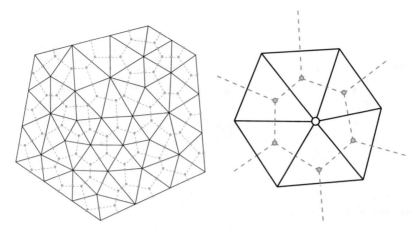

Figure 2.10.: Left: Delaunay grid and the corresponding Voronoi grid. Right: single control volume and associated cells. The control volumes are given by the dashed lines.

2.3. Local Grid Adaptivity

The underlying assumption of all grid-based methods for PDEs is that the approximation error can be decreased by increasing the grid resolution, i.e., by making the element size h smaller and smaller. The following result for p-th-order Lagrangian finite elements is prototypical. It states that the H^1 error decreases like the p-th power of h.

Theorem 2.1 *Let $p \geq 1$ and $u \in H^1_\Gamma(\Omega) \cap H^{p+1}(\Omega)$. Then if the grid size h is small enough we get the a priori bound*

$$\|u - u_h\|_{H^1} \leq C h^p |u|_{H^{p+1}} \tag{2.33}$$

with a constant C independent of u.

Corresponding results exist for many other finite element discretizations [37, 42, 63], and for finite volume methods [66].

Theorem 2.1 implies that u_h converges to u for h going to zero. However, for the practitioner, the convergence *order* is crucial: higher orders of the exponent translate into more accuracy for a given number of degrees of freedom, i.e., for a given cost.

Unfortunately, bounds like (2.33) hold only if the solution u of the PDE is sufficiently smooth. For example, to obtain linear convergence with first-order finite elements, u has to be an H^2 function. Situations where u is not H^2-regular are easily constructed, and sublinear convergence is indeed observed in these cases. The classic example are domains with inward corners such as the one in Figure 2.11, where singularities in the solution develop in these corners. Figure 2.11 also shows the solution of a boundary

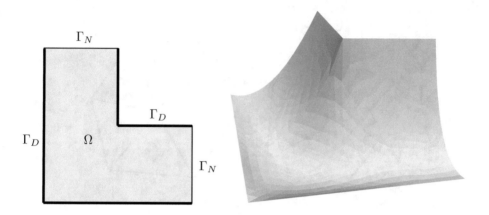

Figure 2.11.: Corner singularity in the solution of the equation $-\Delta u = -5$ (height plot), on an L-shaped domain

value problem for the Poisson equation on this domain. The solution is H^2-regular away from the inward corner, but not on the entire domain.

For a more "visual" explanation of why the convergence is not optimal for this example, observe that near the corner, the graph of the solution has a very high curvature. A grid with many small elements is therefore needed to approximate the function well near the corner. On the other hand, away from the corner, the solution is reasonably flat. There, few elements suffice to approximate the function well, and using a high-resolution grid is wasteful.

2.3.1. Local Adaptivity of Grids and Finite Element Spaces

The convergence order can frequently be improved if local grid adaptivity is used. Local grid adaptivity means automatically using small elements near the singularities, and bigger elements elsewhere. This approach can improve the overall accuracy per work dramatically. We discuss it here, because the need for adaptive grids motivates a number of design decisions of DUNE.

Local adaptivity comprises two tasks: First, for a given discrete solution u_h on a certain grid, one needs to estimate the current error $u - u_h$, and how this error is distributed across the domain. With this information the grid is then locally modified in such a way as to concentrate degrees of freedom in areas of high error, while keeping the overall number of degrees of freedom as low as possible. The first part, which is called *a posteriori error analysis*, is a well-established subfield of numerical mathematics [157]. It involves the differential equations and their discretizations, and is beyond the scope of this book. The second part is purely geometrical, and only concerns the grid data structure. It can nevertheless be quite challenging to

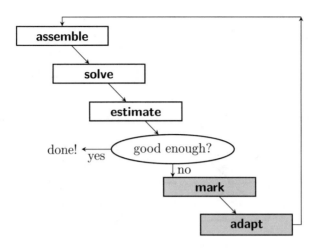

Figure 2.12.: The classic adaptive refinement loop

implement. Together with the actual solving of the PDE, these steps form the classic *adaptive refinement loop*, visualized in Figure 2.12.

There are two ways to increase the grid resolution in particular areas of the domain:

- Split large elements into smaller ones (*h-refinement*). In some cases, it can also be convenient to merge several small elements into a larger one (coarsening).

- Move grid vertices from zones of low interest to zones of high interest, without changing the grid topology (*r-refinement*). This is of interest for certain flow problems, and problems with boundary layers. See [95] for an overview.

A third approach to local adaptivity is *p-refinement*, which leaves the grid untouched, but increases the polynomial order of the finite element spaces locally on elements where this seems appropriate. Frequently, *p*-refinement is combined with *h*-refinement, which together is called *hp*-refinement. Under suitable conditions, the resulting discretization errors drop exponentially with the number of degrees of freedom [145]. *hp*-refinement for finite element spaces of continuous functions is nevertheless rarely used, because the inter-element continuity between polynomials of different orders is difficult to implement. For finite elements without continuity it is more natural [55].

The rest of this chapter focuses on *h*-refinement, as that is what the DUNE grid interface mainly aims at.

2.3.2. *h*-Refinement

In *h-refinement*, the element size h is decreased in areas that need a high resolution. To achieve this, the elements in the affected areas are split and replaced by elements

Figure 2.13.: Edge bisection splits a simplex into two new ones

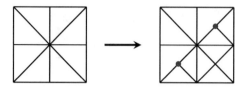

Figure 2.14.: Splitting triangles along one edge will typically lead to hanging nodes

with a smaller size. The challenge is to split the elements in such a way that the element shapes do not deteriorate too much, because the element quality crucially influences the discretization errors and solver convergence speed [37, 147]. There are several different strategies for splitting elements into smaller ones. All of them are available in DUNE.

Edge Bisection

Edge bisection is a refinement technique for grids with simplex elements. In a two-dimensional grid, a given triangle is split by dividing one edge in half, and connecting the midpoint with the opposite vertex. This way, the original triangle is split into two smaller ones. In three-dimensional grids, the midpoint of one edge is connected to the opposite edge by a triangle (Figure 2.13).

The choice of edge to refine matters for preserving the grid quality. For example, repeatedly bisecting elements across their shortest edges leads to sequences of grids with unbounded element aspect ratios. Various strategies to pick appropriate edges have been proposed in the literature. Rivara [137] showed that the overall grid quality remains bounded if elements are always bisected along their *longest* edges. Other criteria are purely combinatorial [128].

Simply splitting elements along one edge typically leads to grids with hanging nodes, i.e., element vertices that do not correspond to vertices of all adjacent elements (Figure 2.14). Grids with hanging nodes are called *nonconforming*; they violate Condition 2 of the Definition 2.1 of a grid. Depending on the discretization method, hanging nodes may or may not be a problem. For example, cell-centered finite volume methods handle grids with hanging nodes easily, and so do certain finite element methods like the Discontinuous Galerkin (DG) method of Chapter 11.2.6. In fact, this ability to work on nonconforming grids is one of these methods' main selling points.

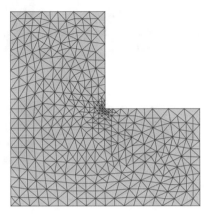

Figure 2.15.: Poisson problem on the L-domain with bisection refinement

To recover conforming grids, edge bisection algorithms will additionally refine elements with a hanging node, until no more such elements exist [128] Figure 2.15 shows a grid for the singular solution shown in Figure 2.11, obtained by local adaptive edge bisection refinement.

Red (Regular) Refinement

While edge bisection splits every element into two new elements, *red* (or *regular*) refinement subdivides any d-dimensional element into 2^d new ones. It was introduced by Bank, Sherman, and Weiser [10] for two-dimensional grids. Figure 2.16 shows red refinement for a quadrilateral and a triangle: New vertices are introduced by bisecting edges, and connecting the edge midpoints forms new elements. For the quadrilaterals, an additional new vertex appears at the element center.

Unlike edge bisection, red refinement of triangles produces elements that are *similar* to the old ones in the sense that each new element can be transformed into its father

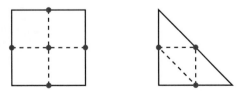

Figure 2.16.: Red (regular) refinement splits two-dimensional elements into four similar new ones

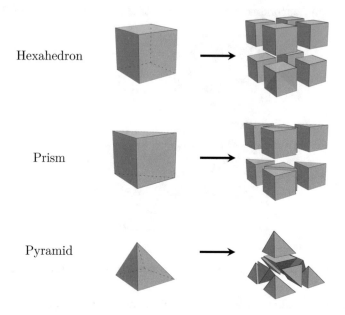

Hexahedron

Prism

Pyramid

Figure 2.17.: Red (regular) refinement of three-dimensional elements

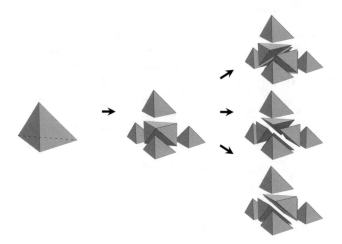

Figure 2.18.: Red (regular) refinement of a tetrahedron. Cutting off the corners leaves an octahedron, which can be split into four tetrahedra in three different ways.

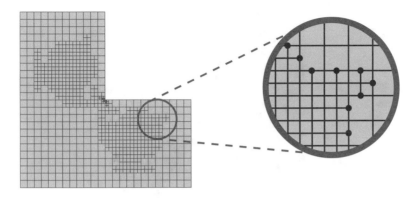

Figure 2.19.: Poisson problem on the *L*-domain with adaptive red (regular) refinement. (Some) hanging nodes are shown as blue dots.

by rotations, translations, and scaling alone. The same holds for quadrilaterals with parallel opposite sides. This implies that a sequence of refinement operations will not influence the element quality of the grids.

Unfortunately, this desirable similarity property does not hold in more general situations; in particular, red refinement of three-dimensional elements is more involved. Figure 2.17 shows the canonical splittings for hexahedra, prisms and pyramids, which are all canonical. However, refining a tetrahedron is problematic. Splitting off the four corners leaves an octahedron, which can be split along any one of the three diagonals into four additional tetrahedra (Figure 2.18). These tetrahedra are *not* similar to the original one, and their shape depends on the choice of diagonal. Various authors remarked that in a sequence of refinements of a single given tetrahedron, there are strategies to pick the diagonals that result in only a finite number of similarity classes appearing [25, 70]. Consequently, the grid quality is impacted by regular refinement of tetrahedral grids, but the grid quality remains bounded. While these algorithms rely on vertex numberings alone, similar results exist for always selecting the shortest diagonal [113, 165].

Red–Green Refinement

Figure 2.19 shows a grid for the singularity function from Figure 2.11 obtained by adaptive red refinement. The figure shows that when regular refinement operations are applied locally, hanging nodes will again appear in the grid. To recover a conforming grid, Bank, Sherman, and Weiser [10] proposed to use so-called *closure* (or *green*) refinement rules at the transition zone (Figure 2.22). Suppose that a subset of elements has been refined using *red* rules. Closure refinement rules are then used to refine elements with hanging nodes without introducing further vertices. While it is obvious that for all 8 combinations of having hanging nodes at the edge midpoints of a triangle

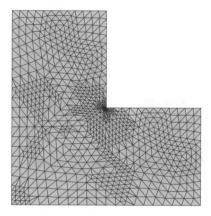

Figure 2.20.: Poisson problem on the *L*-domain with red–green refinement

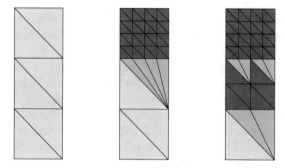

Figure 2.21.: Naive repetition of green closure leads to a degradation of the element quality (center). This is avoided by proper red–green refinement (right).

there is a corresponding refinement rule, the generalization to higher dimensions is more involved [78].

Green closures, however, have a negative impact on the element quality. Two triangles obtained by a green refinement have a worse element quality than the original triangle. Repeated application of green refinement rules to the same triangle will lead to sequences of grids of unbounded grid quality (Figure 2.21). To avoid this scenario, Bank, Sherman, and Weiser [10] proposed, instead of refining green elements, to remove them first and to replace them by a red refinement. This can lead to additional hanging nodes, which are then eliminated by additional steps of green refinement. The result is a grid where no element has a history of being refined by a green rule more than once. Consequently, as red refinement leaves the element quality bounded, the overall grid quality stays bounded even when using green closures to

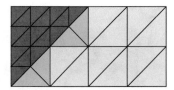

Figure 2.22.: Green closures allow to avoid hanging nodes. In grids consisting of hexahedral elements, non-hexahedral element types appear in the closure.

avoid hanging nodes.

2.3.3. Hierarchical Grids and Refinement Trees

In most texts on the finite element method grids appear as non-hierarchical objects, as in Definition 2.1. In such a context, local grid refinement takes place by removing elements from the grid, and replacing them with the refined elements. However, most implementations of grid refinement take a different view on this. There are several problems with the non-hierarchical view:

- It is difficult to undo refinement steps. The traditional way "forgets" that certain groups of elements where created by splitting a larger element. In principle, such a group could easily be replaced by the original element, if that original element would still be available somehow. In contrast, coarsening of general unstructured grids without such additional information is difficult.

- As a special case of this, removing green closure elements before applying a new round of red–green refinement is difficult, if there is no information on the history of these elements.

- Geometric multi-level methods [37, 83, 155], which form a large and successful family of solvers for linear (and some nonlinear) problems, require a hierarchy of grids of different resolutions.

For these reasons, most grid data structures that support grid refinement do not only store the latest refined grid, but also all coarser refinement stages leading up to the final grid. The data structure of choice is usually a *forest*, i.e., a collection of *trees* [54]. Initially, a grid is a collection of elements (the so-called *coarse grid* or *macro grid*), but each element is seen as a tree, consisting of only a single node. If such an element is now refined into a set of smaller ones, the new elements do not replace the old one. Rather, they are added to the corresponding tree as the children of the old element. The forest that represents the macro grid and the current state of refinement is called a *hierarchical grid* (Figure 2.23).

41

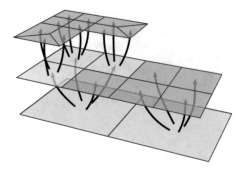 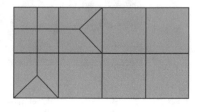

Figure 2.23.: A forest of refinement trees, and the grid of leaf elements

The grid in the old sense of the word is formed by the elements without children, the so-called *leaf elements*. DUNE calls this the *leaf grid view* or *leaf grid*. When the refinement process is continued, some of its elements stop being leaf elements, and disappear from the leaf grid view. In the forest data structure, the current grid always appears as the set of leaf elements, but previous stages of the refinement process are nevertheless available from the data structure, in form of the elements that are not leaves of the forest.

As a consequence of this construction, coarsening of a previously refined grid becomes easy: By removing a leaf element together with all its siblings from the tree, the common father element becomes a leaf element again, and therefore appears in the finite element grid. Also, hierarchical grids contain the information that is necessary to implement geometric multigrid solvers. The different levels of the refinement forest naturally form sets of grids at different resolutions, as required by these methods.

The DUNE grid interface sees grids as collections of trees in the sense described here. This is why a bit of graph terminology appears in the grid interface. Actual implementations of the DUNE grid interface may explicitly store the refinement trees, or they may recompute them when needed from the root elements and the refinement history. The topic is taken up again from a more implementation-centered point of view in Chapter 5.1.

It is interesting to note that refinement trees are not only used in implementations of adaptive finite element and finite volume methods. They also appear as a theoretical tool in the analysis of the approximation properties of adaptive finite element spaces; see, e.g., [26].

3. Getting Started with Dune

This chapter describes the first steps with DUNE. It shows how to install DUNE on a computer, and how to create a first DUNE module. Then, it presents two example applications that give a first impression of how DUNE can be used.

3.1. Installation of Dune

The first step into the DUNE world is its installation. We have tried to make this as painless as possible. The following instructions assume a Unix-style command shell and toolchain installed. This will be easy to obtain on all flavors of Linux and Apple OS X. On Windows there is the WINDOWS SUBSYSTEM FOR LINUX,[1] or the CYGWIN environment.[2]

Running the examples in this chapter requires eight DUNE modules: dune-common, dune-geometry, dune-grid, dune-istl, dune-localfunctions, dune-uggrid, dune-typetree, and dune-functions. These are even sufficient to run all examples from this book, except for the ones from Chapter 11, which additionally need dune-pdelab.

3.1.1. Installation from Binary Packages

Installation is easiest when using precompiled packages. At the time of writing this is the case for the Debian Linux distribution and many of its derivatives, but there may be more. An up-to-date list is available on the DUNE project web page at www.dune-project.org. On a Debian-type system, type

```
sudo apt-get install libdune-functions-dev \
                      libdune-istl-dev \
                      libdune-uggrid-dev
```

to install all those eight DUNE modules. The ones not listed explicitly are added automatically. To additionally install dune-pdelab, append libdune-pdelab-dev to the list. The DUNE modules are then installed globally on the machine, and the DUNE build system will find them.

3.1.2. Installation from Source

If there are no precompiled packages available, or when working without sudo rights, then DUNE has to be installed from the source code. First, download the code from

---

[1] https://docs.microsoft.com/en-us/windows/wsl/about
[2] http://cygwin.com

© Springer Nature Switzerland AG 2020
O. Sander, *DUNE — The Distributed and Unified Numerics Environment*, Lecture Notes in Computational Science and Engineering 140, https://doi.org/10.1007/978-3-030-59702-3_3

the DUNE website at `www.dune-project.org`. It is recommended to download the release tarballs, named

```
dune-<modulename>-X.Y.Z.tar.gz
```

where `<modulename>` is one of `common`, `geometry`, `grid`, etc., and `X.Y.Z` is the release version number. The code examples in this book require at least version 2.7.0.

Those who need very recent features can also get the bleeding edge development versions of the DUNE modules. The DUNE source code is stored and managed using the `git` version control software.[3] The repositories are at `https://gitlab.dune-project.org`. To clone (i.e., download) the source code for one module type

```
git clone https://gitlab.dune-project.org/core/dune-<modulename>.git
```

where `<modulename>` is replaced by `common`, `geometry`, `grid`, `localfunctions`, or `istl`. The remaining three modules are available from the `staging` namespace:

```
git clone https://gitlab.dune-project.org/staging/dune-uggrid.git
git clone https://gitlab.dune-project.org/staging/dune-typetree.git
git clone https://gitlab.dune-project.org/staging/dune-functions.git
```

These eight commands create eight directories, one for each module. Each directory contains the latest development version of the source code of the corresponding module. If desired, this development version can be replaced by a particular release version by calling, e.g.,

```
git checkout releases/2.7
```

in the directory. See the `git` documentation for details.

Suppose that there is an empty directory called `dune`, and that the sources of the eight DUNE modules have been downloaded into this directory. When using the tarballs, these have to be unpacked by

```
tar -zxvf dune-<modulename>-X.Y.Z.tar.gz
```

for each DUNE module. To build them all, enter the dune directory and type[4]

```
./dune-common/bin/dunecontrol cmake : make
```

This configures and builds all DUNE modules, which may take several minutes. For brevity, the two commands `cmake` and `make` can be called together as:

```
./dune-common/bin/dunecontrol all
```

Once the process has completed, DUNE can be installed by typing

```
./dune-common/bin/dunecontrol make install
```

---

[3]`https://git-scm.com`
[4]Replace `dune-common` by `dune-commmon-X.Y.Z` when using the tarballs.

This will install the DUNE core modules to /usr/local, and requires root access.

To install DUNE to a non-standard location, a custom installation path can be set. For this, create a text file dune.opts, which should contain

```
CMAKE_FLAGS="-DCMAKE_INSTALL_PREFIX=/the/desired/installation/path"
```

Then call

```
./dune-common/bin/dunecontrol --opts=dune.opts all
```

and

```
./dune-common/bin/dunecontrol --opts=dune.opts make install
```

The dunecontrol program will pick up the variable CMAKE_FLAGS from the options file and use it as a command line option for any call to cmake, which in turn is used to configure the individual modules. The particular option of this example will tell cmake that all modules should be installed to /the/desired/installation/path.

Unfortunately, it has to be mentioned that as of DUNE Version 2.7, working with installed modules is still not very mature, and may lead to build failures. As a consequence of this, the installation of DUNE modules is optional, and the DUNE build system will also accept non-installed modules as build dependencies.

When using the dunecontrol program to manage further modules, it has to be told where to find the DUNE core modules. This is done by prepending the path /the/ desired/installation/path to the environment variable DUNE_CONTROL_PATH. Note, however, that if DUNE_CONTROL_PATH is set to anything, then the current directory is *not* searched automatically for DUNE modules. If the current directory should be searched then the DUNE_CONTROL_PATH variable has to contain :.: somewhere.[5]

3.2. A First Dune Application

DUNE is organized in modules, where each module is roughly a directory with a predefined structure.[6] Each such module implements a C++ library that can be used from other code. Although not necessary, however, it can be convenient to write new code into a new DUNE module, because that simplifies dependency tracking, and allows further DUNE modules to easily depend on the new code. The first step for this is to create a new DUNE module template.

3.2.1. Creating a new Module

In the following we assume again that there is a Unix-type command shell available, and that DUNE has been installed successfully either into the standard location, or that DUNE_CONTROL_PATH contains the installation path. To create a new module, DUNE provides a special program called duneproject. To invoke it, simply type

---

[5]The dot denotes the current directory, and the colons are separators.

[6]See Appendix A for more details on modules.

```
duneproject
```

in the shell. If the program is not installed globally, use the version from dune-common/
bin.

The duneproject program will ask several questions before creating the module.
The first is the module name. Any Unix file name without whitespace is admissible,
but customarily module names start with a dune- prefix. To be specific we will call
the new module dune-foo.

The next question asks for other DUNE modules that the new module will depend
upon. To help, duneproject has already collected a list of all modules it sees on
the system. These are the globally installed ones, and the ones in directories listed
in the DUNE_CONTROL_PATH environment variable. After the installation described
above one should see at least dune-common, dune-geometry, dune-grid, dune-istl,
dune-localfunctions, dune-typetree, dune-uggrid, and dune-functions. The
required ones should be entered in a white-space-separated list (and it is easy to add
further ones later). For the purpose of this introduction please select them all.

Next is the question for a module version number. These should start with X.Y (X
and Y being numbers), and can optionally end with a third number .Y or an arbitrary
string. This is followed by a question for an email address. This address will appear in
the file dune.module of the module (and nowhere else), and will be a point of contact
for others with an interest in the module. After this, the duneproject program exits
and there is now an blank module dune-foo:

```
~/dune: ls
dune-foo
```

A tool like the tree program can be used to see that dune-foo contains a small
directory tree:

```
~/dune> tree dune-foo
dune-foo
|-- cmake
|   '-- modules
|       |-- CMakeLists.txt
|       '-- DuneFooMacros.cmake
|-- CMakeLists.txt
|-- config.h.cmake
|-- doc
|   |-- CMakeLists.txt
|   '-- doxygen
|       |-- CMakeLists.txt
|       '-- Doxylocal
|-- dune
|   |-- CMakeLists.txt
|   '-- foo
|       |-- CMakeLists.txt
|       '-- foo.hh
```

```
|-- dune-foo.pc.in
|-- dune.module
|-- README
'-- src
    |-- CMakeLists.txt
    '-- dune-foo.cc
```

```
7 directories, 15 files
```

This tree contains:

- The cmake configuration files for the DUNE build system,

- A text file dune.module, which contains some meta data of the module,

- a small example program in dune-foo.cc.

The content of a typical DUNE module is described in more detail in Appendix A.

3.2.2. Testing the new Module

The new module created by duneproject contains one C++ source code file src/dune-foo.cc. This is a small test program that allows to verify whether the module has been built properly. Configuring and building the module is controlled by the dunecontrol program again.

The process is hardly any different from configuring and building the DUNE core modules as described in the previous section. Just move to the dune directory again, and type

```
dunecontrol all
```

in the shell. This will output lots of information while it runs, but none of it should be of any concern right now (unless actual error messages appear somewhere). Once dunecontrol has terminated there is a new executable dune-foo in dune-foo/build-cmake/src. Start it with

```
~/dune: ./dune-foo/build-cmake/src/dune-foo
```

and it will print

```
Hello World! This is dune-foo.
This is a sequential program.
```

Congratulations! You have just run your first DUNE program.

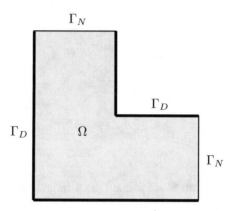

Figure 3.1.: A simple domain Ω with Dirichlet boundary Γ_D (thick lines) and Neumann boundary Γ_N

3.3. Example: Solving the Poisson Equation Using Finite Elements

To get started with a real example we will solve the Poisson equation with the finite element method.[7] More specifically, we will explain the program that produced the function shown in Figure 2.11. The function in that figure is the weak solution of the Poisson equation

$$- \Delta u = -5 \tag{3.1}$$

on the L-shaped domain $\Omega = (0,1)^2 \setminus [0.5, 1)^2$, with Dirichlet boundary conditions

$$u = \begin{cases} 0 & \text{on } \{0\} \times [0,1] \cup [0,1] \times \{0\}, \\ 0.5 & \text{on } \{0.5\} \times [0.5, 1] \cup [0.5, 1] \times \{0.5\}, \end{cases} \tag{3.2}$$

and zero Neumann conditions on the remainder of the boundary. The domain and boundary conditions are shown in Figure 3.1.

The example program solves the Poisson equation and outputs the result to a file which can be opened with the PARAVIEW visualization software.[8] The program code is contained in a single file. We will not show quite the entire source code here, because complete C++ programs take a lot of space. However, the entire code is printed in Appendix B.2. Also, readers of this document in electronic form can access the source code file through the little pin icon in the page margin. The easiest way to build the example is to copy the program file into dune-foo/src/, and then either to

---

[7]Readers that are unsure about how the finite element method works will find a short introduction in Chapter 2.

[8]http://www.paraview.org

adjust the file dune-foo/src/CMakeLists.txt, or to simply replace the existing file dune-foo.cc with the example, leaving CMakeLists.txt intact.

A DUNE example program for solving the Poisson equation can be written at different levels of abstraction. It could use many DUNE modules and the high-level features. In that case, the user code would be short, but there would be no detailed control over the inner working of the program. On the other hand, an example implementation could be written to depend on only a few low-level modules. In this case more code would have to be written by hand. This would mean more work, but also more control and understanding of how the programs works exactly.

The example in this chapter tries to strike a middle ground. It uses the DUNE modules for grids, shape functions, discrete functions spaces, and linear algebra. It does not use a DUNE module for the assembly of the algebraic system—that part is written by hand. Chapter 11.3 contains an alternative implementation that uses dune-pdelab to assemble the algebraic problem.

At the same time, it is quite easy to rewrite the example to not use the DUNE function spaces or a different linear algebra implementation. This is DUNE—the user is in control. The finite volume example in Section 3.4 uses much less parts from DUNE than the Poisson example here does.

3.3.1. The `main` Method

We begin with the main method, i.e., the part that sits in

```
int main (int argc, char** argv)
{
  [...]
}
```

It is located at the end of the file, and is preceded by a few classes that make up the finite element assembler. As a first action, it sets up MPI [9] if available:

```
275    // Set up MPI, if available
276    MPIHelper::instance(argc, argv);
```

Note that this command is needed even if the program is purely sequential, because it does some vital internal initialization work.

Remember that all DUNE code resides in the Dune namespace. Hence type names like MPIHelper need to be prefixed by Dune::. Since that makes the code more difficult to read, the example file contains the line

```
28    using namespace Dune;
```

near the top. This allows to omit the Dune:: prefixes.

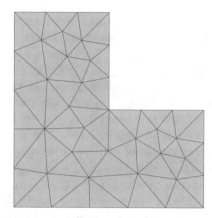

Figure 3.2.: Unstructured coarse grid for the domain used by the example. The actual simulation happens on a grid that results from this one after two steps of uniform refinement.

Creating a Grid

The first real action is to create a grid. The example program will use the triangle grid shown in Figure 3.2. A grid implementation in DUNE that supports unstructured triangle grids is the UGGrid grid manager.

 The grid itself is read from a file in the GMSH format [74].[10] The file can be obtained by clicking on the annotation icon in the margin. The following code sets up a two-dimensional UGGrid object with the grid from the GMSH file 1-shape.msh:

```
284    constexpr int dim = 2;
285    using Grid = UGGrid<dim>;
286    std::shared_ptr<Grid> grid = GmshReader<Grid>::read("1-shape.msh");
287
288    grid->globalRefine(2);
289
290    using GridView = Grid::LeafGridView;
291    GridView gridView = grid->leafGridView();
```

The first two lines of this code block define the C++ data structure used for the finite element grid, and the third line loads the grid from the file into an object of this data structure. Note that the grid dimension dim is a compile-time parameter. Line 288 refines the grid twice uniformly, to get a result with higher resolution. For this part of the code to compile, it needs to have the lines

```
8    #include <dune/grid/uggrid.hh>
9    #include <dune/grid/io/file/gmshreader.hh>
```

---

[9]The Message Passing Interface, used for distributed computing (see www.mpi-forum.org).
[10]http://gmsh.info

at the top. The final two lines extract the result of the second refinement step as a non-hierarchical grid. This so-called *leaf grid view* is where the actual finite element computation will take place on.

Line 285 has introduced the type variable `Grid` to store the type of the grid data structure that is used. This hints at one of the strengths of DUNE: It is easy to use the same code with different grid data structures. For the rest of the code, whenever the type of the grid is needed, we will only refer to `Grid`. This allows to change to, say, a structured cube grid by replacing only the definition of `Grid` and the subsequent constructor call. Indeed, to replace the unstructured grid by a structured cube grid for the unit square, replace Lines 285–286 by

```
using Grid = YaspGrid<dim>;
auto grid = std::make_shared<Grid>({1.0,1.0},    // Upper right corner,
                                                 // the lower left one
                                                 // is implicitly (0,0) here
                             {10, 10});           // Number of elements
                                                 // per direction
```

The YaspGrid class,[11] from the file `dune/grid/yaspgrid.hh`, is the standard implementation of a structured cube grid in DUNE.

Assembling the Stiffness Matrix and Load Vector

Now that we have a grid we can assemble the stiffness matrix and the load vector. For this we first need matrix and vector objects to assemble into. We get these with the lines

```
299    using Matrix = BCRSMatrix<double>;
300    using Vector = BlockVector<double>;
301
302    Matrix stiffnessMatrix;
303    Voctor b;
```

Both `BCRSMatrix` and `BlockVector` are data structures from the `dune-istl` module, and obtained by placing

```
15    #include <dune/istl/bcrsmatrix.hh>
16    #include <dune/istl/bvector.hh>
```

near the top of the program. It is, however, easy to use other linear algebra implementations instead of the one from `dune-istl`. That is another advantage of DUNE.

The next code block selects the first-order finite element space and the volume source term:

```
311    Functions::LagrangeBasis<GridView,1> basis(gridView);
312
313    auto sourceTerm = [](const FieldVector<double,dim>& x){return -5.0;};
```

---

[11]The name means "Yet another structured parallel Grid", for historical reasons.

The space is specified by providing a *basis* for it. The closed-form volume source term is written as a C++ lambda object.

To make the `main` method more readable, the actual assembly code has been put into a subroutine, which is called next:

```
316    assemblePoissonProblem(basis, stiffnessMatrix, b, sourceTerm);
```

The `assemblePoissonProblem` method will be discussed in Chapter 3.3.2.

Incorporating the Dirichlet Boundary Conditions

After the call to `assemblePoissonProblem`, the variable `stiffnessMatrix` contains the stiffness matrix A as defined in (2.11), but we still need to incorporate the Dirichlet boundary conditions. We do this in the way it has been described in Section 2.1.2, viz. if the i-th degree of freedom belongs to the Dirichlet boundary we overwrite the corresponding matrix row with a row from the identity matrix, and the entry in the right hand side vector with the prescribed Dirichlet value.

The implementation proceeds in two steps. First we need to figure out which degrees of freedom are Dirichlet degrees of freedom. Since we are using Lagrangian finite elements, we can use the positions of the Lagrange nodes to determine which degrees of freedom are fixed by the Dirichlet boundary conditions. We define a predicate class that returns **true** or **false** depending on whether a given position is on the Dirichlet boundary implied by (3.2) or not. Then, we evaluate this predicate with respect to the Lagrange basis to obtain a vector of booleans with the desired information:

```
322    auto predicate = [](auto x)
323    {
324      return x[0] < 1e-8
325          || x[1] < 1e-8
326          || (x[0] > 0.4999 && x[1] > 0.4999);
327    };
328
329    // Evaluating the predicate will mark all Dirichlet degrees of freedom
330    std::vector<bool> dirichletNodes;
331    Functions::interpolate(basis, dirichletNodes, predicate);
```

In general, there is no single approach to the determination of Dirichlet degrees of freedom that fits all needs. The simple method used here works well for Lagrange spaces and simple geometries. However, DUNE also supports other ways to find the Dirichlet boundary.

Now, with the bit field `dirichletNodes`, the following code snippet does the corresponding modifications of the stiffness matrix:

```
338    // Loop over the matrix rows
339    for (size_t i=0; i<stiffnessMatrix.N(); i++)
340    {
341      if (dirichletNodes[i])
342      {
```

```
343        auto cIt    = stiffnessMatrix[i].begin();
344        auto cEndIt = stiffnessMatrix[i].end();
345        // Loop over nonzero matrix entries in current row
346        for (; cIt!=cEndIt; ++cIt)
347          *cIt = (cIt.index()==i) ? 1.0 : 0.0;
348      }
349    }
```

Line 339 loops over all matrix rows, and Line 341 tests whether the row corresponds to a Dirichlet degree of freedom. If this is the case then we loop over all nonzero matrix entries of the row, using the iterator loop that starts in Line 346. Note how this loop is very similar to iterator loops in the C++ standard library. Finally, Line 347 sets the matrix entries to the corresponding values of the identity matrix, by comparing column and row indices.

Modifying the right hand side vector is even easier. The previous loop could be extended to also overwrite the appropriate entries of the b array, but it is equally possible to use the interpolation functionality of the dune-functions module a second time:

```
354    auto dirichletValues = [](auto x)
355    {
356      return (x[0]< 1e-8 || x[1] < 1e-8) ? 0 : 0.5;
357    };
358    Functions::interpolate(basis,b,dirichletValues, dirichletNodes);
```

The code defines a new lambda object that implements the Dirichlet value function, and computes its Lagrange interpolation coefficients in the b vector object. The fourth argument of the Functions::interpolate method restricts the interpolation to those degrees of freedom where the corresponding entry in dirichletNodes is set. All others are untouched.

At this point, we have set up the linear system

$$Ax = b \tag{3.3}$$

derived in Chapter 2.1, and this system contains the Dirichlet boundary conditions. The matrix A is stored in the variable stiffnessMatrix, and the load vector b is stored in the variable b. We seize the occasion and write matrix and load vector into two files. While not needed in this chapter, the matrix and vector files will be used in later examples in this book:

```
365    std::string baseName = "getting-started-poisson-fem-"
366                         + std::to_string(grid->maxLevel()) + "-refinements";
367    storeMatrixMarket(stiffnessMatrix, baseName + "-matrix.mtx");
368    storeMatrixMarket(b, baseName + "-rhs.mtx");
```

The data is written in the MATRIXMARKET format, described in [33].

Solving the Algebraic Problem

To solve the algebraic system (3.3) we will use the conjugate gradient (CG) method with an ILU preconditioner (see [138] for some background on how these algorithms work). Both methods are implemented in the dune-istl module, and require the headers dune/istl/solvers.hh and dune/istl/preconditioners.hh, respectively. The following code constructs the preconditioned solver, and applies it to the algebraic problem:

```
376    // Choose an initial iterate that fulfills the Dirichlet conditions
377    Vector x(basis.size());
378    x = b;
379
380    // Turn the matrix into a linear operator
381    MatrixAdapter<Matrix,Vector,Vector> linearOperator(stiffnessMatrix);
382
383    // Sequential incomplete LU decomposition as the preconditioner
384    SeqILU<Matrix,Vector,Vector> preconditioner(stiffnessMatrix,
385                                        1.0);  // Relaxation factor
386
387    // Preconditioned conjugate gradient solver
388    CGSolver<Vector> cg(linearOperator,
389                        preconditioner,
390                        1e-5, // Desired residual reduction factor
391                        50,   // Maximum number of iterations
392                        2);   // Verbosity of the solver
393
394    // Object storing some statistics about the solving process
395    InverseOperatorResult statistics;
396
397    // Solve!
398    cg.apply(x, b, statistics);
```

After this code has run, the variable x contains the approximate solution of (3.3), b contains the corresponding residual, and statistics contains some information about the solution process, like the number of iterations that have been performed.

Outputting the Result

Finally we want to access the result and, in particular, view it on screen. DUNE itself does not provide any visualization features (because dedicated visualization tools do a great job, and the DUNE team does not want to compete), but the result can be written to a file in a variety of different formats for post-processing. In this example we will use the VTK file format [107], which is the standard format of the PARAVIEW software.[12] This requires the header dune/grid/io/file/vtk/vtkwriter.hh, and the following code:

---

[12]http://www.paraview.org

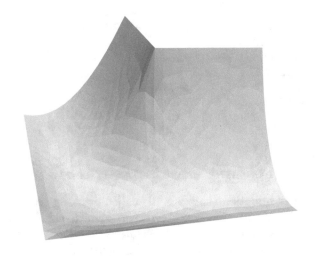

Figure 3.3.: Output of the Poisson example program, visualized as a height field

```
403    VTKWriter<GridView> vtkWriter(gridView);
404    vtkWriter.addVertexData(x, "solution");
405    vtkWriter.write("getting-started-poisson-fem-result");
```

The first line creates a VTKWriter object and registers the grid view. The second line adds the solution vector x as vertex data to the writer object. The string "solution" is a name given to the data field. It appears within PARAVIEW and prevents confusion when there is more then one field. The third line actually writes the file, giving it the name getting-started-poisson-fem-result.vtu.

Running the Program

With the exception of the stiffness matrix assembler (which is covered in the next section), the complete program has now been discussed. It can be built by typing make in the directory build-cmake. The executable getting-started-poisson-fem will then appear in the build-cmake/src directory. After program start one can see the GmshReader object giving some information about the grid file it is reading, followed by the conjugate gradients iterations:

```
Reading 2d Gmsh grid...
version 2.2 Gmsh file detected
file contains 43 nodes
file contains 90 elements
number of real vertices = 43
```

```
number of boundary elements = 22
number of elements = 62
=== Dune::IterativeSolver
 Iter          Defect              Rate
    0         3.26472
    1         0.851622          0.260856
    2         0.510143          0.599025
 [...]
   21         6.52302e-05       0.925161
   22         4.68241e-05       0.717829
   23         2.2387e-05        0.478109
=== rate=0.596327, T=0.0213751, TIT=0.000929351, IT=23
```

After program termination there is a file called `getting-started-poisson-fem-result.vtu` in the `build-cmake/src` directory, which can be opened with, e.g., PARAVIEW. It contains the grid and the solution function u_h, and when visualized using a height field, the result should look like Figure 3.3.

3.3.2. Assembling the Stiffness Matrix

We now show how the stiffness matrix and load vector of the Poisson problem are assembled. The example program here does it "by hand"—it contains a complete assembler loop that uses only the DUNE core modules and `dune-functions`. This will illustrate how to use the grid and discrete function interfaces, and can be used as a starting point for writing assemblers for other PDEs. Several additional DUNE modules provide full frameworks for finite element assemblers. Have a look at the `dune-pdelab`,[13] `dune-fem`,[14] and `dune-fufem`[15] modules, and read Chapter 11.

The Global Assembler

The main assembler loop is contained in the method `assemblePoissonProblem`, which is located above the main method in the example file. It has the signature

```
191   template<class Basis>
192   void assemblePoissonProblem(const Basis& basis,
193                               BCRSMatrix<double>& matrix,
194                               BlockVector<double>& b,
195                               const std::function<
196                                   double(FieldVector<double,
197                                                      Basis::GridView::dimension>)
198                                       > volumeTerm)
```

The method implements the assembly loop explained in Chapter 2.1.3; in particular, it assembles the individual element stiffness matrices, and adds them up to obtain the

---

[13]www.dune-project.org/modules/dune-pdelab
[14]www.dune-project.org/modules/dune-fem
[15]www.dune-project.org/modules/dune-fufem

global stiffness matrix. As the first step, it retrieves the grid object from the finite element basis by

```
202    auto gridView = basis.gridView();
```

The object `gridView` is then the finite element grid that the basis is defined on.

Next, the code initializes the global stiffness matrix. Before a `BCRSMatrix` object can be filled with values, it has to be given its occupation pattern, i.e., the set of all row/column pairs where nonzero matrix entries may appear:

```
208    MatrixIndexSet occupationPattern;
209    getOccupationPattern(basis, occupationPattern);
210    occupationPattern.exportIdx(matrix);
```

For brevity we do not show the code of the `getOccupationPattern` method here, because it is very similar to the actual assembler loop. Consult the complete source code in Appendix B.2 to see it in detail.

For all matrix entries that are part of the pattern, the next line then writes an explicit zero into the matrix:

```
215    matrix = 0;
```

Finally, the vector b is set to the correct size, and filled with zeros as well:

```
219    // Set b to correct length
220    b.resize(basis.dimension());
221
222    // Set all entries to zero
223    b = 0;
```

After these preliminaries starts the main loop over the elements in the grid:

```
228    auto localView = basis.localView();
229
230    for (const auto& element : elements(gridView))
231    {
```

The variable `localView` implements a restriction of the finite element basis to individual elements. Among other things, it provides the local set of shape functions, and how they relate to global degrees of freedom. The **for** loop in Line 230 iterates over the elements of the grid. The free method `elements` from the `dune-grid` module acts like a container of all elements of the grid in `gridView`. At each iteration, the object `element` will be a **const** reference to the current grid element.

Within the loop, we first bind the `localView` object to the current element. All subsequent calls to this `localView` will now implicitly refer to that element. Then, we create a small dense matrix and call the element matrix assembler for it:

```
237    localView.bind(element);
238
239    Matrix<double> elementMatrix;
240    assembleElementStiffnessMatrix(localView, elementMatrix);
```

In this implementation, the element assembler sets the correct matrix size for the current element. After the call to `assembleElementStiffnessMatrix`, the variable `elementMatrix` contains the element stiffness matrix for the element referenced by the `element` variable.

Finally, the element matrix is added to the global one:

```
244    for(size_t p=0; p<elementMatrix.N(); p++)
245    {
246        // The global index of the p-th degree of freedom of the element
247        auto row = localView.index(p);
248
249        for (size_t q=0; q<elementMatrix.M(); q++ )
250        {
251            // The global index of the q-th degree of freedom of the element
252            auto col = localView.index(q);
253            matrix[row][col] += elementMatrix[p][q];
254        }
255    }
```

The two **for**-loops iterate over all pairs of shape functions. The `localView` object knows the corresponding global degrees of freedom, and provides their numbers via its `index` method.

The Element Assembler

Finally, there is the local problem: given a grid element T, assemble the element stiffness matrix A_T for the Laplace operator and the given finite element basis. Remember that an entry $(A_T)_{pq}$ of the element stiffness matrix for the Poisson problem has the form

$$(A_T)_{pq} = \int_T \langle \nabla \phi_i, \nabla \theta_j \rangle \, dx,$$

where ϕ_i and θ_j are the basis functions from the trial and test spaces corresponding to the p-th and q-th local degree of freedom, respectively. For simplicity, the example implementation uses the same basis for both spaces, and we therefore only use the symbol ϕ for basis functions.

The matrix entry is computed by transforming the integral over T to an integral over the reference element T_{ref}

$$(A_T)_{pq} = \int_{T_{\text{ref}}} \langle \nabla F^{-T} \nabla \hat{\phi}_p, \nabla F^{-T} \nabla \hat{\phi}_q \rangle |\det \nabla F| \, d\xi, \tag{3.4}$$

where F is the mapping from T_{ref} to T, and $\hat{\phi}_p$, $\hat{\phi}_q$ are the shape functions on T_{ref} corresponding to the basis functions ϕ_i, ϕ_j on T. We approximate (3.4) by a quadrature rule with points ξ^k and weights ω_k,

$$(A_T)_{pq} \approx \sum_k \omega_k \left\langle \nabla F^{-T}(\xi^k) \nabla \hat{\phi}_i(\xi^k), \nabla F^{-T}(\xi^k) \nabla \hat{\phi}_j(\xi^k) \right\rangle |\det \nabla F(\xi^k)|.$$

This is the formula that the local assembler has to implement.

The corresponding method has the following signature:

```
33  template<class LocalView, class Matrix>
34  void assembleElementStiffnessMatrix(const LocalView& localView,
35                                       Matrix& elementMatrix)
```

The first parameter is the `LocalView` object of the finite element basis. From this view we get information about the current element, in particular its dimension and its shape:

```
39  using Element = typename LocalView::Element;
40  constexpr int dim = Element::dimension;
41  auto element = localView.element();
42  auto geometry = element.geometry();
```

The geometry object contains the transformation F from the reference element T_{ref} to the actual element T. Then, we get the set of shape functions $\{\hat{\phi}_p\}_{p=0}^{n_T-1}$ for this element:

```
47  const auto& localFiniteElement = localView.tree().finiteElement();
```

In DUNE-speak, the object that holds the set of shape functions is called a *local finite element*(details are given in Chapter 8). The need to invoke the method `tree` is a technicality. It exists to support vector-valued or mixed finite element spaces, and can be ignored for the time being.

We can now ask the `localView` object for the number of shape functions for this element, and initialize the element matrix accordingly:

```
52  elementMatrix.setSize(localView.size(),localView.size());
53  elementMatrix = 0;       // Fill the entire matrix with zeros
```

Then we need a quadrature rule. Such rules are provided by the dune-geometry module in the file dune/geometry/quadraturerules.hh:

```
58  int order = 2 * (localFiniteElement.localBasis().order()-1);
59  const auto& quadRule = QuadratureRules<double, dim>::rule(element.type(),
60                                                            order);
```

Line 58 estimates an appropriate quadrature order for simplex grids, and Line 59 gets the actual rule, as a reference to a singleton held by the dune-geometry module. A quadrature rule in DUNE is little more than a `std::vector` of quadrature points, and hence looping over all points is straightforward:

```
65  for (const auto& quadPoint : quadRule)
66  {
```

Now, with `quadPoint` the current quadrature point, we need its position ξ^k, the inverse transposed Jacobian $\nabla F^{-T}(\xi^k)$, and the factor $|\det \nabla F(\xi^k)|$ there. This information is provided directly by the DUNE grid interface via the geometry object:

```
70    // Position of the current quadrature point in the reference element
71    const auto quadPos = quadPoint.position();
72
73    // The transposed inverse Jacobian of the map from the reference element
74    // to the grid element
75    const auto jacobian = geometry.jacobianInverseTransposed(quadPos);
76
77    // The determinant term in the integral transformation formula
78    const auto integrationElement = geometry.integrationElement(quadPos);
```

Then we compute the derivatives of all shape functions $\{\nabla\hat{\phi}_p\}$ on the reference element, and multiply them from the left by ∇F^{-T} to obtain the gradients of the basis functions $\{\nabla\phi_i\}$ on the element T:

```
82    // The gradients of the shape functions on the reference element
83    std::vector<FieldMatrix<double,1,dim> > referenceGradients;
84    localFiniteElement.localBasis().evaluateJacobian(quadPos,
85                                                     referenceGradients);
86
87    // Compute the shape function gradients on the grid element
88    std::vector<FieldVector<double,dim> > gradients(referenceGradients.size());
89    for (size_t i=0; i<gradients.size(); i++)
90      jacobian.mv(referenceGradients[i][0], gradients[i]);
```

Note how the gradients of the $\{\hat{\phi}_p\}$ are stored in an array of *matrices* with one row in Line 83. This is because dune-localfunctions regards all shape functions as vector-valued functions, with a vector size of 1 for scalar-valued spaces. In the scalar case, getting the gradient $\nabla\hat{\phi}_p$ as a vector requires the suffix [0] in Line 90, which returns an object of type FieldVector<**double**,dim>.

Finally we compute the actual matrix entries:

```
95     for (size_t p=0; p<elementMatrix.N(); p++)
96     {
97       auto localRow = localView.tree().localIndex(p);
98       for (size_t q=0; q<elementMatrix.M(); q++)
99       {
100        auto localCol = localView.tree().localIndex(q);
101        elementMatrix[localRow][localCol] += (gradients[p] * gradients[q])
102                              * quadPoint.weight() * integrationElement;
103      }
104    }
```

By operator overloading, gradients[p]*gradients[q] implements the scalar product between two vectors. The expressions localView.tree().localIndex(p) and localView.tree().localIndex(q) compute the element matrix indices from the shape function numbers. In this simple case they simply map p to p and q to q, respectively. See Chapter 10 for more details.

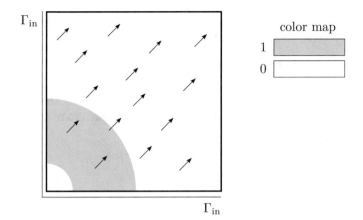

Figure 3.4.: Domain, velocity field (not to scale), and initial condition of the finite volume example

3.4. Example: Solving the Transport Equation with a Finite Volume Method

The second example program in this introductory chapter will show how to implement a simple first-order finite volume method. This will demonstrate a few more features of the DUNE grid interface, e.g., how to obtain face normals and volumes.

Compared to the Poisson solver of the previous section, the presented finite volume implementation uses much less features of the DUNE libraries. Instead of using a dedicated linear algebra library, C++ standard library types are used for coefficient vectors. Similarly, while cell-centered finite volume methods may be implemented using the function space basis objects from the dune-functions module, this would not make the code much simpler. The example therefore does not depend on dune-functions at all.

3.4.1. Discrete Linear Transport Equation

As the example problem we will use the linear scalar transport equation already encountered in Chapter 2.2.1. Let $\mathbf{v} : \Omega \times (0, t_{\text{end}}) \to \mathbb{R}^d$ be a given velocity field, and $c : \Omega \times [0, t_{\text{end}}] \to \mathbb{R}$ an unknown concentration. Transport of the concentration along the velocity flow lines is described by the equation

$$\frac{\partial c}{\partial t} + \text{div}(c\mathbf{v}) = 0 \qquad \text{in } \Omega \times (0, t_{\text{end}}).$$

For this example, we choose the domain $\Omega = (0,1)^2$, and the final time $t_{\text{end}} = 0.6$. As velocity field we pick

$$\mathbf{v}(x, t) = (1, 1),$$

which is stationary and divergence-free (Figure 3.4). By the choice of this field, a part of the boundary becomes the inflow boundary

$$\Gamma_{\text{in}}(t) := \{x \in \partial\Omega \; : \; \langle \mathbf{v}(x,t), \mathbf{n}(x) \rangle < 0\},$$

where $\mathbf{n}$ is the domain unit outer normal. In the current example, the inflow boundary consists of the lower and left sides of the square, and remains fixed over time. On the inflow boundary we prescribe the concentration

$$c(x,t) = 0 \qquad x \in \Gamma_{\text{in}}, \quad t \in (0, t_{\text{end}}).$$

Finally, we provide initial conditions

$$c(x,0) = c_0(x) \qquad \text{for all } x \in \Omega,$$

which we set to

$$c_0(x) = \begin{cases} 1 & \text{if } |x| > 0.125 \text{ and } |x| < 0.5, \\ 0 & \text{otherwise.} \end{cases}$$

For the discretization we cover the domain with a uniform grid consisting of $n = 80 \times 80$ quadrilateral elements. The time interval $[0, t_{\text{end}}]$ is split into uniform substeps

$$0 = t_0 < t_1 < t_2 < \cdots < t_m = t_{\text{end}},$$

with step size $\Delta t_k := t_{k+1} - t_k = 0.006$. We write T_i for the i-th grid element and $|T_i|$ for its volume. Likewise, γ_{ij} will denote the element facet common to elements T_i and T_j, $|\gamma_{ij}|$ the area of that facet, and $\mathbf{n}_{ij}$ its unit normal pointing from T_i to T_j. The velocity field $\mathbf{v}$ evaluated at the center of γ_{ij} will be called $\mathbf{v}_{ij}$. We use a cell-centered finite volume discretization in space, full upwind evaluation of the fluxes and an explicit Euler scheme in time as explained in Section 2.2.1. In particular, we approximate the unknown concentration c by a piecewise constant function, and identify the value $\bar{c}_i$ of this function on element T_i, $i = 0, \ldots, n-1$ by the mean value of c over that element. We obtain the following equation for the unknown element averages $\bar{c}_i^{k+1}$ at time t_{k+1}:

$$\bar{c}_i^{k+1}|T_i| - \bar{c}_i^k|T_i| + \Delta t_k \sum_{\gamma_{ij}} |\gamma_{ij}| \, \phi(\bar{c}_i^k, \bar{c}_j^k, \langle \mathbf{v}_{ij}, \mathbf{n}_{ij} \rangle) = 0 \qquad \forall i = 0, \ldots, n-1. \quad (3.5)$$

The flux function ϕ is an approximation of the flux $\langle c\mathbf{v}, \mathbf{n}_{ij} \rangle$ across the element boundary γ_{ij}. In Section 2.2.1 we have derived the expression

$$\phi(\bar{c}_i^k, \bar{c}_j^k, \langle \mathbf{v}_{ij}, \mathbf{n}_{ij} \rangle) := \bar{c}_i^k \max(0, \langle \mathbf{v}_{ij}, \mathbf{n}_{ij} \rangle) - \bar{c}_j^k \max(0, -\langle \mathbf{v}_{ij}, \mathbf{n}_{ij} \rangle). \quad (3.6)$$

Observe that it effectively switches between two cases, depending on whether there is flux from T_i to T_j or vice versa.

Inserting the flux function (3.6) into (3.5) and rearranging terms, we can solve (3.5) for the unknown coefficients $\bar{c}_i^{k+1}$ at time t_{k+1}. The resulting formula is a simple vector update

$$\bar{c}^{k+1} = \bar{c}^k + \Delta t_k \delta^k \tag{3.7}$$

with the update vector $\delta^k \in \mathbb{R}^n$ given by

$$\delta_i^k := -\sum_{\gamma_{ij}} \frac{|\gamma_{ij}|}{|T_i|} \left(\bar{c}_i^k \max(0, \langle \mathbf{v}_{ij}, \mathbf{n}_{ij} \rangle) + \bar{c}_j^k \max(0, -\langle \mathbf{v}_{ij}, \mathbf{n}_{ij} \rangle) \right). \tag{3.8}$$

3.4.2. The `main` Method

The implementation of the finite volume example is again contained in a single file. The complete file is printed in the appendix (Chapter B.3), and users of an electronic version of this text can get it by clicking on the icon in the margin. Do not forget that this program again has

```
13   using namespace Dune;
```

at the top, to avoid having to write the `Dune::` prefix over and over again.

The implementation is split between two methods: the `main` method and a method `evolve` that computes and applies the update vector δ defined in (3.8). We first discuss the `main` method. As in the finite element case, it begins by setting up the `MPIHelper` variable:

```
112   int main(int argc, char *argv[])
113   {
114     // Set up MPI, if available
115     MPIHelper::instance(argc, argv);
```

The `MPIHelper` instance sets up the MPI message passing system if it is installed. Even though this example does not use MPI, some parts of DUNE call it internally, and not initializing it would lead to run-time errors.

The first real code block sets up the grid:

```
119     constexpr int dim = 2;
120     using Grid = YaspGrid<dim>;
121     Grid grid({1.0,1.0},      // Upper right corner, the lower left one is (0,0)
122              { 80, 80});     // Number of elements per direction
123
124     using GridView = Grid::LeafGridView;
125     GridView gridView = grid.leafGridView();
```

Unlike the previous example, the finite volume implementation uses a structured grid. Therefore there is no need to read the grid from a file; giving the bounding box and the number of elements per direction suffices.

We then set up the vector for the element concentration averages $\bar{c}_i$:

```
130    MultipleCodimMultipleGeomTypeMapper<GridView>
131    mapper(gridView, mcmgElementLayout());
132
133    // Allocate a vector for the concentration
134    std::vector<double> c(mapper.size());
```

The `MultipleCodimMultipleGeomTypeMapper` object constructed in Line 130 is a device that assigns numbers to grid elements. These numbers are then used to address arrays that hold the actual simulation data. The mapper plays a similar role as the function space basis in the previous example, but it is a more low-level construct with less functionality. It is provided by the `dune-grid` module.

Line 134 creates the array that is used to store the concentration values $\bar{c}_i$. Observe that an array type from the C++ standard library is used. There is no dependence on the DUNE linear algebra module `dune-istl`, or any other dedicated linear algebra library.

The array `c` is then filled with the values of the initial-value function c_0 at the element centers. First, the function c_0 is implemented as a lambda object:

```
139    auto c0 = [](const FieldVector<double,dim>& x)
140    {
141      return (x.two_norm()>0.125 && x.two_norm()<0.5) ? 1.0 : 0.0;
142    };
```

Then, the code loops over the elements and samples c0 at the element centers. These one-point evaluations are used as approximations of the element averages that the algebraic variables $\bar{c}_i^k$ represent:

```
146    // Iterate over grid elements and evaluate c0 at element centers
147    for (const auto& element : elements(gridView))
148    {
149      // Get element geometry
150      auto geometry = element.geometry();
151
152      // Get global coordinate of element center
153      auto global = geometry.center();
154
155      // Sample initial concentration c0 at the element center
156      c[mapper.index(element)] = c0(global);
157    }
```

Loops over the elements have already appeared in the previous example. Note the special method `center` used in Line 153 to obtain the coordinates of the center of an element. While there is a more general mechanism to obtain coordinates for any point in an element (the `global` method of the geometry object), the element center is so frequently used in finite volume schemes that a dedicated method for it exists. The center coordinate is then used as the argument for the function object c0, which returns the initial concentration c_0 at that point. Line 156 shows how the mapper object is used: Its `index` method returns a nonnegative integer for the given element, which is used to access the data array `c`.

The next code block constructs a writer for the VTK format, and writes the discrete initial concentration to a file:

```
162    auto vtkWriter = std::make_shared<Dune::VTKWriter<GridView> >(gridView);
163    VTKSequenceWriter<GridView>
164        vtkSequenceWriter(vtkWriter,
165                          "getting-started-transport-fv-result");   // File name
166
167    // Write the initial values
168    vtkWriter->addCellData(c,"concentration");
169    vtkSequenceWriter.write(0.0);  // 0.0 is the current time
```

The `VTKWriter` constructed in Line 162 writes individual concentration fields to individual files. The `VTKSequenceWriter` in the following line ties these together to a time series of data. In addition to the individual data files, it writes a *sequence file* (with a `.pvd` suffix) that lists all data files together with their time points t_k. This information allows to properly visualize time-dependent data even if the time steps are not uniform.

The final block in the `main` method is the actual time loop:

```
174    double t=0;                      // Initial time
175    const double tend=0.6;           // Final time
176    const double dt=0.006;           // Time step size
177    int k=0;                         // Time step counter
178
179    // Inflow boundary values
180    auto inflow = [](const FieldVector<double,dim>& x)
181    {
182      return 0.0;
183    };
184
185    // Velocity field
186    auto v = [](const FieldVector<double,dim>& x)
187    {
188      return FieldVector<double,dim> (1.0);
189    };
190
191    while (t<tend)
192    {
193      // Apply finite volume scheme
194      evolve(gridView,mapper,dt,c,v,inflow);
195
196      // Augment time and time step counter
197      t += dt;
198      ++k;
199
```

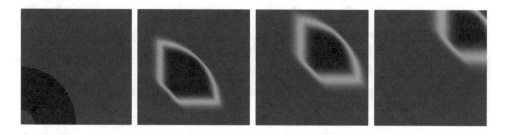

Figure 3.5.: Evolution of the concentration c at times $t = 0$, $t = 0.204$, $t = 0.402$, $t = 0.6$

```
200     // Write data. We do not have to call addCellData again!
201     vtkSequenceWriter.write(t);
202
203     // Print iteration number, time, and time step size
204     std::cout << "k=" << k << " t=" << t << std::endl;
205   }
```

Lines 174–177 initialize several variables, and two further lambda objects inflow and v for the inflow boundary condition and the velocity field, respectively. The loop starting in Line 191 iterates until the current time t has exceeded the specified end time tend. Most of the actual work is done in a separate method evolve, which we discuss below. The main loop then writes the concentration vector to a file, and proceeds to the next time step.

When run, the example program produces 101 output files, and the sequence file getting-started-transport-fv-result.pvd. A visualization of the simulation result is given in Figure 3.5. One can clearly see how the initial condition is transported along the velocity field v. The noticeable diffusion is caused by the crude numerical method.

3.4.3. The evolve Method

The evolve method does the main part of the work: After each call to evolve, the current iterate has advanced to the next time step.

The method signature is:

```
17  template<class GridView, class Mapper>
18  void evolve(const GridView& gridView,
19              const Mapper& mapper,
20              double dt,        // Time step size
21              std::vector<double>& c,
22              const std::function<FieldVector<double,GridView::dimension>
23                               (FieldVector<double,GridView::dimension>)> v,
24              const std::function<double
```

```
25        (FieldVector<double,GridView::dimension>)> inflow)
```

The first two arguments are the grid and the mapper. The third argument is the array of element concentration values c. This argument is a non-**const** reference, because the array is modified in-place. The arguments v and inflow are the velocity field and the inflow boundary condition functions, respectively.

The method starts by a bit of initialization code:

```
29    // Grid dimension
30    constexpr int dim = GridView::dimension;
31
32    // Allocate a temporary vector for the update
33    std::vector<double> update(c.size());
34    std::fill(update.begin(), update.end(), 0.0);
```

The array set up in Line 33 is the correction δ^k defined in (3.8).

The code then loops over all grid elements T_i:

```
39    for (const auto& element : elements(gridView))
40    {
41      // Element geometry
42      auto geometry = element.geometry();
43
44      // Element volume
45      double elementVolume = geometry.volume();
46
47      // Unique element number
48      typename Mapper::Index i = mapper.index(element);
```

This is the same kind of loop already seen in the global finite element assembler in Chapter 3.3. For each element T_i, the loop computes the update δ_i^k defined in (3.8). At the top of the loop, the element volume $|T_i|$ and its index i are precomputed.

The formula (3.8) for the correction δ_i^k for element T_i consists of a sum over all elements T_j whose boundaries intersect with the boundary of T_i in a $d-1$-dimensional set. Such neighborhood relations are represented in the DUNE grid interface by objects of type Intersection. These provide all relevant information about the relationship of an element with a particular neighbor or the domain boundary. The concept is deliberately general enough to allow for nonconforming grids, i.e., grids where the intersection of two elements is not necessarily a common facet. The sum in (3.8) is therefore coded as a loop over all intersections of the current element:

```
53    for (const auto& intersection : intersections(gridView,element))
54    {
55      // Geometry of the intersection
56      auto intersectionGeometry = intersection.geometry();
57
58      // Center of intersection in global coordinates
59      FieldVector<double,dim>
60          intersectionCenter = intersectionGeometry.center();
```

```
61
62          // Velocity at intersection center v_ij
63          FieldVector<double,dim> velocity = v(intersectionCenter);
64
65          // Center of the intersection in local coordinates
66          const auto& intersectionReferenceElement
67              = ReferenceElements<double,dim-1>::general(intersection.type());
68          FieldVector<double,dim-1> intersectionLocalCenter
69              = intersectionReferenceElement.position(0,0);
70
71          // Normal vector scaled with intersection area: n_ij|γ_ij|
72          FieldVector<double,dim> integrationOuterNormal
73              = intersection.integrationOuterNormal(intersectionLocalCenter);
74
75          // Compute factor occuring in flux formula: ⟨v_ij,n_ij⟩|γ_ij|
76          double intersectionFlow = velocity*integrationOuterNormal;
```

The loop itself uses the convenient range-based **for** syntax already seen when looping over the grid elements. It then computes the velocity $\mathbf{v}_{ij}$ and the product $\mathbf{n}_{ij}|\gamma_{ij}|$. To compute $\mathbf{v}_{ij}$, the value of the velocity field $\mathbf{v}$ at the center of the common intersection between T_i and its current neighbor T_j, Line 56 first acquires the geometry (i.e., the shape) of the intersection between T_i and T_j. Just like an element geometry, this intersection geometry has a center method which is called in Line 60. Line 63 then evaluates the velocity field at that position.

To evaluate $\mathbf{n}_{ij}$, we need the center of the intersection in local coordinates of the intersection. A direct method for this does not exist. On the other hand, there is a reference element corresponding to the intersection, which knows its center. That reference element is acquired in Line 67, and its center is evaluated in Line 69.

Rather than computing $|\gamma_{ij}|$ and $\mathbf{n}_{ij}$ separately, the code then calls a dedicated DUNE grid interface method called integrationOuterNormal that directly yields the product of the two. This product is used frequently in finite volume methods, so that a dedicated method makes sense. Even more importantly, it can be much more efficient to evaluate the scaled normal $|\gamma_{ij}|\mathbf{n}_{ij}$ directly, rather than computing the two separate factors first.

The second half of the intersection loop computes the actual update δ_i^k for the element T_i:

```
80          // Outflow contributions
81          update[i] -= c[i]*std::max(0.0,intersectionFlow)/elementVolume;
82
83          // Inflow contributions
84          if (intersectionFlow<=0)
85          {
86            // Handle interior intersection
87            if (intersection.neighbor())
88            {
89              // Access neighbor
```

```
90              auto j = mapper.index(intersection.outside());
91              update[i] -= c[j]*intersectionFlow/elementVolume;
92            }
93
94            // Handle boundary intersection
95            if (intersection.boundary())
96              update[i] -= inflow(intersectionCenter)
97                                 * intersectionFlow/elementVolume;
98          }
```

Line 81 adds $-\bar{c}_i \frac{|\gamma_{ij}|}{|T_i|} \max(0, \langle \mathbf{v}_{ij}, \mathbf{n}_{ij} \rangle)$, which covers the case that the current intersection is an outflow boundary of the element T_i. If $\langle \mathbf{v}_{ij}, \mathbf{n}_{ij} \rangle < 0$, i.e., if there is flow into element T_i, we need to distinguish between whether γ_{ij} really is the intersection with a second element T_j, or whether γ_{ij} is part of the domain boundary $\partial\Omega$ (in which case it is on the inflow boundary Γ_{in}). Lines 84–98 cover these two cases. Observe how the intersection knows whether there is an adjacent element T_j (through the neighbor method), and whether we are on the domain boundary (through the boundary method). If the grid is distributed across several processors, both methods may return **false** at the same time. However, in this simple sequential example this cannot happen.

This ends the loop over the intersections, and the loop over the elements ends as well:

```
100        }  // End loop over all intersections
101      }  // End loop over the grid elements
```

Finally, the concentration vector is updated:

```
105      // Update the concentration vector
106      for (std::size_t i=0; i<c.size(); ++i)
107        c[i] += dt*update[i];
108    }
```

This ends the discussion of the evolve method.

4. The Design and Structure of Dune

DUNE is a large software system. As with all such systems there is always the danger of program complexity increasing uncontrollably, making program maintenance more and more difficult. DUNE tries to keep this problem in check with a modular design, and a set of rules on how the programming interface (API) should look like. This chapter explains and motivates this design. It also gives some background on writing interfaces in C++, and discusses a few conventions that DUNE has adopted regarding such interfaces.

4.1. Software Functionality for Finite Element and Finite Volume Methods

From the brief introduction to finite element and finite volume methods in Chapter 2 it can be seen that the software functionality needed to implement such methods can be grouped into several fields. Here is a list of the more prominent ones.

Grid Handling   All methods of Chapter 2 rely on a computational grid.[1] A software system for such methods therefore needs to provide a data structure for such a grid. The data structure needs to allow efficient access to those grid aspects relevant for finite element and finite volume methods. For example, looping over all grid elements must be possible to assemble stiffness matrices, and the maps F_T from the reference elements to the grid elements (2.18) must be available to perform numerical quadrature. Grids also need to be read from files, and written back later for the visualization of results.

Grids come in various kinds, ranging from simple triangle grids like the one in Figure 2.2 to the exotic ones in Figure 2.3. A good grid software must handle all these cases, while retaining maximum efficiency. Grid data structures should support the various types of local refinement and coarsening, and possibly allow for vertex repositioning (r-refinement, see [95]) as well. Ideally, they implement the concept of refinement trees of Chapter 2.3.3, which then also permits to implement geometric multigrid methods [37, 155].

In this book we differentiate between the *handling* of a grid, and the *construction* of a grid. We believe that the latter problem is beyond the scope of DUNE, and best left to dedicated special-purpose software. See [153] for an overview.

---

[1] There are also methods for the numerical solution of PDEs that do not make use of a grid, but such methods are not the focus of DUNE.

© Springer Nature Switzerland AG 2020
O. Sander, *DUNE — The Distributed and Unified Numerics Environment*, Lecture Notes in Computational Science and Engineering 140, https://doi.org/10.1007/978-3-030-59702-3_4

Finite Element Basis Functions and Shape Functions As seen in Chapter 2.1.2, the assembly of finite element problems requires basis functions that span given finite element spaces. These functions can be used directly, but they are frequently pulled back to the reference elements, and are then called *shape functions*. A finite element code therefore has to provide implementations of finite element basis functions, either directly or in form of shape functions. While a few standard bases are predominantly used, there is also a wide range of rather exotic finite element bases for special problems [106].

In a finite element assembly routine, shape functions typically appear in integral terms. For second order PDEs, basis function values and first derivatives are needed, but for higher order equations higher derivatives of the basis functions can become necessary. Also, a mechanism is needed to associate individual shape functions with global degrees of freedom.

Numerical Quadrature The integrals (2.14) and (2.15) making up the entries of the stiffness matrices and load vectors cannot generally be evaluated in closed form. Therefore, computer codes have to provide a mechanism for numerical quadrature on the reference elements. A quadrature rule is a fairly simple object, consisting of little more than a set of points in a reference element and corresponding scalar weights. Finite element codes are expected to provide a library of such rules, for different element types, up to high orders, and possibly with a high precision. This usually means storing large tables of point coordinates and weights.

Assembly There needs to be a part in the code that, for a given equation and finite element space, assembles the stiffness matrices and load vectors, using the infrastructure described so far. In the simplest case, this code implements the loop in Chapter 2.1.3. While the loop itself is the same for all PDEs, the assembly of the element stiffness matrices depends on the problem. Therefore, finite element software will usually provide a collection of local assembly routines for different PDEs.

Matrices and Vectors The result of the assembly routine is usually a linear or nonlinear algebraic system of equations. A computer code therefore needs to provide data structures to handle such systems. Global stiffness matrices are usually sparse, i.e., they contain a large number of zero entries. There is a variety of data structures for sparse matrices, each with advantages and disadvantages [133]; a potent computer code will give the user a choice. Also, many finite element discretizations introduce some form of block structure in the degrees of freedom. Linear algebra data structures may try to reflect this, in order to increase efficiency.

Algebraic Solvers Finally, the algebraic problems resulting from the problem discretization need to be solved. In most cases this involves solving large sparse linear systems of equations. Many different algorithms are used for this, including direct methods [47], Krylov methods [138], and multigrid methods [37, 155]. For nonlinear problems, Newton-type methods are common. Such methods have to interact tightly with the discretization because tangent stiffness matrices have to be reassembled at each iteration, and it can be helpful to couple the solver precision and damping to the discretization error [53].

Concurrency and Communication Many of today's application PDE problems get so large and challenging that every aspect of modern hardware needs to be exploited to obtain acceptable simulation times. This includes concurrency on various levels, like vectorized computation, multi-threading, and distributed computing. Several of these techniques require communication between threads or processes running in parallel. While good general purpose libraries for inter-process communication exist (MPI being the most prominent one), they do not usually directly support the type of more structured communication patters that are used in simulation software for partial differential equations. For example, communication in a finite element code will usually send data near the boundary of a subdomain grid to the processors holding the neighboring subdomain grids. A good finite element software should offer such functionality directly, and help to translate these communication patterns from the grid-based view to the grid-agnostic view of MPI.

4.2. The Structure of Dune

Most finite element or finite volume codes offer the functionality of the previous list in some way or another. However, they usually combine it in a monolithic program. This means that there is a *single program* to download and install. This program contains all functionality tightly connected. It may be controlled by a graphical user interface (e.g., ABAQUS[2]) a scripting language (FENICS[3] [120] or pdelib[4]), or it may offer programming interfaces like a library.

In contrast, DUNE is not a single program. Instead, it is a set of libraries, which are as independent from each other as possible. In DUNE terminology, each such library is called a *module*. Each module contains a distinct and well-defined part of the functionality. This separation is based on the list of requirements given in the previous section: there is a module for grid handling, a module for shape functions, one for linear algebra and several modules for discretizations. Figure 4.1 shows a diagram of all DUNE modules that are covered in this book, together with the graph of their dependencies. Chapter 4.3 gives an overview over the central components (termed the *core modules*), and several chapters in later parts of this book are dedicated to individual modules.

Each DUNE module is an individual entity that has to be downloaded and installed separately. It contains a text file called dune.module which formally specifies the dependencies on other modules. Most dependencies will be hard requirements, meaning

---

[2]www.simulia.com
[3]fenicsproject.org
[4]www.wias-berlin.de/software/pdelib

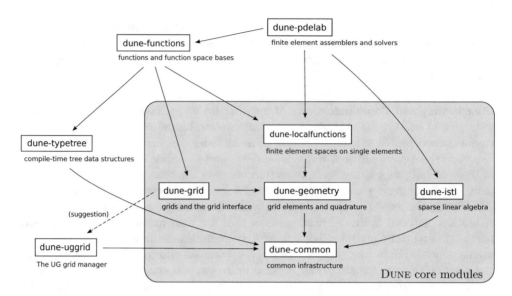

Figure 4.1.: Dependency graph of the DUNE modules covered in this book

that a module cannot be built at all without the modules it depends on. Alternatively, dependencies can be suggestions. In that case a module will work without the dependency, but with reduced functionality.

A set of modules with dependency relations forms a directed acyclic graph. A special program called dunecontrol (provided by the dune-common module) tracks and resolves these dependencies. In other words, if a specific module needs to be built, dunecontrol makes sure that all its dependencies are built and available when they are needed. In this sense, dunecontrol acts very much like the package managers used by the different Linux distributions. The dunecontrol program is described in Appendix A.1.1.

Once a set of DUNE modules is installed on a system, end users can use it just like any other C++ library. For example, the DuMu^X [109] and KASKADE 7 [76] projects[5,6] have both built PDE simulation toolboxes on top of some DUNE modules. On the other hand, many application developers write their application codes in form of DUNE modules, too. This gives them a head start for new projects (an empty new DUNE module with all build system boilerplate code can be created easily with the help of the duneproject program), and handles the dependency tracking automatically. Also, it is not uncommon for something that started as an application to eventually turn into something that other modules will want to depend on. This is then trivially possible if the application is a DUNE module.

---

[5]https://dumux.org
[6]www.zib.de/en/numerik/software/kaskade-7.html

Designing DUNE as a set of separate libraries rather than as one monolithic code has quite a number of advantages.

- Using DUNE is not an all–or–nothing decision. A monolithic code is either used completely, or not at all. This can be problematic if a given code excels in one field, but is suboptimal in another. With DUNE, on the other hand, only those parts have to be installed and used that are really desired, while something else can be used for the rest. For example, there are very good linear algebra libraries available. While DUNE does offer its own linear algebra (in the dune-istl module, see Chapter 7), it is easy to use DUNE for grids but something else (say, PETSc[7]) for linear algebra. This increases the user flexibility tremendously.

- A related point is that a modular structure makes it easier to migrate to (and away from) DUNE. Suppose that the aim is to port a finite element code that implements its entire infrastructure like grids, shape functions, etc. itself, to run on top of DUNE instead. Rather than having to rewrite the entire lower level and switch to DUNE in one go, it is possible to proceed module by module. For example, a first step could be to keep the code as it is and only replace the part implementing shape functions by calls to dune-localfunctions. Once this runs in a stable way, it is then possible to, as a next step, replace the grid data structure by the ones from dune-grid. This way makes transitions much less disruptive.

- The modular structure of DUNE allows for a decentralized development model. Third-party developers can provide additional functionality in separate modules, without ever having to touch the DUNE core modules. This does indeed happen: the DUNE project webpage lists a collection of modules that extend DUNE in various ways, but that are implemented and hosted by people not directly associated with the DUNE project. Examples are the opm grid grid manager for corner point grids [117, 135] currently hosted by the Open Porous Media Initiative,[8] the dune-fufem discretization module[9] and the dune-curvilinear module for grids with polynomial geometries.[10]

  The advantage of such a decentralized model is that many people can work on a common software system while keeping the necessary communication at a minimum. This allows to beat *Brook's law*, which states that as a software engineering team grows in size, the time lost in communication will eat up the additional productivity gained by having more team members [39]. Instead, small agile teams can focus on single modules, without having to care much about the larger DUNE ecosystem.

---

[7]www.mcs.anl.gov/petsc
[8]www.opm-project.org
[9]https://git.imp.fu-berlin.de/agnumpde/dune-fufem
[10]www.curvilinear-grid.org

- A side effect of the decentralized development style has been the appearance of several competing implementations of the same functionality. Prime examples are the `dune-pdelab`[11] and `dune-fem`[12] modules, which both implement finite element and finite volume discretization routines (`dune-pdelab` is presented in Chapter 11). Both are mature projects with a considerable user base. Since both have their individual strengths, having both gives users additional choice and flexibility.

- The design of DUNE as a set of independent modules allows for rapid prototyping: To try something quickly, create a new module and try what you want to try. If it doesn't work out, simply remove the module directory and it is gone completely. This helps to avoids the feature creep sometimes seen in research codes, where small features are occasionally added "for a short time only", never to be removed again.

- Last but not least, modular design is good software design. Modular code is much easier to read and understand, because modules can only access functionality from other modules that they have explicitly declared dependence on. In contrast, monolithic codes sometimes have the tendency to develop a "spaghetti structure" [161], which makes them more difficult to understand and maintain.

DUNE was among the first projects to implement this radical modularization in a PDE context, and it has worked out very well.

4.3. The Dune Core Modules

Due to its decentralized style of development, different DUNE modules are available from different sources. There is a set of six modules though that form the core of DUNE, or DUNE proper. These modules a called the DUNE *core modules*. They are hosted on the main project website at `www.dune-project.org`, and their development is overseen by the DUNE steering committee.[13] The functionality of individual modules follows the thematic blocks that we have described in Chapter 4.1. The correspondence is not always perfect, with deviations sometimes for technical and sometimes for historical reasons.

We briefly present the core modules here, but they are also described in detail in later chapters.

- The module `dune-common` is the foundation of everything. It contains basic infrastructure for the build system, the `dunecontrol` program to track inter-module dependencies, and the `duneproject` program to create new modules (see Section 3.2.1). It further contains lots of helper code like special container

---

[11]`www.dune-project.org/modules/dune-pdelab`
[12]`www.dune-project.org/modules/dune-fem`
[13]`www.dune-project.org/community/people`

classes, help for template metaprogramming, dense matrix and vector classes, an implementation of distributed indices, and much more. dune-common is depended upon by all other modules, and DUNE cannot be used without it.

As dune-common is basically a collection of independent helper classes with little interrelation there is no separate chapter about it in this book. The easiest way to find out about what dune-common has to offer is to look into its online reference documentation at www.dune-project.org.

- dune-grid provides computational grids for finite volume and finite element computations. It contains the abstract C++ interface for grids described in mathematical terms in [15]. This includes things like iterators over vertices and elements, mechanisms to handle adaptive grid refinement, and partitioning of grids on distributed machines. It does *not* include mechanisms for the creation of unstructured grids, which is left to external software. The dune-grid interface is explained in detail in Chapter 5.

  In addition to the interface, the dune-grid module also contains several grid implementations of the interface. Some of them are native, like YaspGrid, a structured parallel grid, or OneDGrid, a fully adaptive one-dimensional grid. Others are wrappers to grid managers of existing finite element libraries. Currently, dune-grid contains adapters to the grids of ALBERTA [141] and of UG3 [13]. The actual code of the latter is contained in the dune-uggrid module, which is a suggested dependency of dune-grid. The separation between abstract interface and implementation allows to access the features of those grid managers in a simple, efficient and unified way.

- The module dune-geometry contains a part of the grid interface, split off from dune-grid into a separate module. In DUNE terminology, a *geometry* is the mapping F_T from a reference element into the world space, which describes individual grid element geometries. dune-geometry provides various implementations for such mappings, e.g., for multilinear or affine mappings. These are not part of the dune-grid module, because they are sometimes needed in a situation where the rest of the grid interface is unnecessary. One example are the non-affine finite elements implemented in the dune-localfunctions module (Chapter 8).

  In addition to the geometries, the dune-geometry module provides quadrature rules for the different reference elements. These are documented in Chapter 9.

- dune-localfunctions provides an interface to sets of finite elements, i.e., basis functions of finite element spaces, and associated interpolation rules. The module currently focusses on functions that can be expressed in form of shape functions on a reference element. dune-localfunctions is independent of dune-grid, and only needs dune-common and dune-geometry to work. It also provides a collection of different finite elements, ranging from simple Lagrangian elements to advanced vector-valued types. The `dune-localfunctions` module is explained in Chapter 8.

- Finally, `dune-istl` (with "istl" abbreviating "iterative solver template library") is a library for linear algebra. It provides implementations for various types of dense and sparse matrices, and corresponding vector types. As a special feature, these types can be nested in an almost arbitrary fashion. This allows to exploit a lot of the blocking and sparsity structure inherent in many finite element problems. `dune-istl` also implements several basic iterative solvers for linear systems. These run both on sequential and on distributed machines. Particularly noteworthy is the implementation of an algebraic multigrid algorithm, which scales to very large processor numbers [31, 100]. The module `dune-istl` is explained in Chapter 7.

The book also discusses three modules that are not part of the official DUNE core:

- `dune-uggrid` contains a fork of the grid data structure of the UG3 finite element software [13]. It provides unstructured, locally adaptive, multi-element grids in two and three space dimensions. It is used in this book because `dune-grid` itself does not provide an unstructured grid implementation. The `dune-uggrid` module is presented in some detail in Chapter 5.10.1.

- The `dune-functions` module provides interfaces for functions. This includes finite element functions defined on a grid, but also general functions defined by closed-form expressions. `dune-functions` also describes spaces of finite element functions by providing an interface for bases of such spaces. A powerful mechanism allows to easily construct complicated spaces from simpler ones. The `dune-functions` module is covered in Chapter 10.

- Finally, `dune-pdelab` provides finite element assemblers on top of these function space bases. It implements the general global element loop of Chapter 2.1.3, and a wide variety of local element assemblers for different partial differential equations, with different types of boundary conditions. Mechanisms for local grid adaptivity are included, and all this can run on distributed architectures. For size reasons we can only cover parts of the functionality of `dune-pdelab`. We do so in Chapter 11.

4.4. Designing Efficient Interfaces

Arguably the most important aspect of DUNE is the set of interfaces it provides for different pieces of infrastructure used by finite element and finite volume codes. These interfaces must be general enough to cover (almost) all possible use cases. On the other hand they should be simple and concise: large complicated interfaces are difficult to implement and difficult to learn. They must be easy to document, and, last but not least, their impact on run-time performance should be negligible.

The C++ language offers several different approaches for constructing interfaces. They differ in flexibility, ease of use, and the amount of run-time overhead incurred by the interface layer. Several of these approaches are used in different parts of DUNE; sometimes for factual, and sometimes for historical reasons. To allow people a better understanding of why certain interfaces in DUNE are designed in a particular way, this section discusses three ways to construct interfaces in C++.

4.4.1. Abstract Base Classes

The classical way to build interfaces in C++ is based on *abstract base classes* [150]. An interface is an abstract base class, which is a base class that declares all required methods as *pure virtual* methods, for example,

```
class AbstractInterface
{
public:
  virtual void doSomething() = 0;
};
```

The compiler will not allow objects of type AbstractInterface to be constructed. Only derived classes may be constructed, and only if they overload all pure virtual functions. Implementations of the interface are therefore classes that derive from the interface class, and overload all pure virtual methods in the base class:

```
class InterfaceImplementationA
: public AbstractInterface
{
public:
  virtual void doSomething();
};

void InterfaceImplementationA::doSomething()
{
  // do something
}
```

Objects that implement the interface are referred to by pointers to the base class. As the compiler will not allow construction of class objects with pure virtual methods, it is guaranteed that all methods of the interface are implemented.

Interfaces based on abstract base classes have many advantages. They are fairly easy to understand and teach. Compiler support is good, which means in particular that compilers emit short, helpful error messages if code is malformed. Also, the approach puts a lot less stress on the compiler than corresponding template techniques. Finally, the interface specification is explicitly encoded in the source code: The class AbstractInterface *is* the interface. This is a big advantage compared to techniques like duck typing (Chapter 4.4.2), where the interface is only prescribed implicitly, and must be documented separately.

Unfortunately, abstract base classes also have disadvantages, in particular for high-performance libraries such as DUNE. First of all, they cannot be used for methods that have member template parameters

```
template<class T>
virtual void foo(const T& t) = 0;    // error
```

Such methods, however, are required in various DUNE interfaces.

Secondly, virtual function calls incur a run-time overhead. Although very small, this overhead can become noticeable when the called method is very short and called very frequently. This is best explained by extending the example above by a second interface implementation:

```
class InterfaceImplementationB
: public AbstractInterface
{
public:
  virtual void doSomething();
};

void InterfaceImplementationB::doSomething()
{
  // do something
}
```

Interface methods are typically called via the base class:

```
void someMethod(AbstractInterface* object)
{
  object->doSomething();
}
```

In this situation, the compiler cannot decide which of the two implementations of the doSomething method is going to be called—It is the run-time system which has to determine this, which usually means at least one additional pointer access. What is worse, not knowing the called method at compile-time prevents compiler optimizations such as inlining. Together, these two effects can lead to a measurable slowdown [59].

To improve upon this situation, all major compilers implement an optimization technique called *devirtualization*. In certain situations the compiler is able to gather enough information to conclude at compile time which method is being called by a virtual function call. This call can then be replaced by a direct call, and inlined if desired. If the precise call destination is not known with absolute certainty, but with a high probability, *speculative devirtualization* allows to decrease the price of that call almost to the price of a direct call [96].

Because of the potential run-time costs, and because it mingles badly with template techniques, textbook-style abstract base classes are only rarely used in DUNE. They are used internally on some occasions, but not for any major interfaces.

4.4.2. Duck Typing

With the previous approach, an interface was declared in form of an explicitly given abstract base class. Classes that claim to implement the interface had to explicitly state this by deriving from the base class. The compiler then made sure that the implementation class actually does implement all required methods.

This approach is quite rigid. Even classes that do implement all methods of the interface specification are not accepted as interface implementations if they do not derive from the correct base class. Also, in many situations classes should implement more than a single interface. It is possible to derive from more than one base class in C++, but this approach has its own share of pitfalls, and is generally discouraged.

A different approach called *duck typing*[14] works without any interface classes. In duck typing, an interface is merely an informal list of methods, types, and semantic requirements that each implementation has to fulfill. Any class that implements all requirements is accepted as an interface implementation, without having to derive from particular interface classes [110]. This approach is heavily used in the C++ standard library. For example, the std method for sorting the elements of a container is declared as

```
template<class RandomIt>
void sort(RandomIt first, RandomIt last)
```

Objects of any type can be passed as arguments, as long as they implement the methods of a RandomAccessIterator.[15]

The main advantage of duck typing is the vast increase in flexibility compared to traditional dynamic polymorphism. Any class can implement any number of interfaces, without having to worry about complicated inheritance hierarchies. What is more, classes can get away with implementing only parts of an interface, if it is known that only that part is required in a particular situation. C++ introspection techniques even allow the calling code to test whether individual parts of the interface exist or not, and adjust its behavior accordingly.

The second advantage of duck typing in C++ is run-time efficiency. Interface implementations are typically passed around as template arguments. There are therefore no virtual function calls, and no corresponding performance penalty. Compilers can apply their full range of optimization techniques.

The downside of duck typing in C++ is that the interface cannot be explicitly enforced by the compiler. The compiler does report errors when the implementation does not meet all requirements of the calling code, but the error messages are frequently cryptic and a lot less clear than the corresponding messages for dynamic polymorphism. This is no surprise, because the interface specification is only given implicitly, and hence unknown to the compiler. It never appears anywhere in the code, and is merely

---

[14]This expression is inspired by American poet James Whitcomb Riley, who reportedly gave a rule on how to recognize ducks: "When I see a bird that walks like a duck and swims like a duck and quacks like a duck, I call that bird a duck." It does not matter what the bird *is*, it is what the bird *does* that counts.

[15]There are a few other technical restrictions, but you get the idea.

a convention among the programmers. If a formal interface specification is required, then it must be written in form of a separate text document. Such a text is difficult to create and maintain, because there is no technical mechanism that verifies whether the implementation matches the specification document.

Concept checking [97, 151] has been proposed as a way to allow for a bit of formal interface verification. Aspects of it have been included in the 2020 revision of the C++ language, and at the time of writing they are starting to appear in major compilers. Until they become fully established, DUNE provides a manual implementation of light-weight concept checks, based on the techniques proposed by Niebler [126] and implemented in the RANGE-V3 library.[16] For example, the following method evaluate checks whether its argument type is a function that maps a **double** to a **double**, i.e., whether **operator**() can be called for it with a **double** argument and returns a **double**:

```
template<class F>
void evaluate(F&& f)
{
  // Concept check that F has method 'double operator()(double)'
  static_assert(models<Functions::Concept::Function<double(double)>, F>(),
      "Type does not model function concept");

  std::cout << "Value of f(1): " << f(1) << std::endl;
}

int main(int argc, char *argv[])
{
  evaluate([](double x){return x*x;});    // Works
  //evaluate(42);                          // Error, '42' is not a function
}
```

If evaluate is instantiated with a type that is not a function, then a readable error message is produced. For example, for the code

```
evaluate(42);      // The integer 42 is not a function object
```

GCC-8.2 prints the error message (file paths and line numbers removed)

```
In instantiation of 'void evaluate(F&&) [with F = int]':
  required from here
error: static assertion failed: Type does not model function concept
    static_assert(models<Functions::Concept::Function<double(double)>, F>(),
    ~~~~~~~~~~~~~~~~~~~~~~~~~~~~~~~~~~~~~~~~~~~~~~~~~~~~~~~~~~~~~~~^~
error: expression cannot be used as a function
    std::cout << "Value of f(1): " << f(1) << std::endl;
                                     ~^~~
```

---

[16]https://github.com/ericniebler/range-v3

The dune-common module provides the general concept checking facility in the header dune/common/concept.hh. The concept definitions themselves are contained in the respective modules.

In DUNE, duck typing is used in a lot of places, e.g., the dune-istl, dune-localfunctions, and dune-functions modules. In some cases, interface classes do exist, but they have no functional purpose and only exist to document the interface.

4.4.3. Wrappers and Engines

Several places in the DUNE core modules, in particular the DUNE grid interface, use a third technique known as the *wrapper–engine* approach. We have not found this technique described in the literature, but it is related to, e.g., the *bridge* pattern of [72]. As a special form of static polymorphism, it has the run-time efficiency of pure template duck typing, but uses an explicit interface class which enforces and documents the interface.

In the wrapper–engine approach, the interface class is not a base class in an inheritance tree. Instead, it is a wrapper that takes the actual implementation class (the engine) as a template parameter, and forwards all calls to it. In the following example, Interface is the interface class, and InterfaceImp is one implementation. The user is prevented to see a separate object of type InterfaceImp, typically by making all constructors private. Instead, she will always deal with objects of type Interface<InterfaceImp>.

```
template<class Imp>
struct Interface
{
  void doSomething() const
  {
    imp_.doSomothing();
  }

private:
  Imp& imp_;
};

struct InterfaceImp
{
  void doSomething() const
  {
    std::cout << "This is InterfaceImp doing something!" << std::endl;
  }
};
```

There are several advantages to this approach. The first is encapsulation: The user only gets to see the interface class Interface, with a fixed set of methods and exported static information. Secondly, given a good optimizing compiler, the efficiency of wrapper–engine interfaces is as high as for duck typing. There is no overhead due to virtual functions. Furthermore, since all types are known at compile time, the method calls can be inlined where appropriate. Support for template member functions is not a problem. Finally, there exists a class which *is* the interface. This automatically provides a verbose description of the interface, unlike in pure duck typing where the interface is only implicit and needs to be described in a separate text document. In particular, this documentation is up-to-date by construction, and will not diverge from the actual code. Furthermore, the interface is enforced by the compiler. This makes it easier to detect errors if implementation classes do not implement the correct interface.

Unfortunately, the approach also has its problems: The support by the language and compilers is not as good as for dynamic polymorphism. While the compiler would report an error if `InterfaceImp` did not implement the method `doSomething` in a dynamically polymorphic setting *at object declaration time*, in static polymorphism that error only occurs if the method is actually used.[17] Secondly, debugging becomes more difficult. As the type of the implementation class is part of the type of the interface class, compiler error messages become longer and more difficult to read. (Not *very* difficult, though, as there are no recursive constructions.) Also, due to the additional redirection, stepping through a program with a debugger becomes more cumbersome.

For the case of the DUNE grid interface (Chapter 5) with its focus on efficiency, it has been decided that the advantages of the wrapper–engine concept outweigh the costs.

4.5. Coding Style

The code in the different DUNE modules follows certain style rules. These are a lot less thorough and formalized than the rules for some other projects, and only a few of them are enforced automatically. The advantage of such rules is that it becomes easier to use and talk about DUNE code if all modules adhere to a common set of rules regarding their formal structure. This does not only concern people that work on the core DUNE modules themselves, but also most users. In many cases, using DUNE involves writing new DUNE modules, and it is then convenient (even if not required) to follow the same style rules.

In this chapter, we focus on style rules for the C++ code in DUNE modules. There are also rules concerning the file naming and directory layout of such modules. These are discussed in Appendix A.2.1.

---

[17] Actually, some consider this as an advantage.

4.5.1. Rules Regarding the Code in a Dune Module

Consistent naming rules make libraries easier to use. Unfortunately, while DUNE does have such rules, they are not consequently adhered to throughout the code base. This has historical reasons and is difficult to change, because these names are part of the public interface, and are used in a lot of user code.

- Class- and method names use camel case [159]. Types and classes start with a capital letter, objects and method names start with a lower-case letter. For example, the class implementing the ALBERTA grid manager [141] is called AlbertaGrid, and a grid object of this class could be called albertaGrid. There are a few exceptions to these rules, to increase the interoperability with the standard template library. These are explained in the following section.

- If a method or class name contains an acronym, then this acronym is either spelled in all upper- or all lower-case letters (depending on the context). For example, the type of a block-compressed-row-storage matrix in dune-istl is BCRSMatrix, and a method returning such a matrix would be called bcrsMatrix().

- The code in each DUNE module is contained in the Dune namespace (but do not forget that the Dune:: prefix is omitted in all code examples in this book). Furthermore, the code of each module is contained in nested namespaces with a module-specific name. This is usually the module name without the dune- prefix, in camel case. For example, the code in the dune-functions module is contained in the Dune::Functions namespace, and the code in dune-pdelab is contained in Dune::PDELab. As a historical exception, the code in the core modules dune-common, dune-geometry, dune-grid, dune-istl, and dune-localfunctions is placed directly with the Dune namespace, and is *not* contained in a further module-specific namespace.

- Code that is not to be used from outside the module that it appears in should be placed in a namespace called Impl, nested within the above-mentioned namespaces. Such code is allowed to be changed at any moment.

- Names of private class data members end with an underscore.

There are further rules specifying indentation, the use of curly braces, etc. A list is given at www.dune-project.org/dev/codingstyle.

4.5.2. Compatibility with the C++ Standard Library

One idea underlying the DUNE design and implementation is that existing standards should be followed as much as possible. Such practice makes it easier for new users to get acquainted with DUNE, because existing standards may already be known, and if not there is usually good documentation available to learn about them. Also, established standards and standard implementations have usually seen a lot of attention by dedicated teams of programmers. A reimplementation is therefore unlikely to be better in all regards. In general, a reinvented wheel is usually not as round as the wheel from a well-established standard.

For the C++ language, the established standard is the C++ Standard Library. It offers containers, algorithms, and a plethora of other small useful things, and is specified by the C++ standardization committee together with the language itself [98]. It follows a set of well-established design rules and is well-documented.

Many concepts in DUNE are directly inspired from the standard library. For example, the idea of a grid as a set of elements that can be iterated over is closely related to standard library containers with forward iterators. It would therefore make sense to follow the standard library conventions also in the naming rules. However, for historical reasons this is not done. For example, while the standard library uses lowercase letters and underscores for method names (like `stl_method_name`), DUNE uses the camel-case style (`duneMethodName`). While this may seem like a purely aesthetic problem, it does become inconvenient in certain situations. For example, several classes in `dune-common` such as `ArrayList`, `ReservedVector`, and `lru` implement container types. It is natural to expect to be able to use such containers with standard library algorithms such as `std::fill` or `std::sort`. However, as the standard library uses duck typing (Chapter 4.4.2), this can only work if the DUNE container classes follow the standard library naming conventions; for example, containers need to export the type of the objects they contain under the name of `value_type`.

Therefore, as an exception to the rules of Chapter 4.5.1, there are some classes that follow pure standard library naming rules. These are, e.g., the above-mentioned `ArrayList`, `ReservedVector`, and `lru`, but also a few others. Such classes behave like standard library classes in all respects, except for the class name itself. In particular, standard library algorithms like `std::sort`, `std::find`, and `std::fill` will operate on these classes.

Furthermore, quite a number of classes offer at least a partial standard interface in addition to the DUNE interface. For example, the `BlockVector` class from `dune-istl`, which is basically a dynamic array for (arrays of) floating point numbers, exports the type of its iterator both as `iterator` (the standard library name) and `Iterator` (the DUNE name). The same holds for the `FieldVector` class of `dune-common`, which is a fixed-size container of numbers, and widely used for coordinates and the like. Conversely, some DUNE classes are parametrized with classes that need to implement an standard library interface. For example, various classes in `dune-common` and `dune-istl` take an allocator type that needs to implement the standard library concept of an allocator [150].

While this compromise may not be as clear and elegant as using standard library names exclusively in DUNE, it does allow to benefit from the existing standard, without having to completely rewrite the DUNE programming interface.

4.6. Interface Stability and Backward Compatibility

Designers and developers of libraries are always confronted with a dilemma. As they work on the library, ponder its design, and possibly even use it themselves, they frequently come up with improvements and new features. This is no surprise: on the one hand, library designers are human beings just like other people, and keep learning new things. On the other hand, the requirement for a library can change over time. If the library provides infrastructure for numerical mathematics, then a new requirement may come up because somebody has developed a new numerical method which cannot properly be supported by the current design. For example, the `Intersection::centerUnitOuterNormal()` method in dune-grid was introduced because there was an interest to use first-order finite volume methods on a grid with degenerate elements. Alternatively, a new hardware platform may appear that may be impossible to benefit from without some library changes. For example, interface changes that allow to make better use of multi-core architectures and vectorization units are being discussed [19].

In the simplest case evolving the interface means adding new methods which bring additional functionality. However, such changes may also involve removing certain functionality. This may happen if an alternative superior way to provide the same functionality has been introduced. Then, in the interest of a small interface, the old functionality is removed.

Such changes will make many users happy, as they benefit from new features and a polished interface design. On the other hand, users value stability. Library interface changes always mean work for the users, since the code using the libraries needs to be updated. It is a rather harmless problem if code that compiled with an old version of a library fails to compile with a newer version. The situation is more severe if the dependent program continues to build, but starts to malfunction in subtle ways.

Library writers solve this dilemma in various ways. Some declare complete stability of all existing interfaces, meaning in particular that methods can only be added, but never be removed. This means that user code that works with a given library version will continue to work with all future versions. On the other extreme, some library designers may not care at all about the woes of their users, and add and remove interfaces at any given moment.[18]

DUNE positions itself somewhere in the middle. On the one hand, the DUNE developers recognize the value of interface stability, and make an effort to not change the interface without a good reason. On the other hand, it is assumed that there are relatively few DUNE users (as compared to, say, users of glibc), and that many of those users will be relatively tech-savvy. Therefore things are changed in non-backward-compatible ways, if this is necessary to improve the overall design.

To make this process not too painful for users, there are a few promises.

---

[18]This discussion tacitly omits the important distinction between stability of the programming interface (API stability) and binary stability (ABI stability). The DUNE project does not have any rules at all concerning the latter, and may break the ABI at any moment. However, as DUNE is mostly a template library with little precompiled code, this is of little practical importance.

- The programming interface (API) will not change between different maintenance releases (i.e., code that built and worked with version $X.Y.n$ will continue to build and work with version $X.Y.(n+1)$). There is no guarantee of ABI stability; you will need to rebuild your code against the newer version.

- Code that compiles without warnings and runs with a given feature release $X.n$ will compile and run with the next feature release $X.(n+1)$. This promise is to be taken with a rather largish grain of salt. See the discussion below.

- A feature that is scheduled for removal (be it because it is truly removed, or because it is renamed) is first marked as "deprecated". The deprecation must appear in at least one feature release, to allow users to take notice of it. Only then the features can be actually removed in the next release (or later).

  The precise way of the deprecation marking depends on the feature. Methods, classes, and other things can make the compiler emit a deprecation warning when outfitted with the `[[deprecated( "Custom text here" )]]` attribute. Files can be deprecated by placing

  `#warning` This file is deprecated. Use [...] instead!

  somewhere in the file.

While the DUNE project is strict about the first rule, there are situations when exceptions to the other rules are needed. For example, the deprecation compiler attribute is notorious for not working reliably in various situations. Then the use of a deprecated feature will not trigger a warning, and the user is not informed that the feature is scheduled for removal. Also, it is sometimes technically impossible to provide a feature both in the old way and the new way, for example, changing an exported constant

`static const int` foo = 42;

to a method

`int` foo() `const {return` 42;`}`

in the same class. Users of DUNE are therefore strongly advised to thoroughly read the release notes before upgrading to a newer version.

Part II.

The Core Modules

5. Grids and the Dune Grid Interface

The dune-grid module and the interface to finite element grids it contains are the historical and conceptual core of the DUNE software system. It is one of the central components, and displays the design ideas of DUNE at their best.

This chapter will explain the abstract concepts underlying DUNE grids, and show how they are used in practice. This will allow to write assembly routines, error estimators, and I/O-methods based directly on the grid itself. Later chapters show prefabricated solutions for some of this, but programming directly at the grid interface level gives the most control and understanding.

The founding idea of DUNE was the observation that there cannot be one implementation of a grid data structure that fits all needs. Take, for example, the grid manager built into the UG3 finite element software system [13]. This is an extremely powerful and flexible data structure, handling two- and three-dimensional grids of different element types, with full local adaptivity, on single- and multi-processor machines. However, this power has its price: UG3 grids need quite a bit of memory, and are comparatively slow. This is inconvenient for problems having a lot of structure, e.g., for flow problems in a rectangular domain, where a uniform grid can be used. In such a situation, only very little memory should be needed to store the grid itself, and the grid manager should be very fast, because information like the element volumes and surface normals can be precomputed. Of course, an unstructured grid data structure like UG3 could be used for a structured grid, but its memory and run-time performance would never match the ones of a dedicated structured-grid implementation.

Continuing this train of thought, the dilemma encompasses more than just the difference between structured and unstructured grids. Beyond those, there is another vast range of constructs that count as computational grids, with varying amounts of structure. Figure 2.3 shows two-dimensional grids in $\mathbb{R}^3$, grids without a manifold topology, grids with curvilinear elements, and structured grids with adaptive multiresolution (AMR) refinement [134]. Even powerful grid managers such as UG3 do not support all these features, and therefore applications that require them frequently involve a hand-coded dedicated grid manager.

DUNE proposes to solve this dilemma by offering not just one but a whole selection of grid data structures. These all implement the same interface, and it is therefore very easy to exchange one for another in a given code. This allows a flexibility previously unheard of, without compromising run-time and memory efficiency. The mechanism is implemented in the dune-grid module, whose central part is an *abstract interface* for grid implementations, i.e., a standard way to access all these different grid implementations. The interface describes the functionality that a grid has to provide, such as iteration over vertices and elements, the mapping of grid elements

© Springer Nature Switzerland AG 2020

O. Sander, *DUNE — The Distributed and Unified Numerics Environment*, Lecture Notes in Computational Science and Engineering 140, https://doi.org/10.1007/978-3-030-59702-3_5

to reference elements, and methods for local refinement. Users of a grid (like a finite element assembly code) only need to know the interface. Since all grid implementation implement the same interface, they can then be exchanged with minimal changes to the client code. This allows, e.g., to easily switch between structured and unstructured grids. It also allows an easier access to more exotic grids, which can be written as implementations of the DUNE grid interface rather than as special-purpose implementations hard-wired into particular simulation codes. Indeed, any grid manager that implements the DUNE grid interface becomes very easy to (re-)use, because the interface is the same for all DUNE grids, and is well-documented.

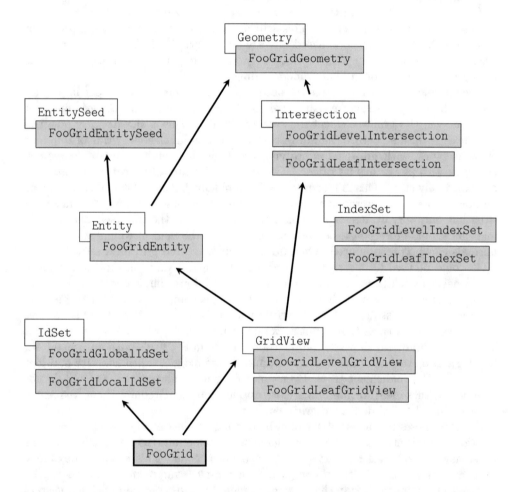

Figure 5.1.: Schematic view of the DUNE grid interface. White boxes are interface classes, and grey boxes are implementation classes. Arrows signify that an object of one class hands out interface objects of another class.

The abstract grid interface is based on a mathematically rigorous definition of the concept of a finite element or finite volume grid. As hinted at in Chapter 2.1 already, this is a surprisingly difficult task. Text books on finite elements usually contain fairly restrictive definitions of what a "grid" is (like, e.g., partitions into triangles and quadrilaterals in [37]). However, practitioners regularly use more general grids containing, e.g., hanging nodes, or elements with polynomial boundaries (Figure 2.3). While people in the field "know" what a finite element grid "is", a precise definition encompassing all grids in use gets more and more complex the closer one looks.

Few people have tried to formalize the notion of a general computational grid. Noteworthy are the work of Berti [24], its generalization to non-planar geometries by Benger [22], and the very abstract *relation-based computations* of Botta et al. [36]. For this book it is assumed that the reader will have an intuitive idea of what a grid is, and the details of an abstract grid definition will be skipped. The DUNE grid interface is based on the formal definitions given in [16], which readers may want to consult for a deeper understanding of the design ideas of the DUNE grid interface.

The concepts from [16] map fairly directly to a set of C++ interface classes in the dune-grid module. These interface classes and some of their interrelations are shown in Figure 5.1. They are implemented using the wrapper–engine concept of Chapter 4.4.3, which appeared to offer the best balance between efficiency, expressiveness, and ease of use. Therefore, in Figure 5.1, there is an implementation class corresponding to each interface class. The implementation classes are visible in compiler error messages, but they are never handled directly by the user.

The largest part of this chapter explains the interface. It focuses on single process applications, and leaves the grid functionality needed for parallel grids to the next chapter. Chapter 5.10 presents the implementations of the grid interface that are contained in the dune-grid module. Some of them are standalone implementations, whereas others rely on external finite element libraries. Indeed, all the power of the UG3 [13], ALBERTA [141], and ALUGRID [40, 144] grid managers is available through the DUNE grid interface. This is an important feature, because it allows the use of powerful mature codes with a uniform simple way to access them.

5.1. Hierarchical Grids and Grid Views

Starting point of the DUNE grid interface is the notion of a *hierarchical grid* introduced in Chapter 2.3.3. While numerical analysis typically sees grids as flat partitions of a given domain, research in multigrid methods and adaptive grid refinement has shown that it is advantageous to think of grids as hierarchical structures. These are constructed by starting with one flat grid, called the "coarse grid" or "macro grid", and refining some or all elements, possibly repeatedly. Rather than having the new elements replace the old ones, they are placed "above" them, leading to a hierarchical tree structure, called a *refinement forest* (a collection of trees; Figure 5.2).

Hierarchical grids are called "grids" in DUNE terminology. When the word "grid" appears in the DUNE code, it usually means "hierarchical grid". The class for hierar-

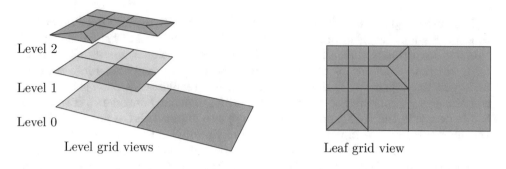

Figure 5.2.: Hierarchical grid and corresponding leaf grid view

chical grids appears at the bottom of the diagram in Figure 5.1.[1] Simulation codes will typically construct an object of this type somewhere in the code, and take all grid-related information from it or from the objects it hands out. Grid construction is discussed in Chapter 5.7, which one can read before proceeding further, if desired.

The main functionality of hierarchical grid objects is to provide access to non-hierarchical grids. After all, these are the ones that numerical discretization methods operate on.[2] There are two canonical ways to get a non-hierarchical grid from a refinement forest. The more important one consists of collecting all leaf elements of the forest. These form a grid of the domain which is called the *leaf grid*. It is the canonical choice for non-hierarchical numerical methods on adaptively refined grids. For multigrid methods one also needs access to non-leaf elements. These come organized in levels, the so-called *level grids*.

In the language of Dune, the leaf and level grids are *views* onto a hierarchical grid. Such views are represented in the Dune grid interface by a class called `GridView`. The (hierarchical) grid implementation class hands out the aforementioned two types of grid views: the level views and the leaf view. They can be obtained by the methods

```
LevelGridView levelGridView(int level) const
```

and

```
LeafGridView leafGridView() const
```

Both the level grid view and the leaf grid view implement the interface class `GridView`. The interface methods are listed in Table 5.1. These methods offer access to the grid elements, element neighborhood information, and methods to control grids that are distributed across several processors.

---

[1]While the grid interface uses the wrapper–engine concept of Chapter 4.4.3, the hierarchical grid implementation class is handled directly in user code. That is the reason why only the implementation class FooGrid appears in Figure 5.1. While an interface class Grid does exist (in dune/grid/common/grid.hh), its main purpose is documentation.

[2]The grid class offers more than just that. The methods used to control grid adaptivity are covered in Chapter 5.9, and the methods for load-balancing in Chapter 6.

	`GridView(`**`const`** `GridView& other)`
	Copy constructor
`GridView&`	**`operator=(const`** `GridView& other)`
	Assignment operator
`const` `Grid&`	`grid()` **`const`**
	The underlying hierarchical grid
`const` `IndexSet&`	`indexSet()` **`const`**
	The index set, which assigns numbers to entities
`int`	`size(`**`int`** `codim)` **`const`**
	Number of entities of a given codimension
`int`	`size(`**`const`** `GeometryType& type)` **`const`**
	Number of entities with a given reference element
`template<class` `Entity>`	
`bool`	`contains (`**`const`** `Entity& entity)` **`const`**
	Whether the given entity is contained in this grid view
`template<int` `cd>`	
`Codim<cd>::Iterator`	`begin()` **`const`**,
	`end()` **`const`**
	Iterators over the entities of codimension `cd` for this view
`IntersectionIterator`	`ibegin(`**`const`** `Codim<0>::Entity& entity)` **`const`**
	`iend(`**`const`** `Codim<0>::Entity& entity)` **`const`**
	Iterators over the intersections of the given element with
	its neighbors
`template<int` `cd,PartitionIteratorType pitype>`	
`Codim<cd>::Partition<pitype>::Iterator`	
	`begin()` **`const`**, `end()` **`const`**
	Entity iterators for this view
`const`	
`CollectiveCommunication&`	`comm()` **`const`**
	Obtain collective communication object
`int`	`overlapSize(`**`int`** `codim)` **`const`**
	Size of the overlap region for a given codimension on the
	grid view
`int`	`ghostSize(`**`int`** `codim)` **`const`**
	size of the ghost region for a given codimension on the
	grid view
`template<class` `DataHandleImp,`**`class`** `DataType>`	
`void`	`communicate(`
	`CommDataHandleIF<DataHandleImp,DataType>& data,`
	`InterfaceType iftype,`
	`CommunicationDirection dir)` **`const`**
	Communicate numerical data on this grid view

Table 5.1.: Interface methods of the `GridView` class. The bottom block shows the methods for distributed grids (see Chapter 6)

int	dimension
	Dimension of the grid
int	dimensionworld
	Dimension of the world space
class	ctype
	Number type used for coordinates
bool	conforming
	True, if the grid view is conforming, i.e., has no hanging nodes
class	Grid
	Type of the underlying hierarchical grid
class	IndexSet
	The corresponding IndexSet type (see Chapter 5.6.1)
class	Intersection
	Type of intersections between elements in this grid view (see Chapter 5.4)
class	IntersectionIterator
	Type of iterator over intersections between elements in this grid view (see Chapter 5.4)
class	CollectiveCommunication
	For parallel computations: encapsulate several communication methods (see Chapter 6.4)
template<int codim>	
class	Codim
	Exports dimension-dependent types like entity iterators

Table 5.2.: Static information exported by the GridView class

Element access is discussed in Chapter 5.2 below. The list of GridView methods for distributed computations is shown here only for completeness, and it will not be discussed until Chapter 6.

Note that the levelGridView and leafGridView methods return the grid view *by value*, and the return value can be copied around freely. This copying is expected to be cheap both in terms of run-time and of memory consumption. In other words, GridView objects have reference semantics: They do not store data itself, but rather provide views onto the actual data in the hierarchical grid object. The downside of this is that in general a GridView object will not contain anything meaningful once the hierarchical grid object has disappeared. Also, as GridView objects may cache data, it cannot be guaranteed that they still hold correct information after the hierarchical grid has been modified. The leafGridView and levelGridView methods have to be called again in such a case, to obtain up-to-date GridView objects.

In addition to the public methods, the GridView interface also consists of a set of statically exported numbers and types. These are listed in Table 5.2. Each GridView object exports the dimension of the grid and the dimension of the space it lives in (the *world space* or *physical space*). It also exports the return types of all methods of the class, and some more. The convention to do this originates from the days before the C++11 standard, which did not have the **auto** keyword and introspection features like **decltype**. While the exported types are less relevant today, dune-grid still follows the policy that an interface class should export all types that appear as return types of public methods.

5.2. Iterating over Vertices and Elements

The main purpose of the grid view is to provide access to its grid elements and vertices. In this context, DUNE terminology uses the word *entity* to denote grid vertices, edges, elements, etc. of all dimensions. So, for example, a 0-dimensional entity is a grid vertex, a 1-dimensional entity is an edge, etc. There are two ways for accessing entities: The range-based **for** mechanism introduced in the C++11 standard allows to write loops over all elements and vertices very concisely, and will be the most useful for many situations. If flexibility is needed, there is a way to access all elements using iterators. While this section will mainly speak of grid vertices and elements, some DUNE grids offer the same iteration facilities also for edges and facets.[3]

It is important to realize that access to elements and vertices is by *forward iteration* exclusively. There is a mechanism that hands out a "first" element of a grid view, and a way to proceed from one entity to the next. There is no other way to access an element or vertex; in particular, there is no random access operator that allows to access a vertex by its number.

The reason for this is that many existing unstructured grid managers store vertices and elements, edges, etc. internally in data structures that do not support random access. For example, linked lists are common, because they allow easy insertion and removal of list items [43], which is needed for local grid adaptivity. At the same time, virtually all algorithms in finite element and finite volume computations visit vertices and elements in a linear fashion only. Random access is never really needed, and hence the restriction to forward-iterating access is not severe. In the rare situation that random access to elements or vertices is indeed needed, the relevant grid data will have to be copied into an array beforehand.

We illustrate entity access by some small examples. Given a grid view object in a variable called gridView, the following two lines print all vertex coordinates:

```
for (const auto& vertex : vertices(gridView))
  std::cout << vertex.geometry().corner(0) << std::endl;
```

The expression vertices(gridView) in the first line creates a *linear range* containing all vertices in the grid view. The **for** loop then iterates over this range. At each iteration, the variable vertex is a **const** reference to the current vertex. The second

---

[3]However, only iteration over vertices and elements is mandatory for all grid implementations. Consult the boolean constant hasEntityIterator<Grid,codim>::v from the Capabilities namespace to find out whether a given grid supports iteration over edges and facets.

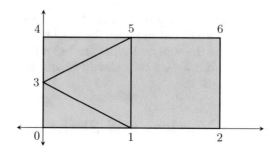

Figure 5.3.: Two-dimensional unstructured example grid

line then prints the vertex coordinates. For DUNE, all entities are polytopes, and a vertex is simply a polytope with a single corner. The expression `vertex.geometry()` accesses the shape of the entity, and `corner(0)` returns the position of its only, i.e., zero-th, corner. When run for the example grid in Figure 5.3, the program will print

```
0 0
1 0
2 0
0 0.5
0 1
1 1
2 1
```

which is the list of the coordinates of the grid's seven vertices.

Iterating over all elements in a grid view works just the same way. The following example prints the types of all elements of a grid view:

```
for (const auto& element : elements(gridView))
  std::cout << element.type() << std::endl;
```

The output produced by this for the same grid is

```
(simplex, 2)
(simplex, 2)
(simplex, 2)
(cube, 2)
```

It is a list of the reference element types of all elements contained in the grid. The number 2 in the output refers to the fact that all entities looped over have two-dimensional reference elements. See Chapter 5.5.2 for more details on what the `type` method actually returns.

As mentioned previously, iteration over grid entities other than vertices and elements is optional, and is not implemented by all DUNE grids.[4] If implemented, ranges of edges and facets can be obtained through `edges(gridView)` and `facets(gridView)`, respectively. If the corresponding iteration is not implemented, the behavior of these ranges is undefined.

Many times accessing vertices and elements as described above is just what is needed, but sometimes a bit of additional flexibility is necessary. Switching between vertices and elements using a parameter is easily possible; indeed, the line

```
for (const auto& entity : entities(gridView,Dim<0>()))
{
  // Do something with 'entity'
}
```

loops over all grid vertices, (i.e., 0-dimensional objects). Correspondingly, if the grid in `gridView` is is `dim`-dimensional, then

```
for (const auto& entity : entities(gridView,Dim<dim>()))
{
  // Do something with 'entity'
}
```

loops over the elements. In a three-dimensional grid, `Dim<1>()` and `Dim<2>()` may work, depending on the actual grid implementation.

For symmetry there is also a way to access vertices, elements, etc. by their *codimension* with respect to the grid. The codimension of an entity is the entity dimension subtracted from the grid dimension. In particular we have the relations

$$
\begin{aligned}
\text{codim(vertex)} &= \text{grid dimension} - \text{vertex dimension} &= \text{grid dimension} \\
\text{codim(facet)} &= \text{grid dimension} - \text{facet dimension} &= 1 \\
\text{codim(element)} &= \text{grid dimension} - \text{element dimension} &= 0.
\end{aligned}
$$

Hence

```
for (const auto& vertex : entities(gridView,Codim<dim>()))
{
  // Do something with 'vertex'
}
```

loops over the vertices in a `dim`-dimensional grid, and

```
for (const auto& element : entities(gridView,Codim<0>()))
{
  // Do something with 'element'
}
```

loops over the elements. Indexing vertices and elements by their codimension is widely used in the DUNE grid interface, but there is no technical reason to prefer the codimension over the dimension.

---

[4]For grids that do not support iteration over edges and facets directly, such loops have to be simulated by iterating over each edge/facet of each element, cf. Chapter 5.3.1.

Iterating over elements using ranges works very well in most situations. For even more control there is also access to vertices and elements using iterators. The rest of this section may be skipped on first reading.

Iterators are objects that point to other objects. They are heavily used in the C++ standard library [103, 150], and the DUNE grid interface tries to follow that library as good as possible. Iterators are obtained via methods of the GridView class, and can be dereferenced to obtain the objects they point to. Additionally, iterators can be incremented to make them point to the next object. Indeed, the iterators in the DUNE grid interface are *forward iterators*, that is they can only be used to access the vertices of a grid view in a linear fashion, starting from the first one, without ever going backwards. Iterators in the DUNE grid interface behave exactly like forward iterators in the C++ standard library.

We revisit the examples used above, and rewrite them using iterators. Just as it is done in the standard template library, the grid view objects hands out an iterator to the first vertex, and one that points to an imaginary vertex right past the last one, to be used as a loop end marker:

```
auto it = gridView.begin<dim>();
auto endIt = gridView.end<dim>();
```

The number dim is the co(!)dimension of the entities that are to be iterated over. For this example suppose that dim is the dimension of the grid. Hence, calling the method gridView.begin<dim>() produces an iterator to the first *vertex*. A loop over all vertices is then as simple as

```
for (; it!=endIt; ++it)
{
  // Do something with the vertex that 'it' points to
}
```

In particular, here is the example again that prints a list of all vertex coordinates on the console:

```
auto it = gridView.begin<dim>();
auto endIt = gridView.end<dim>();

for (; it!=endIt; ++it)
  std::cout << it->geometry().corner(0) << std::endl;
```

Notice how unlike in the loop using range-based **for**, the loop variable it is not the vertex itself, but an iterator pointing to the vertex. To get the actual vertex the iterator is dereferenced by writing it->. For the example grid shown in Figure 5.3, the program output is again

```
0 0
1 0
2 0
0 0.5
```

```
0 1
1 1
2 1
```

Similarly, the following example prints the element types of all elements in the grid view

```
auto it = gridView.begin<0>();
auto endIt = gridView.end<0>();

for (; it!=endIt; ++it) {
  std::cout << it->type() << std::endl;
```

The first two lines request iterators for entities of codimension 0, i.e., iterators for grid elements. For the grid of Figure 5.3, this will again print

```
(simplex, 2)
(simplex, 2)
(simplex, 2)
(cube, 2)
```

Note how in these examples, the **auto** keyword allows to write iterator loops without ever explicitly stating the type of the iterator. This is very helpful, because the exact type specifications can be unwieldy. The iterator type is exported by the grid view as

```
GridView::Codim<codim>::Iterator
```

The argument `codim` is a compile-time integer that specifies the type of objects to iterate over (vertices, elements, edges, etc.), and it does so by giving the object *codimension* with respect to the grid. Consequently, the type of an iterator over all elements in a grid view is

```
GridView::Codim<0>::Iterator
```

and the vertex iterator type is

```
GridView::Codim<dim>::Iterator
```

Both depend on the grid view, because iteration over a leaf grid view will do something different than iteration over a level grid view. Indirectly, the iterator type will also depend on the grid implementation itself. This dependence, an aspect of static polymorphism, allows to make sure internally that the iterator implementation is optimally tuned for the grid implementation, without unnecessary run-time checks.

5.3. Entities and Geometries

Explaining the functionality of DUNE grid entities requires to say a little bit more about the design ideas of the DUNE grid interface. In DUNE, constituents of a (non-hierarchical) grid such as elements, vertices, and edges are conceptually split

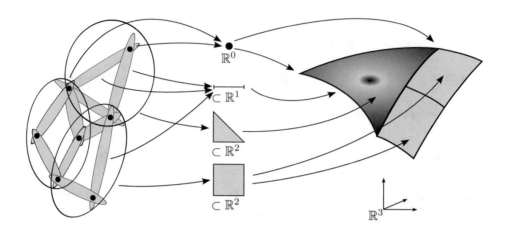

Figure 5.4.: Split of an example grid into a topological part and its geometric realization. Left: the topology is given by a set of abstract vertices, and subsets
of these vertices. Center: list of reference elements. To each set on the
left corresponds a reference element. Right: Images of the geometric
realizations in the world space $\mathbb{R}^3$. Picture taken from [16].

into two parts. The first part, which is called *entity*, is a purely topological object.
Entities of dimension zero, i.e., entities corresponding to grid vertices, are seen as
elements of an abstract finite set, devoid of anything resembling a position. Entities
of higher dimension consist of sets of entities of lower dimension. For example, an
element-entity consists of its vertex-entities and edge-entities, and from this one can
infer its neighboring element-entities, or all edges or elements adjacent to a given
vertex. This concentration on neighborhood information is what makes the entity a
purely topological object. An example set system for a grid consisting of one triangle
and two quadrilaterals is shown in Figure 5.4 on the left.

   The second part of each element, vertex, edge, etc. is the geometrical information.
In algebraic topology, this is called a geometric realization, and in DUNE it is simply
called *geometry*. It specifies the entity shape by embedding it into a manifold. This
manifold, which in the vast majority of cases will be a low-dimensional Euclidean space,
is called the *world space*.[5] A geometric realization of an entity is a C^1-diffeomorphism
from a reference element to the world space. This diffeomorphism is precisely the map
$F : T_{\mathrm{ref}} \to T$ that appears in the description of the finite element assembly algorithm
in Chapter 2.1.3. The intuitive idea of the shape of an entity as a point set is obtained
as the image of F, but in addition this shape is equipped with a parametrization or

---

[5]Boundary value problems with periodic boundary conditions can be interpreted as problems with
a toroidal world space. More general world spaces may be useful for applications in general
relativity (cf. Benger [22]).

coordinate system. This construction is also shown in Figure 5.4.

The idea of splitting grid elements in a topological and a geometrical part is not new, and has already been driven to considerable sophistication in the works of Berti [24] and Benger [22]. The mathematical roots can be found in combinatorial and algebraic topology, which has introduced the notion of *complex*, of which there are different flavors of varying complexity [90]. Pure abstract simplicial complexes correspond precisely to the topological information of a conforming simplex grid, whereas (non-abstract) pure simplicial complexes, i.e., abstract simplicial complexes with a geometrical realization correspond to the grids themselves. Polyhedral complexes allow elements to be any convex polyhedron, but prohibit curved element boundaries. The widely used CW-complexes allow curved boundaries, but are too general as a model for finite element grids [90]. The concept most closely resembling the grids used for finite element and finite volume computations appear to be the *regular cell complexes with the intersection property* [27]. They do allow polyhedral elements with curved boundaries, but preclude a lot of pathological cases.

The one serious disadvantage of all these concepts from topology is that none of them satisfactorily deals with the ideas of nonconforming grids and hanging nodes, such as the ones in Figures 5.4 or 5.8. This was the motivation for the work in [16], which modifies the idea of a complex in this direction. This general definition forms the basis for the DUNE grid interface.

5.3.1. Entities

The DUNE grid interface explicitly incorporates the distinction between topology and geometry. The topological part of each element, vertex, edge, etc. is given by an interface class called `Entity`, while the geometrical part is given by a separate interface class `Geometry`. This section discusses the `Entity` interface class, while the `Geometry` interface class will be considered in Chapter 5.3.2.

The `Entity` interface class comes in two flavors. While all entities share some common functionality, there is a specialization for entities of codimension zero (i.e., grid elements), which has additional features. We address both interfaces in turn, and begin with the one that is available for all codimensions. For that, Tables 5.3 and 5.4 list the public methods and types, respectively. Even though not mentioned on those tables, all `Entity` objects can be default- and copy-constructed, and assigned from other such objects.

The first interface method is called

```
int level() const
```

It simply returns the level of the entity in the grid hierarchy. The level is a natural number, and 0 signifies that the entity is part of the coarsest grid. The next method

```
GeometryType type() const
```

int	level() const
	Refinement level of this entity
Geometry	geometry() const
	Geometric realization of the entity
GeometryType	type() const
	Type of the reference element
unsigned int	subEntities(unsigned int codim) const
	Number of subentities for a given codimension
EntitySeed	seed() const
	Return memory-optimal specifier for the entity
PartitionType	partitionType() const
	For parallel computations; see Chapter 6
Implementation&	impl()
const Implementation&	impl() const
	The underlying implementation object

Table 5.3.: Interface methods of the generic Entity class

returns a marker object which denotes the reference element corresponding to the entity. The marker is an object of type GeometryType,[6] and is presented in detail in Chapter 5.5.2.

The actual geometric realization, or *geometry*, is available via the method

```
Geometry geometry() const
```

Its return value, an object of the interface class Geometry, implements the mapping F from the reference element into the world space, and consequently all geometric aspects of the entity. It is a complex class with quite a bit of functionality, and is discussed in Chapter 5.3.2. The method

```
unsigned int subEntities(unsigned int codim) const
```

returns the number of all subentities of the entity of a given codimension. If entity is an entity representing a quadrilateral, the code

```
std::cout << entity.subEntities(2)
          << " " << entity.subEntities(1)
          << " " << entity.subEntities(0) << std::endl;
```

will print "4 4 0", the number of vertices, edges, and elements of a quadrilateral.
Finally, the method

```
PartitionType partitionType() const
```

is only needed when using DUNE on distributed machines, and will be explained in Chapter 6.

---

[6]The naming is unfortunate, because reference elements are also topological objects.

`int`	codimension
	Entity codimension with respect to the grid it is part of
`int`	dimension
	Dimension of the grid(!), not of the entity
`int`	mydimension
	Dimension of the entity
`class`	Geometry
	Type of the corresponding geometric realization
`class`	EntitySeed
	The corresponding entity seed (for storage of entities)
`class`	Implementation
	Type of the underlying implementation class

Table 5.4.: Static information exported by the generic `Entity` class

Like several other parts of the DUNE grid interface, `Entity` objects allow to access the actual implementation class behind the `Entity` interface class by means of the method

```
Implementation& impl()
```

(A **const** version of the method is available as well.) The method allows to circumvent the grid interface abstraction layer, and to control special functionality of particular grid implementations. This is not without danger, though. Code that uses the `impl` methods will only run with one grid implementation, and it may break when updating from one release of DUNE to another. It should only be used if really necessary.

The last method of the generic `Entity` interface is called

```
EntitySeed seed() const
```

It returns a so-called *entity seed*. An entity seed is a light-weight storage mechanism for entities. Its raison d'être is subtle: Some applications need to store entities outside of a grid. In principle, `Entity` objects can be stored directly in a container (they have the usual constructors and assignment/move operators). For example, the following code snippet stores every boundary element of a given grid view in a `std::vector` called boundaryElements, and then prints their reference element types:[7]

```
std::vector<GridView::Codim<0>::Entity> boundaryElements;
for (const auto& element : elements(gridView))
  if (element.hasBoundaryIntersections())
    boundaryElements.push_back(element);

for (const auto& boundaryElement : boundaryElements)
  std::cout << boundaryElement.type() << std::endl;
```

---

[7]Even though this code looks like the entity objects have value semantics, their implementations usually reference the underlying grid. This means that all elements copied out of a grid cease to contain meaningful information once the grid object gets destroyed.

This code does what it is supposed to do; however, there can be one problem: Depending on the grid implementation, an Entity object can use quite a bit of memory. Therefore, storing large quantities of Entity objects can be problematic, and this is where the EntitySeed class comes into play. Objects of type EntitySeed can be stored with a very low memory footprint, because they cannot really be used for anything except for obtaining an Entity back when combined with the corresponding grid. The following example code does the same as the previous one, but it uses EntitySeed objects to store entities:

```cpp
std::vector<Grid::Codim<0>::EntitySeed> boundaryElements;
for (const auto& element :  elements(gridView))
  if (element.hasBoundaryIntersections())
    boundaryElements.push_back(element.seed());

for (const auto& boundaryElement : boundaryElements)
  std::cout << grid.entity(boundaryElement).type() << std::endl;
```

Depending on the grid implementation this can save a considerable amount of memory.

template<**int** codim> Codim<codim>::Entity	subEntity(**int** i) **const** Obtain a subentity of given codimension
Entity	father() **const** Access to father entity on the next-coarser grid
bool	hasFather() **const** True if entity has a father entity
HierarchicIterator	hbegin(**int** maxLevel) **const** Access to descendant elements in the refinement tree (not higher than maxLevel)
HierarchicIterator	hend(**int** maxLevel) **const** Iterator to one past the last descendant element (not higher than maxLevel)
bool	isLeaf() **const** True if the entity is part of the leaf grid view
LocalGeometry	geometryInFather() **const** The embedding of this element into its father element
bool	isNew() **const** True if the entity has been created during the last call to adapt
bool	mightVanish() **const** True if entity might disappear during the next call to adapt
bool	hasBoundaryIntersections() **const** True if entity has intersections with boundary (see Chapter 5.4)

Table 5.5.: Additional methods of the codimension-0 specialization of the Entity class

Besides these methods, the interface of `Entity` consists of several pieces of static information (Table 5.4). In particular, the class exports the precise return types of all of its methods, and three integer numbers: its own dimension in `mydimension`, its codimension with respect to the grid in `codimension`, and the dimension of the grid itself in `dimension`. The naming of these integers is a bit unfortunate; it has historic reasons and is unlikely to change.

The specialization of the `Entity` class for grid elements (codimension 0) contains a number of additional methods. These are listed in Table 5.5. The more powerful interface of codimension 0 entities reflects an important design principle of the DUNE grid interface: Elements are the main objects, and most access to entities of other dimensions should happen via the elements. For example, even though many grid implementations allow to directly iterate over all grid edges, this is not required. Instead, the standard way to do so is by iterating over the elements, and for each element to iterate over the edges. This mirrors the fact that the main assembly loop in finite element and finite volume codes is almost always element-based (see Section 2.1.3), and that even operators that contain facet contributions (as, e.g., for DG methods; see Chapter 11.2.6) usually have an easy element-wise reformulation. The focus on elements can be seen in various other places of the grid interface as well.

The first additional method of the codimension-0 `Entity` interface hands out element subentities like edges and vertices:

```
template<int codim>
Codim<codim>::Entity subEntity(int i) const
```

The requested codimension is a template parameter, because it influences the return type. The number of the requested subentity, on the other hand, is a run-time argument. The numbering follows the conventions described in Chapter 5.5. For example, to get the edge with number 2 of the quadrilateral entity of the previous example, write

```
auto edge = entity.subEntity<1>(2);   // '1' is the codimension of edges
                                       // in two-dimensional grids
```

This returns an object of type `Entity` of the correct dimension. The number of subentities of a given codimension is computed by the `subEntities` method, which is available for entities of all codimensions, not just for the grid elements (see above).

The further methods of `Entity<0>` can be grouped into two sets. The methods of the first set connect an element with the elements above and below it in the grid refinement hierarchy. Access to coarser elements differs from access to finer elements, because elements only have a single father (at most), but they may have an arbitrary number number of sons (Figure 5.5). Access to the father element is provided by the method

```
Entity<0> father() const
```

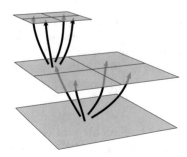

Figure 5.5.: Grid element with descendants

If the entity does not have a father, then the behavior of the method is undefined. This may happen for two reasons. Either the element is already part of the coarsest grid level. Or, if the grid is distributed across several processes (Chapter 6), then the father element may exist only on a different process (in which case the `father` method cannot find it). There is an additional method

```
bool hasFather() const
```

which returns whether the father element exists on the local process, and which can be used to test whether it is safe to call `father`.

Accessing child elements works differently. Since there is generally more than one child there is a special iterator to iterate over the children. Indeed, this iterator, called *hierarchic iterator*, not only iterates over the children, but (if requested) also over descendants more than one generation away. An iterator is obtained by calling the method

```
HierarchicIterator hbegin(int maxLevel) const
```

The resulting forward iterator will traverse all descendants of the entity on which hbegin was called, up to and including those on level `maxLevel`. The corresponding method to obtain a loop-ending iterator is

```
HierarchicIterator hend(int maxLevel) const
```

The type of the iterator is implementation-dependent, and is exported by the interface class. So, for example, to loop over all direct sons of an element called `element`, write

```
auto it    = element.hbegin(element.level()+1);
auto endIt = element.hend(element.level()+1);

for (; it != endIt; ++it)
{
  // Do something with the son available through it->
}
```

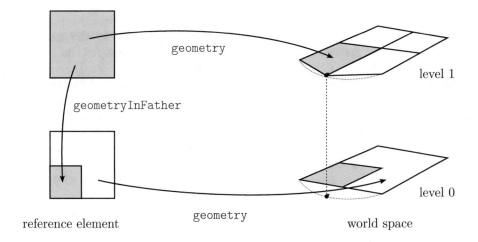

geometry

geometryInFather

level 1

level 0

reference element

geometry

world space

Figure 5.6.: The geometryInFather method returns an embedding of the element into its father element. The dotted line is the domain boundary.

As this syntax is unnecessarily verbose in modern C++, the range-based-**for** syntax can be used as well:

```
for (const auto& son : descendantElements(element,element.level()+1))
{
    // Do something with the son available in 'son'
}
```

To determine whether an element is a leaf element is similarly difficult as determining whether an element has a father. On a single process, an element is a leaf if and only if hbegin and hend return the same iterator. If the grid is distributed across multiple processes, however, there may be children on a different process. Therefore, the method

```
bool isLeaf() const
```

is provided which returns **true** if and only if the element is a leaf entity of the global refinement hierarchy.

The methods explained above allow to determine father–son relationships between elements, but their purely topological nature is usually not sufficient. To be able to implement the transfer of numerical data between different grid levels, e.g., for geometric multigrid methods, it is necessary to obtain information on how the sons partition the father element geometrically. This information is represented in the grid interface in form of a second geometric realization. This second realization can be obtained by calling the method

```
LocalGeometry geometryInFather() const
```

Like the geometry method, the `geometryInFather` method provides information about the geometrical shape of the element. However, unlike the former, the `geometryInFather` method returns the embedding of the element into its own father element.[8] That is, the object returned by `geometryInFather` implements a map from the element's reference element into the reference element of its father. This is illustrated in Figure 5.6. The object returned by the method implements the full Geometry interface explained in the following section, but it is not necessarily of the same type as the object returned by the geometry method (see below).

To show how the `geometryInFather` method is used to implement data interpolation from coarser to finer grid levels and back, we give the following example: Let `element` be an `Entity` of codimension 0 on grid level 1. Suppose that on grid level 0 there is a scalar function `f` given which can be evaluated by calling

```
auto value = f(level0Element, localCoordsInLevel0Element);
```

On `element` we suppose a set of Lagrange points (in local coordinates of `element`), in a `std::vector<FieldVector<double,dim> > lp`, and we want to interpolate `f` at these Lagrange points. The challenge is that the Lagrange points are defined with respect to the level 1 grid element, whereas the function f needs to be evaluated in terms of coordinates on the father. The `geometryInFather` method provides the necessary coordinate transformations:

```
for (size_t i=0; i<lp.size(); i++)
  values[i] = f(element.father(),
                element.geometryInFather().global(lp[i]));
```

While `lp[i]` is the position of the i-th Lagrange point in coordinates of the level-1 element, the expression `element.geometryInFather().global(lp[i])` is the position in coordinates of the father element. The `geometryInFather` method produces a map from the element to its father, and the method `global` evaluates this map. Note that

```
element.geometry().global(x);
```

is not necessarily the same point as

```
element.father().geometry().global(element.geometryInFather().global(x));
```

because the finer grid may cover a domain that is slightly different from the coarse grid one; due to the presence of a boundary parametrization (Figure 5.6).

The codimension-0 `Entity` interface also provides functionality needed for grid refinement. As explained in Chapter 2.3, grid refinement is typically performed as a sequence of steps: first, a subset of the (leaf) grid elements is marked for refinement or

---

[8]If there is no father element on the same process, then the behavior of `geometryInFather` is undefined.

coarsening, and only after all these marks have been set, a subsequent step performs the actual grid modifications. Numerical data attached to the grid entities may need to be transported across this sequence of steps.

Marking elements for refinement is not done through a method on the entity, but on the grid itself. However, the `Entity` class provides a few auxiliary methods needed for data handling during grid refinement. In particular, the

```
bool mightVanish() const
```

method tells whether the entity may disappear during the next round of grid modifications (triggered by calling the method `adapt` of the hierarchical grid), given the current state of refinement marking. Knowing that the element may disappear allows to save all data attached to it.

If `mightVanish` returns **false**, then the element is guaranteed to stay, but if it returns **true** the element may either stay or disappear. The reason for this asymmetry is the fact that it is quite difficult to determine for sure whether an element will disappear or not. So, in a sense, `mightVanish` is just a hint; one may as well ignore it and always save all the data, but that would be wasteful.

Conversely, the method

```
bool isNew() const
```

is to be called after a call to `adapt` has triggered a set of grid modifications. It returns whether a given entity has been newly created during the last adaptivity cycle. If so, data for this element needs to be constructed by interpolation from data from coarser entities. A detailed presentation of local adaptivity in DUNE and examples of how the `mightVanish` and `isNew` methods should be used is given in Chapter 5.9.

5.3.2. Geometries

Geometric realizations of entities are implemented by the interface class

```
class Geometry
```

Unlike the `Entity` class, it has the same interface for objects of all codimensions. The geometry is what determines the "shape" of an element, edge, etc. in physical space, and when embedded into other elements.

Conceptually, the `Geometry` class implements a mapping F from the reference element T_{ref} into Euclidean space. The image of this map is *defined* to be the shape of the element T. Therefore, one can alternatively say that a `Geometry` maps the reference element T_{ref} to the element T. By doing so, F induces a coordinate system on T which is called the *local* coordinate system of T (Figure 5.7). It is assumed that F is continuously differentiable on T_{ref}, with a continuously differentiable inverse. This condition holds for an overwhelming part of the geometries of interest in finite element and finite volume methods, including isoparametric and isogeometric elements of all orders.

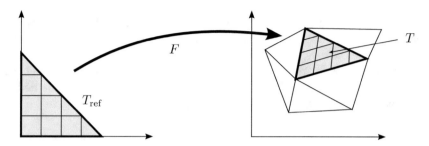

Figure 5.7.: The Geometry class implements the map F from the reference element T_{ref} to the grid element T.

The target space of a geometry is not always the world space of the grid. Indeed, the two may even have different dimensions. This happens when a Geometry object is not used to embed an entity into the grid world space, but into another entity. For example, the method Entity::geometryInFather returns a geometry object that embeds an element into its father in the grid refinement hierarchy. Hence, the target space of that geometry is the reference element of the father. If the grid consists of a triangle surface embedded into $\mathbb{R}^3$, then the grid world space will be three-dimensional, but the target space of the geometry returned by geometryInFather will only be two-dimensional. Geometric realizations into other entities are sometimes called LocalGeometry objects in DUNE, but they are represented by the same interface class as realizations that map into the grid world space.

The interface of the Geometry class reflects the requirements of finite element and finite volume methods. A lot of the concepts from Chapter 2.1.3 on how to assemble stiffness matrices appear in the Geometry interface. A list of all public methods of the class can be found in Table 5.6.

The discussion in Chapter 2.1.3 established that standard finite element methods require the evaluation of F, its first derivatives, and the Jacobian factor $\sqrt{\det(\nabla F^T \nabla F)}$ required for the integral transformation formula. Therefore, the largest part of the interface consists of methods that allow to evaluate these quantities. The method

```
GlobalCoordinate global(const LocalCoordinate& xi) const
```

evaluates F at a given point ξ in the reference element T_{ref}. Given a vector of coordinates in the reference elements (in "local coordinates"), it returns the corresponding point in the world space (the "global coordinates"). If global is called for a point outside of T_{ref}, then its behavior is undefined. The types LocalCoordinate and GlobalCoordinate are arrays of sizes dim T_{ref} and the dimension of the world space, respectively. Together with all other method return types, they are exported by the Geometry interface class (see Table 5.7).

GlobalCoordinate	global(**const** LocalCoordinate& xi) **const**
	Evaluate the map $F : T_{\mathrm{ref}} \to T$
LocalCoordinate	local(**const** GlobalCoordinate& x) **const**
	Evaluate the inverse map $F^{-1} : T \to T_{\mathrm{ref}}$
GlobalCoordinate	corner(**int** i) **const**
	Position of the i-th corner
int	corners() **const**
	Number of corners of the reference element
ctype	integrationElement(
	const LocalCoordinate& xi) **const**
	The factor $\sqrt{\det(\nabla F^T \nabla F)}$ appearing in the integral transformation formula
JacobianInverseTransposed	jacobianInverseTransposed(
	const LocalCoordinate& xi) **const**
	Inverse of transposed of the Jacobian: $(\nabla F)^{-T}$
JacobianTransposed	jacobianTransposed(
	const LocalCoordinate& xi) **const**
	Transposed of the Jacobian: $(\nabla F)^T$
bool	affine() **const**
	True if the geometry mapping is affine
GeometryType	type() **const**
	Type of the reference element
ctype	volume() **const**
	Volume of the image of the geometry
GlobalCoordinate	center() **const**
	Center of the image of the geometry
Implementation&	impl()
	Access to internal implementation class (also as **const** method)

Table 5.6.: Methods of the Geometry class. ctype is the number type used by the grid for coordinates.

Typically, the LocalCoordinate and GlobalCoordinate arrays are instances of the FieldVector class (from the file dune/common/fvector.hh), which implements an array class of static size with vector-space operations. Observe that the two arrays may not have the same length.

While the method global allows to evaluate the map $F : T_{\mathrm{ref}} \to T$, the method

```
LocalCoordinate local(const GlobalCoordinate& x) const
```

evaluates the inverse $F^{-1} : T \to T_{\mathrm{ref}}$. The single parameter of this method is a GlobalCoordinate, and it returns a LocalCoordinate object with the coordinates in the reference element. Again, the behavior is undefined if the input coordinates are not actually contained in T. Evaluating local can be difficult and costly if T is highly nonlinear, or if T has positive codimension in the world space. In the latter case it is unlikely that input points are exactly on T, given that only finite-precision arithmetic is available. Implementations of local are therefore expected to do something reasonable even when called for points in the vicinity of, but not quite on T.

In principle, the corners of an element can be obtained by using the corners of the reference element as input to the global method. As a convenient shortcut there is the method

```
GlobalCoordinate corner(int i) const
```

which takes an integer argument i and returns the position of the i-th entity corner in a GlobalCoordinate. The number of corners is available through the method

```
int corners() const
```

As an example, the following code snippet computes the center of mass of the element corners

```
GlobalCoordinate centerOfMass(0);  // Initialize with zeros
for (int i=0; i<geometry.corners(); i++)
  centerOfMass += geometry.corner(i);
centerOfMass /= geometry.corners();
```

Note that the Geometry class provides the interface for the geometric realizations of *all* dimensions. This includes the extreme case of zero-dimensional entities (i.e., vertices), whose reference element is $\mathbb{R}^0$, which we interpret as a point. In this case, LocalCoordinate is an array of size zero, and effectively arguments of type LocalCoordinate disappear. Hence to obtain the position of a vertex stored in an object called vertex of type Geometry, one may either write

```
Geometry::LocalCoordinates dummy;  // Array of size zero
Geometry::GlobalCoordinate p = vertex.global(dummy);
```

or

```
Geometry::GlobalCoordinate p = vertex.corner(0);  // Zero-th (only) corner
                                                  // of a zero-dimensional
                                                  // polytope
```

The latter has already appeared in Chapter 5.2.

Several methods provide the derivative of F in various forms. The method

```
ctype integrationElement(const LocalCoordinate& xi) const
```

returns the Jacobian determinant $\sqrt{\det \nabla F^T \nabla F}$ at the point given by the argument xi. The method

```
JacobianInverseTransposed
    jacobianInverseTransposed(const LocalCoordinate& xi) const
```

returns the matrix ∇F^{-T} at the same point. Both quantities appear in the expressions for element stiffness matrices in Chapter 2.1.3. Example programs in Chapters 3, 6, and others show these methods in context.

The object returned by the method jacobianInverseTransposed has the semantics of a matrix, but the precise type depends on the grid implementation. In many cases the type is a FieldMatrix (a dense matrix of static size, implemented in dune/common/fmatrix.hh), but this is not mandatory. For example, the YaspGrid implementation (Chapter 5.10.2) implements an axis-aligned structured cube grid, and hence it is known a priori that ∇F^{-T} is always a diagonal matrix. Consequently, the method jacobianInverseTransposed returns objects of type DiagonalMatrix (from dune/common/diagonalmatrix.hh), which store only the diagonal matrix entries. Hence, multiplications with a DiagonalMatrix are much faster than regular matrix multiplications.

For symmetry, there is also a method

```
JacobianTransposed jacobianTransposed(const LocalCoordinate &xi) const
```

which returns the matrix ∇F^T. It is used much less frequently than its inverse analog. One important use are simulations on surface grids, where the rows of $(\nabla F(\xi))^T$ span the tangent space of the grid at $F(\xi)$.

The rest of the interface consists of various auxiliary methods. For example, calling

```
bool affine() const
```

tells whether the geometric realization $F : \mathbb{R}^d \to \mathbb{R}^w$ has the form

$$F(\xi) = B\xi + c \qquad B \in \mathbb{R}^{w \times d}, \ c \in \mathbb{R}^w.$$

If this is the case (which includes all simplex and uniform grids), the results of the methods integrationElement, jacobianTransposed, and jacobianInverseTransposed are independent of their arguments, and need to be computed only once for each element. This can save a considerable amount of time when using high-order integration methods.

The method

```
GeometryType type() const
```

returns a token that identifies the reference element, that is, the domain of definition of F. It can be used to get more information about the reference element itself (Chapter 5.5), or to request finite element shape functions (Chapter 8) or quadrature rules (Chapter 9.2).

Two further methods exist to help implementing finite volume methods. Such methods frequently involve functions that are constant on each element. Integrals over one element can then be computed by multiplying the element volume by the function value at one fixed point. To facilitate this, the Geometry interface class provides two methods

int	mydimension	
	Dimension of the reference element T_{ref}	
int	coorddimension	
	Dimension of the geometry world space	
class	ctype	
	Number type used for coordinates	
class	Volume	
	Number type used for the geometry volume	
class	LocalCoordinate	
	Array type for local coordinates	
class	GlobalCoordinates	
	Array type for global coordinates	
class	JacobianInverseTransposed	
	Matrix type returned by jacobianInverseTransposed	
class	JacobianTransposed	
	Matrix type returned by jacobianTransposed	

Table 5.7.: Static information exported by the Geometry class

```
Volume volume() const
```

and

```
GlobalCoordinate center() const
```

which return the element volume and barycenter. They work for objects of all dimensions; by convention, the volume of a vertex is 1. For an example of how to use these methods see the finite volume example in Chapter 3.4.

As the very last methods, the Geometry interface class provides

```
Implementation& impl()
```

(and the corresponding **const** method), which allows access to the underlying implementation class. This method is very powerful, but it requires a lot of knowledge, too. Use it only with a good reason.

5.4. Intersections Between Elements

The Geometry class allows to compute integrals over elements, and this is a central part of what is needed for conforming finite element methods. However, for finite volume methods, and also for some finite element methods such as DG methods, integrals over the interface between two adjacent elements are needed. Such integrals typically combine shape functions from both elements with the normal of the interface. Similarly, Neumann boundary conditions involve integrals over the domain boundary. DUNE grids provide the relevant information through the Intersection interface.

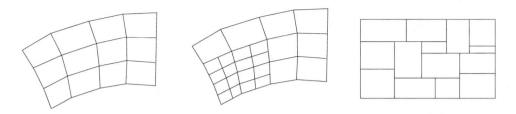

Figure 5.8.: Conforming and nonconforming grids. Left: conforming grid. Center: nonconforming grid obtained by red refinement without closure. Right: general nonconforming grid

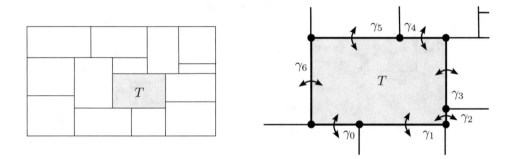

Figure 5.9.: Nonconforming grid, and close-up of one element with its intersections. The element has four faces of codimension 1, but seven intersections with neighboring elements.

5.4.1. Intersections

Providing information about adjacent elements is fairly straightforward if the grid is conforming. Remember that a grid is called *conforming* if the intersection of any two elements is a common subface of them, or empty (Figure 5.8, left). Such grids are required in many situations; e.g., for Lagrangian finite element methods. However, DG methods (Chapter 11.2.6) and finite volume methods work just as well on nonconforming grids, and it can be convenient to use this extra flexibility. Nonconforming grids frequently arise from adaptive red refinement without a closure (Figure 5.8, center), but more general situations are equally possible (Figure 5.8, right).

DUNE, aiming for maximum flexibility, does not make any assumptions about the conformity of the grids. The intersection between two neighboring elements is *not* restricted to be a common face. Rather, it can be an almost arbitrary subset of the element boundaries. In particular, it does not necessarily have to come about by refinement of a facet (as in Fig. 5.8, center).

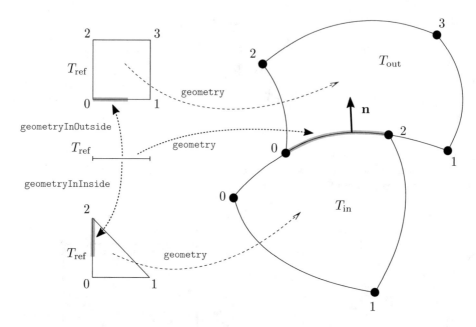

Figure 5.10.: The three different geometries of an intersection

DUNE handles such general situations by the concept of *intersections*.[9] Mathematically, intersections in the DUNE sense are set-theoretic intersections of the closures of two elements, and must have positive $d-1$-dimensional measure. For example, consider Figure 5.9, which shows the general nonconforming grid of Figure 5.8, and a close-up of one quadrilateral element T with its direct neighborhood. The element T meets its neighbors in a nonconforming way. The (set-theoretic) intersections with the neighboring elements are marked as bold lines, and labelled $\gamma_0, \ldots, \gamma_6$. These objects, which in DUNE parlance are called simply "intersections", carry the information of how T relates to its neighbors.

By convention, element facets on the domain boundary (or, more generally, intersections of an element with the exterior of the domain), are also DUNE intersections. Hence, there are two types of intersections: Intersections of one element with another, and intersections of one element with the domain boundary.

The intersections are represented in the DUNE grid interface by the interface class Intersection. It offers the functionality needed to evaluate the boundary and interface integrals occurring in finite element and finite volume methods. For example, there are methods to obtain element normal vectors **n**. Since intersections may be nonplanar, the normal vectors depend on the position in the intersection

---

[9]It is easy to get confused by the language here: The DUNE notion of *intersection* models a set-theoretic intersection (a subset of $\overline{\Omega}$) between two elements, and has certain functionality. The implementation represents the *intersection* concept by the interface class Intersection.

(Figure 5.10). The direction of the normal also gives an orientation to the intersection. This orientation is not definite, however. Rather, it depends on the way the intersection was obtained in the program. The next section will explain that individual intersections are obtained by iterating over all intersections of a given element. For all intersections obtained that way, this element is called the *inside* element. The element on the other side of the intersection is called the *outside* element, and normals always point from the inside to the outside. The names *inside* and *outside* will appear several times in the Intersection interface class.

Intersections have a shape as objects in a Euclidean space. This shape is encoded in the same way as the shape of grid elements: each intersection has a reference element, and its shape is a map F from the reference element into physical space (Figure 5.10). Since this is so similar to element shapes, it is actually implemented using the very same interface class Geometry already used for elements. So after having understood how element shapes work, intersection shapes will be easy.

There is more than one intersection shape, however. When integrating over intersections, the integrand frequently involves shape functions from the two elements that meet at this intersection. To evaluate those, one needs to be able to go from local coordinates on the intersection to local coordinates in the two elements. This coordinate change is provided by the interface by offering two additional geometric realizations of the intersection: one in the reference element of the *inside* element, and one in the reference element of the *outside* element (Figure 5.10). In other words, there are maps from the intersection reference element into the reference elements of the two elements. Going from local coordinates of the intersection to local coordinates of one of the two elements is as easy as evaluating these maps. This construction is very similar to the way the geometryInFather method of Chapter 5.3.1 produces an element shape in the reference element of its father.

5.4.2. Iterating over Intersections

Intersections are accessed by iterating over all intersections of a given element. For all intersections thus retrieved, this element is then the *inside* element. As always, there is forward iteration only, and range-based **for** loops can be used as well as direct iterator loops.

It is important to realize that the intersections of an element depend on whether that element is interpreted as part of a level grid view or as part of the leaf grid view. This is best explained by an example. Consider the hierarchical grid shown in Figure 5.11. In the level 0 view, element T_1 has three intersections with the grid boundary, and one with element T_2. When seen as part of the leaf view, however, it has *two* intersections with neighboring elements, namely with T_{11} and T_{12}, in addition to the three boundary intersections. Stating the desired grid view is therefore necessary to obtain the appropriate intersections.

Given a DUNE GridView object called gridView, and a codimension-zero entity named element from that grid view, the corresponding intersections can be accessed easily using range-based **for**. The following example loops over all intersections of

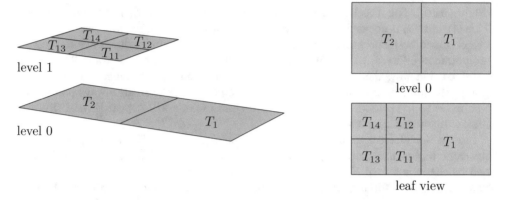

Figure 5.11.: Element intersections depend on the grid view.

element, and prints the normal at the barycenter of each intersection:

```
for (const auto& intersection : intersections(gridView, element))
  std::cout << intersection.centerUnitOuterNormal() << std::endl;
```

For the element in Figure 5.9 this will print

```
0 -1
0 -1
1 0
1 0
0 1
0 1
-1 0
```

To do the same thing using iterators, write

```
auto it    = gridView.ibegin(element);
auto endIt = gridView.iend(element);

for (; it!=endIt; ++it)
  std::cout << it->centerUnitOuterNormal() << std::endl;
```

The iterators completely follow the specification of a standard library forward iterator [103, 150]. Their type is available as `GridView::IntersectionIterator`.

If an element has m intersections, then the intersection iterator will stop *at least* m times. It may stop more than m times because grid implementations are allowed to subdivide intersections. Consider a nonconforming three-dimensional simplex grid, allowing very general neighborhood configurations. The intersection between two adjacent tetrahedra is a convex n-gon with n between 3 and 6. To integrate over such an intersection requires a quadrature rule. However, quadrature rules for penta- and hexagons are actually difficult to construct—the most pragmatic way is to triangulate the *n*-gon. To hide this complexity from the user, the DUNE grid interface allows the grid implementations to do the triangulation internally. If two tetrahedra intersect in a pentagon, the grid implementation may return three triangles in place of a single pentagonal intersection.

5.4.3. The `Intersection` Interface Class

The DUNE grid interface represents intersections by objects of the interface class `Intersection`. These objects provide all the information about the relationship between neighboring elements. The list of all public methods and all exported static information of the `Intersection` class is given in Tables 5.8 and 5.9, respectively.

The first group of methods provides information about the type of the intersection. The method

```
bool boundary() const
```

returns **true** if it is called for an intersection that is a part of the domain boundary. If that is the case, all methods that refer to the *outside* element have undefined behavior. On the other hand, boundary returning **false** does not imply that there is an *outside* element. For an example have a look at Figure 5.11 again. In the level 1 grid view, the element T_{11} has four intersections. The one corresponding to the right vertical edge is not part of the domain boundary, but there is no *outside* element in that grid view, either. A similar situation occurs at inter-process boundaries in a distributed grid. To really know whether the *outside* element exists (on the same processor), there is the method

```
bool neighbor() const
```

which returns **true** if and only if that is the case.

The method

```
GeometryType type() const
```

returns the type of the reference element for the intersection, i.e., the domain for its parametrization. As for grid entities, the reference element is encoded by the marker class GeometryType (see Chapter 5.5.2). Finally, the method

```
bool conforming() const
```

returns whether the intersection is conforming, i.e., whether it is equivalent to an entire element facet shared by the *inside* and *outside* elements. Having this information sometimes allows to switch between different code paths for "simple" conforming situations and more challenging nonconforming ones.

bool	boundary() **const**
	True if intersection is part of the domain boundary
bool	neighbor() **const**
	True if intersection is shared with another element on the same process
GeometryType	type() **const**
	Type of reference element for this intersection
bool	conforming() **const**
	True if intersection is conforming
Entity	inside() **const**
	The *inside* entity for this intersection
Entity	outside() **const**
	The *outside* entity for this intersection
int	indexInInside() **const**
	Local index of facet in the *inside* entity that contains the intersection
int	indexInOutside() **const**
	Local index of facet in the *outside* entity that contains the intersection
LocalGeometry	geometryInInside() **const**
	Geometry of this intersection in local coordinates of the *inside* entity
LocalGeometry	geometryInOutside() **const**
	Geometry of this intersection in local coordinates of the *outside* entity
Geometry	geometry() **const**
	Geometry of the intersection in the grid world space
GlobalCoordinate	outerNormal(**const** LocalCoordinate& xi) **const**
	An outer normal vector (length not necessarily 1)
GlobalCoordinate	unitOuterNormal(**const** LocalCoordinate& xi) **const**
	Unit outer normal
GlobalCoordinate	centerUnitOuterNormal() **const**
	Unit outer normal at the center of the intersection
GlobalCoordinate	integrationOuterNormal(**const** LocalCoordinate& xi) **const**
	Outer normal scaled with the functional determinant of the intersection
size_t	boundarySegmentIndex() **const**
	Index of the boundary segment within the macro grid

Table 5.8.: Public methods of the Intersection class

Assembling contributions for a Neumann boundary term involves only those intersections for which the boundary method returns **true**. However, even though this is a frequent use-case, there is no way to iterate over only these intersections. Instead, one has to loop over all elements, and for each element loop over all intersections of that element, skipping those intersections not on the boundary. Needless to say that this can be inefficient.

int	mydimension	
	Dimension of the intersection	
int	dimensionworld	
	Dimension of the grid world space	
class	Entity	
	Type of entity that this intersection belongs to	
class	LocalCoordinate	
	Type for vectors of coordinates in the intersection	
class	GlobalCoordinate	
	Type for normal vectors	
class	ctype	
	Number type used for coordinates	
class	Geometry	
	Geometry returned by the geometry method	
class	LocalGeometry	
	Geometry returned by geometryInInside and geometryInOutside	

Table 5.9.: Static information exported by the Intersection class

As a compromise, the Entity class offers the method

```
bool hasBoundaryIntersections() const
```

which will return **true** if at least one of the intersections of the element is part of the domain boundary. If this is true for a given element, one needs to loop over all intersections of that element to get more details. However, if the methods returns **false**, iteration over the element intersections can be skipped.

The next group of methods returns information about the two elements that make up the intersection. Remember that they are called *inside* and *outside* element, with *inside* being the element that was used to obtain the intersection. The methods

```
Entity inside() const
Entity outside() const
```

return these two elements. If the intersection has no *outside* element, then the behavior of the outside method is undefined. The methods

```
int indexInInside() const
int indexInOutside() const
```

return the local number of the element facet (i.e., element face of codimension 1) that the intersection is part of, the second method again only if the intersection has a neighbor. The numberings of the facets for the different element types are given in Chapter 5.5. For example, for the intersection in Figure 5.10, if T_1 is the *inside* element, indexInInside and indexInOutside return 1 and 2, respectively. These numbers can be used directly as input arguments for the method `Entity<0>::subEntity` of Chapter 5.3.1. Indeed, the line

```
auto facet = intersection.inside().subEntity<1>(intersection.indexInInside());
```

returns the facet of the *inside* entity that the intersection is part of.

A third group of methods gives information about the actual shape of the intersections. The grid interface assumes that intersections are C^1-diffeomorphic images of a reference element, just as grid elements are. Therefore the geometry of intersections can be represented by the `Geometry` interface class that has already been used for entity geometries in Chapter 5.3.2.

For each intersection, the `Intersection` interface offers three different geometries (Figure 5.10). The first, given by the method

```
Geometry geometry() const
```

gives the shape of the intersection in the grid world space. This can be used, for example, to evaluate globally defined functions on the intersection. Suppose that the function is implemented by an object f that supports evaluation with `operator()(FieldVector x)`, where x is a coordinate in the domain Ω. Then at a point xi on the intersection f is evaluated by

```
auto value = f(intersection.geometry().global(xi));
```

More often than not, however, functions are given in local coordinates on the grid elements. To evaluate such functions on an intersection, there needs to be a way to go from local coordinates on the intersection to local coordinates on the element. This is what the

```
LocalGeometry geometryInInside() const
```

and

```
LocalGeometry geometryInOutside() const
```

methods do. They return geometries whose world spaces are the reference elements of the *inside* and *outside* elements, respectively (see Figure 5.10). The result is a *local geometry* just like the geometry returned by the geometryInFather method. The dimension of their target space is equal to the grid dimension rather than to the dimension of the grid world space. However, both local and global (regular) geometries are implemented by the same Geometry class of Chapter 5.3.2, hence they both behave the same way. As an example for the use of the geometryInInside and geometryInOutside methods, the following code computes the jump $[u_h] := u_{h,1} - u_{h,2}$ of a piecewise continuous finite element function at a point ξ on an intersection:

```
auto jump = u(intersection.geometryInInside().global(xi))
                - u(intersection.geometryInOutside().global(xi));
```

The method `global` transforms the coordinates on the intersection to coordinates on the *inside* and *outside* elements.

Care must be taken when comparing intersection geometries with the corresponding subentity geometries. The situation is particularly confusing if the intersection is conforming, i.e., if it covers an entire facet. Such an intersection has a geometry, which can be obtained with

```
intersection.geometry();
```

Likewise, the facet has a geometry, which can be retrieved by

```
intersection.inside().subEntity<1>(intersection.indexInInside()).geometry();
```

These two geometry objects describe the same shape in space, so one would expect them to be equal. However, the grid interface specification does *not* require them to be represented by the same parametrization, and indeed for many grid implementations they are not. They do both describe the same shape by maps $T_{\text{ref}}^{\text{facet}} \to \mathbb{R}^w$, but the two maps can differ by a map from $T_{\text{ref}}^{\text{facet}}$ onto itself. Overlooking this can be the cause of very subtle bugs.

We point out a second pitfall: While the methods geometry, geometryInInside, and geometryInOutside return their results *by value*, the returned objects may not have true value semantics. The DUNE grid interface only guarantees that the objects will remain valid until either the grid object is deleted, localBalance is called, or adapt is called. In particular, incrementing intersection iterators will not change an intersection geometry. In other words, the following code snippet is valid:

```
std::vector<Geometry> geometries;
for (const auto& intersection : intersections(gridView, element))
  geometries.push_back(intersection.geometry());

for (const auto& geometry : geometries)
  std::cout << geometry.center() << std::endl;
```

It will print the centers of all intersections of the element given in the object `element`.

The last group of `Intersection` interface methods returns various normal vectors of the intersection. Remember that these always point away from the *inside* element. Given local coordinates ξ on the intersection, the method

```
GlobalCoordinate outerNormal(const LocalCoordinate& xi) const
```

returns a normal vector at the point ξ, of unspecified length, in an object of type `FieldVector`. This method is to be used if all that is need is the normal direction.

To get a *unit* outer normal the return value of `outerNormal` can simply be normalized. However, in some cases (structured grids, in particular), unit outer normals can be computed more efficiently by the grid itself, because the grid may have certain values precomputed. To benefit from this, the `Intersection` class offers the method

```
GlobalCoordinate unitOuterNormal(const LocalCoordinate& xi) const
```

which returns a normal vector at ξ of unit length.

The related method

```
GlobalCoordinate centerUnitOuterNormal() const
```

also returns a unit outer normal, but it does so without requiring an argument. The normal is always computed at the intersection center, more precisely at the point returned by `intersection.geometry().center()`. This is a convenience method for finite volume algorithms, which need precisely this information very frequently. It is used, e.g., in the finite volume example in Chapter 3.4.

There is a third method offering normal vectors of specific length. It is a common task in finite volume and DG methods to compute fluxes across element boundaries, i.e., terms of the form

$$\int_{\partial T} \langle \nabla u, \mathbf{n} \rangle \, ds = \sum_{\gamma \subset \partial T \text{ an intersection}} \int_{\gamma} \langle \nabla u, \mathbf{n} \rangle \, ds, \tag{5.1}$$

where u is a scalar function, and $\mathbf{n}$ is the unit outer normal to the element T. Written as an integral over the intersection reference element T_{ref}^{γ} this involves the expression

$$\mathbf{n} \sqrt{\det(\nabla F_{\gamma}^{T} \nabla F_{\gamma})},$$

where F_{γ} is the map represented by the intersection geometry, i.e., the map from the intersection reference element to the grid world space. The term $\mathbf{n} \sqrt{\det(\nabla F_{\gamma}^{T} \nabla F_{\gamma})}$ can be interpreted as a scaled normal vector, and this is what the method

```
GlobalCoordinate integrationOuterNormal(const LocalCoordinate& xi) const
```

returns. Calling this method can be faster than computing $\mathbf{n}$ and $\sqrt{\det(\nabla F_{\gamma}^{T} \nabla F_{\gamma})}$ separately, because the grid implementation may be able to use precomputed information. As an example, the following code snippet implements the expression (5.1) for a given function u that can be evaluated in global coordinates:

```
// Loop over all intersections
for (const auto& intersection: intersections(gridView, element))
{
  // Quadrature loop
  const auto& quadRule
      = Geometry::QuadratureRules<double, dim>::rule(intersection.type(),
                                                     quadOrder);

  for (const auto& quadPoint : quadRule)
  {
```

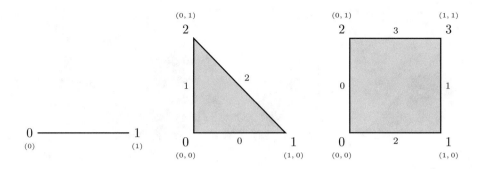

Figure 5.12.: The one- and two-dimensional reference elements, with vertex and edge numbers

```
GlobalCoordinate uGradient
    = derivative(u)(intersection.geometry().global(quadPoint.position()));
GlobalCoordinate scaledNormal
    = intersection.integrationOuterNormal(quadPoint.position());
result
    += quadPoint.weight() * (uGradient * scaledNormal);  // Scalar product
    }
}
```

See Chapter 9 for details on the numerical quadrature, and Chapter 10.5.3 on the syntax for computing the gradient ∇u.

There is one final method in Table 5.8 that has not been discussed yet. The method boundarySegmentIndex is needed to attach data to the domain boundary. Chapter 5.6 is dedicated to this issue, and the method will be explained there.

5.5. Reference Elements

The DUNE grid interface defines element geometries as maps from a reference element into Euclidean space, and this idea is reflected in the interface of the Geometry class. In this interface, the reference elements appear only indirectly, e.g., as the set of valid arguments for the method global, which evaluates the map.

Sometimes it is necessary, however, to obtain some information about the reference elements themselves. For example, it may be helpful to know whether a given point is contained in a given reference element. Or, in a grid containing prism elements, whether the i-th face is a triangle or quadrilateral, and what the local numbers of its vertices are. While this may sound like information to be obtained from the grid element, it is more elegant to have it stored separately, once and for all prisms.

Similarly to grid elements, reference elements have a geometrical and a topological aspect to them. Geometrically, a reference element is a fixed polytope. In large

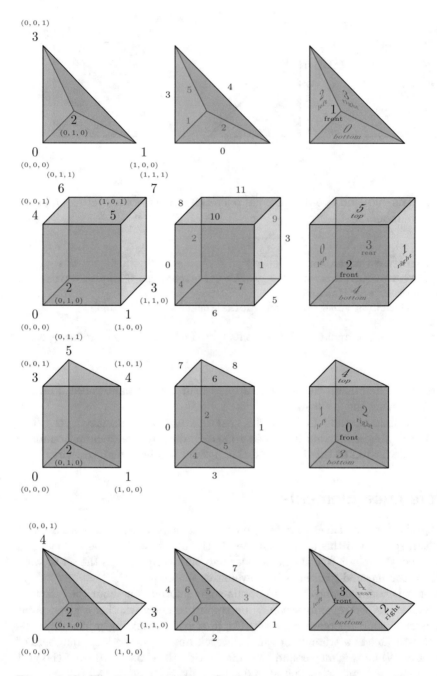

Figure 5.13.: The most important three-dimensional reference elements, with their subentity numberings

parts of the finite element literature, the reference triangle is the triangle with corners $(0,0)$, $(1,0)$, $(0,1)$. For the reference quadrilateral both $[0,1]^2$ and $[-1,1]^2$ are used. In the more rigorous presentation in [16], a reference element is therefore defined as an *equivalence class* of polytopes.

Topologically, reference elements define a local numbering of the faces of an element. There are various ways to define these numberings, and every finite element implementation needs to make a choice. For the one-, two-, and three-dimensional reference elements available in DUNE, these numberings are illustrated in Figures 5.12 and 5.13.

While there are infinitely many topologically distinct convex polytopes of any dimension larger than one, few of them are actually used as reference elements. The finite element method works with simplices and cubes almost exclusively, while for red–green refinement in three-dimensional cube grids pyramids and prisms are required (Figure 2.22). Advanced methods like the virtual element method or mimetic finite differences can accommodate more general element types as well [21, 119].

The DUNE grid interface does not offer a complete set of reference elements. Rather, it offers the important types mentioned above, and a set of others that can be constructed by certain rules (explained in Section 5.5.2). For more general reference elements one would have to extend the dune-geometry module code itself.

5.5.1. Using the Dune Reference Elements

The DUNE interface for reference elements is contained in the dune-geometry module, in the file dune/geometry/referenceelements.hh. Unlike the rest of the DUNE grid interface, the reference elements use duck typing, i.e., user code gets to handle implementation classes directly (Chapter 4.4.2). At the time of writing there is a single implementation of reference elements that is used by all grid implementations, but the infrastructure exists to let particular grids implement their own reference elements.

Reference elements do not have public constructors. Instead, they can be produced by the factory class [72]

```
template<class ctype, int dim>
ReferenceElements<ctype,dim>
```

It hands out reference element objects through static methods

```
auto ReferenceElements<ctype,dim>::simplex()
auto ReferenceElements<ctype,dim>::cube()
auto ReferenceElements<ctype,dim>::general(const GeometryType& type)
```

The first two of these hand out the simplex and cube reference elements of the corresponding dimension. The general method provides any implemented reference element of the given dimension. The requested type is specified by an object of type GeometryType (see Section 5.5.2 for details). For example, to obtain the reference element for three-dimensional prism elements, write

```
auto referencePrism
    = ReferenceElements<double,3>::general(GeometryTypes::prism);
```

To use this to get, e.g., the volume of that reference prism write

```
std::cout << referencePrism.volume() << std::endl;
```

This will print the value 0.5, which is correct given the definition of the DUNE reference prism in Figure 5.13.

As an alternative to the factory, there are free functions

```
auto referenceElement(...)
```

that allow to get reference element objects for a variety of arguments. For example, if geometryType is GeometryType object, dim is a dimension, and ctype a number type suitable for point coordinates, then the following three lines all produce reference elements:

```
auto refElement = referenceElement<ctype,dim>(geometryType);
auto refElement = referenceElement<ctype>(geometryType, Dim<dim>());
auto refElement = referenceElement(ctype(), geometryType, Dim<dim>());
```

Furthermore, the referenceElement method allows to get reference elements for Geometry objects:

```
auto refElement = referenceElement(geometry);
```

This is a bit shorter than

```
auto refElement = referenceElement<ctype,dim>(geometry.type());
```

and can be more efficient, too.

The list of public methods of the ReferenceElement class can be seen in Table 5.10.[10] Corresponding to the topological and geometrical aspects of grid elements, it can be split into a topological and a geometrical part. In the topological part, the first two methods inform about the types of the reference element faces. The method

```
GeometryType type(int i, int c) const
```

returns the type of the i-th subface of codimension c. Setting i and c to zero returns the type of the reference element itself. For convenience, there is also the method

```
GeometryType type() const
```

which returns the same thing.

---

[10] As usual all return types are exported by the ReferenceElement interface class as well.

GeometryType	type() **const** Type of the reference element
GeometryType	type(**int** i, **int** c) **const** Type of the reference element of subentity (i,c)
int	size(**int** c) **const** Number of subentities of codimension c
int	size(**int** i, **int** c, **int** cc) **const** Number of subentities of codimension cc of subentity (i,c)
SubEntityRange	subEntities(**int** i, **int** c, **int** cc) **const** Local numbers of the subentities with codimension cc of subentity (i,c)
FieldVector<ctype, dim>	position(**int** i, **int** c) **const** Position of the barycenter of subentity (i,c)
bool	checkInside(**const** FieldVector<ctype, dim>& local) **const** Check if a point is contained in the reference element
template<**int** codim> Codim< codim >::Geometry	geometry(**int** i) **const** Embedding of subentity (i,codim) into the reference element
Volume	volume() **const** Volume of the reference element
FieldVector<ctype, dim>	integrationOuterNormal(**int** i) **const** Outer normal of the i-th facet, scaled with the Jacobian determinant of that facet

Table 5.10.: Public methods of the ReferenceElement class. The expression "subentity (i,c)" denotes the i-th subentity of codimension c.

The method

```
int size(int c) const
```

returns the number of faces of codimension c. So, for example, if dim is the dimension of the reference element, then size(dim) will return the number of corners, size(1) will return the number of facets, and size(0) will always return 1.

The following example uses both the type and the size method to print the types of all facets of a pyramid:

```
auto referencePyramid
   = ReferenceElements<double,3>::general(GeometryTypes::pyramid);

for (int i=0; i<referencePyramid.size(1); i++)    // 1 is the codimension
                                                  // of facets
  std::cout << referencePyramid.type(i,1) << std::endl;
```

More generally, the reference elements know about the sizes of their faces. The method

```
int size(int i, int c, int cc) const
```

gives the number of subfaces of codimension cc of the i-th face of codimension c.
Both codimensions are with respect to the reference element. So, for example,
size(0,1,dim) returns the number of vertices of the 0-th facet, and size(0,1,dim-1)
returns the number of edges of this facet. If dim==3, then size(0,2,dim) returns
the number of vertices of the 0-th edge (always 2, of course), and so on. If cc is less
than c then the method returns zero.

Yet another method provides the actual subfaces. Calling

```
SubEntityRange subEntities(int i, int c, int cc) const
```

gives information about the i-th face of codimension c: It returns a range over the
local numbers of the subfaces of codimension cc of that face. Both codimensions are
with respect to the reference element. The return value of the subEntities method
is a subclass of Dune::IteratorRange from the dune-common module. As such it
provides methods begin and end that allow to iterate over the range. In addition,
SubEntityRange objects have methods

```
std::size_t size() const
```

and

```
bool contains(std::size_t i) const
```

While the first returns the number of elements in the range, the second one can be
used to test whether a range contains a particular subface index. Hence, to get the
vertex numbers of all facets of a tetrahedron write

```
// Loop over facets
for (int i=0; i<referenceTetra.size(1); i++)
  // Loop over vertices of current facet
  for (auto v : referenceTetra.subEntities(i,1,3))
    std::cout << "facet: " << i << ",  vertex: " << v << std::endl;
```

This will print

```
facet: 0,  vertex: 0
facet: 0,  vertex: 1
facet: 0,  vertex: 2
facet: 1,  vertex: 0
facet: 1,  vertex: 1
facet: 1,  vertex: 3
facet: 2,  vertex: 0
facet: 2,  vertex: 2
facet: 2,  vertex: 3
facet: 3,  vertex: 1
facet: 3,  vertex: 2
facet: 3,  vertex: 3
```

which matches the numbering illustrated in Figure 5.13.

The remaining methods of the ReferenceElement class are of geometrical nature. For example, the method

```
bool checkInside(const FieldVector<ctype,dim>& local) const
```

returns **true** if the point with coordinates given in local is contained in the reference element. The method

```
template<int codim>
Codim<codim>::Geometry geometry(int i) const
```

returns the embedding of a reference element face into the reference element itself. This is very similar to the geometryInInside method of the Intersection class, which returns the embedding of (parts of) facets into the reference element (Chapter 5.4). Unlike the case of the geometryInInside method, however, faces of any dimension can be handled here; in particular, calling this method for codimension 0 will return the geometry of the reference element itself. The object returned by ReferenceElement::geometry implements precisely the interface of the Geometry class explained in Chapter 5.3.2, and its type is exported by the ReferenceElement class.

The last three methods are simple but convenient. The method

```
Volume volume() const
```

returns the volume of the reference element, in the number type Volume. The method

```
FieldVector<ctype,dim> position(int i, int c) const
```

computes the center of mass of the vertices of the i-th subface of codimension c. This can be useful for finite volume methods, or when simply needing a point that is guaranteed to be in the reference element. If c equals dim, the return value is the position of the i-th vertex; if c is 0, it is the center of mass of the reference element itself. Both volume and position information is also available via the geometry method, but it can be convenient to have them directly on the reference element itself.

Finally, the method

```
FieldVector<ctype,dim> integrationOuterNormal(int i) const
```

returns an outer normal vector of the i-th facet. The length of the vector is equal to the integration element of that facet, i.e., to the term $\sqrt{\det \nabla F_{\text{facet}}^T \nabla F_{\text{facet}}}$ that appears in the integral transformation formula (Chapter 5.3.2). This allows for very efficient numerical quadrature of vector fields across element facets.

Unlike in the case of the Intersection class, which also returns normal vectors, no position argument is needed here. As all reference elements are polytopes, their facets are planar, and hence normal vectors of facets are independent of position.

5.5.2. `GeometryType` and the Topology Id

Several chapters have already mentioned the `GeometryType` class, which implements a token to specify a particular reference element. It is used by the `Entity` class to report the reference element corresponding to the entity (Chapter 5.3.1), and in turn it is accepted by the `ReferenceElements` container class which hands out specific reference element objects. In Chapter 9.2, it will be used to request quadrature rules for a given reference element.[11]

The `GeometryType` class is easier to understand with a bit of background knowledge about how DUNE constructs its reference elements. Ideally, a token class for reference elements would enable to label all convex polytopes there are. Unfortunately, such a labeling is difficult for all but the one- and two-dimensional polytopes, and besides, very few of these types are ever needed in finite element computations.[12] DUNE opts for a pragmatic approach, and implements those reference elements that are found in the standard finite element methods. This includes simplices and hypercubes of all dimensions, prisms and pyramids needed for red–green adaptive hexahedral grids, and various prism- and pyramid-like types in higher dimensions. For situations that are not covered by the list of available reference elements there is a special none-type (one for each dimension), which can signal to applications that some sort of special handling is necessary.

DUNE constructs the reference elements systematically, in a process explained in [49]. The basic idea is induction over the dimension. The only one-dimensional reference element is the unit interval (up to homeomorphisms of $\mathbb{R}$), which can be interpreted either as a one-dimensional cube or as a one-dimensional simplex. Starting with the unit interval, DUNE considers two ways to construct a two-dimensional polytope. Tensor-multiplying the unit interval with a second unit interval yields the unit square $[0,1]^2 = [0,1] \times [0,1]$. Conical multiplication, on the other hand, leads to the unit triangle. Formally, for a given reference element $T_{\text{ref}} \in \mathbb{R}^{d-1}$ we define the tensor product

$$T_{\text{ref}}^{|} := \big\{ (x,z) \ : \ x \in T_{\text{ref}}, z \in [0,1] \big\}$$

and the conical product

$$T_{\text{ref}}^{\circ} := \big\{ (x(1-z),z) \ : \ x \in T_{\text{ref}}, z \in [0,1] \big\}.$$

Continuing this way to three dimensions, the tensor product of the unit square with an interval results in the unit cube, and the conical product of a triangle with the interval leads to the unit tetrahedron. Conical multiplication of the unit square with the unit interval leads to the pyramid, and tensor multiplication of the triangle with a unit interval leads to a prism. This process is illustrated in Figure 5.14. Polytopes of

---

[11] The `GeometryType` class is not particularly well-named: It only specifies information about the topology/combinatorial structure of the reference element, and no geometrical information in the sense of Chapter 5.3 at all.

[12] Methods like mimetic finite differences [119] and virtual element methods [21] allow more general elements, but typically do not use reference elements.

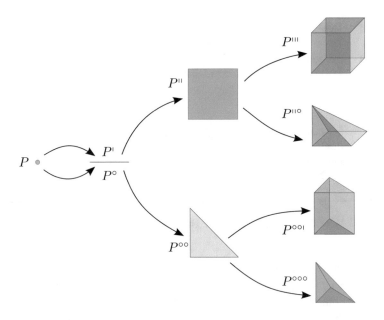

Figure 5.14.: Construction of reference elements by tensor and conical multiplication starting from a single point P

arbitrary dimension can be constructed this way, including hypercubes and simplices of all dimensions.

While this construction evidently yields all relevant reference elements as point sets, its true strength lies in the fact that it also produces consistent numberings of all the faces of these reference elements. The precise nature of the algorithm is beyond the scope of this text, but it allows to implement the subEntity method of the ReferenceElement class (Chapter 5.5.1) even for high-dimensional reference elements.

DUNE encodes the defining sequence of tensor and conical products in an integer number. Denoting a tensor product by "1" and a conical product by "0", a sequence of such products gives a sequence of binary digits. This sequence of digits, read as a number, is called the *topology id* of the reference element. Together with the element dimension (which specifies the number of bits of the *topology id* bit sequence), such a *topology id* specifies a reference element. The *topology id* is unique, with one exception: Since the unit segment can be constructed both by tensor- and by conical multiplication of a point with the interval $[0, 1]$ (it is both a simplex and a cube), the one-digit numbers 0 and 1 both denote the same reference element. This has to be kept in mind when interpreting the *topology id* bits.

The GeometryType class (from the file dune/geometry/type.hh) hides the complexity of the *topology id* idea behind a convenient interface. It is a C++ "literal type",

`bool`	`isTriangle() const`
	Return **true** if it is a triangle
`bool`	`isQuadrilateral() const`
	Return **true** if it is a quadrilateral
	$\ldots$
	$\ldots$
`bool`	`isNone() const`
	Return **true** if it is a singular reference element
`unsigned int`	`dim() const`
	Dimension of the corresponding reference element
`unsigned int`	`id() const`
	Corresponding topology id of the type

Table 5.11.: Methods of the `GeometryType` class

and can be used in **constexpr** contexts if created with a **constexpr** constructor.

There are two ways how `GeometryType` objects can be obtained. The first way is to call one of its constructors. The most important one is

```
constexpr GeometryType(unsigned int topologyId,
                       unsigned int dim,
                       bool none=false)
```

but `GeometryType` objects can also be default-constructed.

In many situations it is more convenient to use the second construction method, however. The namespace `GeometryTypes` provides a set of static objects and factory methods that can be used to set up `GeometryType` objects. For example,

```
GeometryType gt = GeometryTypes::triangle;
```

will set up a `GeometryType` object representing the two-dimensional reference triangle, and

```
GeometryType gt = GeometryTypes::prism;
```

will do the same for the three-dimensional reference prism. The lines

```
GeometryType gt = GeometryTypes::simplex(4);
GeometryType gt = GeometryTypes::cube(4);
```

will set up four-dimensional reference simplex and cube elements, respectively, and

```
GeometryType gt = GeometryTypes::none(3);
```

gives a token for the none type in three dimensions. Of all these types, `pyramid` and `prism` are only available for three-dimensional elements, but the others can be used with any space dimension. For a complete list of these objects and methods refer to the online documentation.

Given a `GeometryType` object, its properties can be queried. For example,

```
bool isHexahedron() const
```

returns **true** if the given GeometryType object is a three-dimensional hexahedron, and

```
bool isCube() const
```

returns whether it is a hypercube of any dimension. The full list of these methods is given in the online documentation. `GeometryType` objects provide their topology id representation by

```
unsigned int id() const
```

an their dimension by

```
unsigned int dim() const
```

For anything deeper you need to ask the corresponding `ReferenceElement` object.

Finally, `GeometryType` implements several comparison operators, that allow it to be used as key type in associative standard library containers [103]. Observe that all none objects compare equal, even though they may have different dimensions. For debugging convenience, **operator**<< is also implemented.

5.6. Attaching Data to Grids

It is the purpose of finite element grids to serve as the basis for the definition of the discrete function spaces used in finite element methods. Functions from such spaces are expressed as vectors of coefficients with respect to some basis of the discrete space, and once a grid and a function space basis are chosen, storing a discrete function from that space means storing the vector of coefficients. In DUNE, function space bases are represented by `GlobalBasis` objects. These are implemented in the `dune-functions` module, and explained in Chapter 10.

For most finite element spaces, the basis functions and corresponding coefficients can be associated to individual grid entities. For example, the nodal basis functions of a first-order Lagrange space can be associated to the grid vertices (Chapter 2.1.2). Similarly, the standard basis functions of Crouzeix–Raviart elements can be associated to grid edges, and the basis functions of Discontinuous Galerkin (DG) spaces are naturally associated to elements. Not *all* functions used as basis functions for finite element spaces can be sensibly assigned to entities of the grid; one example are the functions of a B-spline basis, with a much larger support than Lagrangian basis functions. For the standard finite element spaces, though, such an assignment is possible and useful.

Since data coefficients are associated to basis functions, and basis functions are associated to grid entities, the coefficients are also associated to entities of the grid.

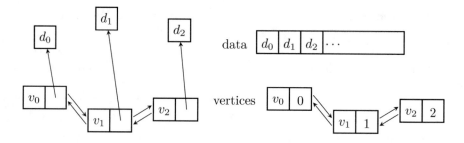

Figure 5.15.: Two ways to attach vertex data to a grid that stores its vertices in a doubly-linked list. Left: each vertex $v_0, v_1, \dots$ stores a pointer to the data item on the heap. Right: The data $d_0, d_1, \dots$ is stored consecutively in an array, and each vertex stores the corresponding array index.

A software that implements grid data structures for numerical simulations therefore needs to provide a way to attach numerical data to the individual grid entities. For any given entity, this data should be easily and efficiently retrievable. Also, when the grid changes due to adaptive refinement or load-balancing, there must be a way to retain discrete grid function data across such grid modification steps as much as possible. Finally, the type and amount of data allowed per entity should be arbitrary. Applications can have arbitrary requirements here, and good software infrastructure should impose no restrictions.

It is tempting to store the numerical data in the entity data structures themselves, either by storing the data directly, or by storing a pointer to the data (Figure 5.15, left). This approach, used, e.g., by the UG3 software [13], allows to easily and cheaply access the data for any given entity. When the grid changes locally, those entities not affected by the change naturally keep their data.

However, the approach comes with a considerable price in execution speed. The coefficient vectors will serve as input and output data of linear algebra algorithms, in particular of linear solvers for sparse systems. Since the algebraic problems are notoriously big, the corresponding algorithms and their implementations must be maximally efficient. Storing the data for each vertex separately on the heap leads to data fragmentation, i.e., the data is scattered all over the computer memory. On modern computer architectures this leads to a severe increase in execution time.

To optimally use the memory hierarchies of modern computer architectures, the data needs to be placed contiguously in memory. An array-like data structure is mandatory, with the individual entries being addressable by numbers. This is the approach taken by DUNE. All numerical data must be stored in array-like data types completely outside of the grid data structure.[13] To access this data, the grid entities

---

[13]This is actually one of the big advantages of the DUNE grid data interface: It does not force the user to use one particular data structure for coefficient vectors and sparse matrices. For coefficient vectors, any data structure with random access interface is usable. DUNE proposes its own set of

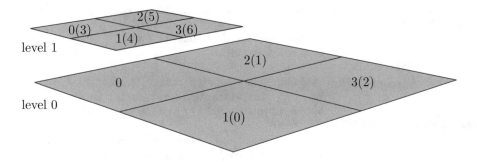

Figure 5.16.: Entity indices depend on the grid view. Here, for each element, the level grid view index is shown. For those elements in the leaf grid view, the leaf grid view index is shown additionally in parentheses.

provide integer numbers that allow to address the arrays. In DUNE terminology, such a number is called an *index*. For the set of all entities of a given GeometryType of a grid view, the corresponding set of indices is guaranteed to be consecutive and to start at zero. Hence the corresponding data can be stored in an array data structure without unused space. For vertex data, this is illustrated in Figure 5.15, right.

The approach using arrays and indices fails, however, as soon as the grid changes. After a step of adaptive refinement or load-balancing, the old indices necessarily cease to be consecutive, and new indices have to be assigned. These new indices then fail to address the existing data arrays, unless those are properly reordered. To do this reordering, the DUNE grid interface offers a second type of number called *id*, which will be explained in Chapter 5.6.3.

5.6.1. Index Sets and the Boundary Segment Index

The following section explains the interface used to obtain indices for given entities. Since these indices do not only depend on the entity itself, but also on the grid view (Figure 5.16), the index is not available from the Entity class directly. Instead, assignment of indices to entities for a given grid view is provided by a dedicated interface class called IndexSet, available from the GridView class through the method

```
const IndexSet& indexSet() const
```

Note that the return value is a **const** reference—IndexSet objects cannot be copied. The public methods of the IndexSet class are listed in Table 5.12. Additionally, the

linear algebra data structures in the dune-istl module (Chapter 7), but these are by no means mandatory. Coefficient vectors may equally well be stored in std::vectors, C-style arrays, or the PETSc data structures. Indeed, the dune-pdelab discretization toolbox (Chapter 11) proposes a set of linear algebra backends, which allow to assemble the algebraic problems into different linear algebra data structures.

```
template<class Entity>
        IndexType    index(const Entity& entity) const
                     Map entity to index
template<class Entity>
        IndexType    subIndex(const Entity& entity,
                              int i, unsigned int codim) const
                     Map a subentity of entity to an index
template<class Entity>
             bool    contains(const Entity& entity) const
                     True if index is available for the given entity
        IndexType    size(GeometryType type) const
                     Number of entities of given geometry type in the grid view
        IndexType    size(int codim) const
                     Number of entities of given codimension in the grid view
            Types    types(int codim) const
                     Return iterator range over the set of all GeometryTypes in
                     this grid view
```

Table 5.12.: Public methods of the `IndexSet` class

class statically exports the grid dimension in `dimension` and, among other things, the type `IndexType` used by the grid implementation to store indices (Table 5.13).

The following examples always assume that there is an `IndexSet` object for a given grid view, in a variable called `indexSet`, and an `Entity` object of arbitrary dimension called `entity`. Then, `indexSet.index(entity)` will return the index of that entity. As an example use, suppose there is a `std::vector<double>` called `data` holding vertex data. The following loop will print out that data together with the corresponding vertex coordinates:

```
for (const auto& vertex : vertices(gridView))
    std::cout << "index: " << indexSet.index(vertex) << "  "
              << "position: " << vertex.geometry().corner(0) << "  "
              << "data: " << data[indexSet.index(vertex)] << std::endl;
```

Depending on the grid implementation, this loop may or may not traverse the list of vertices in order of increasing index. For example, the `UGGrid` implementation numbers the vertices in the order they were inserted into the grid, but when iterating, the boundary vertices will appear first. This discrepancy may appear strange at first, but there is hardly ever a need for the two orderings to be the same.

In addition to the index of a given entity, the `IndexSet` interface also provides the indices of its subentities. This is of importance in assemblers which, for a given element, need to access the data of that element's vertices, edges, facets, etc. The method for this is

```
template<class Entity>
IndexType subIndex(const Entity& entity, int i, unsigned int codim) const
```

int	dimension
	Grid dimension
class	IndexType
	Number type used for indices
class	Types
	Iterator range used as return value of the types method

Table 5.13.: Static information exported by the IndexSet class

The second integer parameter is a codimension (with respect to the grid), and the first one is the local number for a subentity of that codimension. So for example, the following loop prints all edge data for a hexahedron given in entity:

```
for (size_t i=0; i<entity.subEntities(2); i++)      // Loop over all 12 edges
  std::cout << data[indexSet.subIndex(entity,i,2)] << std::endl;
```

If entity is an element, and the subentity codimension is known at compile time, then instead of

```
indexSet.subIndex(entity,i,2)
```

it is also possible to use the Entity::subEntity method and write

```
indexSet.index(entity.subEntity<2>(i))
```

which returns the same number. However, using subIndex not only leads to shorter code, it also tends to be faster than constructing the full subentity with the subEntity method.

The resulting behavior of calling either of the last two methods for an entity that is not contained in the grid view corresponding to the index set is not specified. To test whether a given entity is part of the grid view, the GridView class provides the method

```
template<class Entity>
bool contains(const Entity& entity) const
```

For convenience, this method is replicated in the IndexSet class. Hence, if calling indexSet.contains(entity) returns **true**, then it is safe to also call the method indexSet.index(entity). There is no guarantee regarding the run-time complexity of the contains method, but for most reasonable grid implementations it should take constant time.

The next two methods inform about how many objects of a given type are in the domain of the IndexSet. This implies in particular the largest index that can appear. Both methods are called size:

```
IndexType size(GeometryType type) const
IndexType size(int codim) const
```

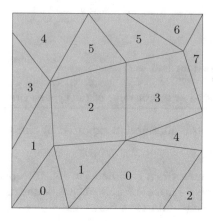

Figure 5.17.: Element indices of a two-dimensional grid with two element types. Note how triangles and quadrilaterals are counted separately.

The first takes a `GeometryType` object as its only argument, and returns the number of entities of that `GeometryType` (e.g., the number of all triangles in the grid view). The second method takes an integer argument with the meaning of a codimension with respect to the grid, and returns the number of entities of that codimension. Hence, with an `IndexSet` object for a two-dimensional grid view,

```
indexSet.size(GeometryTypes::simplex(2))
```

is the number of triangles, and

```
indexSet.size(0)    // Elements have codimension 0
```

is the number of all elements.

Finally, the method

```
Types types(int codim) const
```

allows to determine what types of entities of a given codimension are actually present in the grid view. The method returns an iterator range, that is, a lightweight container type that can be forward-iterated over. For example, the code

```
for (const auto& type : indexSet.types(0))
  std::cout << type << std::endl;
```

will print the types of all elements in the grid view. For the grid in Figure 5.17, the result will be

```
(simplex, 2)
(cube, 2)
```

There is one important pitfall for people working with grids that contain more than one element type: Indices are counted *separately* for each `GeometryType`. For

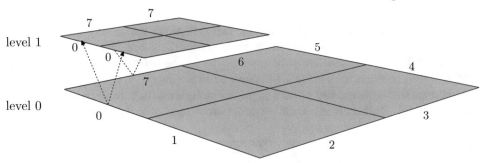

Figure 5.18.: Boundary segment indices enumerate the boundary intersections on level 0. Boundary intersections on higher levels are assigned the index of their level 0 ancestors.

example, in the grid in Figure 5.17, the element index 0 occurs twice, once for a triangle element and once for a quadrilateral element. The reason for this is run-time speed. Many finite element spaces assign different numbers of degrees of freedom to different element types. Computing the global number of a degree of freedom from a given element and shape function number is much easier and faster if separate element types have separate indices. For a single consecutive index for all entities of a given codimension, use the `MultipleCodimMultipleGeomTypeMapper` class of Chapter 5.6.2.

So far the text has only talked about how to assign numbers to grid entities. These numbers are needed to attach user data to these entities. However, it is also necessary to be able to attach user data to the boundary. Such data can consist of tags or enumeration values to specify different types of boundary conditions, or it can be scalar or vector-valued boundary data.

The basic idea of how to store such data in DUNE is the same as for entity data: The data itself is stored in an array, and the boundary segments can provide a number suitable to index the array. However, there are also a few differences. First of all, the objects being assigned numbers are not the entities but intersections. Chapter 5.4 explained how intersections exist between pairs of adjacent elements, and on the domain boundary. Only the latter carry indices. For this set of intersections, the index is consecutive and starts at 0. However, unlike the indices for entities, the boundary segment index does not distinguish between different boundary segment types. There is a single consecutive index starting at zero, not one for each `GeometryType`. This is a small inconsistency in the DUNE grid interface, but arguably it makes boundary segment indices easier to use.

There is a second difference between boundary segment indices and entity indices: boundary segment indices are only defined for boundary segments on the level 0 grid view. However, the index can be asked from *all* boundary intersections. Intersections on level 0 will return their index. Intersections on higher levels will return the index of their ancestor on level 0. The reason for this design is that usually the input grid consists of a level 0 view only, and the boundary conditions are specified with respect to that. Higher level grid views only appear later in the course of the simulation, by adaptive grid refinement. Figure 5.18 shows an example.

For historical reasons, the boundary segment index is not available through the index set class. Rather, it is a member function of the `Intersection` interface class (see Table 5.8):

```
size_t boundarySegmentIndex() const
```

Additionally, the total number of level 0 boundary segments can be obtained by calling the method

```
size_t numBoundarySegments() const
```

of the (hierarchical) grid object.

5.6.2. The `MultipleCodimMultipleGeomTypeMapper` Class

Using the `IndexSet` interface it is easy to write applications for simple finite element spaces like first-order Lagrangian finite elements, DG spaces on single-element-type grids, or Crouzeix–Raviart elements. These all have in common that they attach data only to a single type of entity. For more general spaces, the global numbers of the degrees of freedom must be *computed* given the information provided by the `IndexSet`. Depending on the actual space, implementing this computation can range from "quite simple" to "pretty difficult". The `GlobalBasis` classes of Chapter 10 do such computations, and should be used unless there is a good reason not to.

For some tasks, however, using a full `GlobalBasis` may not be feasible. In particular, writing new `GlobalBasis` classes requires manual implementations of the index computations, which can be tedious. To assist in these computations, dune-grid provides a helper class called `MultipleCodimMultipleGeomTypeMapper`, available from the file `dune/grid/common/mcmgmapper.hh`. It fulfills a task very similar to index sets: it assigns indices (i.e., elements of a set of consecutive integers starting from 0) to the entities of a grid view. These assignments can be more general than what index sets provide, without reaching the full generality of `GlobalBasis` objects: While an index set assigns one index to each entity of the grid, the `MultipleCodimMultipleGeomTypeMapper` allows to assign one *or more* indices to entities *of particular geometry types*. The indices are then consecutive and zero-starting for all entities of one of these geometry types. For example, to get the numbering for a third-order Lagrange space on a triangle grid, there should be one index for each vertex and triangle, and two for each edge.

An important special case is the construction of consecutive element numbers for grids with more than a single element type. Remember that index sets assign indices for each `GeometryType` separately (Figure 5.17), i.e., in a grid view with both triangle and quadrilateral elements there will be two elements with index 0. The canonical way to obtain a single consecutive index for all elements is by using a `MultipleCodimMultipleGeomTypeMapper`.

The `MultipleCodimMultipleGeomTypeMapper` is a standalone class with the signature

```
template<class GridView>
class MultipleCodimMultipleGeomTypeMapper
```

Objects of this type are obtained by calling the class constructor

```
MultipleCodimMultipleGeomTypeMapper(const GridView& gridView,
                                    const MCMGLayout& layout)
```

The `layout` parameter is a small helper class that specifies which geometry types should get indices. It is explained in detail below.

Since the mapper functionality is similar to index set functionality, the two programming interfaces are similar, too. Table 5.14 lists the public methods of the `MultipleCodimMultipleGeomTypeMapper` class. The methods

```
size_type size() const
size_type size(GeometryType type) const
```

return the total number of indices, and the number of indices assigned to one entity of a given `GeometryType`, respectively. The methods

```
template<class Entity>
bool contains(const Entity& entity, Index& result) const
bool contains(const GridView::Codim<0>::Entity& entity,
              int i,
              int cc,
              Index& result) const
```

return **true** if a given entity or subentity is assigned at least one index by the `MultipleCodimMultipleGeomTypeMapper` class. This is very similar to the `contains` method of the `IndexSet` class. However, unlike those, if the entity is assigned a set of indices, then the `contains` methods of the `MultipleCodimMultipleGeomTypeMapper` return the first such index in the `result` field. This is helpful in the case where it is known a priori that this index is the only one—it can then be used directly, without further calls to the mapper.

To get all indices for a given entity, there are the methods

```
template<class Entity>
IntegralRange<Index> indices(const Entity& entity) const
IntegralRange<Index> indices(const GridView::Codim<0>::Entity& entity,
                             int i,
                             unsigned int codim) const
```

145

	`MultipleCodimMultipleGeomTypeMapper(`
	`        const GridView& gridView`
	`        const MCMGLayout& layout)`
	Construct mapper for a given layout object
`size_type`	`size() const`
	Total number of indices for the grid view
`size_type`	`size(GeometryType type) const`
	Number of indices assigned to this geometry type
`template<class Entity>`	
`bool`	`contains(const Entity& entity, Index& result) const`
	True if at least one index is assigned to the given entity
`bool`	`contains(const GridView::Codim<0>::Entity& entity,`
	`        int i, int cc, Index& result) const`
	True if at least one index is assigned to the given subentity
`template<class Entity>`	
`IndexRange<Index>`	`indices(const Entity& entity) const`
	Get all indices assigned to entity e
`IndexRange<Index>`	`indices(const GridView::Codim<0>::Entity& entity,`
	`        int i, unsigned int codim) const`
	Get all indices assigned to the requested subentity of entity
`template<class Entity>`	
`Index`	`index(const Entity& entity) const`
	Get the first index assigned to entity entity
`Index`	`subIndex(const GridView::Codim<0>::Entity& entity,`
	`        int i, unsigned int codim) const`
	Get the first index assigned to the requested subentity of entity
`void`	`update()`
	Recalculate mapper after grid adaptation
`const std::vector<GeometryType>&`	
	`types(int codim) const`
	Geometry types of the given codimension that have at least one index
`const MCMGLayout&`	`layout() const`
	Retrieve the layout object
`const GridView&`	`gridView() const`
	Retrieve the underlying grid view

Table 5.14.: Public methods of the `MultipleCodimMultipleGeomTypeMapper` class. Index is the type used by the grid view's index set.

They return light-weight container-like objects that can be random-access-iterated over. They can be used directly in range-based **for** loops. For example, the following code prints all global indices attached to a given entity:

```
for (auto i: mapper.indices(entity))
  std::cout << i << std::endl;
```

For the important case that there is only one degree of freedom per entity there are also the methods

```
template<class Entity>
Index index(const Entity& entity) const
Index subIndex(const GridView::Codim<0>::Entity& entity,
               int i,
               unsigned int codim) const
```

The method

```
void update()
```

finally, recomputes the internal state. It has to be called whenever the grid has changed.

Selecting which entities get how many indices works using a predicate class. In this context, the predicate class is called the *layout* of the mapper. It must have a public method

```
std::size_t operator()(GeometryType type, int gridDim)
```

which shall return the number of indices that the given geometry type carries in a grid of dimension gridDim. For example, the following lambda implements the aforementioned third-order Lagrange layout for a triangle grid:

```
auto p3Layout = [](GeometryType type, int gridDim)
{
  if (type.isVertex())
    return 1;
  if (type.isLine())
    return 2;
  assert(type.isTriangle());
  return 1;
};
```

Several standard layout objects are implemented directly in the mcmgmapper.hh file. For example, a layout object for one index per element can be obtained by calling the global method

```
MCMGLayout mcmgElementLayout()
```

Similarly, there is mcmgVertexLayout for vertex-based data layouts, or, more generally,

```
template<int dim>
MCMGLayout mcmgLayout(Dim<dim>);
```

to select all entities of a given dimension. The class `MCMGLayout` returned by these methods is defined as

```
using MCMGLayout = std::function<std::size_t(GeometryType,int)>;
```

It can hold objects of any type that implement

```
size_t operator()(GeometryType,int)
```

The following short example sets up a mapper with the p3Layout object defined above and prints the resulting indices:

```
MultipleCodimMultipleGeomTypeMapper<GridView> mapper(gridView, p3Layout);
unsigned int dim = GridView::dimension;

for (const auto& element : elements(gridView))
{
  std::cout << "element index: " << mapper.index(element) << std::endl;

  for (size_t i=0; i<element.subEntities(dim-1); ++i)
    for (const auto& index : mapper.indices(element,i,dim))
      std::cout << "  vertex index: " << index << std::endl;

  for (size_t i=0; i<element.subEntities(dim-1); ++i)
    for (const auto& index : mapper.indices(element,i,dim-1))
        std::cout << "  edge index: " << index << std::endl;
}
```

When running this code, edge and vertex indices are printed several times, because the main loop iterates over the elements, and hence edges and vertices are visited once for each element they are part of.

5.6.3. Persistent Numberings

By construction, the indices of the previous section become invalid when the grid changes. If entities are added to the grid, these can be indexed in a natural way. More general grid changes, however, will also involve entities being removed. It is then impossible to maintain consecutive indices without changing the indices of entities that existed before the grid change already.

There are basically two ways how a DUNE grid can change: adaptive grid refinement and load-balancing.[14] Adaptive grid refinement primarily adds new entities, but entities can also disappear when using red–green refinement rules, or when their

---

[14]Besides these two, the FoamGrid grid implementation (www.dune-project.org/modules/dune-foamgrid) allows the grid to grow and shrink [140]. This was introduced originally to model growth of root- and fracture networks, but it is not part of the official DUNE grid interface.

removal is explicitly requested. Load-balancing on a distributed machine leaves the overall grid unchanged, but changes the grids on the individual processes.

To transfer data across a grid change, the *index* mechanism is not sufficient. There needs to be a way to attach data to a grid that is persistent across modifications of the grid, keeping the data in arrays indexed by entity indices. The dune-grid approach to this problem is to offer a second type of entity numbering that does not change when the grid is modified. Such a number is called *id* in DUNE. Each entity has two of them, and these two do not change throughout the life-time of the entity.

Naturally, such ids cannot be used to address data that is kept in arrays. However, they can be used for more general, associative containers such as search trees or hash maps [43]. Before changing the grid, all data needs to be copied from their arrays to separate containers that can be indexed by the persistent ids. This copying is costly; however, it is argued that grid refinement and load-balancing operations are even more costly, and that the price of copying is not relevant in comparison. After the grid modification, the data is copied back into arrays. Chapter 5.9.2 shows a detailed example of how to do this for adaptive grid refinement.

While ids are similar to indices in many regards, there are also a few differences. Most important are the different injectivity properties. Let E be the set of all entities of all codimensions, reference element types, and refinement levels together, and let I be a set of ids. Then the id map $E \to I$ is injective, with one important exception: If an entity on a given grid level also exists on a different level, then these two copies map to the same id. This is illustrated in Figure 5.19: Observe how several edges and vertices have the same id there. The reason for this rule is that interpolating coarse grid data onto new, finer grid elements can be an expensive operation. If the finer entity is a copy of another one on a lower-level grid, this interpolation step can be skipped, and the data only has to be copied. Detecting and implementing this case is particularly easy if the two entities share the same id.

The two ids that entity has are called the *local* and the *global* one. On a single-processor system, there is no functional difference between the two, although they are not required to be identical. On a multi-processor system, local ids are bijective in the sense described above if E is the set of all entities on one process (with the copy-rule applying within that process). In contrast, the global id map is bijective for the set of all entities of the entire distributed grid. The motivation for this double construction is the claim that full global bijectivity is not always required, and that a local id map can be easier to construct and maintain than a global one.

Similarly to indices, entity ids are not obtained directly from the entities, but rather through a separate class. This class, which goes by the name of IdSet, corresponds to the IndexSet class for indices. However, as ids are not constructed with respect to a specific grid view, the IdSet classes are directly obtained from the hierarchical grid, rather than from a grid view. To obtain an IdSet for the local ids, write

```
const auto& localIdSet = grid.localIdSet();
```

Correspondingly, write

```
const auto& globalIdSet = grid.globalIdSet();
```

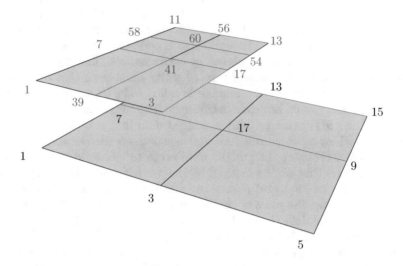

(a) Vertex ids

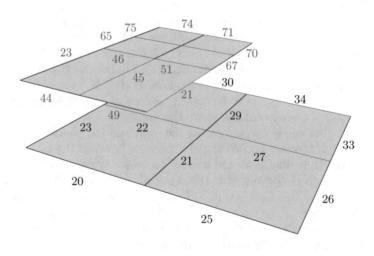

(b) Edge ids

Figure 5.19.: Grid hierarchy with vertex and edge ids. Black and red numbers are ids for entities on level 0 and level 1, respectively. Note how all entities have different ids, except for pairs of entities that are the same on two levels. Not all ids are shown, to avoid overcrowding the illustration.

for the global ids. `IdSet` objects cannot be copied, and have to be passed around as **const** references. Consequently, such references always point to the up-to-date `IdSet` object, even when the grid has been modified.

The interface of the `IdSet` class resembles a slimmed-down version of the `IndexSet` interface. It practically contains only two methods, namely

```
template<class Entity>
IdType id(const Entity& entity) const
```

and

```
IdType subId(const Grid::Codim<0>::Entity& entity,
             int i,
             unsigned int codim) const
```

These correspond to the two methods `index` and `subIndex` of the `IndexSet` class, respectively: `id` returns the id for the entity given in the argument `entity`, `subId` returns the id of the `i`-th subentity of codimension `codim` of `entity`. The grid adaptivity example of Chapter 5.9.2 makes ample use of these methods.

The return type `IdType` is the implementation type for ids used by the grid implementation, and can be different for local and global ids. The type can be retrieved from the `IdSet` class as `IdSet::IdType`. There are only very weak requirements on these types. Conceptually, an `IdType` can be pretty much anything, but it has to be usable as key for the `std::map` and `std::unordered_map` containers of the standard library [103]. In particular, for `std::map` this means that `IdType` implements

```
bool operator<(const IdType&, const IdType&)
```

and that this operator implements a strict weak ordering. To be usable as a key for `std::unordered_map`, there additionally has to be

```
bool operator==(const IdType&, const IdType&)
bool operator!=(const IdType&, const IdType&)
```

and `std::hash<IdType>` has to fulfill the requirements of a standard library hash object. For completeness, it is required that `IdType` be default-constructible, copy-constructible, and copy-assignable. Finally, for debugging purposes it must be possible to write objects of type `IdType` into standard C++ streams using **operator**<<.

5.7. Creating Grids

So far we have talked about the grid interface and how existing grids can be accessed and manipulated. This chapter will show how to create grids in the first place. There are several ways to do this. For example, unstructured grids can be read from files, and there are ways to set up structured and unstructured grids directly from your code.

The central idea of DUNE is that all grid implementations behave equally, and conform to the same interface. However, this concept reaches its limits when it comes to constructing grid objects, because grid implementations differ too much to make one construction method work for all. For example, setting up an unstructured grid requires explicitly specifying all vertex positions, whereas for a structured grid this would be a waste of effort. On the other hand, constructing a structured grid may require the bounding box, which, for an unstructured grid, is not necessary. As a consequence, while there is an official interface for the construction of *unstructured* grids (Chapter 5.7.2), there is no such interface for *all* grids. In practice, the differences do not pose serious problems.

In this chapter, creating a grid always means creating a grid with only a single level. Such a grid is alternatively called a *level-0 grid*, a *coarse grid*, or a *macro grid*. There is currently no functionality in DUNE to set up truly hierarchical grids directly. The underlying assumption is that the user will create a coarse grid first and then generate a hierarchy using refinement. Despite the name (and grid implementations permitting), the coarse grid can of course be as large and fine as desired.

To avoid misunderstandings let us point out that dune-grid does not provide a way to *generate* grids. By generation of grids we mean the process of constructing a partition of a domain into elements from a description of the boundary. For this task, many good implementations exist,[15] and DUNE does not attempt to compete with them. For this chapter we assume that unstructured grids are provided from outside sources, typically in form of a grid file.

The chapter begins by showing how to construct *structured* grids. This is the easiest task, because there is no external grid file involved. Next, Chapter 5.7.2 describes the GridFactory, a lower-level interface that allows to construct unstructured grids directly in C++ code. It is the natural basis for implementing alternative file readers. Finally, Chapter 5.7.3 explains how grid files and associated data can be read from files in the GMSH format. This is currently the principal way to load unstructured grids into DUNE.

5.7.1. Creating Structured Grids

Creating uniform structured grids is comparatively simple, as little information needs to be provided. In general, for uniform structured grids, the grid dimension, bounding box, and number of elements in each direction suffices.

There are two ways in DUNE to work with structured grids: either use a dedicated structured grid data structure such as YaspGrid, or use an unstructured grid manager such as UGGrid, and set it up with a structured grid. Both are reasonable things to do. If the grid is known to remain fixed throughout the computation, the first way yields a time- and memory-efficient grid implementation. The second one may be useful as the initial step towards an adaptively refined grid.

---

[15]Good open-source grid generators are GMSH (http://gmsh.info), NETGEN (https://ngsolve.org), and TETGEN (http://tetgen.org), and there is a long list of commercial ones, too.

In dune-grid, the YaspGrid class is the standard implementation of a structured uniform cube grid. All relevant information is handed directly to one of its constructors. For example, to create a two-dimensional YaspGrid for the domain $\Omega = [0,1]^2$ with 10 elements in each direction call

```
FieldVector<double,2> upper = {1.0, 1.0};
std::array<int,2> elements = {10, 10};
YaspGrid<2> grid(upper, elements);
```

or, shorter,

```
YaspGrid<2> grid({1.0, 1.0}, {10, 10});
```

In this example, the lower left corner is hard-wired to $(0,0)$ within YaspGrid. For more flexibility, YaspGrid can be extended with several template policy classes. The following code creates a grid for the domain $[0.1, 2.25] \times [-0.5, 0.5] \subset \mathbb{R}^2$:

```
FieldVector<double,dim> lower = {0.1, -0.5};
FieldVector<double,dim> upper = {2.25,  0.5};
std::array<int,dim> elements = {15, 10};
YaspGrid<dim, EquidistantOffsetCoordinates<double,dim> >
                                     grid(lower, upper, elements);
```

Further ways of setting up YaspGrid objects are described in the online documentation.

The dune-grid module also has infrastructure to set up structured grids using *unstructured* grid implementations. The header file structuredgridfactory.hh in dune/grid/utility/ provides the class

```
template<class Grid>
class StructuredGridFactory
```

whose sole interface consists of the two static methods

```
static std::unique_ptr<Grid>
        createCubeGrid(const FieldVector<ctype,dimworld>& lowerLeft,
                       const FieldVector<ctype,dimworld>& upperRight,
                       const std::array<unsigned int,dim>& elements)
```

and

```
static std::unique_ptr<Grid>
        createSimplexGrid(const FieldVector<ctype,dimworld>& lowerLeft,
                          const FieldVector<ctype,dimworld>& upperRight,
                          const std::array<unsigned int,dim>& elements)
```

Both methods create structured grids in the bounding box given by the two vectors lowerLeft and upperRight. The entries of the array elements specify the number of basic cells in each coordinate direction. The grid implementation given as the template parameter Grid is used to store the grid. The first method constructs a grid consisting of hypercubes, whereas the second method decomposes each cube into $d! := d \cdot (d-1) \cdot \ldots \cdot 2 \cdot 1$ simplices, using the Coxeter–Freudenthal–Kuhn triangulation [3]. So, for example, the lines

```
FieldVector<double,2> lowerLeft = {0.0, 1.0};
FieldVector<double,2> upperRight = {2.0, 2.0};
std::array<unsigned int,2> elements = {8, 17};
auto grid
    = StructuredGridFactory<AlbertaGrid<2> >::createSimplexGrid(lowerLeft,
                                                               upperRight,
                                                               elements);
```

create a two-dimensional AlbertaGrid object for the rectangular domain $\Omega = [0, 2] \times [1, 2]$ discretized by $8 \times 17 \times 2$ triangles. The code

```
FieldVector<double,3> lowerLeft = {0.0, 0.0, 0.0};
FieldVector<double,3> upperRight = {1.0, 1.0, 1.0};
std::array<unsigned int,3> elements = {10, 10, 10};
auto grid
    = StructuredGridFactory<UGGrid<3> >::createCubeGrid(lowerLeft,
                                                        upperRight,
                                                        elements);
```

creates a grid of the unit cube with $10 \times 10 \times 10$ hexahedral elements. The algorithms used within the StructuredGridFactory work for any dimension, and are limited only by the available grid implementations.

Internally, the StructuredGridFactory class uses the GridFactory class described in Chapter 5.7.2 to actually create the grids; hence the StructuredGridFactory can be used with all grids that implement the GridFactory interface. In addition, there are specializations even for structured grid implementations. For example, the YaspGrid example from the beginning of this section can alternatively be written as

```
FieldVector<double,2> lowerLeft = {0.0, 0.0};
FieldVector<double,2> upperRight = {1.0, 1.0};
std::array<unsigned int,2> elements = {10, 10};
auto grid
    = StructuredGridFactory<YaspGrid<2> >::createCubeGrid(lowerLeft,
                                                         upperRight,
                                                         elements);
```

This can make life easier when needing to switch between different grid implementations frequently.

5.7.2. The Grid Construction Interface

To set up unstructured grids, there is a low-level C++ interface. This interface, which goes by the name of GridFactory, provides methods to, e.g., insert vertices and elements one by one. It is the basis of grid file readers provided by DUNE, in particular the GMSH file reader described in the next section. In some cases it is also convenient to be able to specify a coarse grid entirely in the C++ code.

The GridFactory is programmed as a factory class (hence the name, cf. [72]). An object of the factory class is fed with all necessary information, and it will create and hand over a grid. The following will first demonstrate the steps necessary to simply construct a grid using the GridFactory, and then explain a few additional features. There are some subtle numbering issues involved if the input grid already has data associated with it. These will be discussed in Chapter 5.7.2.

Setting up a Grid

The GridFactory interface exists for all unstructured grids in DUNE. It is implemented as a class that takes a (hierarchical) grid type as a template parameter. The different grid implementations specialize this class. In addition, each GridFactory specialization inherits from the abstract base class GridFactoryInterface. The base class mainly serves as documentation. A list of its public methods is given in Table 5.15.

Suppose there is a list of vertices and elements given somehow, and the goal is to create a new grid of type, say, AlbertaGrid from it. The first step is to create a GridFactory object. GridFactory objects are default constructible, but may have additional constructors as well. For the AlbertaGrid example, simply call

```
GridFactory<AlbertaGrid> factory;
```

No additional header needs to be included for this—The GridFactory specialization of AlbertaGrid comes included in the header for AlbertaGrid itself. The resulting factory object is empty.

Into this empty GridFactory object the vertex positions can now be entered. For this, there is the method

```
void insertVertex(const FieldVector<ctype,dimworld>& position)
```

The type ctype is the number type used by the grid for coordinates, and dimworld is the dimension of its world space. The insertion order of the vertices does not necessarily reflect their indexing by the IndexSet class, or the order they are iterated over (see below).

Once all vertices have been inserted, the elements can be entered as well. For each element the GridFactory method

```
void insertElement(GeometryType type,
                   const std::vector<int>& vertices)
```

has to be called. The first parameter is the element type encoded as a GeometryType object (Chapter 5.5.2). The grid implementation is expected to throw an exception if an element type that cannot be handled is encountered. The second parameter is the set of indices of the vertices of this element. The size of the array must necessarily match the number of vertices of the polytope specified by the type parameter. The order of the vertex numbers must follow the conventions for vertex numberings of DUNE reference elements shown in Figures 5.12 and 5.13. The vertex numbers themselves must refer to the order of insertion of the vertices into the GridFactory.

155

	GridFactory()
	Default constructor
void	insertVertex(**const** FieldVector<ctype,dimworld>& pos)
	Insert a grid vertex
void	insertElement(**const** GeometryType& type,
	const std::vector<**unsigned int**>& vertices)
	Insert a grid element
void	insertElement(**const** GeometryType& type,
	const std::vector<**unsigned int**>& vertices,
	const std::shared_ptr<VirtualFunction>&
	elementParametrization)
	Insert a grid element with a geometry parametrization
void	insertBoundarySegment(
	const std::vector<**unsigned int**>& vertices)
	Insert a boundary segment
void	insertBoundarySegment(
	const std::vector<**unsigned int**>& vertices,
	const std::shared_ptr<BoundarySegment>&
	boundarySegment)
	Insert a boundary segment with a geometry parametrization
std::unique_ptr<Grid>	
	createGrid()
	Actually create the grid
unsigned int	insertionIndex(**const** Codim<0>::Entity& element) **const**
	Get the number of the element in the order in which the elements were inserted
unsigned int	insertionIndex(**const** Codim<dim>::Entity& vertex) **const**
	Get the number of the vertex in the order in which the vertices were inserted
unsigned int	insertionIndex(
	const Grid::LeafIntersection& intersection) **const**
	Get the number of the intersection in the order in which the boundary segments were inserted
bool	wasInserted(
	const Grid::LeafIntersection& intersection) **const**
	Whether this boundary segment was explicitly inserted
Communication	comm() **const**
	Communication object used by the grid factory

Table 5.15.: Methods of the GridFactory interface class

The factory object now contains all required information for a simple finite element grid. To finish off the construction of the grid object, there is the method

```
std::unique_ptr<Grid> createGrid()
```

This creates the grid object on the heap and hands its ownership to the caller. As a complete example, here is code that constructs the simple grid of Figure 5.21. It uses the UGGrid implementation because it contains triangles and quadrilaterals together.

```
GridFactory<UGGrid<2> > factory;

factory.insertVertex({0, 0, 0});
factory.insertVertex({1, 0, 0});
factory.insertVertex({2, 0, 0});
factory.insertVertex({0, 0.5, 0});
factory.insertVertex({0, 1, 0});
factory.insertVertex({1, 1, 0});
factory.insertVertex({2, 1, 0});

factory.insertElement(GeometryTypes::simplex(2), {0, 1, 3});
factory.insertElement(GeometryTypes::simplex(2), {1, 5, 3});
factory.insertElement(GeometryTypes::simplex(2), {3, 5, 4});
factory.insertElement(GeometryTypes::cube(2), {1, 2, 6, 5});

std::unique_ptr<UGGrid<2> > grid = factory.createGrid();

// do something with grid
...
```

In addition to the methods to insert vertices and elements, there is an optional method that allows to explicitly insert boundary segments:

```
void insertBoundarySegment(const std::vector<unsigned int>& vertices)
```

It may be called for some or all of the boundary segments of the grid, but doing so does not make any difference to the grid as such. Rather, explicitly inserting boundary segments determines an enumeration of these segments, which is later available through the insertionIndex method (Chapter 5.7.2). This is the canonical way to attach externally given data to boundary segments. If boundary segments are not explicitly given to the GridFactory, the grid implementation itself determines their indices.

Parametrized Elements and Domains

There are two more optional methods that allow to provide additional geometric information to the grid factory. Let T_{ref} be the reference element of a grid element or of a boundary segment. We call any sufficiently well-behaved map from T_{ref} into the world space $\mathbb{R}^w$ a *parametrization* of the element or boundary segment.

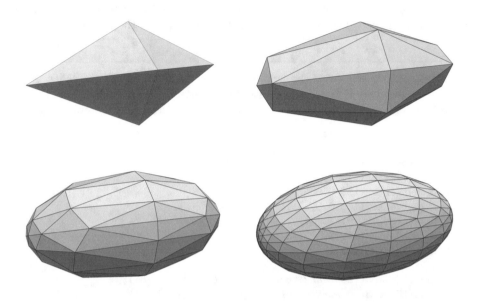

Figure 5.20.: Grid with a boundary parametrization, and three grids produced from it by uniform refinement

Parametrizations can be used by grid refinement when the error of piecewise polynomial approximation of the domain geometry is problematic. For example, Figure 5.20 shows a grid that is an octahedron, but which is intended to approximate an ellipsoid. The exact shape of the ellipsoid is given in form of parametrizations for the eight coarse grid boundary segments. As the grid gets refined, the new boundary vertices are not placed on edge midpoints, but rather their new positions are determined using the geometry information from the boundary parametrizations. That way, increasingly finer grids will approximate the true geometry better and better. Not all DUNE grids implement this feature. The example in Figure 5.20 was constructed using UGGrid.

Similar mechanisms can be implemented for element parametrizations. An example is given in the article [140] on the FoamGrid grid manager. Other grid managers may use the additional geometric information differently. For example, grids implementing higher-order polynomial element shapes may use it to set up the element geometries.

Using boundary or element parametrizations requires implementing an object for each boundary segment or element that computes the parametrization map. These objects are then handed over to the grid factory. This is one of the rare places where DUNE uses abstract base classes. Implementations of boundary parametrizations must be derived from

```
template<int dim, int dimworld = dim, class ctype = double>
struct BoundarySegment
```

where dim is the dimension of the reference element, dimworld is the dimension of the world space, and ctype is the number type used for coordinates. This abstract base class mandates an overload of the method

```
virtual FieldVector<ctype, dimworld>
    operator()(const FieldVector<ctype, dim-1>& local)
```

This method must compute the world coordinates from the local ones on the reference element. For each boundary segment with a parametrization, an object deriving from this class must be handed to the grid factory by calling

```
void insertBoundarySegment
    (const std::vector<unsigned int>& vertices,
     const std::shared_ptr<BoundarySegment<dim,dimworld> >& boundarySegment)
```

In addition, calling this method influences the numbering of the boundary segments in the same way as the insertBoundarySegment method with only a single parameter.

For grids that support parametrized elements,[16] there is the method

```
void insertElement
    (const GeometryType& type,
     const std::vector<unsigned int>& vertices,
     const std::shared_ptr<VirtualFunction>& elementParametrization)
```

where VirtualFunction is an abbreviation of

```
VirtualFunction<FieldVector<ctype,dimension>, FieldVector<ctype,dimworld> >
```

The method works just like the insertElement method described in Chapter 5.7.2, but it allows to hand over a parametrization of the element as the third argument. Unlike boundary parametrizations, element parametrizations have to be derived from

```
template<class Domain, class Range>
class VirtualFunction
```

defined in dune/common/function.hh. Descendants of this class must implement the method

```
void evaluate(const Domain& x, Range& y) const
```

which returns its result in the second argument. Lambdas and other callables need to be wrapped by calling

```
template<class F>
auto makeVirtualFunction(F&& f)
```

---

[16]The only grid that currently implements parametrized elements is FoamGrid. An example is shown in [140].

The design inconsistencies between boundary and element parametrizations have historical reasons, and may be removed in later versions of `dune-grid`.

Element and Vertex Numberings

Problem specifications of boundary value problems typically include a number of data fields defined on the domain or on the domain boundary. Just as the computational grids, these data fields are frequently given in files. They can be included in the file of the grid itself, or they can reside in separate files that reference the main grid file explicitly or implicitly.

In practice, input grid data is usually given in the form of sets of data per grid vertex, per grid element, or per boundary segment. Data on the grid boundary is sometimes also given as a set of data for all vertices or elements, of which all but the restriction to the domain boundary is ignored. The association of data to, e.g., vertices can happen implicitly, i.e., data and vertices are given in corresponding order, or explicitly. In the latter case each data field comes with an index that specifies the vertex or element it belongs to.

As explained previously in Chapter 5.6, DUNE grids cannot have data attached to them directly. Rather, each entity of a grid view has a number (its *index*) attached to it, and this index is used to address all data attached to that entity, stored in containers with fast random access. In order to attach data from outside sources to DUNE grids, the indices of the grid entities must be put in a relationship to the input data. This is a surprisingly difficult problem. The relation hinges on the order in which the vertices, elements, and possibly boundary segments are inserted into the grid factory. We say that a vertex has *insertion index i*, $i = 0, 1, 2, \ldots$, if it was the i-th vertex to be inserted into the grid factory (the same definition holds, mutatis mutandis, for elements and boundary segments). Code that uses the grid factory to load grids and possibly additional data into the DUNE grid factory is expected to organize this additional data such that it references the grid vertices and elements in insertion order.

Accessing such data from within DUNE would be easy if the indices of the level-0 `IndexSet` would directly correspond to the insertion indices. However, this is not possible for two reasons: First, for grids that contain more than one element type, the element *insertion index* is by definition a single consecutive sequence of integers for the set of all elements. The level-0 *index*, however, counts the different element types separately (Chapter 5.6.1). As a consequence, the two cannot coincide.

The second reason is a potential loss of speed. It has been argued that the ordering of the vertices that is natural for a given grid implementation may not always coincide with the insertion order of the vertices. A vertex renumbering would therefore be necessary, and this, so the argument continues, cannot be done once and for all when the grid is set up. It has to be recomputed each time a level-0 vertex is accessed, and that may be too expensive.

For these reasons, the `GridFactory` class provides three methods that provide the *insertion index* for vertices, elements, and boundary segments. The methods are:

```
unsigned int insertionIndex(const Codim<0>::Entity& entity) const
unsigned int insertionIndex(const Codim<dim>::Entity& entity) const
unsigned int insertionIndex(const Grid::LeafIntersection& intersection) const
```

For vertices and elements, these indices always exist. If boundary segments were not given explicitly, the behavior of the corresponding insertionIndex method is undefined.

The insertionIndex methods can be used from the time where GridFactory:: createGrid has been called, until the time when the grid is modified for the first time (because grid modifications are allowed to alter the level-0 indices, even if the level-0 grid view itself stays the same).

To make this discussion a bit less abstract, the following example shows how input vertex and element data can be attached to DUNE grids. The example supposes that a grid has just been inserted into a grid factory object called gridFactory, and that this grid came with vertex and element data. The input data is stored in two containers inputVertexData and inputElementData, and the data in these arrays is ordered in correspondence to the order that the vertices and elements were entered in into the grid factory, respectively. The following code reorders the data such that it becomes indexable by the indices provided by the leaf grid index set:

```
using GridView = Grid::LeafGridView;
GridView gridView = grid.leafGridView();

// Reorder vertex data
duneVertexData.resize(inputVertexData.size());

for (const auto& vertex : vertices(gridView))
{
  auto inputIndex = gridFactory.insertionIndex(vertex);
  auto duneIndex  = gridView.indexSet().index(vertex);
  duneVertexData[duneIndex] = inputVertexData[inputIndex];
}


// Reorder element data
MultipleCodimMultipleGeomTypeMapper<GridView>
            elementMapper(gridView, mcmgElementLayout());

duneElementData.resize(inputElementData.size());

for (const auto& element : elements(gridView))
{
  auto inputIndex = gridFactory.insertionIndex(element);
  auto duneIndex  = elementMapper.index(element);
  duneElementData[duneIndex] = inputElementData[inputIndex];
}
```

After this code has been executed, the arrays `duneVertexData` and `duneElementData` contain the desired output. Note how vertex data can be accessed directly by vertex indices, but how a `MultipleCodimMultipleGeomTypeMapper` object is needed to construct an array index from an element index. This additional redirection is not necessary if the grid is known to contain only a single element type.

5.7.3. Reading Unstructured Grids from Gmsh Files

The `GridFactory` interface of the previous section forms the basis for code that reads grid and data files in a variety of formats. DUNE provides support for some of these formats,[17] but no interface specification exists for grid file readers in DUNE.

Arguably the most important file format currently supported by DUNE is the GMSH format. [18] GMSH[19] is an open-source geometry modeler and grid generator. It allows to define geometries either interactively or via its own modeling language. From such a geometry representation, it creates simplicial grids in two and three space dimensions. The element geometries can be described by polynomials of up to fifth order. Additionally, GMSH is able to import CAD geometries, e.g., from IGES or STEP files, and it can generate grids for them. The generated grids can be stored in a native file format, with file names typically ending in .msh. The files are ASCII text and can be inspected with a text editor. The format is described in the GMSH manual [74]. In addition to grids, GMSH files can contain subdomain partitionings for the elements and boundary segments.

Reading GMSH files into dune-grid is done by the class

```
template<class Grid>
class GmshReader
```

in the header dune/grid/io/file/gmshreader.hh. The public interface consists of three static methods, all of which are called `read`, and all of which read .msh files. The most basic method loads the grid but disregards any partitioning data. It has the signature

```
static std::unique_ptr<Grid> read(const std::string& fileName,
                                  bool verbose=true,
                                  bool insertBoundarySegments=true)
```

The file given in `fileName` is interpreted as a .msh file, and a new grid object of type `Grid` is constructed from it. The responsibility for memory management is handed over to the caller. Hence, the simplest way to read a grid from a GMSH file is

```
std::unique_ptr<FooGrid> grid = GmshReader<FooGrid>::read(filename);
[...]   // Code that uses the grid
```

---

[17]This text only discusses reading from the GMSH file format. Please consult the online documentation of dune-grid for other formats.

[18]Currently, dune-grid only supports Version 2 of the GMSH format. Support for newer versions may appear in the future.

[19]http://gmsh.info

Parameters 2 and 3 have default values and can therefore be omitted when calling the read method. The second of the three parameters controls the screen output of the GmshReader. When set, the reader will print a few lines of information while reading the file. Otherwise, the read method will remain completely silent.

The third parameter controls whether the GmshReader reads the boundary segment information that may be present in the file. When reading a d-dimensional grid from a file, that file may contain $d - 1$-dimensional elements in addition to the d-dimensional ones. These lower-dimensional elements are interpreted as boundary segments. When insertBoundarySegments is set to **false**, the boundary segments in the file are simply ignored. If they are not ignored, the GmshReader will explicitly insert the boundary segments into the grid using the insertBoundarySegment method of the grid factory that is used internally. Explicitly inserting the boundary segments has three consequences:

- It sets the insertion index of the boundary segment,

- it allows to read boundary condition tags associated to boundary segments (see below),

- it allows to read boundary segment geometries of higher order. GMSH partially supports polynomial elements of orders up to five. Higher-order boundary representations are handed to the grid factory as boundary segment parametrizations, to be used as described in Chapter 5.7.2.

The read method has a twin that uses a grid factory that is given from the outside:

```
static void read(GridFactory<Grid>& factory,
                 const std::string& fileName,
                 bool verbose = true,
                 bool insertBoundarySegments=true)
```

Compared to the previous method, the advantages are two-fold: On the one hand, it allows to use grid factories that have non-standard constructors or need additional setup code before the actual grid can be inserted. On the other hand, having the grid factory object after the call to read has finished is necessary to call the insertionIndex methods of the factory. This is turn is required to load additional data associated to the input grid.

The third variant of the read method extracts additional information from the grid file. GMSH files can contain general purpose integer values for each element. These integers are used by GMSH to group elements into subdomains by assigning them corresponding subdomain numbers. In GMSH terminology, these numbers are called *physical entity* numbers. Similarly, integer numbers can be attached to boundary segments. These can be used to specify different types of boundary conditions on different parts of the boundary. The GmshReader allows to read this information into arrays. The corresponding read method is

163

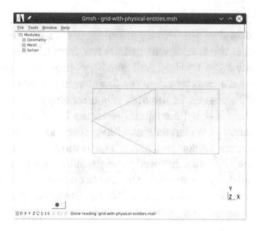

Figure 5.21.: The GMSH example file as shown by GMSH itself

```
static void read(GridFactory<Grid>& factory,
                 const std::string& fileName,
                 std::vector<int>& boundarySegmentToPhysicalEntity,
                 std::vector<int>& elementToPhysicalEntity,
                 bool verbose = true,
                 bool insertBoundarySegments=true)
```

New in this signature are the two array boundarySegmentToPhysicalEntity and elementToPhysicalEntity. After a successful call to the method, they will contain the integer values associated in the file with the boundary segments and elements, respectively. The numbering is the insertion index of the boundary segments and elements. These in turn are determined by the order of the boundary segments and elements in the file, not by the explicit numbering given in the file.

To show how this works here is a short GMSH example file. Figure 5.21 shows how the file looks like when opened in GMSH.

```
1   $MeshFormat
2   2.1 0 8
3   $EndMeshFormat
4
5   $Nodes
6   7
7    1 0    0    0
8    2 1    0    0
9    3 2    0    0
10   4 0    0.5 0
11   5 0    1    0
```

```
12    6 1   1    0
13    7 2   1    0
14   $EndNodes
15
16   $Elements
17   11
18    1 2 1 0   1 2 4
19    2 2 1 0   2 6 4
20    3 2 1 0   4 6 5
21    4 3 1 1   2 3 7 6
22
23    5 1 1 1   1 2
24    6 1 1 1   2 3
25    7 1 1 1   3 7
26    8 1 1 1   7 6
27    9 1 1 1   6 5
28   10 1 1 1   5 4
29   11 1 1 1   4 1
30   $EndElements
```

Most of this is self-explanatory, and only the Elements section needs some discussion.[20] The first four lines of that section are the three triangles and one quadrilateral in the grid, and the remaining seven lines are the seven boundary segments.

In each line, the first number is a running number,[21] and the second number is the type of the element (2 for triangles, 3 for quadrilaterals, and 1 for segments). In a line for a triangle, the last three numbers are the vertex numbers. Correspondingly, in a line for a quadrilateral, the last four numbers are the vertices. This explains all numbers in the Elements section except for those in the third and forth column. These two columns describe the integer data that is attached to the elements. The first number simply says how many more data members there are to come. In this file this number is simply 1. The GMSH file format allows higher numbers here, but the DUNE GMSH reader currently only reads the first one. The following number (the one in the fourth column) is the actual integer data. This is the number that will be read into the arrays elementToPhysicalEntity and boundarySegmentToPhysicalEntity.

5.8. Writing Grids and Data to VTK Files

The simulation data produced by finite element codes needs post-processing to be of use. DUNE itself does not provide any post-processing facilities itself. Just as in the case of grid generation, the DUNE team believes that others do a much better job at data post-processing than DUNE ever could. Therefore, DUNE does not try to compete, but rather simply writes out the simulation data to files. Post-processing

---

[20]Observe that GMSH vertex coordinates have always three components, even if the grid is only two-dimensional.

[21]This number is disregarded by GmshReader.

and data visualization must then be done by other programs that are able to read the files.

While DUNE supports writing grid and grid data in several formats, we focus here on VTK files exclusively, because they are the most relevant. The VTK format is the native format of the VTK toolkit[22] [143], and of the PARAVIEW visualization software.[23] The format is publicly documented [107]. It is based on the eXtensible Markup Language (XML),[24] and can hold binary and ASCII data. The format can accommodate general unstructured grids, but it also has special representations for various topologically structured and even uniform cube grids. Unstructured grids can mix element types such as tetrahedra and hexahedra, together with obscurer elements such as prisms and pyramids. Even certain second-order element geometries are possible.

Almost all of this functionality can be accessed for reading and writing from DUNE using the module `dune-vtk`. [25] At the same time, the `dune-grid` module offers partial support for writing grids in the VTK format, and to add vertex and element data to the files. This support is explained in this chapter: Section 5.8.1 will show how to write vertex and element data, and Section 5.8.2 then covers time-dependent data. Handling grid functions is covered in Chapter 10.7.

5.8.1. Writing Grids and Data

The VTK writing capabilities of DUNE are implemented in a class `VTKWriter`, located in the header `dune/grid/io/file/vtk/vtkwriter.hh`. The public interface of the method is short, and can be found in Table 5.16.

The class `VTKWriter` takes a single template parameter, which must be a `GridView`:

```
template<class GridView>
class VTKWriter
```

A VTK file contains exactly one (non-hierarchical) grid, and therefore a `VTKWriter` object stores exactly one object of type `GridView`. In addition, a VTK file can contain an arbitrary number of data sets defined on this grid. Each data set is either associated to the grid vertices, or to the grid elements (which VTK calls "cells"). Data sets are handed to the `VTKWriter` object using dedicated methods. Finally, the grid and data sets are written to a file when a particular method is called.

The `VTKWriter` class has only one constructor. Its signature is

```
VTKWriter(const GridView& gridView,
          VTK::DataMode dm = VTK::conforming,
          VTK::Precision precision = VTK::Precision::float32)
```

---

[22] www.vtk.org
[23] www.paraview.org
[24] www.w3.org/XML
[25] https://dune-project.org/modules/dune-vtk

```
                    VTKWriter(const GridView& gridView,
                            VTK::DataMode dm = VTK::conforming,
                            VTK::Precision coordPrecision
                                        = VTK::Precision::float32)
                    Constructor for given grid view, data mode, and precision
template<class Container>
          void    addCellData(const Container& v,
                            const std::string& name,
                            int ncomps = 1,
                            VTK::Precision prec = VTK::Precision::float32)
                    Attach element-based data
template<class Container>
          void    addVertexData(const Container& v,
                            const std::string& name,
                            int ncomps = 1,
                            VTK::Precision prec = VTK::Precision::float32)
                    Attach vertex-based data
template<class F>
          void    addCellData(F&& f, VTK::FieldInfo vtkFieldInfo)
                    Interpret a function as element data (Chapter 10.7)
template<class F>
          void    addVertexData(F&& f, VTK::FieldInfo vtkFieldInfo)
                    Interpret a function as vertex data (Chapter 10.7)
          void    clear()
                    Remove all attached functions
  VTK::Precision  coordPrecision() const
                    A tag for the data type used for numerical data (float32 or
                    float64)
     std::string  write(const std::string& name,
                            VTK::OutputType type = VTK::ascii)
                    Write grid and data to file
     std::string  pwrite(const std::string& name,
                            const std::string& path,
                            const std::string& extendpath,
                            VTK::OutputType type = VTK::ascii)
                    Write grid and data to file at a given path (Chapter 6.6)
```

Table 5.16.: Interface methods of the VTKWriter class

The first parameter is the grid view to be written to the VTK file. The second parameter provides support for data fields that are discontinuous across element boundaries. By default, (i.e., when dm is set to VTK::conforming), neighboring elements in a VTK file share common vertices. Since VTK only allows single degrees of freedom on a vertex, piecewise linear fields are then necessarily continuous. To have discontinuous fields, set dm to VTK::nonconforming. In that case, each vertex has a copy for each adjacent element, and each element will have its own set of vertices. This trick allows for multiple degrees of freedom (one per element) on the "same" vertex, at the price of an increased output file size. The parameter precision, finally, allows to set the data type used in the VTK file for numerical values. The most important choices are VTK::Precision::float32 and VTK::Precision::float64, but there are a few others, too.

Once the VTKWriter object is set up, it can receive data via several different methods. The first two of them add element and vertex data:

```
template<class Container>
void addCellData(const Container& v, const std::string& name, int ncomps=1)

template<class Container>
void addVertexData(const Container& v, const std::string& name, int ncomps=1)
```

Both methods take a container v that must contain cell (i.e., element) or vertex values for the grid view that the VTKWriter object was created with. The container type can be arbitrary, as long as it supports random access by implementing **operator**[]. A std::vector will do, for example, and so will the vector data types from the dune-istl module in Chapter 7. The data value for a given vertex or element is accessed by the VTKWriter implementation by first computing an index using the MultipleCodimMultipleGeomTypeMapper class. That index is then used to call **operator**[] of the container v. Note that the DataMode parameter of the constructor is ignored here— if it is set to VTK::nonconforming the output will be conforming nevertheless. To actually get discontinuous output requires a function as data, see Chapter 10.7.

The second parameter name of the two methods allows to associate an arbitrary string to each data set, which is stored in the VTK file. ParaView, for example, uses this string to label the data sets in its user interface.

Finally, the parameter ncomps allows to write multi-component data. By default, ncomps is set to 1, and scalar data fields are written. If ncomps is set to a larger value m, then each group of m values in the container v is interpreted as vector data for either one cell or one vertex. Consequently, the input container v must be m times as long as for scalar data. The output appears in ParaView as m independent scalar fields. Writing actual VTK vector fields is only possible when writing functions (Chapter 10.7).

When all data fields have been added to the VTKWriter object, the output file can be written by the method

```
std::string write(const std::string& name,
                  VTK::OutputType type = VTK::ascii)
```

The first parameter is the file name. Currently, all data is written as unstructured data (UnstructuredGrid in VTK terminology), and therefore the suffix .vtu is appended to the string in name to get the actual filename. One-dimensional grids are written as PolyData and receive the suffix .vtp. The return value of the method write is the complete filename with the suffix.

The second parameter determines how the data is stored in the file. There are currently four options:

- VTK::ascii: All numerical data is written as ASCII, the file is hence completely human-readable. This can be helpful for debugging; however, writing ASCII data leads to large files, and floating-point precision may be lost due to truncation.

- VTK::appendedraw: The data is written in raw binary, and the binary data is not written into the correct XML locations, but rather appended after the XML text. The leads to compact file with no loss of precision. However, such files can only be moved between machines with the same byte order [160].

- VTK::base64, VTK::appendedbase64: These two options use base64 encoding [102], which is a compromise between file size and portability. Data is either inserted into the XML segments (VTK::base64), or appended at the end of the file (VTK::appendedbase64).

The behavior of the write method changes somewhat when called for a distributed grid. File names are then prefixed with s####-p####-, where the first set of #### encodes the total number of processes and the second one the rank of the local writing process. For additional control, there is another method pwrite (for: "path write") that allows more control of where the resulting files are placed. See Chapter 6.6 for details.

5.8.2. Writing Time-Dependent Data

So far there has been a single fixed grid, and data fields on this grid. The simulation output of evolution problems consists instead of sequences of grids and data fields, associated to certain points in time. While the grid may stay the same for all time points, it may also change due to grid adaptivity.

In a limited sense, post-processing for time-dependent data is possible even without further infrastructure. If a sequence of data sets is written into files with file names that consist of a common prefix followed by consecutive numbers

```
my_data_file_0000.vtu
my_data_file_0001.vtu
my_data_file_0002.vtu
...
```

then the file-opening dialog of the PARAVIEW visualization software will interpret the whole sequence as a time series of data, and load it into a single data object. All PARAVIEW features for time-dependent data are then available for this object; for example, individual files can be selected by dragging the "time" slider in the PARAVIEW GUI, and animations can be created with only a few clicks.

This approach to time-dependent data is simple and convenient, but it only works if the data is associated to uniformly distributed time points. With simulation algorithms that perform adaptive time-stepping, however, the precise time points must be recorded in the file, so that visualization and post-processing software can use it.

The dune-grid module offers the VTKSequenceWriter class to write sequences of grids and data with associated time points (from the file dune/grid/io/file/vtk/vtksequencewriter.hh). It has already been used in the finite volume example in Chapter 3.4. The VTKSequenceWriter class writes the individual data sets just as individual invocations of the VTKWriter class would (with filenames that have integer numbers at the end). Additionally, it writes a *sequence file* with file extension .pvd, which contains all data file names and the corresponding time points. If the grid is distributed across several processes in an MPI-parallel setting, then only rank 0 will write the sequence file. When reading this sequence file into PARAVIEW, it will automatically read all associated data files, and it will associate the given time points with them.

To write a time series of data, first create a VTKWriter object[26] with the grid view that the data is defined on. This VTKWriter object will write the data files, and hence the choice of data mode in the constructor will be respected. Then, create a VTKSequenceWriter object. There are two constructors

```
VTKSequenceWriter(std::shared_ptr<VTKWriter<GridView> > vtkWriter,
                  const std::string& name)
```

and

```
VTKSequenceWriter(std::shared_ptr<VTKWriter<GridView> > vtkWriter,
                  const std::string& name,
                  const std::string& path,
                  const std::string& extendpath)
```

which correspond to the write and pwrite methods of the VTKWriter class. Either constructor receives the VTKWriter object, which will do the writing of the individual data files. Both constructors also have a string argument name, which is the string prefix used for all time step files and the sequence file. If the name prefix is my_series, for example, then the sequence file will be

```
my_series.pvd
```

and the individual time steps will be

```
my_series-00000.vtu
my_series-00001.vtu
my_series-00002.vtu
...
```

(if the grid is one-dimensional, then the file extension will be .vtp instead of .vtu).

---

[26]Or a SubsamplingVTKWriter; see Chapter 10.7.

As the next step, register all the data fields or discrete functions for the first time step with the VTKWriter. This works just as described in the previous section. To write a data set, call the method

```
void write(double time, VTK::OutputType type = VTK::ascii)
```

of the VTKSequenceWriter object. A call to this method does two things. First, it calls the method write (or pwrite, depending on which constructor was used) of the VTKWriter object. This writes the grid and all currently registered data sets to a single VTK file. The filename is what has been given in the name argument of the VTKSequenceWriter constructor, together with an integer number. The parameter type has the same purpose as when writing a single data set. Secondly, a call to the write method of the VTKSequenceWriter object appends a line to the sequence file. There is no need to explicitly inform the VTKWriter and VTKSequenceWriter classes when the data has changed between one time step and the next. The VTKWriter object stores references to the data containers, and it will therefore always write their current content. For a complete example have a look at the finite volume code in Chapter 3.4.

5.9. Local Grid Adaptivity

One of the strengths of the dune-grid module is the support for a wide variety of different local grid adaptivity mechanisms. Chapter 2.3 has explained how adaptive grid refinement works; here we show how to implement it with the DUNE grid interface.

Local grid adaptivity is sometimes shied away from, because it can be difficult to program. Indeed, getting red–green refinement of three-dimensional grids to work reliably can be a challenge, especially if the code is supposed to run on a parallel computer. Luckily, dune-grid does all the heavy lifting here. It already contains several grid implementations that provide various forms of adaptive mesh refinement, which are controlled through an abstract interface. For example, the AlbertaGrid grid manager does edge bisection refinement of simplex grids (Figure 2.13). UGGrid implements red and red–green refinement in two and three space dimensions, for grids containing more than one element type, and even on distributed machines. The ALUGrid grid manager, available as a separate DUNE module,[27] does bisection and nonconforming red refinement as well [2]. Relocation of vertices (r-refinement [95]) can be applied to all grids using the GeometryGrid meta grid. In the future, further DUNE grid data structures may implement even more adaptivity strategies.

It is difficult to unify all these different ways to locally adapt grids, and make them usable through a precise and specific interface. An interface that is too detailed will leave most grids unable to fully implement it, but an interface that is too vague does not provide enough control. DUNE offers a compromise. The part of the adaptivity interface that is codified and mandatory for all grids is fairly vague, and grid managers

---

[27]www.dune-project.org/modules/dune-alugrid

have a lot of freedom to interpret the modification requests. To access more specific adaptivity features, most grid managers that implement some form of local adaptivity offer certain non-interface methods to control the refinement behavior. For example, while all grids have a method to mark elements for refinement, UGGrid has additional element marking methods for certain anisotropic refinement rules (see Figure 5.25). Since these methods are not part of the official interface, they can be tailored to each grid implementation, and allow to access very specific features. On the downside, users of such features need to be aware that they cannot simply swap their grid implementation for another one when they use them.

We focus in this chapter on the general interface that is common to all grids. Some implementation-specific features are discussed when a few grid implementations are presented in Chapter 5.10. For a complete list of features it is best to consult the online documentation of the grid classes.

5.9.1. Local Grid Adaptivity Without Data Transfer

There are two levels of difficulty to local grid adaptivity in DUNE. If there is no data attached to the grid, local grid adaptivity really only means adapting the grid structure to one's liking. This is what is described in this section. The more challenging task of how to preserve grid-based data while modifying the grid will be covered afterwards.

Adaptive grid modification in DUNE is a two step procedure. First, a subset of the leaf grid elements is marked for either refinement or coarsening. Then, a call to the method

```
bool adapt()
```

of the grid does the entire grid restructuring automatically.[28] It is up to the grid implementation to interpret the refinement and coarsening marks. For example, the same code will do edge-bisection refinement with AlbertaGrid, and red–green-refinement with UGGrid. Structured grids may decide to refine the entire grid when a single refinement marker is encountered, or they may decide to not do anything at all. The adapt method returns **true** if the grid changed at all, and **false** otherwise.

Elements are marked for refinement or coarsening by calling the method

```
bool mark(int refCount, const Codim<0>::Entity& element)
```

on the hierarchical grid object. The second of the two parameters is the leaf element to be marked. The first one is an integer specifying the kind of modification requested. It is the number of subdivisions that should be applied to the element. So, a value of 1 means *refinement*, while a value of -1 means *coarsening*. A value of 0 means to not do anything with the element. Most grids only implement at most these three values.

---

[28]Strictly speaking, the dune-grid interface specification requires that two additional methods preAdapt and postAdapt have to be called before and after the call to adapt, respectively (see Chapter 5.9.2 below). However, in practice no current grid implementation actually requires this.

The method `mark` returns **true** if the marking has been successful. This allows grid implementations to signal early that certain marking operations are not allowed. The current marking status of each element can be accessed by calling the method

```
int getMark (const Codim<0>::Entity& element) const
```

again on the grid, which returns one of 1, 0, or -1.[29] For example, the following code snippet refines all elements for which a precomputed indicator exceeds the value 10, and coarsens those for which the indicator is below 0.1:

```
MultipleCodimMultipleGeomTypeMapper<GridView> mapper(gridView,
                                             mcmgElementLayout());

for (const auto& element : elements(gridView))
{
  auto idx = mapper.index(element);      // Get element index

  if (indicator[idx] < 0.1)
    grid.mark(-1, element);              // Mark for coarsening
  else if (indicator[idx] > 10)
    grid.mark(1, element);               // Mark for refinement
}

grid.preAdapt();
grid.adapt();                            // Actual grid modification happens here
grid.postAdapt();
```

Note how we have used a `MultipleCodimMultipleGeomTypeMapper` to get a unique index for each element, since the element indices provided by the `IndexSet` interface are not unique if there is more than one element type in the grid. As mentioned, `preAdapt` and `postAdapt` are required by the grid interface. They have no real purpose here, and will be explained in Chapter 5.9.2.

While this short example shows pretty much all there is to know about grid adaptivity without data transfer, we nevertheless follow it up with a larger, complete example. This larger example produces the grids shown in Figure 5.22, and it should allow to play with the power of DUNE adaptivity without much preparation. The complete source code is available in the appendix (Appendix B.4), and also by clicking on the icon in the text margin, if reading this text in electronic form.

Consider the following scenario:[30] Given a sphere of radius 1 moving slowly upwards in a rectangular domain of dimensions 6×15 units, we want a grid that is highly resolved near the circle, and less so away from it. Since the sphere is moving this means that certain parts of the grid have to be refined, only to be coarsened again later when the sphere has moved on.

The program code starts by defining a helper class `Sphere`. The sole purpose of this class is to compute distances of points to a sphere of given radius and center. We

---

[29]The reason why the methods `mark` and `getMark` are part of the interface of the grid class, rather than being member methods of the `Entity` class, is that technically `mark` modifies the element (it stores the marker). Hence a non-**const** element would be needed to call it. However, it has turned out that the grid interface is easier to implement when only **const** elements are available.

[30]The example is inspired by the two-phase flow simulations of Gross and Reusken [86].

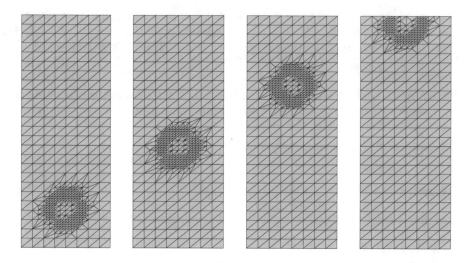

Figure 5.22.: Local grid adaptivity without data transfer. This example uses red–green refinement and a two-dimensional simplex grid.

implement a sphere of arbitrary dimension right away. That way it becomes easier to also try out the program with three-dimensional grids:

```
16  template<int dim>
17  class Sphere
18  {
19    double radius_;
20    FieldVector<double, dim> center_;
21  public:
22    Sphere(const FieldVector<double, dim>& center, const double& radius)
23    : radius_(radius),
24      center_(center)
25    {}
26
27    double distanceTo(const FieldVector<double, dim>& point) const
28    {
29      return std::abs((center_ - point).two_norm() - radius_);
30    }
31
32    void displace(const FieldVector<double, dim>& increment)
33    {
34      center_ += increment;
35    }
36  };
```

Then the `main` method follows. It is short—just over a hundred lines of code including blank lines and comments. We split it up in parts, and discuss the parts one by one. The first part starts MPI if it is installed. Then, a structured two-dimensional simplex grid is constructed using the `UGGrid` grid manager. Note how we are using the `StructuredGridFactory` utility class from Chapter 5.7.1 to create such a grid:

```
41  int main(int argc, char *argv[])
42  {
43    // Set up MPI if available
44    MPIHelper::instance(argc, argv);
45
46    constexpr int dim = 2;    // Grid and world dimension
47    using Grid = UGGrid<dim>;
48
49    // Start with a structured grid
50    const std::array<unsigned, dim> n = {8, 25};
51
52    const FieldVector<double, dim> lower = {0, 0};
53    const FieldVector<double, dim> upper = {6, 15};
54
55    std::shared_ptr<Grid> grid
56        = StructuredGridFactory<Grid>::createSimplexGrid(lower, upper, n);
57
58    using GridView = Grid::LeafGridView;
59    const GridView gridView = grid->leafGridView();
```

The next block constructs a sphere with center $(3, 2.5)$ and radius 1.0, and sets various parameters, like the number of time steps:

```
64    Sphere<dim> sphere({3.0, 2.5}, 1.0);
65
66    // Set parameters
67    const int steps = 30;                 // Total number of steps
68    const FieldVector<double, dim>
69        stepDisplacement = {0, 0.5};       // Sphere displacement per step
70
71    const double epsilon = 0.4;           // Thickness of the refined region
72                                          // around the sphere
73    const int levels = 3;                 // Number of refinement levels
```

Then follows the main loop over the 30 time steps. First, in Lines 82–90, the code coarsens any refined element left over from previous iterations. Then, the loop in Lines 93–103 does the actual refinement. Any element whose center is within `epsilon` of the sphere is marked for refinement. Then, `adapt` is called to do the actual grid refinement. The whole procedure is repeated `levels−1` times, to obtain element refinement trees of up to `levels` levels:

```
77    for (int i = 0; i < steps; ++i)
78    {
79      std::cout << "Step " << i << std::endl;
80
81      // Coarsen everything
82      for (int k = 0; k < levels-1; ++k)
83      {
84        for (const auto& element : elements(gridView))
85          grid->mark(-1, element);
86
87        grid->preAdapt();
88        grid->adapt();
89        grid->postAdapt();
90      }
91
92      // Refine near the sphere
93      for (int k = 0; k < levels-1; ++k)
94      {
95        // Select elements that are close to the sphere for grid refinement
96        for (const auto& element : elements(gridView))
97          if (sphere.distanceTo(element.geometry().center()) < epsilon)
98            grid->mark(1, element);
99
100       grid->preAdapt();
101       grid->adapt();
102       grid->postAdapt();
103     }
```

Finally, the grid is written to a VTK file, and the sphere is moved. The closing parenthesis in Line 113 marks the end of the main loop:

```
108     VTKWriter<GridView> vtkWriter(gridView);
109     vtkWriter.write("refined_grid_"+std::to_string(i));
110
111     // Move sphere
112     sphere.displace(stepDisplacement);
113   }
```

This concludes the example. Consult the text on the different grid managers in Chapter 5.10 to learn how to modify the code to make it use non-conforming or bisection refinement.

5.9.2. Preserving Data Across Grid Changes

Local grid adaptivity gets more challenging when there is data attached to the grid. Usually, at least some of this data is not thrown away, but must instead be preserved as good as possible while the grid changes. For example, a discrete function may exist

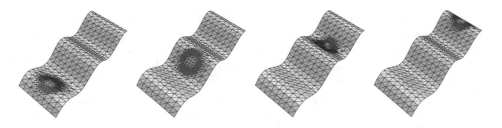

Figure 5.23.: Local grid adaptivity with data transfer. The values of a scalar function, visualized as a height field, are preserved across several steps of grid modifications.

on the old grid, and should be used as the initial iterate of an iterative solver on the new grid.

Of course such discrete functions cannot be preserved perfectly: Since the grid changes, the corresponding discrete function spaces change along with it. Therefore, generally, the old function cannot be represented exactly on the new grid. Roughly speaking, information is lost wherever elements get removed. Refinement of elements does not lead to information loss if the corresponding finite element space are nested, but the new degrees of freedom must be determined by suitable interpolation from known ones.

Implementing this is reasonably easy when the numerical data is stored within the data structures of the grid elements and vertices itself. However, DUNE keeps the grid and numerical data rigorously separate, and the only connection between them is through the *index* and *id* numbers described in Chapter 5.6: For computations on a fixed grid view, each entity provides an *index*, i.e., a non-negative integer number that can be used to address the array data types that contain the numerical data. However, when the grid is locally modified, these indices change a fortiori. It is generally not possible to modify the set of entities and preserve the existing enumeration. To preserve data across grid modifications, the DUNE grid interface additionally offers persistent entity numbers, the so-called *id*s (Chapter 5.6.3). Ids are neither required to be positive nor to be consecutive, but they offer enough functionality to work as keys in an associative container such as a `std::map` or a `std::unordered_map` [103]. The idea is to archive numerical data in such an associative container before changing the grid, and to copy it from there into new arrays afterwards, once the new indices are available. Of course data storage in a `std::map` is much slower than using arrays. However, it is expected that the time needed for the actual grid modification dominates the time needed to copy the data to and from a `std::map`.

To show how this works in practice we extend the previous example. Again, we implement grid refinement following a moving circle. However, additionally, a discrete function is set up by interpolating a given closed-form function at the beginning, and this discrete function is then transferred from each grid to the next. For simplicity, a first-order Lagrangian finite element space is used. The result is shown in Figure 5.23, which corresponds to Figure 5.22, but now uses a height field to visualize the function values. It can be seen that at each time step, exactly the same function is defined on the grid. No detail is lost at all, because in this example the initial interpolation happens on the coarsest grid level.

As usual the complete source code is contained in a single file. It is listed in Appendix B.5, and it is also available through the icon in the text margin. The code is very similar to the previous one; in particular, it reuses the Sphere class which is therefore not printed here again. Almost the complete code is placed in the main method. Besides the Sphere class, the only exception to this is a small helper function for linear interpolation of scalar values on a simplex:

```
37  template<int dim>
38  double interpolate(const std::vector<double> values,
39                     FieldVector<double,dim> p)
40  {
41    assert(values.size() == dim+1);
42    double result = values[0];
43    for (std::size_t i=0; i<p.size(); i++)
44      result += p[i]*(values[i+1]-values[0]);
45    return result;
46  }
```

It is used to compute function values at new vertices by interpolation between vertex values on a coarser grid element. In production code, shape function implementations like the ones from the dune-localfunctions module (Chapter 8) would be used for this, but here the interpolation is implemented directly to keep the code self-contained.

The main method is structured very much like the main method of the previous example. In particular, its beginning reads identically, with setting up MPI, and constructing a structured triangle grid using the UGGrid grid manager:

```
50  int main(int argc, char *argv[])
51  {
52    // Set up MPI if available
53    MPIHelper::instance(argc, argv);
54
55    constexpr int dim = 2;    // Grid dimension
56    using Grid = UGGrid<dim>;
57
58    // Create UGGrid from structured triangle grid
59    const std::array<unsigned, dim> n = {8, 25};
60
61    const FieldVector<double, dim> lower = {0, 0};
62    const FieldVector<double, dim> upper = {6, 15};
63
64    std::shared_ptr<Grid> grid
65        = StructuredGridFactory<Grid>::createSimplexGrid(lower, upper, n);
66
```

```
67    using GridView = Grid::LeafGridView;
68    const GridView gridView = grid->leafGridView();
69    const auto& indexSet = gridView.indexSet();
70    const auto& idSet    = grid->localIdSet();
```

Note, however, that unlike in the previous example, the code asks the grid for an IndexSet and a local IdSet in Lines 69–70. The index set will be used to address the array (here: a std::vector) that holds the data. The ids will be used to index the std::map that will preserve the data while the grid is being changed.

The next few lines set up a few parameters like the number of time steps, the sphere radius, and so on. This part is a verbatim copy of the previous example from page 175:

```
75    Sphere<dim> sphere({3.0, 2.5}, 1.0);
76
77    // Set parameters
78    const int steps = 30;
79    const FieldVector<double, dim> stepDisplacement = {0, 0.5};
80
81    const double epsilon = 0.4;   // Thickness of the refined region
82                                  // around the sphere
83    const int levels = 2;
```

More interestingly, the code then sets up and interpolates the example grid function:

```
88    auto dataFunction = [](const FieldVector<double,dim>& x)
89    {
90      return std::sin(x[1]);
91    };
92
93    std::vector<double> data(gridView.size(dim));
94    for (auto&& v : vertices(gridView))
95      data[indexSet.index(v)] = dataFunction(v.geometry().corner(0));
```

The assignment starting at Line 88 stores the function

$$(x_0, \ldots, x_{d-1}) \mapsto \sin(x_1)$$

as a C++ lambda. Lines 94–95 then loop over all vertices of the leaf grid view, evaluate the function at these vertices, and store the values in the data array. Note how the indexSet object is used to access the data associated to the vertex.

At this point, everything is prepared and we can proceed to the main loop. As in the example without data transfer, each time step consists of two substeps: First, all refined elements are removed. Then, the grid is refined again near the sphere at the new position, and the sphere is moved. The following code does the coarsening:

```
99    for (int i = 0; i < steps; ++i)
100   {
101     std::map<Grid::LocalIdSet::IdType, double> persistentContainer;
```

179

```
102
103      // Coarsen everything
104      for (int j = 0; j<levels; ++j)
105      {
106        for (const auto& element : elements(gridView))
107          grid->mark(-1, element);      // Mark element for coarsening
108
109        grid->preAdapt();
110
111        for (const auto& vertex : vertices(gridView))
112          persistentContainer[idSet.id(vertex)] = data[indexSet.index(vertex)];
113
114        grid->adapt();
115
116        data.resize(gridView.size(dim));
117
118        for (const auto& v : vertices(gridView))
119          data[indexSet.index(v)] = persistentContainer[idSet.id(v)];
120
121        grid->postAdapt();
122      }
```

In this loop, the same calls to mark and adapt appear that have already been used in the example without data transfer, but additionally, the discrete function is handled. Remember that its coefficients are stored in the array called data. To keep the function while the grid is being modified, the code additionally creates a std::map that can hold **double** values and is indexed by the local IdType of the grid. The map is called persistentContainer, because it persists across grid changes. Before actually removing one level from the grid hierarchy in Line 114, all vertex data is copied from that level into the persistent container (Lines 111–112). Then, in Line 114, the grid is modified, and the data is copied back into the data array. After the loop over j has terminated, what is left is a single-level grid, and the point values of the original function attached to the vertices.

In this academic example we know beforehand that we are going to remove all but the coarsest grid level. Copying the data back and forth for each level separately is therefore not necessary. However, in actual simulations the decision whether to remove elements of a certain level is based on the data on that level, and for this it needs to be copied into an array. An adaptivity example that uses an actual error estimator is shown in Chapter 11.5.2.

To organize the data transfer, the code has to use two further methods on the grid object. The method

```
bool preAdapt()
```

needs to be called after the elements have been marked, and before the call to adapt. It sets the mightVanish flags of the elements. These flags, available through the method

```
bool mightVanish()
```

of the Entity class for codimension 0, return **true** if the element may disappear during the next call to adapt. If the method returns **false**, then the corresponding element is guaranteed to stay, and its data does not have to be projected to coarser levels. The method is not used in this example. The preAdapt method returns **false** if no leaf element will be removed from the hierarchical grid during the subsequent call to adapt.

The call to adapt sets the isNew flags. These can be accessed by the method

```
bool isNew()
```

of Entity<0>. It is used below to interpolate data only onto those elements that have been newly created.

The method

```
void postAdapt()
```

is to be called after the grid has been adapted and the data has been processed. It removes the isNew flags of the elements from the last call to adapt. The dune-grid interface specification requires both the preAdapt and the postAdapt methods to be called alongside adapt, even if no data is being transferred. This is why they also appeared in the example in the previous section.

After the coarsening, the grid is refined around the sphere (which has been moved at the end of the previous iteration of the time loop). Again, the code loops over the number of desired levels, marks the correct elements, and modifies the grid:

```
127    for (int j = 0; j<levels; ++j)
128    {
129      // Select elements that are close to the sphere for grid refinement
130      for (const auto& element : elements(gridView))
131        if (sphere.distanceTo(element.geometry().center()) < epsilon)
132          grid->mark(1, element);
133
134      grid->preAdapt();
135
136      for (const auto& vertex : vertices(gridView))
137        persistentContainer[idSet.id(vertex)] = data[indexSet.index(vertex)];
138
139      grid->adapt();
```

Lines 136–137 save the grid data in the persistent container, and the following line triggers the actual grid modification. The subsequent reconstruction of the discrete function is a bit more involved, because values for newly created vertices have to be computed by interpolation from father elements:

```
143    data.resize(gridView.size(dim));
144
```

```
145        for (const auto& element : elements(gridView))
146        {
147          if (element.isNew())
148          {
149            for (std::size_t k=0; k<element.subEntities(dim); k++)
150            {
151              auto father = element;
152              auto positionInFather
153                  = ReferenceElements<double,dim>::general(element.type())
154                                                      .position(k,dim);
155
156              do
157              {
158                positionInFather
159                    = father.geometryInFather().global(positionInFather);
160                father = father.father();
161              } while (father.isNew());
162
163              // Extract corner values
164              std::vector<double> values(father.subEntities(dim));
165              for (std::size_t l=0; l<father.subEntities(dim); l++)
166                values[l] = persistentContainer[idSet.subId(father,l,dim)];
167
168              // Interpolate linearly on the ancestor simplex
169              data[indexSet.subIndex(element,k,dim)]
170                  = interpolate(values, positionInFather);
171            }
172          }
173          else
174            for (std::size_t k=0; k<element.subEntities(dim); k++)
175              data[indexSet.subIndex(element,k,dim)]
176                  = persistentContainer[idSet.subId(element,k,dim)];
177        }
178
179        grid->postAdapt();
```

The code loops over all elements on the leaf grid. For each element it is checked whether it has been created by the last call to adapt (Line 147). If this is not the case, then its data must exist in the persistentContainer, and can simply be copied back out into the data array in Lines 174–176. Otherwise, the data for the elements needs to be constructed by interpolation. Since the data is vertex-based we first loop over the element corners in Line 149. Then, for each corner, Lines 151–161 find its local coordinates in the finest ancestor of the element for which data is available. Note that with red–green refinement being used, this may not be the direct father. The data of the ancestor is then interpolated linearly in Lines 164–170, using the short interpolate helper method shown above. This part exploits that all ancestors of a vertex share the same id (Chapter 5.6.3), which implies that the persistentContainer can be accessed for non-leaf vertices. When the element loop terminates, all leaf elements of the new leaf grid view have the correct data.

The main loop closes identically to the previous example: The current grid is saved (this time with data), and the sphere is moved:

```
187    VTKWriter<GridView> vtkWriter(gridView);
188    vtkWriter.addVertexData(data, "data");
189    vtkWriter.write("refined_grid_"+std::to_string(i));
190
191    // Move sphere
192    sphere.displace(stepDisplacement);
193  }
```

This concludes the example. When this program is run, it will write 30 files with a grid and data in each, four of which are the ones shown in Figure 5.23.

5.10. Some Existing Grid Managers

Various implementations exist of the DUNE grid interface. Each implementation has its particular strengths. Some are more flexible than others, but use more memory or run-time. Some can be used on distributed architectures, or they can do adaptive refinement, while others cannot. This chapter gives a short overview over the grid implementations directly provided by the dune-grid module. It also briefly presents ALUGrid[31] [2], which needs to be installed from a separate module, but which is powerful and widely used.

At the time of writing there are six grid implementations in the dune-grid module. Two of them, YaspGrid and OneDGrid are native DUNE grids, with their code contained entirely in dune-grid. Another two, UGGrid and AlbertaGrid, are adapter implementations that rely on external libraries to do the real work. The last two, GeometryGrid and IdentityGrid, are so-called *meta grids*. These are grids that get parametrized with other DUNE grids, extending and modifying their behavior. In this section, we will briefly present each of these grids.

While the six grid managers in dune-grid together form a powerful set that allows to handle many situations, there are special situations not covered by any implementation in dune-grid. Luckily, the modular structure of DUNE makes it easy for third-party developers to provide additional grid managers in separate DUNE modules. We mention CpGrid, the implementation of degenerate hexahedral grids by the OPM Project,[32] CurvilinearGrid as an unstructured grid with polynomial element geometries[33] [69], and FoamGrid [140] for unstructured one- and two-dimensional surface and network grids in $\mathbb{R}^d$.[34] There is also a variety of additional meta grids available, e.g.,

---

[31]www.dune-project.org/modules/dune-alugrid
[32]http://opm-project.org
[33]www.curvilinear-grid.org
[34]www.dune-project.org/modules/dune-foamgrid

SubGrid [82], MultiDomainGrid [124], and PrismGrid [73]. A more extensive list is given on the DUNE project homepage.

5.10.1. External Grid Managers

Two grid managers in dune-grid connect external libraries for unstructured grids to the DUNE grid interface.

UGGrid

The UGGrid grid implementation uses the grid manager of the UG3 finite element software system [13]. Developed since the early 1990ies at the Universities of Heidelberg and Stuttgart, UG3 was one of the most powerful and widely used finite element research code of its time. While direct usage of UG3 has basically ceased (among other reasons because it was difficult to use and poorly documented), its grid manager remains one of the most flexible ones even today. The relevant parts of UG3 were forked and moved into a dedicated DUNE module called dune-uggrid. This module is an optional dependency for dune-grid. The UGGrid class contained in the dune-grid module allows to harness the flexibility of the UG3 grid manager through the modern and easy-to-use DUNE grid interface. It is used in many examples in this book.[35]

UGGrid implements two- and three-dimensional unstructured grids. Two-dimensional grids can contain triangles and quadrilaterals together, and three-dimensional ones can contain tetrahedra, hexahedra, prisms, and pyramids. The grid dimension is a template parameter of the UGGrid class. While UG3 itself only allowed grids of a single dimension in a program, within DUNE two- and three-dimensional grids can be freely mixed. Memory permitting, any number of UGGrid objects can be alive at the same time.

One of the strengths of UGGrid is the very flexible set of refinement rules. By default, it uses the classical red–green strategy described in Chapter 2.3.2. Additionally, UGGrid provides a non-interface method

```
void setClosureType(ClosureType type)
```

with possible values ClosureType::GREEN (the default), or ClosureType::NONE. Choosing the latter switches off the green closure, which leads to nonconforming red refinement. There is no restriction on the order of the resulting hanging nodes. Figure 5.24 shows several example grids obtained by running the adaptivity example of Chapter 5.9.1 on different initial grids, and with or without the green closure. Boundary parametrizations can be specified, and are taken into account when refining the grid. The boundary parametrization example in Figure 5.20 was computed using UGGrid.

---

[35]Eventually, UG3 was completely reimplemented in a more modern style. The new implementation, called UG4, has many advantages over the old code, but it is also lacking some of the flexibility in grid handling. Its code is available at http://gcsc.uni-frankfurt.de/simulation-and-modelling/ug4.

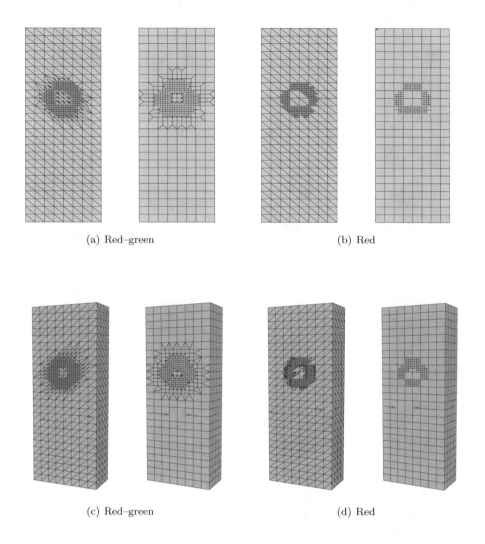

(a) Red–green (b) Red

(c) Red–green (d) Red

Figure 5.24.: Example adaptive refinements obtained with UGGrid

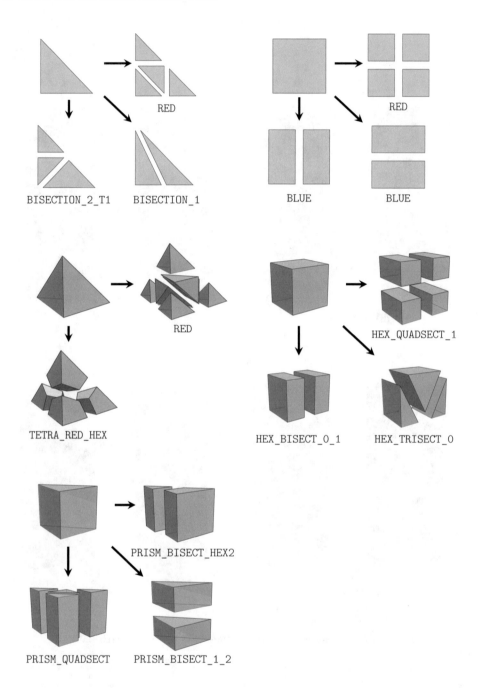

Figure 5.25.: Some (but not all) anisotropic refinement rules of UGGrid

UGGrid supports anisotropic refinement, i.e., marking elements explicitly with nonstandard refinement rules. Figure 5.25 shows a few possibilities. Since such flexibility cannot be controlled with the simple marking mechanism of the standard DUNE grid interface described in Chapter 5.9, the UGGrid class offers an additional mark method that takes a UG refinement rule as parameter:

```
bool mark(const Codim<0>::Entity& element,
          UG_NS<dim>::RefinementRule rule,
          int side=0)
```

The first parameter is the element to mark for anisotropic refinement. Specific anisotropic refinement rules are selected through the second parameter. Several possible values of the enumeration type are shown in Figure 5.25, and a complete list is given in the online documentation. The third parameter only has a meaning for two-dimensional grids, where it selects the orientation of the rules (The orientation of three-dimensional rules is encoded in the enumeration value name). As this interface to anisotropic grid refinement is not particularly pretty, it may change eventually.

Another strength of UGGrid are its features for distributed computing. Grids can be distributed across an arbitrary number of processes, and redistributed if so desired. This dynamic load-balancing works well together with the advanced refinement rules described above. A simple partitioning algorithm is contained in the grid itself; for more advanced algorithms the grid can be connected to external grid partitioners. Unlike many other grid implementations, UGGrid supports even hierarchical load-balancing, i.e., coarser levels of the refinement hierarchy may reside on other processes than finer ones. This combination of features makes UGGrid one of the most flexible grid implementations currently available.

AlbertaGrid

ALBERTA is a complete finite element software that was developed starting in the early 1990ies by Alfred Schmidt and Kunibert Siebert in Freiburg [141],[36] with a particular interest in adaptive multilevel methods. The grid manager of ALBERTA is available through the DUNE grid interface in form of the AlbertaGrid class. Unlike the case of UG3, there is no DUNE-specific fork of ALBERTA. If the actual ALBERTA program is installed, the DUNE build system will pick it up and build the AlbertaGrid bindings.

With its focus on adaptive multilevel methods, Alberta is centered around a grid data structure that implements hierarchical grids as sets of refinement trees. Only simplex elements are allowed, and grids can be one-, two-, or three-dimensional. The adaptation mechanism used is newest vertex bisection in two dimensions, and recursive edge bisection following Kossaczký [112] in three dimensions. Hence, all refinement trees are binary trees. Boundary parametrizations are taken into account when the grid is refined. The adaptive Poisson example in Figure 2.15 was computed using AlbertaGrid.

---

[36]www.alberta-fem.de

As a peculiarity, a lot of entity information is not actually stored at the tree nodes, but rather it is recomputed on the fly from the coarse grid elements and refinement information. This makes the data structure very lean, but access to grid elements is available solely via routines that traverse the hierarchical trees. As this maps badly to the DUNE grid interface (which uses level- and leaf-wise traversals), access to AlbertaGrid entities can be a bit slower than for other grids. Also, AlbertaGrid is restricted to sequential problems. The grid manager is mainly of interest for its bisection refinement, which is sometimes preferable to red or red–green refinement.

ALUGrid

Even though not in dune-grid at all, this text would be incomplete without at least briefly mentioning ALUGrid [2]. ALUGrid is another unstructured grid implementation, geared towards very large distributed problems, where grid performance is important. ALUGrid is provided in a separate module dune-alugrid, available from www.dune-project.org/modules/dune-alugrid. It is mentioned here because it is very powerful and mature, and widely used.

ALUGrid implements unstructured grids in two and three space dimensions. Elements can be hypercubes or simplices, but grids cannot contain both element types at once. Two-dimensional grids can be embedded into a three-dimensional world space.

Its maintainers developed ALUGrid primarily for explicit time-stepping schemes. As such methods do not solve algebraic systems of equations, the performance of the grid itself becomes critical. Much effort was therefore put into data structures that are both time- and space-efficient. ALUGrid supports nonconforming red refinement and conforming bisection refinement (Chapter 2.3.2), but no red–green refinement.

The ALUGrid data structure supports distributed grids, and scales to a large number of processes. To accommodate time-dependent adaptive methods, grids can be repartitioned if adaptive refinement has lead to a load imbalance. A wide range of load-balancing algorithms are available, including a built-in algorithm based on space-filling curves, but also external load-balancing libraries such as ZOLTAN [35].[37] Further special features for large-scale computing include an interface for the direct construction of distributed grids (something that the standard GridFactory class of Chapter 5.7.2 does not do yet), backup and restore of complete grid hierarchies, and the possibility to overlap communication and computations in time. For some of these features, the grid interface had to be extended.

ALUGrid has traditionally been a playground to experiment with extensions to the grid interface, mainly in the interest of parallel performance. The article [2] describes these extensions in detail. Some of them are likely to become part of the official DUNE grid interface eventually.

---

[37] www.cs.sandia.gov/zoltan

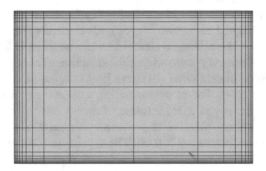

Figure 5.26.: A tensor product grid consists of axis-aligned cube elements arranged in planes of varying thickness

5.10.2. Built-in Standalone Grid Managers

Currently, the dune-grid module contains two grid implementations that do not rely on external libraries. These two are YaspGrid, a fast and light-weight implementation of a distributed structured grid, and OneDGrid, which implements one-dimensional, locally adaptive grids.

YaspGrid

The three external grid managers have all implemented unstructured grids. In contrast, the YaspGrid grid manager[38] implements a production-grade *structured* grid. The domain of a YaspGrid is a d-dimensional hypercube in $\mathbb{R}^d$, and the elements are d-dimensional hypercubes with axis-aligned sides. The elements can all be of the same size, or they can be arranged in a tensor product structure, that is, with the elements arranged in parallel planes of varying thickness (Figure 5.26). Switching between these two modes happens at compile-time. YaspGrid does not support adaptive grid refinement at all; grids can only be refined globally.

Since YaspGrid is a dedicated structured grid implementation, it achieves the space- and time-efficiency of such a grid, combined with the convenience of the common DUNE grid interface. Indeed, a YaspGrid object with uniform element size needs only a constant amount of memory regardless of the number of grid elements. Individual elements and vertices are not stored, as all element information can be computed from the index alone. For the tensor grid mode, the space requirement is still only proportional to the sum of the number of elements in each direction, which is much less than the total number of elements.

---

[38] YaspGrid: Yet another structured parallel Grid

The implementation uses various tricks to make the grid access as fast as possible. For example, the implementation of the Geometry interface class (Chapter 5.3.2) is hand-tuned for axis-aligned elements. [39] For such elements the Jacobi matrix of the transformation F from the reference element to the element is known a priori to be diagonal. Hence, the Geometry method jacobianInverseTransposed returns objects of a dedicated diagonal matrix class (DiagonalMatrix, see also Chapter 7.3.2). As a consequence, all subsequent matrix–vector-multiplications only need d multiplications.

Unfortunately, the dune-grid grid interface does not offer extensions for structured grids. For example, it is not possible to access an element T_{ijk} of a three-dimensional grid by its integer coordinates $i, j, k \in \mathbb{N}$. There is no particular reason not to have such an interface, but there has never been a concrete move to introduce one. At the time of writing, YaspGrid does not offer non-interface methods for this kind of access either, but this may change in the future. The indexing of the elements and vertices by YaspGrid is lexicographic, with the indices varying fastest along the x_0-direction. With that knowledge, at least the data for each element and vertex can be accessed from integer coordinates in constant time.

The YaspGrid class has the signature

```
template<int dim, class Coordinates = EquidistantCoordinates<double, dim> >
class YaspGrid
```

The parameter dim is the dimension of the grid and the dimension of the world space. In principle it may be any natural number, but larger values of dim can be a compiler stress test. The second parameter is a policy object that encapsulates the way coordinates are stored. There are three choices:

1. EquidistantOffsetCoordinates: All elements have the same size.

2. EquidistantCoordinates (the default): All elements have the same size, and the lower-left corner of the grid is hard-wired to $(0, \ldots, 0) \in \mathbb{R}^d$.

3. TensorProductCoordinates: Each plane of elements can have its individual thickness.

YaspGrid objects are set up directly through their constructors.[40] There are various of these, either taking a coordinate policy object together with additional information, or all required information without direct use of a coordinates object. For example,

```
EquidistantCoordinates coordinates({1.0, 1.0, 1.0}, {10,10,10});
YaspGrid<3> grid(coordinates);
```

constructs a structured uniform grid for the unit cube $\Omega = [0, 1]^3$ with $10 \times 10 \times 10$ elements of equal size. The code

---

[39] This special implementation can be reused for other grids; see the class AxisAlignedCubeGeometry in the dune-geometry module.

[40] Alternatively, one can use the StructuredGridFactory class of Chapter 5.7.1.

```
14    const int dim = 2;
15    using Coordinates = TensorProductCoordinates<double,dim>;
16    using GridType = YaspGrid<dim,Coordinates>;
17
18    std::array<std::vector<double>, dim> coordinates;
19
20    for (int j=-10; j<10; j++)
21    {
22      coordinates[0].push_back(1.6*std::atan(j));
23      coordinates[1].push_back(std::atan(j));
24    }
25
26    GridType grid(coordinates);
27
```

creates the two-dimensional tensor grid of Figure 5.26. Consult the class documentation for all the details on how to construct YaspGrid objects.

YaspGrid provides many features for large-scale parallel computing. Grids can be distributed over any number of processors. Indeed, to accommodate for very large problems, YaspGrid numbers its entities with a custom integer type that allows an adjustable bit length. Subdomains must be block-shaped, but their sizes and distributions can be controlled using simple load-balancing classes. On the other hand, dynamic load-balancing, i.e., the repartitioning of a grid during a simulation, is not possible. As YaspGrid does not support adaptive grid refinement, the static nature of the load-balancing is not much of a problem. As the only grid implementation in DUNE (currently), YaspGrid supports overlapping partitions of the grid. Such partitions are, e.g., naturally suited to implement overlapping Schwarz preconditioners [148].

OneDGrid

The OneDGrid grid manager is geared towards a niche application: it only provides sequential one-dimensional grids in $\mathbb{R}^1$. It is, however, fully unstructured and supports element refinement and coarsening. Besides the real-world PDE applications that are posed on one-dimensional domains, a one-dimensional unstructured grid manager can be helpful for debugging, prototyping, and teaching. In particular, the OneDGrid grid manager is probably the simplest unstructured and adaptive grid manager possible. It is therefore easy to understand its code and inner workings. This can be helpful for people who are interested in implementing their own grid data structures for the DUNE grid interface.

5.10.3. Meta Grids

The last two grid implementations in dune-grid are so-called *meta grids*. Meta grids are grids that are parametrized by another DUNE grid (called the *host grid*), which they modify in some way. Since meta grids implement the complete DUNE grid interface, they can be used just like any other grid. There are further interesting

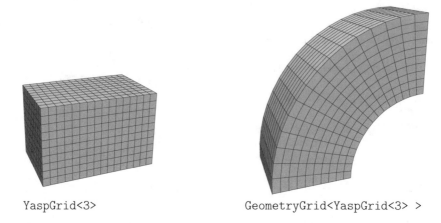

YaspGrid<3>	GeometryGrid<YaspGrid<3> >

Figure 5.27.: Left: a three-dimensional YaspGrid. Right: The GeometryGrid meta grid changes the geometry of the YaspGrid object.

meta grids beyond the two in the dune-grid module, for example SubGrid [82], MultiDomainGrid [124], and PrismGrid [73].

GeometryGrid

The first meta grid in DUNE is GeometryGrid. It modifies the geometry of the host grid. More precisely, it allows to assign new coordinates to each grid vertex, and it replaces the host grid element geometries by multilinear/affine ones. See Figure 5.27, where this is demonstrated with a YaspGrid that is warped into a quarter-circle. The resulting grid implementation is topologically but not geometrically structured.[41] Also, GeometryGrid can be used to implement r-refinement and moving-mesh methods [95]— the grid interface itself does not allow to move or deform existing grids.

Setting up a GeometryGrid requires a host grid object and a function object that produces the new vertex coordinates from either the positions or the indices of the old ones. The following function object produces the deformation from Figure 5.27:

```
struct AnalyticalDeformation
  : public AnalyticalCoordFunction<double, 3, 3, AnalyticalDeformation>
{
  FieldVector<double,3> operator()( const FieldVector<double, 3>& x) const
  {
    return {(1+x[2])*std::cos(x[0]),
            x[1],
            (1+x[2])*std::sin(x[0])};
```

---

[41] For maximum performance it may be better to write a dedicated grid manager for a structured grid with non-uniform geometry. While the run-time overhead of meta grids is small [108], it is not zero.

```
  }
};
```

A deformed YaspGrid can then be set up by the following code:

```
using AnalyticallyDeformedGrid = GeometryGrid<HostGrid,
                                              AnalyticalDeformation>;
AnalyticalDeformation analyticalDeformation;
AnalyticallyDeformedGrid analyticallyDeformedGrid(hostGrid,
                                                  analyticalDeformation);
```

Alternatively, the new vertex coordinates can be provided in a container. For this the deformation function object has to inherit from Dune::DiscreteCoordFunction:

```
template<class HostGridView>
struct DeformationFunction
  : public DiscreteCoordFunction<double,
                                 HostGridView::dimension,
                                 DeformationFunction<HostGridView> >
{
  static constexpr int dim = HostGridView::dimension;

  DeformationFunction(const HostGridView& gridView,
                      const std::vector<FieldVector<double,dim> >&
                                                  deformedPosition)
    : indexSet_(gridView.indexSet()),
      deformedPosition_(deformedPosition)
  {}

  template<class HostEntity>
  void evaluate(const HostEntity& hostEntity, unsigned int corner,
                FieldVector<double,dim> &y) const
  {
    auto idx = indexSet_.subIndex(hostEntity, corner,dim);
    y = deformedPosition_[idx];
  }

private:
  const typename HostGridView::IndexSet& indexSet_;
  const std::vector< FieldVector<double,dim> > deformedPosition_;
};
```

Setting up the YaspGrid example of Figure 5.27 with such a discrete deformation takes the form:

```
std::vector<FieldVector<double,3> > vertexPositions(hostGridView.size(dim));

for (const auto& vertex : vertices(hostGridView))
{
  FieldVector<double,3> x = vertex.geometry().corner(0);
```

```
vertexPositions[hostGridView.indexSet().index(vertex)]
                            = {(1+x[2])*std::cos(x[0]),
                               x[1],
                               (1+x[2])*std::sin(x[0])};
}

using DiscretelyTransformedGrid
    = GeometryGrid<HostGrid, DeformationFunction<HostGridView> >;
DeformationFunction<HostGridView> discreteTransformation(hostGridView,
                                                vertexPositions);
DiscretelyTransformedGrid discretelyTransformedGrid(hostGrid,
                                        discreteTransformation);
```

GeometryGrid implements the entire grid interface, and can be used like any other grid. In particular, it implements all features for parallel distributed computing, and can be applied to distributed host grids.

IdentityGrid

The second meta grid available in dune-grid is IdentityGrid. As the name says, it produces a meta grid that behaves identically to the host grid. IdentityGrid exists for documentation purposes only. It can be instructive to look at, because it consists only of the boilerplate code that DUNE grids have to implement, without any functionality. Indeed, to start writing a new grid manager it is recommended to start by copying IdentityGrid, renaming everything to a new name, and then start implementing the desired functionality one by one.

6. Dune Grids on Distributed Machines

The previous chapter has always assumed that there is a single process only, which knows the entire grid. However, a considerable portion of today's applications needs multi-processor machines to run, either to keep the run-time acceptable or to handle the large storage requirements, frequently both. The `dune-grid` interface therefore provides facilities for distributing grids across parallel machines, and for accessing them concurrently. It also provides means to exchange data between the individual subgrids. These means are designed to cover all common communication patterns usually found in distributed algorithms for partial differential equations. More special needs can always be satisfied by using MPI directly, on top of the DUNE grid interface.

The interface (and this chapter along with it) currently only covers distributed-memory multi-processing. Explicit support for shared-memory architectures using OpenMP[1] or similar tools is being worked on [18, 19], but little has materialized in an official DUNE release yet.

6.1. The Dune Data Decomposition Model

The underlying idea of parallel computations with the DUNE grid interface is grid partitioning. The complete grid is split up into subdomains, and each subdomain is

---

[1] http://openmp.org

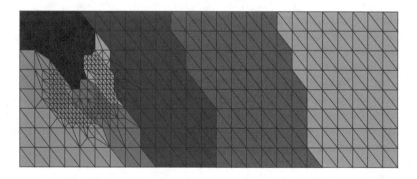

Figure 6.1.: Decomposition of an adaptively refined grid into eight subdomains

© Springer Nature Switzerland AG 2020
O. Sander, *DUNE — The Distributed and Unified Numerics Environment*, Lecture Notes in Computational Science and Engineering 140, https://doi.org/10.1007/978-3-030-59702-3_6

assigned to a separate process. The subdomains may or may not overlap. Each process works only with the data attached to its subdomain grid. The processes exchange data, but only with their direct neighbors on the subdomain partition (Figure 6.1).

This paradigm corresponds directly to the classical one-level domain decomposition methods known from the numerical analysis literature [148, 154]. But DUNE can go further: As DUNE grids are hierarchical (Chapter 5.1), some grid implementations such as UGGrid can also have coarse grids assigned to different processes than the corresponding fine grids. This allows to implement distributed two-grid and multigrid methods, where it is frequently useful to distribute the coarser grids across fewer processes than the finer ones.

A distributed DUNE grid is a collective object: all processes participating in the computations on the grid instantiate the grid object at the same time. Each process stores a subset of all the grid elements, but each element may be stored on more than one process. If this is the case, it is called a *distributed* element. The same holds for the grid entities of lower dimension: Any such entity may be stored in more than one process, and it is then called a *distributed* entity.

In the DUNE data decomposition model, each entity in a process has a *partition type* assigned to it. There are five different possible partition types:

$$\textit{interior, } \textbf{border, } \textit{overlap, } \textbf{front, } \text{and } \textbf{ghost.}$$

We first discuss the partitioning of grid elements, which is simpler than the partitioning of lower-dimensional entities. Grid elements are restricted to the three partition types **interior**, **overlap**, and **ghost**. Consider all elements of a given grid view, either of a level grid view or the leaf grid view. Each element can exist on several processes, and it has a separate partition type on each process. Its partition type must be **interior** on exactly one of them. The **interior** leaf elements in process number i form a subdomain $\Omega_i \subseteq \Omega$, and all the Ω_i, $0 \leq i < P$, form a nonoverlapping decomposition of the computational domain Ω. That is, for each leaf element there is exactly one process where this element has partition type **interior**.

This approach works well for algorithms like the Neumann–Neumann method [148]. For this method, a decomposition of the computational domain into *nonoverlapping* subdomains is required, and such a decomposition is provided by the **interior** elements. In principle, Neumann–Neumann methods can be implemented assigning each element to a single process only (where it is then **interior** a fortiori).

However, in order to manage data like the subdomain boundary residuals, it can be convenient to have an extra layer of elements around each subdomain. Elements in this layer exist on more than one process; they are not **interior**, but correspond to **interior** elements on a neighboring subdomain. Such elements that exist only to facilitate data exchange are called *ghost* elements. Consequently, in DUNE, these elements have the partition type **ghost**.

On the other hand, when the numerical algorithm is an overlapping domain decomposition method like, e.g., the classical Schwarz method [148], the algorithm itself operates on an overlapping partition of the computational domain. Then two types of distributed elements are needed. The first type are the elements in the algorithmic

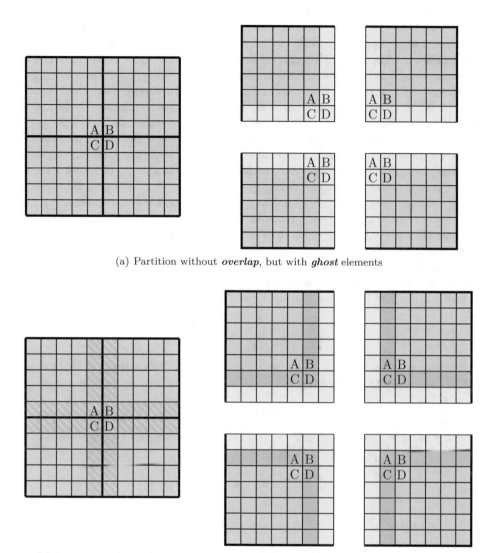

(a) Partition without **overlap**, but with **ghost** elements

(b) Partition with **overlap** and **ghost** elements. The hatched elements show the overlap.

Figure 6.2.: Grid partitioning into four subdomains, without and with overlap. **Interior** elements are grey, **ghost** elements are yellow, and **overlap** elements are green. The letters just exemplify some corresponding sets of elements.

overlap. If an element is contained in the overlap of p subdomains, then it is ***interior*** in one of them, and ***overlap*** in the $p - 1$ others. This is illustrated in Figure 6.2. The second type of distributed elements are again the ***ghost*** elements already known from the nonoverlapping methods, which form a layer around the subdomain grids. Interface-wise, there is no difference between ***overlap*** and ***ghost*** elements, and the distinction exists only for conceptual purposes.[2]

The existence, thickness, and exact structure of the ***overlap*** and ***ghost*** layers is determined by the grid implementations. For example, a distributed YaspGrid has no ***ghost*** elements, but it may have ***overlap***. The thickness of the ***overlap*** layer (in number of elements) is arbitrary, and can be controlled by a dedicated parameter of the YaspGrid constructor. UGGrid and ALUGrid, on the other hand, do not implement ***overlap*** layers at all. Instead, they provide a one-element layer of ***ghost*** elements around each subdomain, with no other options.

The assignment of partition types to grid entities of lower dimensions (vertices, edges, etc.) is more involved. In particular, the remaining two partition types ***border*** and ***front*** come into play. Roughly speaking, given partition types for the elements of a subdomain grid, ***border*** entities separate ***interior*** elements from ***overlap*** elements, and ***front*** entities separate ***overlap*** elements from ***ghost*** elements. If there is no overlap, ***interior*** and ***ghost*** elements are separated by ***border*** entities. More formally, for any process i let Ω_i be the domain determined by the ***interior*** elements, i.e.,

$$\Omega_i := \text{int}\Big(\bigcup_{T \text{ is } \textbf{\textit{interior}}} T \Big).$$

Each entity in Ω_i or in $\partial\Omega_i \cap \partial\Omega$ is an ***interior*** entity. The ***border*** domain is that part of the boundary of the ***interior*** domain that is not also domain boundary

$$B_i := \overline{\partial\Omega_i \setminus \partial\Omega},$$

and all grid entities contained in B_i on process i have partition type border on that process. The ***overlap*** elements form a separate domain

$$\Omega_i^{overlap} := \text{int}\Big(\bigcup_{T \text{ is } \textbf{\textit{overlap}}} T \Big).$$

All points on the boundary of that domain are ***front***, unless they are ***border*** or domain boundary:

$$F_i := \overline{\partial\Omega_i^{overlap} \setminus \partial\Omega} \setminus B_i.$$

Note that $\Omega_i^{overlap}$ may well be empty, in which case F_i is empty, too. All entities that are not contained in one of these four sets will be ***ghost*** entities.

The assignment of grid entities to partition types is illustrated in Figure 6.3 for three different examples. Each example shows a two-dimensional structured grid with

---

[2]Note, however, that as of dune-grid 2.7 the ***ghost*** elements of the ALUGrid implementation do not implement the full interface. Strictly speaking, this is not covered by the grid interface specification, though.

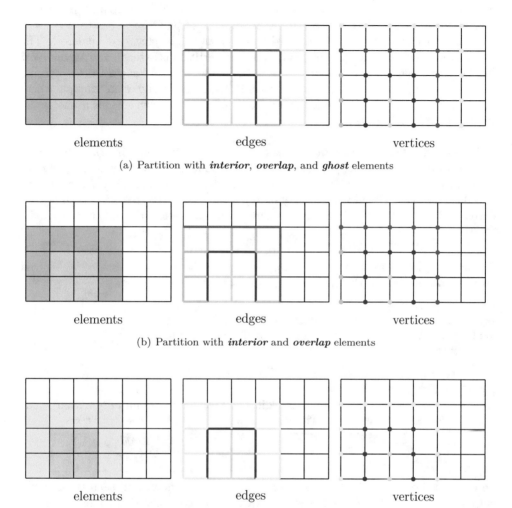

elements   edges   vertices

(a) Partition with *interior*, *overlap*, and *ghost* elements

elements   edges   vertices

(b) Partition with *interior* and *overlap* elements

elements   edges   vertices

(c) Partition with *interior* and *ghost* elements

Figure 6.3.: Color coded illustration of different data decompositions: *interior* (grey), *border* (blue), *overlap* (green), *front* (magenta), and *ghost* (yellow). Elements marked in white are not stored by the process at all.

199

6×4 elements. The entities stored in a fixed process i are shown in color, and the colors indicate the partition type. The first row shows an example where process i has codimension 0 entities of all three partition types *interior*, *overlap* and *ghost* (left picture in that row). The corresponding assignment of partition types to entities of codimension 1 and 2 is then shown in the middle and right picture of that row. The middle row shows an example where an *interior* partition is extended by an *overlap*, but no *ghost* elements are present. This is the model used in YaspGrid (the overlap can be more than one layer of elements thick). The last row shows an example where the *interior* partition is extended by one layer of *ghost* elements. This is the model used in UGGrid and ALUGrid.

The partition type of grid entities is represented in the DUNE grid interface by an enumeration type PartitionType, defined in the file dune/grid/common/gridenums.hh. It can take the values InteriorEntity, BorderEntity, OverlapEntity, FrontEntity, and GhostEntity. Given a grid entity of any codimension, the partition type of that entity with respect to the local process can be obtained using the member method

```
PartitionType Entity::partitionType() const
```

So, for example, the following code snippet will count the number of *interior* elements on the local process on the grid view given in the object gridView:

```
std::size_t counter = 0;
for (const auto& element : elements(gridView))
  if (element.partitionType() == InteriorEntity)
    counter++;
```

The partition type of an entity cannot be set by the user—it is under the exclusive control of the grid implementation.

The DUNE grid interface offers two dedicated methods to obtain information about the *overlap* and *ghost* layers. These are member methods of the GridView

```
int overlapSize(int codim) const
int ghostSize(int codim) const
```

and return the thickness of the *overlap* and *ghost* layers around each subdomain, counted as the number of elements across the transverse direction. The parameter codim exists for historical reasons only, and has no meaning. Unfortunately, a precise definition of the thickness of a layer of grid elements is difficult, and most implementations report an approximation. Also, it is not specified whether a one-element layer of *ghost* elements includes only facet neighbors of *interior* elements, or other neighbors as well. In particular, UGGrid objects claim a *ghost* size of 1, but omit vertex neighbors from the *ghost* layer (Figure 6.4).

Iterating over elements has already been discussed in Chapter 5.2. On a sequential grid, loops like the one in the example above will iterate over all elements of the grid view. In a distributed setting, calling that loop will iterate over all elements of the local process, irrespective of the partition types. To iterate only over certain partition types, there are generalizations of the global `elements` and `vertices` methods that create the appropriate ranges. These generalized methods take a second argument of type `PartitionSet`, which specifies the set of partition types to iterate over. The type `PartitionSet` is defined in the file dune/grid/common/partitionset.hh, and it defines singleton objects `interior`, `overlap`, `ghost`, etc. in the namespace `Dune::Partitions`, which can be combined to sets. So, for example, to iterate over all *interior* elements of a given grid view, write

```
for (const auto& element : elements(gridView,
                                Dune::Partitions::interior))
{
  // Do something with 'element'
}
```

The singletons can be combined to sets by using the addition operator. For example,

```
Partitions::interior + Partitions::border
```

represents the set $\{interior\} \cup \{border\}$. To iterate over all vertices in the *interior* and *border* partitions, write

```
for (const auto& vertex : vertices(gridView,
                             Partitions::interior + Partitions::border))
{
  // Do something with 'vertex'
}
```

The set of all partition types is available as `Partitions::all`. Hence, to iterate over all elements of a subdomain, write

```
for (const auto& element : elements(gridView, Partitions::all))
{
  // Do something with 'element'
}
```

For historical reasons, only the following combinations of partition types are allowed:

interior

interior + border

interior + border + overlap

interior + border + overlap + front

ghost

interior + border + overlap + front + ghost

Since always writing sets as sums can become tedious, the file partitionset.hh also defines a few short cuts. Currently, there is `interiorBorder` for interior + border, and similarly `interiorBorderOverlap`, `interiorBorderOverlapFront`, and `all`. For a complete list check the online documentation of the partitionset.hh file.

It has to be noted that iterating over subsets of the entities of a given subdomain grid does not change the numbering of these entities. In a distributed setting, the

IndexSet objects of Chapter 5.6.1 always enumerate the *all* partition of the local process. Iterating over a true subset, e.g., the *interior* partition, may therefore skip certain indices. Data arrays must always be sized according to the *all* partition, even if some entries, for example the ones for *ghost* elements, are not used.

Most other methods of the grid interface do not change their behavior in a multi-process setting. A notable exception are the methods

```
bool neighbor() const
bool boundary() const
```

of the interface class `Intersection` (see Chapter 5.4). In a single-process setting, `neighbor` returns whether a neighbor element exists across the intersection, and `boundary` returns whether the intersection is on the domain boundary. On the leaf grid view `neighbor` is always the negation of `boundary`, but on level grid views of adaptively refined grids there may be intersections not on the domain boundary that do not have a neighbor.

For a distributed grid, the method `neighbor` returns **true** if there is a neighbor element *on the same process*. Hence it may return **false** even for leaf grid view intersections not on the domain boundary. The boundary method, on the other hand, continues to return **true** only on the *domain* boundary. Hence, for intersections on an inner subdomain boundary, both methods will return **false**.

6.2. Setting up a Distributed Grid

Setting up a distributed grid is easy when using a structured grid implementation: such a grid can be directly constructed as a distributed grid. For an unstructured grid, there are two ways. If the grid is small enough to fit into the master process memory,[3] it can be constructed on the master as described in Chapter 5.7, and then distributed among all processes. Otherwise, the grid needs to be partitioned by external means, and each part has to be fed directly to its corresponding process. This is currently only supported by the ALUGrid grid manager[4] [2].

6.2.1. Distributed Structured Grids

Uniform structured grids need very little memory to store them in. Grid setup, partitioning and repartitioning are therefore operations that do not require a lot of communication. The only implementation in dune-grid of a distributed structured grid is YaspGrid (Chapter 5.10.2), which implements cube grids of any dimension, with axis-aligned element geometries. These can be distributed over an arbitrary number of processes, but the subdomains must have cube form as well.

---

[3]The *master* process is the process with rank (i.e., number) 0.
[4]https://www.dune-project.org/modules/dune-alugrid

Distributed `YaspGrid` objects are set up by means of dedicated constructors. For example, to set up a distributed two-dimensional grid for the unit square with 10×10 elements, call:

```
89    constexpr int dim = 2;
90    FieldVector<double,dim> upper = {1.0, 1.0};
91    std::array<int,dim> elements  = {10, 10};
92    std::bitset<dim> periodic("00");       // Not periodic in either
93                                           // of the two directions
94    int overlapSize = 1;                   // Thickness of the overlap layer
95
96    YaspGrid<dim> yaspGrid(upper, elements,
97                           periodic, overlapSize,
98                           MPI_COMM_WORLD);
```

The first two arguments set the domain size and grid resolution. The last argument is the MPI communicator to be used for the grid. It specifies the set of processes that the grid will be distributed to. The MPI-defined value `MPI_COMM_WORLD` means that all available processes will be used. The argument right before the communicator controls the size of the ***overlap*** layer. Recall from Chapter 6.1 that DUNE allows for overlapping subdomain partitions, and `YaspGrid` is a data structure that actually implements this. The value 1 chosen here means that there will be one layer of ***overlap*** elements around each subdomain. The bitset `periodic` controls periodicity of the grid for each coordinate direction (switched off in this example), which means that corresponding parts of the domain boundary can communicate using the subdomain communication mechanism of Chapter 6.4.1. This is not directly related to distributed computing, but there is currently no `YaspGrid` constructor that allows to specify the MPI communicator without also setting the periodicity bits.

There are more constructors for distributed `YaspGrid` objects, and only the online documentation describes them all. Some of the constructors allow to select custom partitioning algorithms. The default implementation tries to guess a good partitioning by considering the number of processes and the grid dimensions. Alternative partitioning algorithms can be implemented by deriving from the abstract base class `YLoadBalance`, and handing the derived class to the `YaspGrid` constructor. For example, the class `YaspFixedSizePartitioner`, in the file `dune/grid/yaspgrid/partitioner.hh`, allows to explicitly set the requested number of subdomains for each coordinate direction.

6.2.2. Distributed Unstructured Grids

Setting up unstructured distributed grids is more of a challenge, because so much more data is involved. If the initial grid is very large, it may not fit on a single process at all. It then needs to be split up into separate pieces before feeding it to DUNE, and each piece is handed to DUNE directly by the corresponding process. The `ALUGrid` implementation allows this, following a procedure described in [2]. Unfortunately, there is no official support for it in the DUNE grid interface yet.

The grid interface, in form of the GridFactory class described in Chapter 5.7.2, supposes that the initial grid is small enough to fit onto the master process. To construct a distributed grid, the initial grid must be read into the master process first. Then, a partitioning of the elements has to be computed, either by an external partitioning library, or by relying on the grid implementation itself. Finally, a load-balancing step sends each element to the corresponding process in the partition. All Dune grids that support distributed computations are expected to provide this functionality, be it only using very simple partitioning algorithms. The hierarchical grid class implementation must offer a method

```
bool loadBalance()
```

that takes no arguments and returns a boolean. Calling this method for a given grid object should compute an element partitioning, and distribute the grid over the available processes accordingly. The precise way of partitioning is left to the individual grid implementations. For example, UGGrid uses simple coordinate bisection, while ALUGrid uses an approach based on space filling curves [2, 5]. The method returns **true** if the grid has changed at all. For those grid implementations that do not support distributed grids at all, the loadBalance method exists, but is empty and returns **false**. As an example, the following code reads a UGGrid from a Gmsh file and distributes it across all processes in MPI_COMM_WORLD using coordinate bisection:

```
24    using Grid = UGGrid<2>;
25    std::shared_ptr<Grid> grid
26        = GmshReader<Grid>::read("l-shape-refined.msh");
27    grid->loadBalance();
```

The grid partitioning algorithms offered by the grid implementations are frequently too simple. A large variety of such algorithms has been published in the literature [35, 105, 131, 132], each with different strengths and weaknesses. Unfortunately, they all need different sets of parameters, making them difficult to represent in a single interface. Therefore, any grid implementation is allowed to additionally offer non-interface load-balancing methods with additional parameters, which trigger load-balancing algorithms particular to the grid implementation. Both UGGrid and ALUGrid, for example, offer ways to distribute the grids according to arbitrary partition of the leaf grid elements computed by external grid partitioners like ParMETIS [105][5] or Scotch [131, 132].[6]

For UGGrid, given such a partitioning in form of a container with the target process number for each element, the grid can be distributed by calling

```
bool loadBalance(const std::vector<Rank>& targetProcesses,
                 unsigned int fromLevel)
```

on the UGGrid object, where Rank is the type used by MPI for process ranks. The input container targetProcesses is expected to have its size equal to the number

---

[5]http://glaros.dtc.umn.edu/gkhome/metis/parmetis/overview
[6]http://www.labri.fr/perso/pelegrin/scotch

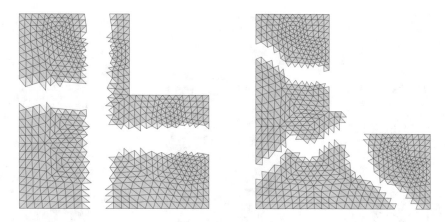

Figure 6.4.: A `UGGrid` object distributed over four processes. Left: the simple partitioner that is built into `UGGrid`; Right: partition created by PARMETIS. The red elements are ***ghost*** elements.

of `AllPartition` elements of the leaf grid. For each such element it must contain the rank that the element is supposed to be sent to. The element numbering is the one implemented by the `MultipleCodimMultipleGeomTypeMapper` class. The parameter `fromLevel` has only meaning when distributing a grid that has been refined before (which is explicitly allowed; see the following section). In such a situation, an entire grid hierarchy needs to be distributed. The number `fromLevel` is the lowest (i.e., coarsest) level to redistribute in the element refinement hierarchy. All elements on lower levels will stay where they are. This feature is typically used to keep the coarsest levels of a multigrid hierarchy on a small number of processes, because if very coarse grids are distributed, the communication costs typically outweigh the savings in computation time.

As one way to compute such partitions `dune-grid` offers a simple interface to the PARMETIS grid partitioner (in the file `parmetisgridpartitioner.hh` from the folder `dune/grid/utility/`). The following code snippet loads a structured `UGGrid` object from a GMSH file and distributes it using PARMETIS:

```
49    using Grid = UGGrid<2>;
50    std::shared_ptr<Grid> grid
51        = GmshReader<Grid>::read("l-shape-refined.msh");
52
53    using GridView = Grid::LeafGridView;
54    const GridView gridView = grid->leafGridView();
55
56    // Create initial partitioning using ParMETIS
57    std::vector<unsigned int> part
58        = ParMetisGridPartitioner<GridView>::partition(gridView, mpiHelper);
```

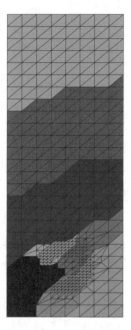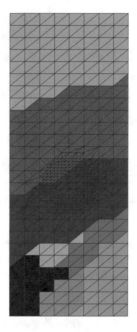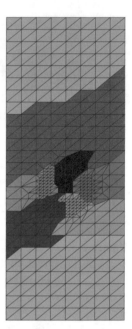

Figure 6.5.: Repartitioning an adaptively refined grid to distribute the load evenly among the processes. Colors show the process numbers.

```
59
60      // Distribute the grid using the partitioning computed by ParMETIS
61      grid->loadBalance(part, 0);
```

The result of this operation can be seen in Figure 6.4. The left image shows the result using the built-in `UGGrid` partitioner, and the right one shows the result produced by PARMETIS.

6.3. Dynamic Load-Balancing

In the previous section grids have been distributed to a set of processes right after they were created. However, one may also want to redistribute an already distributed grid later in the course of a simulation. This is called *dynamic load-balancing*. Figure 6.5 shows an example: the adaptively refined grid on the left has been partitioned such that all processes receive roughly the same amount of degrees of freedom. However, after a few time steps, the refined area has moved to another position, and the partition becomes imbalanced (center). Dynamic load-balancing produces an equilibrated work load again (right).

Triggering a grid load-balancing using the DUNE grid interface differs little from how the initial distribution was set up in the previous section. All load-balancing methods can be called at (almost) any moment to adapt the distribution to the current workload. The semantics of these methods do not change at all. What does change is the use of the PARMETIS interface, which depends on whether an initial distribution is computed, or whether an existing distribution is modified. The following code snipped redistributes an already distributed UGGrid object:

```
float transVsComm=1000;  // Quality parameter,
                         // consult ParMETIS documentation
part = ParMetisGridPartitioner<GridView>::repartition(gridView,
                                                      mpiHelper,
                                                      transVsComm);
grid->loadBalance(part, 0);
```

The variable part is the std::vector of target process ranks already used in the previous section. The floating-point argument transVsComm balances the transportation cost vs. the communication cost of the new partition, and is described in the PARMETIS documentation. Inserting these three lines into the grid adaptivity example of Chapter 5.9 allows to create the grids shown in Figure 6.5.

Dynamic load-balancing gets more interesting when there is numerical data attached to the grid. Such data will typically need to be transported along when parts of the grid move to other processes. As explained already, DUNE leaves the grid transport to the grid implementations. The user may be allowed to say which element is to go where, but the actual element transport together with the corresponding book-keeping is completely hidden. This is convenient, but it makes moving numerical data along with the grid difficult. The problem here is again that grid data structures and numerical data are kept strictly separate:

1. The grids do not know anything about the way user data is stored. For each grid entity there may be an arbitrary number of data fields, of different types, and stored in different containers. The grid implementation cannot be expected to know about this.

2. Each data item may be a custom data type that may be nontrivial to serialize (i.e., converted to a byte stream).

The DUNE grid interface solves this problem by using a call-back mechanism. For a grid redistribution with a particular data payload, the user has to write a specific call-back class, called a DataHandle. The class implements methods that collect and serialize data on the sender side, and distribute it again on the receiving side. The DataHandle knows about the data structures used to store the numerical data, and its methods get called by the grid implementation during the grid transfer. Transferring data along with grid entities is then a three-step process:

Gathering: Data pertaining to entities that will be sent to different processes is gathered from its containers. The entire data to be sent from one process is serialized, and stored in a special memory location called the *message buffer*. The message buffer is provided by the grid implementation.

Communication: The entities and the data in the message buffer are sent to the receiving process.

Scattering: The receiving process places data received from other processes in the correct containers.

As an additional complication observe that redistributing grids must necessarily change the index sets, because entities get added and/or removed on individual processes. Therefore, data that is kept in arrays becomes invalid even if it is associated to entities that are not affected by the load-balancing. The remedy is the same as for grid adaptivity in Chapter 5.9.2: Before the load-balancing, all data needs to be copied into associative containers that can be addressed by an *id*. The DataHandle operates only on the associative containers, and user code must copy the data back into arrays after the load-balancing procedure has terminated, using the updated index sets.

Implementations of the DUNE grid interface are expected to handle the communication step automatically for the user. However, they cannot handle the gathering and scattering, because it depends on the user data layout. For this, the user has to write the data handle class, the interface of which is specified by the CommDataHandleIF class (in dune/grid/common/datahandleif.hh):[7]

```
template<class DataHandleImp, class DataTypeImp>
class CommDataHandleIF
```

Actual implementations must derive from that class, giving their own type as the first template argument DataHandleImp.[8] The second template parameter DataTypeImp is the C++ type of the data to be transported, if there is one such fixed type. For example, if the numerical data consists of one **double** per vertex, then DataTypeImp should be **double**. If it consists of a **double** and a complex number, then

```
std::tuple<double,std::complex<double> >
```

is a suitable argument. If, on the other hand, the data type is not the same for all entities, or if it is not known at compile time, then **unsigned char** is usually the best choice for the DataTypeImp. Note that all size information is then computed in units of **unsigned char**.

To use a data handle implementation for load-balancing, each grid implementation has a load-balancing method that accepts a data handle object:

```
template<class DataHandle>
bool loadBalance(DataHandle& data)
```

---

[7]The name of the interface as such is DataHandle, and it will be referred to by that name in the following.

[8]This technique is known as the Curiously Recurring Template Pattern (CRTP) [1]. It is used here mainly for historical reasons.

bool	contains(**int** dim, **int** codim) **const**	
	Whether data for the given codimension should be communicated	
template<**class** Entity>		
std::size_t	size(**const** Entity& entity) **const**	
	How many objects of type DataTypeImp have to be sent for the given entity	
bool	fixedSize(**int** dim, **int** codim) **const**	
	Whether size of data per entity of given codimension is a constant	
template<**class** MessageBuffer, **class** Entity>		
void	gather(MessageBuffer& buffer, **const** Entity& entity) **const**	
	Pack data from user to message buffer	
template<**class** MessageBuffer, **class** Entity>		
void	scatter(MessageBuffer& buffer, **const** Entity& entity, size_t n)	
	Unpack data from message buffer to user	

Table 6.1.: Interface methods of the CommDataHandleIF class

For such a method to work, a proper implementation of a DataHandle class requires a certain number of public methods, which are listed in Table 6.1. All these methods exist to be called by the grid implementation, and never by the user herself. In addition to the interface methods, the data handle class must have constructors or setup methods that tell it about the location and structure of the data that is to be transported.

The first three interface methods are used to tell the grid implementation about what data needs to be transported along with the grid entities. The method

```
bool contains(int dim, int codim) const
```

must return **true** for the grid to transport data for all entities of codimension codim. When calling this method, the grid implementation always provides the grid dimension in the first argument. The method

```
template<class Entity>
std::size_t size(Entity& entity) const
```

must return the number of data items (of type DataTypeImp) that will be transported with this entity. This method will only be called on the sender side, and only for those entities of a codimension for which the contains method has specified that entities of this codimension carry data. It allows the grid implementation to compute the sizes of the required message buffers.

In principle, each entity can be accompanied by a different number of data items. However, in many cases the number is the same for all entities of a given codimension. To make this important case known to the grid implementation, there is the method

```
bool fixedSize(int dim, int codim) const
```

It is called once by the grid implementation before starting to traverse the grid to gather the numerical data. The first parameter is always the dimension of the grid. If fixedSize returns **true** for a given codimension codim, then all entities of that codimension are expected to have the same number of data items (given by the return value of the size method for any one of these entities). If fixedSize returns **false**, then size is called for each entity before the actual data gathering, which may lead to a small performance loss.

The two central methods of the DataHandle interface, however, are the methods gather and scatter. These transfer the numerical data from the (associative) container(s) used to store the numerical data while the index sets change to the message buffer and back, respectively. The gather method receives a DUNE grid entity and a non-**const** reference to the message buffer:

```
template<class MessageBuffer, class Entity>
void gather(MessageBuffer& messageBuffer, const Entity& entity) const
```

The message buffer is a small container-like object with entries of type DataTypeImp (see below).

The gather method is called before the actual grid change for each entity that will be sent to another process, and that has a codimension for which the contains method returns **true**. The method must find the data attached to the entity, and write it into the message buffer. After all relevant user data has been gathered this way, the grid will sent the content of the message buffer along with the entity.

Conversely, the scatter method is called *after* the grid entities have been transported:

```
template<class MessageBuffer, class Entity>
void scatter(MessageBuffer& messageBuffer, const Entity& entity, size_t n)
```

It is called for those entities for which the contains methods returns **true**, and which have travelled from one process to another as part of the grid redistribution. It is given a message buffer and an entity, and is expected to read the entity's data from the message buffer, and write it into the associative container(s) used to store the numerical data. Since the number of data items may vary from entity to entity, the scatter method is not expected to know how much data to read. Instead, the number of data items is handed to the method by the grid implementation through the third method parameter n.

The message buffer is of unspecified type, and each grid implementation has its own. Following the duck-typing paradigm (Chapter 4.4.2), the interface of the message buffer is not enforced by the compiler. It is very thin, and only consists of the two methods

```
void write(const DataTypeImp& value)
```

and

```
void read(DataTypeImp& value)
```

where `DataTypeImp` is the second type passed to the `CommDataHandleIF` base class. The `write` method must be used by `gather` to write data to the message buffer. It works similarly to an output iterator: when `gather` is called, the message buffer is empty, has the correct length, and there is an internal iterator pointing to the first entry. Calling `write` writes the given value to that entry, and advances the iterator. For example, suppose that an entity needs to transport four values stored in an array called `values`. These four values can be written to the message buffer by

```
for (const auto& v : values)
  messageBuffer.write(v);
```

The data is copied byte-wise into the message buffer. This is a rather crude form of serialization: It works nicely for the standard number types, but it may fail for more exotic types of data. Future versions of `dune-grid` may have an improved interface here.

To read data from the message buffer the `scatter` method must use the method `read`. When `gather` is called for an entity, the buffer is filled with n entries of type `DataTypeImp`, and the internal iterator is pointing to the first entry. A call to `read` copies the current entry into the argument `value`, and advances the iterator by one `DataTypeImp`. The `gather` method is expected to read precisely the number of entries given by the parameter n. Note that the `read` method changes the message buffer, because it advances the iterator. Therefore, `read` is not a **const** method, and the `messageBuffer` argument of the `scatter` method is a non-**const** reference.

The following example implementation of a `DataHandle` class makes the previous explanations more concrete. It is taken from the complete example in Chapter 6.7:

```
228  template<class Grid, class AssociativeContainer>
229  struct LBVertexDataHandle
230    : public CommDataHandleIF<LBVertexDataHandle<Grid, AssociativeContainer>,
231                              typename AssociativeContainer::mapped_type>
232  {
233    LBVertexDataHandle(const std::shared_ptr<Grid>& grid,
234                       AssociativeContainer& dataContainer)
235      : idSet_(grid->localIdSet()), dataContainer_(dataContainer)
236    {}
237
238    bool contains(int dim, int codim) const
239    {
240      assert(dim == Grid::dimension);
241      return (codim == dim);    // Only vertices have data
242    }
```

```
243
244    bool fixedSize(int dim, int codim) const
245    {
246      return true;       // All vertices carry the same number of data items
247    }
248
249    template<class Entity>
250    size_t size(const Entity& entity) const
251    {
252      return 1;          // One data item per vertex
253    }
254
255    template<class MessageBuffer, class Entity>
256    void gather(MessageBuffer& buffer, const Entity& entity) const
257    {
258      auto id = idSet_.id(entity);
259      buffer.write(dataContainer_[id]);
260    }
261
262    template<class MessageBuffer, class Entity>
263    void scatter(MessageBuffer& buffer, const Entity& entity, size_t n)
264    {
265      assert(n==1);  // This data handle implementation
266                     // transfers only one data item.
267      auto id = idSet_.id(entity);
268      buffer.read(dataContainer_[id]);
269    }
270
271  private:
272    const typename Grid::LocalIdSet& idSet_;
273    AssociativeContainer& dataContainer_;
274  };
```

This class handles the transfer of one data element of unspecified type per grid vertex. The data is handed to the class in an associative container, the type of which is determined by the template parameter AssociativeContainer. The class keeps a non-**const** reference to the container, and it can therefore read from it and write to it. The method contains specifies that there is vertex data only, fixedSize tells that each vertex has the same number of data items, and size tells that there is exactly one data item per vertex. The method gather (in Lines 255–260) takes the data of the vertex given in the entity parameter, and writes it into the message buffer. Conversely, the scatter method is called on the processes that receive entities, and the message buffer will contain the corresponding data. The code of that method reads the data from the message buffer, and copies it into the container.

To see the data-handle class in action, suppose that there is an array object dataVector of type std::vector<**double**> that contains one item of data for each vertex of the current leaf grid view. The following code copies this data into an

212

associative container, triggers the grid redistribution, and recuperates the data:

```
380    // Copy vertex data into associative container
381    using PersistentContainer = std::map<Grid::LocalIdSet::IdType, double>;
382    PersistentContainer persistentContainer;
383    const auto& idSet = grid->localIdSet();
384
385    for (const auto& vertex : vertices(gridView))
386      persistentContainer[idSet.id(vertex)]
387          = dataVector[gridView.indexSet().index(vertex)];


391    // Distribute the grid and the data
392    LBVertexDataHandle<Grid, PersistentContainer>
393                                    dataHandle(grid, persistentContainer);
394    grid->loadBalance(dataHandle);


398    // Get gridView again after load-balancing, to make sure it is up-to-date
399    gridView = grid->leafGridView();
400
401    // Copy data back into the array
402    dataVector.resize(gridView.size(dim));
403
404    for (const auto& vertex : vertices(gridView))
405      dataVector[gridView.indexSet().index(vertex)]
406          = persistentContainer[idSet.id(vertex)];
```

This code fragment is also taken from the complete example in Chapter 6.7, where it can be seen in context.

6.4. Communication

The different processes involved in a distributed simulation need to communicate with each other. In a finite element or finite volume computation there are two types of communication. During *subdomain communication*, processes that hold a given subdomain exchange information with the processes that hold adjacent subdomains. In contrast, *collective communication* is sometimes necessary, where all processes talk to all others, or to the master process.

6.4.1. Subdomain Communication

As explained in Chapter 6.1, in distributed computing each process typically stores a subdomain, i.e., a connected part of the complete grid. The subdomains may overlap with their neighbors, and elements in such overlap regions then exist on multiple processes. The elements are of **_interior_** type on one process, and of **_overlap_** or **_ghost_** type on the others.

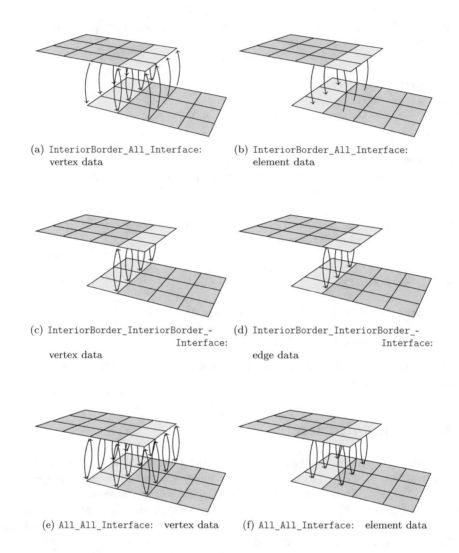

(a) InteriorBorder_All_Interface: vertex data

(b) InteriorBorder_All_Interface: element data

(c) InteriorBorder_InteriorBorder_- Interface: vertex data

(d) InteriorBorder_InteriorBorder_- Interface: edge data

(e) All_All_Interface: vertex data

(f) All_All_Interface: element data

Figure 6.6.: Communication patterns between two subdomains for different interfaces and entity codimensions. Grey elements are ***interior***, and yellow elements are ***ghost***. Setting the CommunicationDirection argument to BackwardCommunication reverses the directions of all arrows.

As part of any numerical algorithm for the solution of a PDE, the subdomain grids will have to communicate with each other. This means that data attached to one subdomain grid needs to be sent to other processes. The fundamental assumption of the DUNE grid interface is that data only needs to be sent for entities that also exist in other subdomains. So, for example, in Figure 6.2 only data on elements that are yellow or green on at least one process can partake in communication. This is justified for virtually all classical one-level algorithms [148]. Distributed two-grid methods, where there is also communication between coarser and finer grids for the same domain, are not directly supported by DUNE.

Subdomain communication is provided by the grid implementations. The grid interface allows to specify the entity sets that need to send and receive data, and a call-back mechanism to get the numerical data to and from their containers. Communication is triggered by a single method `communicate`, which is a member method of the `GridView` interface class. The complete signature is

```
template<class DataHandleImp, class DataType>
void communicate(CommDataHandleIF<DataHandleImp, DataType>& dataHandle,
                 InterfaceType ifType,
                 CommunicationDirection dir) const
```

The first parameter of this method is the call-back class that transfers the numerical data from its containers into a message buffer and back. Observe that the interface class is the same as the one that has been used for data transfer during load-balancing! The second parameter specifies the sets of entities for which to communicate data, and the third one selects the communication direction. Communication triggered by this method is synchronous, i.e., process execution blocks until the entire communication has completed. More efficient asynchronous communication mechanisms have been implemented in ALUGrid [2], and are likely to appear in future releases of dune-grid.

DUNE calls the sets of entities that exchange data *communication interfaces*. Remember from Chapter 6.1 that the entities of a subdomain are classified into the five partition types ***interior***, ***border***, ***overlap***, ***front***, and ***ghost***, and that combinations of such types are called *partition sets*. A communication interface is simply a pair of partition sets P_1 and P_2. In the communication step, process i will send a message to another process j if for any valid codimension c, there exist entities of codimension c that have a partition type in P_1 on process i, and a partition type in P_2 on process j. The message from process i to process j contains data from every entity on process i of a partition type in P_1 that exists on process j and has a partition type in P_2 there. This is illustrated in Figure 6.6.

In the DUNE grid interface, the communication interfaces are implemented as an enumeration type `InterfaceType`, defined in the file dune/grid/common/gridenums.hh.

It can take the values

```
InteriorBorder_InteriorBorder_Interface,
InteriorBorder_All_Interface,
Overlap_OverlapFront_Interface,
Overlap_All_Interface,
All_All_Interface.
```

When the third parameter of the `GridView::communicate` method is set to the value `ForwardCommunication`, the different `InterfaceType` values implement the communication patterns as shown in Figure 6.6. For more choices, the communication direction can be reversed by setting the third parameter of `GridView::communicate` to `BackwardCommunication`.

Specifying the actual data to be sent reuses the call-back idea and the interface class `CommDataHandleIF` already employed for load-balancing. The `communicate` method is given a data handle class that copies the data to and from a message buffer. Unlike in the case of load-balancing, however, the grids themselves do not change. Therefore, the numbering of entities stays intact during communication, and there is no need to copy the data into associative arrays before giving it to the data handle. As an example, we show here a full data handle implementation for vertex data communication, taken from the example in Chapter 6.7:

```
279  template<class GridView, class Vector>
280  struct VertexDataUpdate
281    : public Dune::CommDataHandleIF<VertexDataUpdate<GridView,Vector>,
282                                     typename Vector::value_type>
283  {
284    using DataType = typename Vector::value_type;
285
286    // Constructor
287    VertexDataUpdate(const GridView& gridView,
288                     const Vector& userDataSend,
289                     Vector& userDataReceive)
290      : gridView_(gridView),
291        userDataSend_(userDataSend),
292        userDataReceive_(userDataReceive)
293    {}
294
295    // True if data for this codim should be communicated
296    bool contains(int dim, int codim) const
297    {
298      return (codim == dim); // Only vertices have data
299    }
300
301    // True if data size per entity of given codim is constant
302    bool fixedSize(int dim, int codim) const
303    {
304      return true;           // All vertices carry the same number of data items
305    }
```

216

```
306
307    // How many objects of type DataType have to be sent for a given entity
308    template<class Entity>
309    size_t size(const Entity& e) const
310    {
311      return 1;                    // One data item per vertex
312    }
313
314    // Pack user data into message buffer
315    template<class MessageBuffer, class Entity>
316    void gather(MessageBuffer& buffer, const Entity& entity) const
317    {
318      auto index = gridView_.indexSet().index(entity);
319      buffer.write(userDataSend_[index]);
320    }
321
322    // Unpack user data from message buffer
323    template<class MessageBuffer, class Entity>
324    void scatter(MessageBuffer& buffer, const Entity& entity, size_t n)
325    {
326      assert(n==1);
327      DataType x;
328      buffer.read(x);
329
330      userDataReceive_[gridView_.indexSet().index(entity)] += x;
331    }
332             •
333 private:
334    const GridView gridView_;
335    const Vector& userDataSend_;
336    Vector& userDataReceive_;
337 };
```

There are only minor differences compared to the load-balancing data handle implementation of Chapter 6.3. The methods contains, fixedSize, and size in Lines 296–312 are identical, because both data handle classes deal with single items of vertex data. On the other hand, the data handle operates directly on arrays of numerical data, because the grid does not change during communication, and IndexSet objects can therefore be used. The constructor therefore receives data arrays rather than an associative container as in Chapter 6.3.

The gather method does the same as the gather method of the load-balancing data handle: it takes the single data item of the current entity (which, in this example, is a vertex), and writes it into the message buffer. The scatter method is different, however. Unlike for the load-balancing case the method reads the data item from the message buffer and *adds* it to the corresponding value there. That way, after the call to communicate, each value associated to a vertex that exists on more than one process is the sum of all its values on the different processes. Chapter 6.7 will show that this is exactly what the application requires. Please also refer to that chapter to see how the data handle class is used in code.

6.4.2. Collective Communication

While the data exchange mechanism of the previous section covers most communication patterns occurring in numerical algorithms, sometimes nonlocal communication is necessary. Frequently this means simple things like sending a set of solver parameters to all processes, or a global sum needed to compute a scalar product. Communication that involves all processes together is called *collective communication*.

Collective communication does not directly involve the grid, and therefore any general-purpose distributed-computing library can be used. Nowadays this will most likely be an implementation of MPI. However, hard-coding the use of MPI directly forces the user to have it installed at all times, whereas in many cases it is convenient if the code can also be used if MPI is not installed.

For these reasons the `dune-grid` module offers an interface to commonly needed collective communication functionality. The interface is heavily inspired by MPI, but there may be other implementations for it. In particular, there exists a sequential implementation, which is used when MPI is not installed. This allows to write programs that use a communication layer like MPI, but that gracefully degenerate to a working single-process implementation if no such library is installed at build time.

This abstraction layer of `dune-grid` comes in form of a single class

```
template<class Communicator>
class CollectiveCommunication
```

Different implementations of the collective communication mechanism are given as specializations of this class, and they are selected by the `Communicator` template argument, which should be the communicator type used by the message-passing implementation. For MPI this will be `MPI_Comm`, which is a type defined by the MPI implementation. For sequential codes there is the dummy communicator `No_Comm`, defined in the file `dune/common/parallel/communication.hh`. It is assumed that other message-passing systems have a concept that is equivalent to MPI communicators, which will in principle allow to specialize the `CollectiveCommunication` class for them.

The `CollectiveCommunication` class has no public constructors. An object of type `CollectiveCommunication` is obtained by calling the method

```
const CollectiveCommunication& comm() const
```

provided by each DUNE grid and each DUNE grid view. Grids that do not implement distributed features return the sequential implementation of `CollectiveCommunication`, i.e., `CollectiveCommunication<No_Comm>`, which mimics what the MPI equivalent would do when run on a single process. It is also possible to get a `Collective-Communication` object without a grid, by using the `MPIHelper` class described below in Chapter 6.5.

218

```
template<class T>
         int   broadcast(T* inout, int len, int root) const
               Distribute array from the root process to all other processes
template<class T>
         int   gather(T* in, T* out, int len, int root) const
               Gather arrays on root process
template<class T>
         int   gatherv(T* in, int sendlen,
                       T* out, int* recvlen,
                       int* displ, int root) const
               Gather arrays of variable size on root process
template<class T>
         int   scatter(T* send, T* recv, int len, int root) const
               Scatter array from root process to all others
template<class T>
         int   scatterv(T* send, int* sendlen, int* displ,
                        T* recv, int recvlen, int root) const
               Scatter arrays of variable length from root process to all others
template<class T>
         int   allgather(T* sbuf, int count, T* rbuf) const
               Gather data from all processes and distribute it to all processes
template<class T>
         int   allgatherv(T* in, int sendlen,
                          T* out, int* recvlen, int* displ) const
               Gather data of variable length from all processes and distribute it
               to all processes
template<class BinaryFunction, class T>
         int   allreduce(T* inout, int len) const
               Componentwise reduction of the data in inout with the operation
               BinaryFunction. The result is available on all processes.
template<class BinaryFunction, class T>
         int   allreduce(T* in, T* out, int len) const
               Componentwise reduction of the data in in with the operation
               BinaryFunction. The result is available on all processes.
         int   barrier() const
               Wait until all processes have arrived at this point in the program
```

Table 6.2.: Some (but not all) communication methods of the CollectiveCommunication class. The type T must be a number type supported by MPI. BinaryFunction must be one of std::plus, std::multiplies, Dune::Max, or Dune::Min.

The member methods of `CollectiveCommunication` encapsulate a set of the most important MPI communication routines. First of all, there are the methods

```
int rank() const
int size() const
```

which provide the local process rank and the total number of processes in the communicator. Also, the method

```
operator MPI_Comm() const
```

allows to cast the `CollectiveCommunication` object into an object of type `MPI_Comm` (or `No_Comm` if no MPI is installed), and returns the MPI communicator currently in use:

```
MPI_Comm communicator = gridView.comm();
```

Then, there is a set of arithmetic methods. For example,

```
template<class T>
T max(T& in) const
```

computes the maximum of the argument over all processes, and returns the result in every process. Similarly, the method

```
template<class T>
int max(T* inout, int len) const
```

computes the maximum over all processes for each component of the array `inout` of length `len`, and returns the result in every process. Corresponding methods exist for minima, sums, and products. Check the online documentation for an up-to-date list. The methods all operate on an unspecified number type T. It is assumed that T is a type that can be used in MPI collective communication methods.

Then, there are general global communication methods, listed in Table 6.2. Most of them concern sending or receiving a set of data from or to all processes. For example,

```
template<class T>
int broadcast(T* inout, int len, int root) const
```

sends the content of the array `inout` from the process with number `root` to all other processes. The method

```
template<class T>
int gather(const T* in, T* out, int len, int root) const
```

in turn, collects data from all process on the process with the number given in `root`. These are all thin wrappers around MPI functions with a more convenient syntax, and we refer the reader to the MPI documentation for details. The **int** values returned by the methods are MPI error codes. If all has gone well, the number 0 (which means MPI_SUCCESS) is returned. Nonzero return values signify errors.

6.5. MPI Setup with the `MPIHelper` Class

The MPI message passing system needs to run a setup routine before the start of the actual program. Each C++ program that uses MPI therefore needs to start with the line

```
MPI_Init(&argc, &argv);
```

which does this setup. Likewise, before terminating the program,

```
MPI_Finalize();
```

needs to be called, to properly shut down MPI. Of course, if code contains these lines, MPI has to be installed for the code to build.

The previous section has argued that it is desirable to have distributed code that can build and run even without MPI. It has described a communication mechanism that allows MPI programs to at least run sequentially if MPI is not installed at build time. For that mechanism to work comprehensively it needs to cover MPI startup. DUNE has the `MPIHelper` singleton for this, which is provided in the file dune/common/parallel/mpihelper.hh. Its constructor performs the setup of MPI if MPI is installed, but does nothing if it is not. The constructor is called the first time the singleton instance is requested, which is done by calling

```
MPIHelper::instance(argc, argv);
```

At build time, this call will simply translate to

```
MPI_Init(&argc, &argv);
```

if MPI is installed. The destructor of `MPIHelper` contains the call to `MPI_Finalize`, so there is no need for explicit shutdown code. If MPI is not installed, then the constructor and destructor are empty. Consequently, if

```
MPIHelper::instance(argc,argv);
```

is the first line of the `main` method then the program will build and run in sequential and parallel situations.[9]

The `MPIHelper` can do more, but for this the actual singleton object has to be acquired:

```
const MPIHelper& mpiHelper = MPIHelper::instance(argc,argv);
```

The `mpiHelper` object has methods

```
int rank() const
int size() const
```

---

[9]As a rule, any DUNE program should start with this call to the `MPIHelper` instance, even if it is purely sequential. The reason is that when DUNE is built with MPI, some parts of DUNE do MPI calls even if run on a single process, which fails if `MPI_Init` has not been called.

which allow to determine the rank of the current process, and the total number of processes. These numbers always refer to MPI_COMM_WORLD—beware that this may not be the communicator used by the grids.[10] If there is no MPI installed, rank and size return 0 and 1, respectively.

The MPIHelper also provides its communicator itself. The type is available as MPIHelper::MPICommunicator. The object can be obtained through the method

```
static MPICommunicator getCommunicator()
```

which returns either MPI_COMM_WORLD or an object of type No_Comm. Similarly, the method

```
static MPICommunicator getLocalCommunicator()
```

returns either MPI_COMM_SELF or an object of type No_Comm. Finally, MPIHelper hands out a CollectiveCommunication object for its communicator via the method

```
static CollectiveCommunication getCollectiveCommunication()
```

which provides the full global communication infrastructure described in Chapter 6.4.2.

6.6. Writing Distributed Grids to VTK Files

Writing grids and associated vertex and element data has been discussed at length in Chapter 5.8. Very little of this changes when the grid is distributed. As before, writing a grid to a VTK file has the form

```
VTKWriter<GridView> vtkWriter(gridView);
// Attach numerical data to 'vtkWriter' object
// ...
vtkWriter.write("filename");
```

However, the behavior of the write method changes slightly. Calling write from a process with rank k will only write the grid on that local process (more precisely: the InteriorBorder_Partition). It will also prepend the string s####-p####- to the file name, where the first block of # is replaced with the total number of processes, and the second block with the process rank k (both with leading zeros). As in the sequential case, the complete filename including prefixes and suffixes is returned as the return value of the write method.

Calling write for a distributed grid will leave the user with a separate file for each subdomain. These files are recognized as being parts of a larger whole by an additional file, the so-called *parallel file*. This file is also written by the write method if there is more than one process calling write, and it has the file extension .pvtu. It mainly keeps the list of all the .vtu files that make up the entire grid, and a list of data fields

---

[10]See Chapter 6.4.2 on how to get the communicator of particular grids, and the corresponding rank and size information.

expected to be contained in each subdomain file. When loading the *parallel file* into PARAVIEW, all subdomain files referenced there are loaded as well, and the entire set is treated as a single grid object.

If the number of processes is large, the amount of files can be overwhelming. To keep a little bit of order there is a second writing method, which places the different types of files into different directories. This method is called `pwrite` (for "path write"):

```
std::string pwrite(const std::string& name,
                   const std::string& path,
                   const std::string & extendpath,
                   VTK::OutputType type = VTK::ascii)
```

In addition to the two parameters `name` and `type` known from the `write` method, it accepts two additional string arguments `path` and `extendpath`. The first gives the name of the directory where the `.pvtu` file is to be placed. (As before, the name string itself is not supposed to contain a directory part, and no file type extension either.) The `extendpath` contains the directory path (relative to `path`!) where the individual subdomain files are written to. This second string can contain the rank number, and hence it is possible not only to separate the *parallel file* from the subdomain files, but also the subdomain files for the different ranks from each other. This can be useful when writing time-dependent data, which can lead to a lot of files.

To use the `pwrite`-facilities for the writing of time-dependent data, remember that the VTKSequenceWriter class presented in Chapter 5.8.2 has a separate constructor

```
VTKSequenceWriter(std::shared_ptr<VTKWriter<GridView> > vtkWriter,
                  const std::string& name,
                  const std::string& path,
                  const std::string& extendpath)
```

The `vtkWriter` parameter is the VTKWriter object that will do the writing of the individual data files. The three string parameter have the same meaning as in the `pwrite` method. When the `write` method of the VTKSequenceWriter class is called, it will use the names and paths given in the string parameters: The `.pvtu` file goes into `path`, and the `.vtu` files go into `path/extendpath`. The `.pvd` file is always written to the current directory, hence it is possible to place `.pvtu`, `.vtu`, and `.pvd` files all in separate locations.

6.7. Example: The Poisson Equation on a Distributed Grid

We now combine the features discussed in this chapter in a complete example. As these features are all of a rather low-level nature, the code is a bit wordy. Higher-level interfaces for distributed computations are available in the `dune-istl` and `dune-pdelab` modules, but beyond the scope of this book.

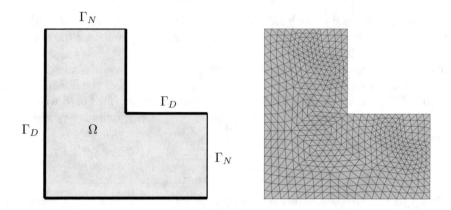

Figure 6.7.: Boundary conditions and grid for the distributed Poisson example

The example program is an extension of the Poisson solver of the introductory Chapter 3. Indeed, both share a lot of common code, and therefore not all parts of the new program will be presented here. As usual, the entire example code is contained in a single file, which appears in its entirety in Appendix B.6. Readers of an electronic copy of this document will also find the file by clicking on the icon in the margin.

The program solves exactly the same problem as the one from Chapter 3.3, namely the Poisson equation

$$-\Delta u = -5$$

on an L-shaped domain shown in Figure 6.7, with boundary conditions

$$u = \begin{cases} 0 & \text{on } \{0\} \times [0,1] \cup [0,1] \times \{0\}, \\ 0.5 & \text{on } \{0.5\} \times [0.5,1] \cup [0.5,1] \times \{0.5\}, \end{cases} \tag{6.1}$$

$$\langle \nabla u, \mathbf{n} \rangle = 0 \quad \text{elsewhere.}$$

Unlike the original code, however, the example here distributes the grid across several processes, and then implements a distributed conjugate gradient (CG) method, with a Jacobi preconditioner. While this is not a very powerful preconditioner for this kind of problems, it is easy to implement using only the low-level features of Dune that have been covered so far. We will use a non-overlapping decomposition, and implement the algorithm such that it produces exactly the same iterates no matter how many processes are used.

The grid, also shown in Figure 6.7, is a twice uniformly refined version of the one in Chapter 3.3, because the original grid is too coarse to allow for a decent distribution across several processes. The Gmsh grid file is embedded in the electronic version of this text, and is available via the icon in the margin.

We discretize the Poisson equation using first-order Lagrange finite elements. Using elements of higher order could be implemented similarly, but would make the code more technical. The example program uses features of the `dune-functions` module for the problem assembly just as the Poisson example of Chapter 3 does, but it implements the actual distributed equation-solving without help from `dune-functions`.

6.7.1. Setting up the Distributed Algebraic Problem

We start by discussing how the distributed algebraic problem is set up. This involves loading and distributing the grid, assembling the subdomain algebraic problems in parallel, and setting up the correct boundary conditions on each process. All this happens in the `main` method, which we step through now. Large parts of it will be familiar from Chapter 3. The method starts with the setup of the `MPIHelper` instance:

```
341  int main(int argc, char *argv[])
342  {
343    // Set up MPI
344    const MPIHelper& mpiHelper = MPIHelper::instance(argc, argv);
```

We store a reference to the `MPIHelper` instance, and use it later to get the process rank. Then, we construct a `UGGrid` object and load the grid shown in Figure 6.7 from the file `1-shape-refined.msh`:

```
352    constexpr int dim = 2;
353    using Grid = UGGrid<dim>;
354    using GridView = Grid::LeafGridView;
355
356    std::shared_ptr<Grid> grid = GmshReader<Grid>::read("1-shape-refined.msh");
357    auto gridView = grid->leafGridView();
```

This looks like the grid setup in a sequential setting; however, internally the `GmshReader` object only loads the grid into the `UGGrid` object on process 0, and leaves all other processes empty. The grid then needs to be distributed explicitly.

At this point the example becomes a bit contrived. To distribute the grid one could simply call

```
grid->loadBalance();
```

and be done with it. However, we want the example to also demonstrate load-balancing with data transport. For this purpose, we construct an initial iterate for the CG solver on the master process, and then distribute it along with the grid. The initial iterate is the pointwise evaluation of the function

$$u_0 : \mathbb{R}^2 \to \mathbb{R}, \qquad u_0(x_0, x_1) = \min\{x_0, x_1\},$$

which complies with the Dirichlet boundary conditions (6.1). The following code block sets up this function and samples it on the vertices on the master process:

```
361    std::vector<double> dataVector;
362
363    if (mpiHelper.rank()==0)
```

```
364    {
365        // The initial iterate as a function
366        auto initialIterate = [](auto x){return std::min(x[0],x[1]);};
367
368        // Sample on the grid vertices
369        dataVector.resize(gridView.size(dim));
370        for (const auto& vertex : vertices(gridView,
371                                        Dune::Partitions::interiorBorder))
372        {
373            auto index = gridView.indexSet().index(vertex);
374            dataVector[index] = initialIterate(vertex.geometry().corner(0));
375        }
376    }
```

The grid and the vertex data in the dataVector array can now be distributed across the available processes. For this, it first needs to be copied into an associative container, because the indices used to address arrays will change during the load-balancing (see Chapter 6.3):

```
380    // Copy vertex data into associative container
381    using PersistentContainer = std::map<Grid::LocalIdSet::IdType, double>;
382    PersistentContainer persistentContainer;
383    const auto& idSet = grid->localIdSet();
384
385    for (const auto& vertex : vertices(gridView))
386        persistentContainer[idSet.id(vertex)]
387            = dataVector[gridView.indexSet().index(vertex)];
```

Next is the actual load-balancing:

```
391    // Distribute the grid and the data
392    LBVertexDataHandle<Grid, PersistentContainer>
393                                dataHandle(grid, persistentContainer);
394    grid->loadBalance(dataHandle);
```

Line 393 sets up an object of the custom data handle class LBVertexDataHandle. Its implementation appears near the top of the example file, and has been discussed in detail in Chapter 6.3. It implements the call-backs for the transfer of the vertex data in persistentContainer during the load-balancing. The constructor of the LBVertexDataHandle class receives the hierarchical grid and the data container, and the entire data handle object is then passed to the loadBalance method of the grid. Note that the constructor of LBVertexDataHandle is not part of the grid interface, but can take any number of arguments to suit the needs of the application.

After the call to loadBalance, each of the available processes has a part of the grid. Figure 6.4, left, shows the decomposition obtained when the program is run on four processes. Note how the subdomain grids consist of *interior* elements and *ghost* elements. In this example, the simple built-in UGGrid partitioner is used.

Finally, the dataVector array is adapted to the new subdomain size, and the data is copied back into it:

226

```
398    // Get gridView again after load-balancing, to make sure it is up-to-date
399    gridView = grid->leafGridView();
400
401    // Copy data back into the array
402    dataVector.resize(gridView.size(dim));
403
404    for (const auto& vertex : vertices(gridView))
405      dataVector[gridView.indexSet().index(vertex)]
406          = persistentContainer[idSet.id(vertex)];
```

Note that the gridView object needs to be refreshed (in Line 399), because it is not guaranteed to remain up to date when the hierarchical grid changes.

The next block is a copy of the code from Chapter 3.3: It selects vector and matrix data types, constructs empty objects b and stiffnessMatrix for the load vector and the stiffness matrix, respectively, and assembles them for a first-order Lagrangian finite element space:

```
415    using Matrix = BCRSMatrix<double>;
416    using Vector = BlockVector<double>;
417
418    Matrix stiffnessMatrix;
419    Vector b;
420
421    auto sourceTerm = [](const FieldVector<double,dim>& x){return -5.0;};
422
423    // Assemble the Poisson system in a first-order Lagrange space
424    Functions::LagrangeBasis<GridView,1> basis(gridView);
425    assemblePoissonProblem(basis, stiffnessMatrix, b, sourceTerm);
```

The method assemblePoissonProblem does the actual assembly. It is identical to the corresponding method in Chapter 3.3, with a small but important exception: In the sequential code, the loop over all elements in source code line 230 on Page 57 has the form

```
for (const auto& element : elements(gridView))
{
  // ...
}
```

In a distributed setting, this loop would iterate over all elements of the subdomain, including the **ghost** elements. The example, however, uses an additive decomposition of matrix and load vector, and has to restrict the loop to the **interior** elements. Therefore, in the parallel code, the loop is replaced by

```
for (const auto& element : elements(gridView,Partitions::interior))
{
  // ...
}
```

The next step is a preparation for the preconditioner. To parallelize the CG method we suppose that the partitioning of the domain Ω into subdomains is *nonoverlapping*. Each process then assembles the matrix and right-hand-side vector on the **interior** elements of its subdomain only. The result is an *additive* decomposition of the load vector b and the stiffness matrix A. Their values corresponding to **interior** vertices are correct—they match the values obtained by assembling on a non-distributed grid. However, entries on vertices on subdomain boundaries are lacking the integral contributions of adjacent elements that are in a different subdomain. Instead, the true value is obtained by summing the values of all copies of the vertex over the different processes.

For other quantities like iterates and corrections, we need *consistent* representations, that is, coefficient vectors that have the correct values even on subdomain boundaries. Consistent representations can be constructed from additive representations by one step of interface communication. The general rule is that primal objects like iterates and corrections need to be available in consistent representation, whereas dual objects like load vectors and matrices need only be available additively.

The stiffness matrix we have just assembled is given in additive representation. The Jacobi preconditioner we will use below, however, will require a consistent representation of the matrix diagonal. Therefore, the following code block triggers a subdomain communication to add the relevant parts of the matrix diagonal together:

```
430    Vector diagonal(basis.size());
431    for (std::size_t i=0; i<basis.size(); ++i)
432      diagonal[i] = stiffnessMatrix[i][i];
433
434    auto consistentDiagonal = diagonal;
435    VertexDataUpdate<GridView,Vector> matrixDataHandle(gridView,
436                                                        diagonal,
437                                                        consistentDiagonal);
438
439    gridView.communicate(matrixDataHandle,
440                         InteriorBorder_InteriorBorder_Interface,
441                         ForwardCommunication);
```

For this, the diagonal is first copied into an array called `diagonal`. This is not strictly necessary, but it avoids having to write a separate data handle for matrix diagonals. The array `diagonal` is then copied into a second array called `consistentDiagonal`, which, despite its name, at this point only contains the matrix diagonal contributions of the local subdomain. The code then constructs an object of type `VertexDataUpdate`. This is a data handle implementation, and has been discussed in Chapter 6.4. It takes vertex data from an input array, and adds it to the vertex data in an output array. Applying this to the arrays `diagonal` and `consistentDiagonal` results in the consistent representation of the matrix diagonal in the `consistentDiagonal` array.

The actual communication happens in the call to the `communicate` method in Line 441. It is given the data handle and the communication interface, i.e., the set of entities that are involved in the communication. In this example we need to send and receive data on the border entities. The smallest communication interface that contains these is precisely `InteriorBorder_InteriorBorder_Interface`.

As the final setup step the algebraic system has to be modified to account for the Dirichlet boundary conditions. As this example uses a position-based criterion to determine the Dirichlet degrees of freedom, the code does not change when the grid is distributed. Just as in the sequential implementation the following code block determines the Dirichlet degrees of freedom:

```
447    auto dirichletPredicate = [](auto p)
448    {
449      return p[0]< 1e-8 || p[1] < 1e-8 || (p[0] > 0.4999 && p[1] > 0.4999);
450    };
451
452    // Interpolating the predicate will mark
453    // all desired Dirichlet degrees of freedom
454    std::vector<bool> dirichletNodes;
455    Functions::interpolate(basis, dirichletNodes, dirichletPredicate);
```

The corresponding Dirichlet values are already contained in the `dataVector` array. The stiffness matrix and the load vector are then modified as described in Chapter 2.1.4:

```
463    // Loop over the matrix rows
464    for (size_t i=0; i<stiffnessMatrix.N(); i++)
465    {
466      if (dirichletNodes[i])
467      {
468        auto cIt    = stiffnessMatrix[i].begin();
469        auto cEndIt = stiffnessMatrix[i].end();
470        // Loop over nonzero matrix entries in current row
471        for (; cIt!=cEndIt; ++cIt)
472          *cIt = (i==cIt.index()) ? 1.0 : 0.0;
473
474        // Modify corresponding load vector entry
475        b[i] = dataVector[i];
476      }
477    }
```

Have a look at Chapter 3.3.1 for more details.

6.7.2. The Distributed Preconditioned CG Method

The remaining part of the code implements the distributed conjugate gradient (CG) method [138, 146]. The implementation is such that each iterate is identical to what a sequential CG implementation would produce. This may not be optimal in terms of efficiency, but it makes the code easier to explain.

For a symmetric positive definite matrix A, the preconditioned CG method solves the linear system of equations

$$Ax = b$$

by successively minimizing the functional $J(x) := \frac{1}{2}x^T A x - b^T x$ along search directions $d_0, d_1, d_2, \ldots$ Under the assumptions on A, minimizers of J are solutions of $Ax = b$. Given an initial iterate x^0, the first search direction is

$$d_0 := W^{-1} r^0 = W^{-1}(b - Ax^0).$$

The quantity $r^0 := b - Ax^0$ is the *residual*, and the matrix W is the preconditioner. It is an approximation of A, and frequently only given in form of an algorithm that computes $W^{-1} r$ from a residual vector r. The present example uses the Jacobi preconditioner $W = \mathrm{diag}(A_{11}, \ldots, A_{nn})$, such that d_0 is computed from r^0 by componentwise division by the matrix diagonal elements. Given a current iterate $x^k \in \mathbb{R}^n$ and a direction $d_k \in \mathbb{R}^n$, the method then performs a line search step

$$x^{k+1} := x^k + \alpha^k d_k,$$

where the step length α^k is given by

$$\alpha^k := \frac{\langle r^k, W^{-1} r^k \rangle}{\langle d_k, A d_k \rangle}.$$

Finally, the method updates the residual

$$r^{k+1} := r^k - \alpha^k A d_k,$$

and computes a new search direction using Gram–Schmidt orthogonalization

$$\beta_{k+1} := \frac{\langle r^{k+1}, W^{-1} r^{k+1} \rangle}{\langle r^k, W^{-1} r^k \rangle}$$

$$d_{k+1} := W^{-1} r^{k+1} + \beta_{k+1} d_k.$$

The iteration continues until r^k becomes suitably small.

To apply this method to a distributed linear system of equations, note that the CG method consists primarily of scalar and matrix–vector products. These are easy to compute in a non-overlapping distributed setting, if matrix and residuals are kept in additive representation, and iterates and directions are kept in consistent representation. For example, in such a situation

$$\langle v, w \rangle = \langle v_{\mathrm{add}}, w_{\mathrm{cons}} \rangle = \sum_{i=1}^{N} \langle v_{\mathrm{add}}^i, w_{\mathrm{cons}}^i \rangle, \tag{6.2}$$

where v_{add}^i and w_{cons}^i are additive and consistent representations, respectively, of the vectors v and w on subdomain i, and N is the total number of subdomains. Hence, computing a distributed scalar product involves only one local scalar product on each process, and one global communication for the sum.

The code that implements the distributed CG algorithm begins by setting up the initial iterate and two parameters:

```
485    // Set the initial iterate
486    Vector x(basis.size());
487    std::copy(dataVector.begin(), dataVector.end(), x.begin());
488
489    // Solver parameters
490    double reduction = 1e-3;   // Desired residual reduction factor
491    int maxIterations = 50;    // Maximum number of iterations
```

The Dirichlet boundary values that have been sampled into the `dataVector` array
are used as the initial iterate, because that is an easy way to obtain an initial iterate
vector that contains the correct Dirichlet boundary values.

Then the code computes the residual $r^0 = b - Ax^0$ from the load vector b and the
initial iterate:

```
496    auto r = b;
497    stiffnessMatrix.mmv(x,r);              // r -= Ax
```

The method `mmv` is part of the matrix interface of the `dune-istl` module (Chapter 7.3).
It implements the two operations `r -= A * x` in a single efficient method. Since x is
given in consistent representation, and A and b are given in additive representation,
the residual is obtained in additive representation.

The code will monitor convergence by tracking the norm of the residual. To
achieve invariance with respect to scaling, the residuals are normalized by the residual of the initial iterate. Computing the norm of this initial residual is the first
operation that requires communication. The norm is computed by locally taking
the scalar product of the additive residual (in r) with its consistent representation,
and then summing the result across all processes as in (6.2). These two steps use
the subdomain communication interface from Chapter 6.4.1, and the collective communication interface from Chapter 6.4.2, respectively. Observe that the consistent
representation of a vector is obtained from its additive representation by taking the
latter, and for each vertex in the **border** partition, add the corresponding entries
from the other subdomains. The appropriate communication interface is therefore
`InteriorBorder_InteriorBorder_Interface` which, for vertex data, only communicates on the **border** partition. For the data handle, the class `VertexDataUpdate`
can be used, which has already been used to compute the consistent matrix diagonal.
The code creates a variable `rConsistent` that will be used to store the consistent
representation of the residual, and initializes it with the content of the variable `rhs`,
which contains the additive representation of the residual. It then sets up a data
handle with input variable r and output variable `rConsistent`, and communicates
once:

```
501    // Construct consistent representation of the data in r
502    auto rConsistent = r;
503
504    VertexDataUpdate<GridView,Vector> vertexUpdateHandle(gridView,
505                                                         r,
506                                                         rConsistent);
```

```
507
508    gridView.communicate(vertexUpdateHandle,
509                          InteriorBorder_InteriorBorder_Interface,
510                          ForwardCommunication);
```

After this, rConsistent contains the consistent representation of the residual r^0. To compute its norm, we take the local scalar product of r with rConsistent, and sum the result over all processes:

```
514    double defect0 = r.dot(rConsistent);    // Norm on the local process
515    defect0 = grid->comm().sum(defect0);
516    defect0 = sqrt(defect0);
```

Note how the collective communication object from Chapter 6.4.2 is used to compute the global sum in Line 515. Using MPI here directly would be more verbose, and with the collective communication object the code even compiles and runs correctly when there is no MPI installed.

As the next step, the code then prints the norm of the initial residual. Note how this only happens on the process with rank 0. Without this restriction there would be screen output for every process involved, which would be unreadable:

```
520    if (mpiHelper.rank()==0)
521    {
522      std::cout << " Iteration            Defect        Rate" << std::endl;;
523      std::cout << "    0" << std::setw(16) << defect0 << std::endl;
524    }
```

The following code then sets the initial search direction d_0 by preconditioning the residual, i.e., by computing $d_0 = W^{-1}r^0$:

```
528    // Construct initial search direction in variable d by applying
529    // the Jacobi preconditioner to the residual in r.
530    Vector d(r.size());
531    for (std::size_t i=0; i<stiffnessMatrix.N(); ++i)
532    {
533      d[i] = 0;
534      if (std::abs(consistentDiagonal[i]) > 1e-5)    // Degree of freedom
535                                                     // is not on ghost vertex
536        d[i] = rConsistent[i] / consistentDiagonal[i];
537    }
```

The i-th entry of d_0 is simply $(r^0)_i/A_{ii}$. However, as the search direction is a primal quantity its consistent representation is required. The code must therefore use the consistent representations of the residual and the matrix diagonal to compute the search direction from. The check for a nonzero diagonal entry weeds out those degrees of freedom that correspond to **ghost** vertices.

Finally, the code computes the quantity $\rho_0 := \langle r^0, W^{-1}r^0 \rangle$, which is used in the formula for the step length α_0 and for the Gram–Schmidt orthogonalization factor β_{k+1}:

```
541    double rho = d.dot(r);
542    rho = grid->comm().sum(rho);
```

As $W^{-1}r^0 = d_0$, this is simply a scalar product between an additive and a consistent vector, which is easily computed with a local scalar product and a global sum.

After these preparations starts the actual iteration loop. It is implemented as a standard **for**-loop with the iteration counter k. The first task within the loop is to compute the line search step length

$$\alpha^k = \frac{\langle r^k, W^{-1}r^k \rangle}{\langle d_k, Ad_k \rangle}$$

for the direction d_k:

```
546    // Current residual norm
547    double defect=defect0;
548
549    for (int k=0; k<maxIterations; ++k)
550    {
551      // Step length in search direction d
552      Vector tmp(d.size());
553      stiffnessMatrix.mv(d,tmp);           // tmp=Ad^k
554      double alphaDenom = d.dot(tmp);      // Scalar product
555      alphaDenom = grid->comm().sum(alphaDenom);
556      double alpha = rho/alphaDenom;
```

Note that the variable rho already contains the numerator $\langle r^k, W^{-1}r^k \rangle$ from the expression for the step size α^k. The denominator $\langle d_k, Ad_k \rangle$ is a scalar product between a primal and a dual quantity, and can be computed in the usual way. Knowing the step size α^k allows to set the new iterate

$$x^{k+1} = x^k + \alpha^k d_k$$

using the line

```
560      x.axpy(alpha,d);           // Update iterate
```

The method axpy of the dune-istl vector data types implements x += alpha*d in a single method. Updating the residual uses the formula

$$r^{k+1} = b - Ax^{k+1} = r^k - \alpha^k Ad_k,$$

coded as

```
563      r.axpy(-alpha,tmp);        // Update residual
```

which is more efficient than recomputing the residual from scratch.

The code then computes the residual norm again, prints it to the screen, and checks whether it is small enough to terminate the iteration. None of this uses any new functionality:

```
568        // Compute residual norm again
569        rConsistent = r;
570        gridView.communicate(vertexUpdateHandle,
571                              InteriorBorder_InteriorBorder_Interface,
572                              ForwardCommunication);
573
574        auto residualNorm = r.dot(rConsistent);
575        residualNorm = grid->comm().sum(residualNorm);
576        residualNorm = sqrt(residualNorm);
577
578        if (mpiHelper.rank()==0)
579        {
580          std::cout << std::setw(5)  << k+1 << " ";
581          std::cout << std::setw(16) << residualNorm << " ";
582                              // Estimated convergence rate
583          std::cout << std::setw(16) << residualNorm/defect << std::endl;
584        }
585
586        defect = residualNorm;          // Update norm
587
588        if (defect<defect0*reduction)   // Convergence check
589          break;
```

The last task is to compute the Gram–Schmidt factor

$$\beta_{k+1} := \frac{\langle r^{k+1}, W^{-1}r^{k+1}\rangle}{\langle r^k, W^{-1}r^k\rangle}$$

and the new search direction

$$d_{k+1} = W^{-1}r^{k+1} + \beta_{k+1}d_k.$$

The denominator of β_{k+1} is already known from the computation of the step length α^k, but the numerator requires invoking the preconditioner again. Note how again degrees of freedom corresponding to **ghost** vertices have to be skipped. We use the variable rhoNext for the numerator of β_{k+1}:

```
594        // Precondition the residual
595        Vector preconditionedResidual(d.size());
596        for (std::size_t i=0; i<stiffnessMatrix.N(); i++)
597        {
598          preconditionedResidual[i] = 0;
599          if (std::abs(consistentDiagonal[i]) > 1e-5)    // Degree of freedom
600                                                         // not on ghost vertex
601            preconditionedResidual[i] = rConsistent[i] / consistentDiagonal[i];
602        }
603
604        double rhoNext = preconditionedResidual.dot(r);
605        rhoNext = grid->comm().sum(rhoNext);
```

```
606      double beta = rhoNext/rho;
607
608      // Compute new search direction
609      d *= beta;
610      d += preconditionedResidual;
611      rho = rhoNext;                  // Remember rho for the next iterate
612    }
```

The curly bracket in Line 612 ends the conjugate gradient loop. All that is left to do is to write the result to a VTK file. For prettier visualization we write an additional element data field that contains the process numbers (Figure 6.8):[11]

```
617    // For visualization: Write the rank number for each element
618    MultipleCodimMultipleGeomTypeMapper<GridView>
619                            elementMapper(gridView,mcmgElementLayout());
620    std::vector<int> ranks(elementMapper.size());
621    for (const auto& element : elements(gridView))
622      ranks[elementMapper.index(element)] = mpiHelper.rank();
623
624    VTKWriter<GridView> vtkWriter(gridView);
625    vtkWriter.addVertexData(x, "solution");
626    vtkWriter.addCellData(ranks, "ranks");
627    vtkWriter.write("grid-distributed-poisson-result");
```

The handling of the VTKWriter object is the same as in a sequential setting, but it adapts to the distributed situation automatically. Each process writes a file containing only the *interior* elements of that process, and the master process additionally writes a pvtu file (Chapter 6.6).

6.7.3. Running the Program

How to invoke the program depends on the MPI implementation that is used. For OPEN MPI,[12] the necessary command is

```
mpirun ⁻np ## ./grid-distributed-poisson
```

where ## is the number of processes to use. It will produce the screen output

```
Reading 2d Gmsh grid...
version 2.2 Gmsh file detected
file contains 541 nodes
file contains 1086 elements
number of real vertices = 541
number of boundary elements = 88
number of elements = 992
  Iteration              Defect      Rate
```

---

[11] We use a MultipleCodimMultipleGeomTypeMapper object to get element indices, even though the grid contains only a single element type (Chapter 5.6.2).

[12] www.open-mpi.org

235

Figure 6.8.: Output of the distributed Poisson example program. Elevation denotes the values of the solution u_h, and colors illustrate the subdomain partitions.

```
     0            0.305176
     1              0.416417              1.36451
     2              0.468409              1.12486
     3              0.438753              0.936687
     4              0.412924              0.941131
     5              0.392524              0.950597
[...]
    36          0.000345697              0.625268
    37          0.000271504              0.785382
```

As the sequence of iterates does not depend on the number of processes, neither does the screen output. With four processes, the program will leave the subdomain files

```
s0004-p0000-grid-distributed-poisson-result.vtu
s0004-p0001-grid-distributed-poisson-result.vtu
s0004-p0002-grid-distributed-poisson-result.vtu
s0004-p0003-grid-distributed-poisson-result.vtu
```

and the *parallel file*

```
s0004-grid-distributed-poisson-result.pvtu
```

which can be opened in PARAVIEW. When colored using the subdomain rank field, the result will look as in Figure 6.8. The elevation is the same as in Figure 3.3 of the sequential Poisson example of Chapter 3.3.

7. Linear Algebra with dune-istl

Besides the grids and their data structures, it is probably the linear algebra that requires most care in a finite element code. This is so because the linear systems resulting from finite element discretizations can get enormously big. Storing large sparse matrices makes up a considerable portion of the overall memory requirements, and solving the linear systems can take up the majority of the overall computing time. Many good software packages for linear algebra are available. Without attempting completeness we mention PETSc[1] [6, 7], MTL[2] [77], Trilinos[3] [93], BLAS[4], and Hypre[5] [68]. A comprehensive list is maintained at [56]. To cope with the ever increasing problem sizes, many of them attempt to scale to very large processor counts.

By construction of the DUNE grid interface, DUNE does not require the user to use any particular linear algebra software package (Section 5.6). Instead, the interface from grid to vectors and matrices is data-type agnostic, and basically only requires that vectors and matrices are random-access containers. This allows to use DUNE grids in combination with most existing linear algebra packages.

In spite of the existing zoo of linear algebra software, however, DUNE also proposes its own linear algebra library. It is contained in the dune-istl module, [6] and sports certain features that make it particularly suited for finite element and finite volume simulations. At the core of it is a powerful nesting mechanism to encode typical blocking and sparsity patterns in the matrix and vector data types themselves. This can make specific finite element computations much more time- and memory-efficient. On top of such matrix and vector data types, dune-istl provides a collection of standard solver algorithms, a lot of them fully parallelized. Some of them are implemented in dune-istl itself, while others are adapter codes to use existing solvers like UMFPACK and CHOLMOD [41, 46].

This chapter begins by describing the nesting mechanism for matrices and vectors (Chapter 7.1). Chapters 7.2 and 7.3 then explain the vector and matrix interfaces in detail, and Chapter 7.4 shows solvers and preconditioners. Finally, Chapter 7.5 presents the work-horse of dune-istl: a fully parallel agglomeration-based algebraic multigrid preconditioner. It can serve to precondition the linear solvers of Chapter 7.4, and has been used successfully even on very large machines [31, 100]. The dune-istl

---

[1] www.mcs.anl.gov/petsc
[2] www.mtl4.org
[3] trilinos.org
[4] www.netlib.org/blas and www.netlib.org/blas/blast-forum
[5] http://computation.llnl.gov/project/linear_solvers/software.php
[6] ISTL meaning Iterative Solver Template Library

© Springer Nature Switzerland AG 2020
O. Sander, *DUNE — The Distributed and Unified Numerics Environment*, Lecture Notes in Computational Science and Engineering 140, https://doi.org/10.1007/978-3-030-59702-3_7

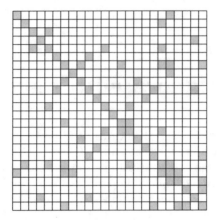

Figure 7.1.: Occupation pattern of a general sparse matrix

module was originally described in [14, 29].

7.1. Constructing Matrix and Vector Types by Nesting

In textbooks on the finite element method, the stiffness matrices are characterized as "sparse", meaning that most entries are zero. Matrix data structures need to exploit this sparsity; otherwise the sheer number of zeros to be stored would rapidly exceed any available memory. Figure 7.1 shows the occupation pattern of such a matrix. The small grey squares represent nonzero entries, and only those need to be stored in the matrix data structure.

A variety of data structures exists for sparse matrices (see [133] for an overview). However, sparse matrices in finite element computations frequently have a lot more structure than simple sparsity. Here are some examples:

- Certain discretizations result in matrices where individual entries are grouped in small dense blocks, say of size 2×2 or 3×3 (Figure 7.2(a)). Such blocks can have one fixed size for the entire matrix; for example, discretizing a three-dimensional elasticity problem will result in a stiffness matrix where all entries are grouped in 3×3 blocks. On the other hand, hp-adaptive DG discretizations [55] lead to matrices with blocks of different sizes, see Figure 7.2(b). There, each block belongs to an element, and the block size depends on the polynomial order of the finite element space on that element. Many iterative methods can be made to operate on such blocks, rather than on scalars, see, e.g., [34].

- When solving a PDE system with m different dependent variables, one possible way of arranging the overall list of coefficients is to order them equation-wise. This results in matrices having an $m \times m$ block structure where the blocks themselves are large and sparse, see Figure 7.2(c). One example is the Stokes problem explained in Chapter 10.8. Solvers such as the Uzawa algorithm and its variants [37, 138] directly operate on such structures.

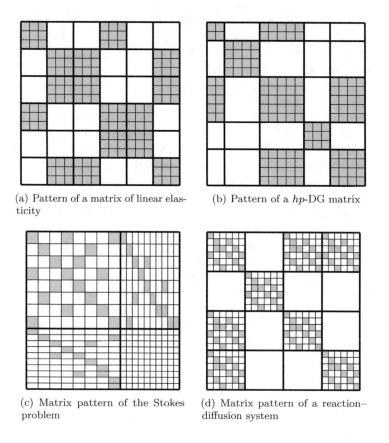

(a) Pattern of a matrix of linear elasticity

(b) Pattern of a *hp*-DG matrix

(c) Matrix pattern of the Stokes problem

(d) Matrix pattern of a reaction–diffusion system

Figure 7.2.: Many finite element discretizations result in sparse matrices with additional structure

- Discretizations of reaction–diffusion systems produce sparse matrices whose blocks are sparse matrices themselves (Figure 7.2(d)). The sparsity pattern is the same for all blocks. Similarly, discretizations of certain vector-valued phase-field equations produce sparse matrices whose blocks are diagonal matrices of fixed size [80].

A lot of this structure is typically known at compile-time and this knowledge can be exploited to produce more efficient code. To get an idea of the savings that using the block structure can bring, consider that a standard compressed-row-storage

(CRS) data structure for a sparse matrix essentially uses 12 bytes per matrix entry: a double-precision number (8 bytes) for the value itself, and one integer (usually 4 bytes, sometimes more) to record the column index of that value. In a dedicated matrix data structure for three-dimensional elasticity problems, each matrix entry is a dense 3×3 matrix, and only one column index per block needs to be stored. Hence, the approximate storage requirement drops to $8 + \frac{4}{9}$ bytes per scalar matrix entry. This is an improvement of about 30 %, which allows to handle larger problems and increases the speed of many linear solvers.

The dune-istl module aims to provide special-purpose data structures for many of these patterns. As usual, the DUNE way is followed in that there is an abstract interface prescribing how matrix and vector data structures can be used, and a number of different implementations of these interfaces. For example, the class BCRSMatrix implements a sparse matrix with compressed row storage and run-time size, and FieldMatrix implements a dense matrix with compile-time size. The class BlockVector implements a vector data structure of run-time size, and FieldVector does the same with compile-time size.

However, considering that Figure 7.2 only shows a subset of the long list of possible patterns, it is clear that implementing a dedicated data structure for each such pattern type is fairly hopeless. Therefore, dune-istl takes a modular approach: It provides a set of data structures for certain basic types of patterns, storage schemes, etc., and it allows to use matrices as matrix entries. Similarly, the entries of dune-istl vectors can be other dune-istl vectors. The entry type of each matrix or vector data structure is a template parameter, which makes the structure visible to the compiler.

This strategy is most easily demonstrated with a few examples.[7] To begin, while the generic sparse matrix is implemented as

```
BCRSMatrix<double>
```

the type

```
BCRSMatrix<FieldMatrix<double,3,3> >
```

is a sparse matrix whose entries are dense 3×3 blocks, as in Figure 7.2(a). The overall size of the matrix can be set at run-time, but the 3×3 block size is hard-coded into the data type, and can therefore be used by optimizing compilers. Similarly

```
BTDMatrix<BCRSMatrix<double> >
```

is a tridiagonal matrix whose entries are sparse matrices with scalar entries, and

```
BCRSMatrix<BCRSMatrix<DiagonalMatrix<double,3> > >
```

is a sparse matrix of sparse matrices of diagonal 3×3 matrices.

Construction of vectors with blocking patterns works in the same way. For example, while

```
BlockVector<double>
```

---

[7] All data types mentioned here will be explained in detail later in this chapter.

is a vector of **double** entries with run-time size, the appropriate vector type for the three-dimensional elasticity system of Figure 7.2(a) would be

```
BlockVector<FieldVector<double,3> >
```

Suitable vector types for the third and fourth example matrix types are

```
BlockVector<BlockVector<double> >
```

and

```
BlockVector<BlockVector<FieldVector<double,3> > >
```

but note that VariableBlockVector is a more efficient replacement for the nested type BlockVector<BlockVector>. The solvers included with dune-istl can work with most of these combinations.

The innermost type always specifies the number type used for scalar entries, and can be real or complex.[8] The fact that the data type used for scalars is never hard-wired anywhere make various interesting applications easily achievable. For example, with changing only a handful of lines of code you can do linear algebra with multi-precision data types such as the ones from boost::multiprecision[9] or GMP[10] [79], or with the "active" number types needed for algorithmic differentiation [85].

The nesting mechanism of dune-istl nicely complements the tree construction of mixed finite element spaces as implemented in the dune-functions module (Chapter 10). Its versatility is demonstrated in the example program given in Chapter 10.8.

7.2. Data Structures for Vectors

The vector and matrix interfaces are based on duck typing (Chapter 4.4.2), i.e., there is no common base class or facade class enforcing them. We first describe the vector interface of dune-istl, because it is a bit shorter and simpler than the matrix interface.

7.2.1. Abstract Interface

A dune-istl vector is a one-dimensional container with additional vector space operations, scalar products, and norms. True to duck typing, a C++ data structure will be called a dune-istl vector, if it implements the container methods listed in Table 7.1, the linear algebra methods of Table 7.3, and if it exports the types listed in Table 7.2. It has to be noted that the dune-istl vector interface supports vectors that are sparse, i.e., vectors that do not store zero entries explicitly. Sparse vectors are rarely used in finite element simulations, but there is no good reason to rule them out, either. In dune-istl they currently only appear in the form of rows of

---

[8]More specifically, it can be any type T for which Dune::IsNumber<T>::value is **true**.
[9]www.boost.org
[10]https://gmplib.org

sparse matrices. We first discuss the container interface of dune-istl vectors, and afterwards the methods for linear algebra.

	`Vector()`
	Default constructor
	`Vector(const Vector& other)`
	Copy constructor
`Vector&`	`operator=(const Vector& other)`
	Assignment operator
`Vector&`	`operator=(const field_type& scalar)`
	Assignment from scalar
`value_type&`	`operator[](size_type i)`
`const value_type&`	`operator[](size_type i) const`
	Random access
`iterator&`	`begin()`, `end()`
	Forward random access iterators
`iterator&`	`rbegin()`, `rend()`
	Reverse random access iterators
`iterator&`	`find(size_type i)`
`const_iterator&`	`find(size_type i) const`
	Iterator to the `i`-th element, or to the end iterator if that element is not contained in the (sparse) vector
`size_type`	`N() const`
	Number of block entries
`size_type`	`size() const`
	Number of nonzero block entries
`size_type`	`dim() const`
	Number of scalar entries

Table 7.1.: Container methods of the vector interface. Default and copy constructors are required, but vector implementations frequently implement further constructors, too.

Container Interface

As a container, a dune-istl vector provides forward and backward sequential access to its entries, and, more importantly, random access. Vector implementations are templated with the type used for the entries in the vector. These are expected to be either number types like **double**, **float**, or std::complex<**double**>, or again vectors in the dune-istl sense. This allows for nested container data structures as shown in the previous section.

The list of all container methods is given in Table 7.1. First and foremost, dune-istl vectors are expected to be random-access containers in the sense of the C++ standard library. That is, there have to be methods

```
reference operator[](size_type i)
```

and

```
const_reference operator[](size_type i) const
```

that provide access to the i-th entry of the container. The method behavior is undefined if that entry does not exist in the data structure.

To be able to do random access on sparse vectors with a well-defined behavior for the case of accessing non-existing entries, dune-istl vectors offer the methods

```
iterator find(size_type i)
const_iterator find(size_type i) const
```

Unlike **operator**[] they return an iterator, which points to the i-th element if it exists, and to the return value of end() otherwise. So, for example, if vec is a sparse array of **double** values that contains only entries at the even positions, then

```
double a = vec[5];
```

will lead to undefined behavior. With the find method, the code can be written in a safe way:

```
auto iter = vec.find(5);
double a = (iter==vec.end()) ? 0 : *iter;
```

The methods

```
iterator begin()
iterator end()
```

and

```
iterator rbegin()
iterator rend()
```

provide access to forward and reverse random-access iterators. For sparse vectors, these iterators stop only at the nonzero entries. The return types have to be exported by the vector class (Table 7.2).

To implement the dune-istl interface, vectors provide the usual constructors and assignment operators expected from a container class. Unlike containers from the standard library, dune-istl vectors must support assignment from a scalar:

```
Vector& operator=(const field_type& scalar)
```

which recursively traverses the nesting tree, and fills each and every scalar entry with the given value.

There are several ways to obtain the size of a dune-istl vector. Most importantly, the method

```
size_type N() const
```

class	block_type (also: value_type)
	Type of the vector entries (if all the same)
class	field_type
	Type used for scalars
class	size_type
	Type used indices and sizes
class	reference
	Reference to a vector entry
class	const_reference
	const reference to a vector entry
class	Iterator (also: iterator)
	Random-access iterator for read/write access
class	ConstIterator (also: const_iterator)
	Random-access iterator for read-only access
class	allocator_type
	Standard-library-compliant allocator used by the vector (optional)

Table 7.2.: Static information exported by vector classes

returns the number of entries of the vector when seen as a container.[11] In other words, size()-1 is the biggest value that can be used as argument for **operator**[]. In contrast, the method with the somewhat unfortunate name

```
size_type size() const
```

only returns the number of nonzero entries. For dense vectors, size will always return the same number as N, but for sparse ones size < N can occur. Finally, the method

```
size_type dim() const
```

returns the total number of scalar entries in the complete nesting hierarchy. For example, for an object of type BlockVector<FieldVector<**float**,2> >, the method dim will return twice the number as size. It is the dimension of the linear space that the vector is an element of, which explains the name.

Several dune-istl vectors can be supplied with custom allocators. An allocator is a C++ object that encapsulates the process of allocating and deallocating memory from the system heap. This allows to instrument memory allocations for debugging, or to insert additional safety checks. The C++ standard describes an interface for allocators in Chapter 20 [98], and dune-istl vectors expect allocators to follow this standard.

Vector Space Operations, Scalar Products, and Norms

Mathematically, a dune-istl vector is an element of a Euclidean vector space. As such, the standard vector space operations must be supported (Table 7.3). Vector addition is supported by

---

[11]This method name starts with a capital letter without any good reason.

```
Vector& operator+=(const Vector& other)
```

and

```
Vector& operator-=(const Vector& other)
```

Vector&	**operator**+=(**const** Vector& y)
	$x \longleftarrow x + y$
Vector&	**operator**-=(**const** Vector& y)
	$x \longleftarrow x - y$
Vector&	**operator***=(**const** field_type& alpha)
	$x \longleftarrow \alpha x$
Vector&	**operator**/=(**const** field_type& alpha)
	$x \longleftarrow \alpha^{-1} x$
Vector&	axpy(**const** field_type& alpha, **const** Vector& y)
	$x \longleftarrow x + \alpha y$
field_type	**operator***(**const** Vector& y)
	Compute $\langle x, y \rangle$
real_type	one_norm()
	Compute $\sum_{i=0}^{n-1} \sqrt{\mathrm{Re}(x_i)^2 + \mathrm{Im}(x_i)^2}$
real_type	one_norm_real()
	Compute $\sum_{i=0}^{n-1} (|\mathrm{Re}(x_i)| + |\mathrm{Im}(x_i)|)$
real_type	two_norm()
	Compute $\sqrt{\sum_{i=0}^{n-1} (\mathrm{Re}(x_i)^2 + \mathrm{Im}(x_i)^2)}$
real_type	two_norm2()
	Compute $\sum_{i=0}^{n-1} (\mathrm{Re}(x_i)^2 + \mathrm{Im}(x_i)^2)$
real_type	infinity_norm()
	Compute $\max_{i=0,\ldots,n-1} \sqrt{\mathrm{Re}(x_i)^2 + \mathrm{Im}(x_i)^2}$
real_type	infinity_norm_real()
	Compute $\max_{i=0,\ldots,n-1} (|\mathrm{Re}(x_i)| + |\mathrm{Im}(x_i)|)$

Table 7.3.: Vector space operations, scalar products, and norms. The type real_type is available as FieldTraits<field_type>::real_type.

Multiplication with a scalar is given by

```
Vector& operator*=(const field_type& alpha)
```

and

```
Vector& operator/=(const field_type& alpha)
```

There must be a direct implementation of the important operation

$$x \longleftarrow x + \alpha y, \qquad \alpha \in \mathbb{K}, \quad x, y \in \mathbb{K}^n,$$

in form of the method

```
Vector& axpy(const field_type& alpha, const Vector& y)
```

In addition to the bare vector space properties, `dune-istl` vectors implement scalar products and norms, which turn the underlying spaces in Hilbert and Banach spaces. There are two different scalar products. The first,

```
field_type operator*(const Vector& y)
```

implements the real scalar product

$$\langle a, b \rangle = \sum_{i=0}^{n-1} a_i b_i.$$

The second one, implemented as a member method

```
real_type dot(const Vector& y)
```

implements the Hermitian product

$$\langle a, b \rangle = \sum_{i=0}^{n-1} \bar{a}_i b_i.$$

If the vector space is real, the two products coincide.

The norms are the same as defined for the sparse BLAS standard [28]. There are three different types:

1. The l_1 norm

   ```
   real_type one_norm()
   ```
   $$|a|_1 := \sum_{i=0}^{n-1} |a_i| = \sum_{i=0}^{n-1} \sqrt{(\operatorname{Re} a_i)^2 + (\operatorname{Im} a_i)^2}$$

   and its real counterpart

   ```
   real_type one_norm_real()
   ```
   $$|a|_{1,\mathrm{r}} := \sum_{i=0}^{n-1} |\operatorname{Re} a_i| + |\operatorname{Im} a_i|.$$

   The second variant is an approximation of the first that avoids the expensive square root that appears in the complex case. For real-valued vectors, the two norms coincide.

2. The l_2 norm

   ```
   real_type two_norm()
   ```
   $$|a|_2 := \sqrt{\sum_{i=0}^{n-1} (\operatorname{Re} a_i)^2 + (\operatorname{Im} a_i)^2},$$

and its square

$$\texttt{real\_type two\_norm2()} \qquad |a|_2^2 := \sum_{i=0}^{n-1} (\operatorname{Re} a_i)^2 + (\operatorname{Im} a_i)^2.$$

Similarly to the l_1 case, the second variant exists to avoid the costly square root.

3. The l_∞ norm

$$\texttt{real\_type infinity\_norm()} \qquad |a|_\infty := \max_{i=0,\dots,n-1} |a_i| \cdot$$

$$= \max_{i=0,\dots,n-1} \sqrt{(\operatorname{Re} a_i)^2 + (\operatorname{Im} a_i)^2}$$

and its real counterpart

$$\texttt{real\_type infinity\_norm\_real()} \qquad |a|_{\infty,\mathrm{r}} := \max_{i=0,\dots,n-1} \big(|\operatorname{Re} a_i| + |\operatorname{Im} a_i| \big),$$

which again avoids the evaluation of the square root.

Note that all three norms work canonically even for nested vector types.

Missing from the interface are the binary operators **operator+**, **operator-**, and **operator***(field_type,Vector). Hence, it is not possible to write

```
BlockVector<double> a = {1.0, 2.0, 3.0};
BlockVector<double> b = {1.0, 4.0, 9.0};
BlockVector<double> c = a+b;  // won't compile
```

The reason for this is that while the first variant looks shorter, it may nevertheless create two loops over all container entries if the compiler is not sufficiently smart. The (in)famous expression templates have been introduced to force the compiler to emit only a single loop for such expressions [1, 156]. However, expression templates are not used in dune-istl, because they lead to very long compile times and compiler error messages, and they make the library a lot more difficult to debug.

Instead, the dune-istl way of adding vectors is

```
BlockVector<double> a = {1.0, 2.0, 3.0};
BlockVector<double> b = {1.0, 4.0, 9.0};
BlockVector<double> c = a;
c += b;
```

This creates two loops, too, but at least it makes this fact explicit. To really make sure that a single loop is created for the expression c = a + b, that loop has to be written by hand.

Pure FieldVector objects are an exception. Indeed, the code

```
FieldVector<double,3> a = {1.0, 2.0, 3.0};
FieldVector<double,3> b = {1.0, 4.0, 9.0};
FieldVector<double,3> c = a+b;
```

does compile and run. In this case it was deemed preferable to offer the binary operators as well, to allow for more readable code. Additionally, anecdotical evidence suggests that modern compilers are smart enough to fuse loops for such arithmetic expressions.

7.2.2. Vector Implementations

The dune-istl module currently provides four vector implementations: FieldVector, BlockVector, MultiTypeBlockVector, and VariableBlockVector. We discuss each of the four vector implementations in turn. Additionally, rows of matrices must implement the vector interface.

FieldVector

The class

```
template<class T, int n>
class FieldVector
```

implements a vector of static size. The data is stored on the stack, and therefore FieldVectors can only be used for small vectors, e.g., the three components of a velocity field. The first template argument T is the type of the vector components, and the second template parameter n is the length of the vector.

The FieldVector class faithfully implements the complete dune-istl vector interface. Objects of type FieldVector can be default- and copy-constructed, they can be constructed from a scalar (which initializes all entries with that scalar), and from initializer lists

```
FieldVector<double,3> v = {1.0, 2.0, 3.0};
```

Of course, short static vectors are useful in a finite element software system far beyond the actual linear algebra. Therefore, even though FieldVector is an implementation of the dune-istl vector concept, it is actually contained in dune-common, in the header dune/common/fvector.hh. As could be seen in Chapter 5, dune-grid and dune-geometry use it for coordinates. The dune-localfunctions module uses it for shape function values (Chapter 8).

As an exception to the general dune-istl nesting mechanism, a FieldVector always ends the nesting recursion. It can therefore only be instantiated with actual real or complex number types for T. The reason for this is partly historical, and partly due to the fact that the FieldVector implementation does not actually reside in the dune-istl module.

As a further exception, FieldVector is the only dune-istl vector class that implements binary arithmetic operators. So, for example, it is valid to write

```
FieldVector<double,3> a, b, c;
double alpha, beta;
```

```
c = alpha * a + beta * b;
```

For all other dune-istl vector types, only the unary operators +=, -=, *= etc. exist.

BlockVector

The class

```
template<class T, class A=std::allocator<T> >
class BlockVector
```

is the work-horse vector type of dune-istl. It implements a vector of dynamic size, with the data stored on the heap. This means that a BlockVector can be as large as desired (physical memory permitting), and BlockVector is the standard choice for finite element coefficient vectors. The template parameter T is the type of the vector entry, and can be any number or dune-istl vector type. The following creates a vector of 10 vectors of length 3 each; useful, e.g., for a displacement field in solid mechanics:

```
BlockVector<FieldVector<double,3> > d(10);
```

Similarly, the following creates a vector of 10 **double** entries:

```
BlockVector<double> v(10);
```

Objects of type BlockVector implement the full container and linear algebra interface. They can be constructed from initializer lists, which can be very convenient for small vectors:

```
BlockVector<float>                fibonacci = {1,1,2,3,5,8};
BlockVector<FieldVector<double,3> > canonical = {{1,0,0}, {0,1,0}, {0,0,1}};
```

Entries of the vector can be accessed using square brackets:

```
v[0] = 1.0;
```

The object returned when indexing the nested vector d with square brackets is a FieldVector<double,3>. Hence individual components as well as entire subblocks can be conveniently addressed:

```
d[0]    = {1.0, 0.0, 0.0};  // Set first block to canonical basis vector
d[0][1] = 1.0;              // d[0] is now {1.0, 1.0, 0.0}
```

The size of a BlockVector can be changed any time using the method

```
void resize(size_type size)
```

As in the standard library, this keeps the data, but invalidates all iterators of the object it is called for.

Note that the BlockVector class is allocator-aware: Its second template parameter defaults to std::allocator, and can be replaced by any standard-library-compliant allocator implementation [150]. This allocator is then used by the BlockVector object for all memory management.

MultiTypeBlockVector

The vector implementation

```
template<typename... Args>
class MultiTypeBlockVector
```

differs from the others because it is a heterogeneous container. This means that it stores a collection of objects of different types, much like `std::tuple` does. In fact, as a container, `MultiTypeBlockVector` is very similar to `std::tuple`. The `MultiTypeBlockVector` and its cousin `MultiTypeBlockMatrix` are very helpful when implementing mixed finite element methods and multi-physics applications. A complete example of such an application is given in Chapter 10.8.

The template parameter **typename**... Args for `MultiTypeBlockVector` is *variadic*, which means that the `MultiTypeBlockVector` class can be instantiated with any number of template arguments. A fortiori, each `MultiTypeBlockVector` has a fixed size of entries. Each argument is expected to be a dune-istl vector, and the `MultiTypeBlockVector` object stores one object of each vector type. Each entry can again be nested, and there is no requirement that the nesting depth should be the same for all entries. As an example, the following code combines a vector of displacements in $\mathbb{R}^3$ for an elastic medium with a scalar variable for a fluid pressure:

```
using DisplacementVector = BlockVector<FieldVector<double,3> >;
using PressureVector =  BlockVector<double>;
using MultiTypeVector = MultiTypeBlockVector<DisplacementVector,
                                             PressureVector>;
```

To construct an object of this type, construct the separate components and then hand them to the constructor of the `MultiTypeBlockVector` class:

```
DisplacementVector displacement = {{0, 0, 0}, {0, 0, 0}, {0, 0, 0}};
PressureVector pressure = {0, 0, 0};
MultiTypeVector multiType(displacement, pressure);
```

All construction methods of `std::tuple` are supported. For example, it is also possible to default-construct the outer vector and then supply the separate components afterwards.

As a heterogeneous container, the `MultiTypeBlockVector` class cannot precisely fulfill all requirements of the dune-istl vector container interface. In particular, since any two vector entries may have differing types, standard run-time iteration over the entries is not possible, and needs to be replaced by static loops. Likewise, the standard random access operator

```
block_type& operator[](size_type i)
```

cannot be implemented, because the return type must be known at compile-time, but it depends on the argument i, which is not known until run-time.

To access the individual entries `MultiTypeBlockVector` implements a special static version of **operator**[] with a compile-time argument:

```
template<std::size_t i>
auto& operator[](const index_constant<i> indexVariable)
```

(and the corresponding **const** operator). The index_constant<i> class (from the file
dune/common/indices.hh) is an alias for std::integral_constant<std::size_t,i>.
The corresponding object encapsulates the number i as a type, and hence forces the
argument to be compile-time information. The return type can then correctly depend
on this information. For example, to copy entry number 2 of a MultiTypeBlockVector
to a separate variable write

```
auto tmp = multiTypeVector[index_constant<2>()];
```

This works, but it is clumsy. To improve readability DUNE provides abbreviations for
the most relevant natural numbers. The file dune/common/indices.hh contains the
definitions

```
constexpr index_constant<std::size_t, 0>  _0 = {};
constexpr index_constant<std::size_t, 1>  _1 = {};
constexpr index_constant<std::size_t, 2>  _2 = {};
...
```

in the namespace Dune::Indices. With this, access to component 2 is concise:

```
using namespace Dune::Indices;
auto tmp = multiTypeVector[_2];
```

Such a line of code looks basically like the corresponding line for a normal container,
and in this sense the heterogeneous MultiTypeBlockVector is hardly more difficult
to use than a BlockVector. If the vector entry to be accessed is not fixed, but only
available as a compile-time variable, then the longer syntax using index_constant
has to be used.

Vector space operations, norms, and scalar products do not pose any problems.
MultiTypeBlockVector implements all the linear algebra methods listed in Table 7.3
For example, given two such vectors a and b of equal type, it is possible to add and
scale them

```
a += b;
a *= 2.0;
```

take the scalar product:

```
auto product = a*b;
```

and compute the norms listed in Table 7.3:

```
auto 12  = a.two_norm();
auto max = a.infinity_norm();
...
```

Based on this functionality, most of the Krylov solvers mentioned in Chapter 7.4.1
work directly with MultiTypeBlockVector objects.

VariableBlockVector

Finally, the class

```
template<class T, class A = std::allocator<T> >
class VariableBlockVector
```

does not offer additional functionality, but allows certain applications to run more efficiently. Sometimes, a vector of vectors is desirable, where both the inner and outer vectors have dynamic sizes. For example, in a p-adaptive DG method, the outer vector would be indexed by the elements, and each inner vector would store the coefficients for one element. Such a data structure can be constructed by nesting `BlockVectors`:

```
BlockVector<BlockVector<double> >;
```

However, such a type uses memory very inefficiently. Indeed, memory is allocated separately for each inner block, leading to memory fragmentation. This is visualized in the center of Figure 7.3. The fragmentation can slow down iterative solvers, which run more efficiently if the vector entries are laid out contiguously in memory.

A solution to this problem is the `VariableBlockVector` class. An object of type

```
VariableBlockVector<V>;
```

(where V is any number or `dune-istl` vector type) will behave just like

```
BlockVector<BlockVector<V> >;
```

However, the implementation of `VariableBlockVector` sees to it that memory is allocated in a single chunk, with no fragmentation, as shown at the bottom of Figure 7.3.

To achieve the continuous memory layout, `VariableBlockVector` is more restrictive than a `BlockVector` of `BlockVectors` regarding when to select the inner vector sizes. Basically, the inner vector sizes can only be set all at once, and setting them will invalidate the entire vector content. Besides the obligatory default- and copy constructors, a `VariableBlockVector` object can be constructed with given outer and inner sizes:

```
VariableBlockVector(size_type nblocks, size_type m)
```

This will allocate an outer vector of size nblocks, and set the size of each inner vector to m. To set the size of each inner vector individually, construct the object giving only the outer size:

```
VariableBlockVector(size_type nblocks)
```

Setting the inner sizes then requires a `CreateIterator` object. Such an iterator is obtained by calling the method

```
CreateIterator VariableBlockVector::createbegin()
```

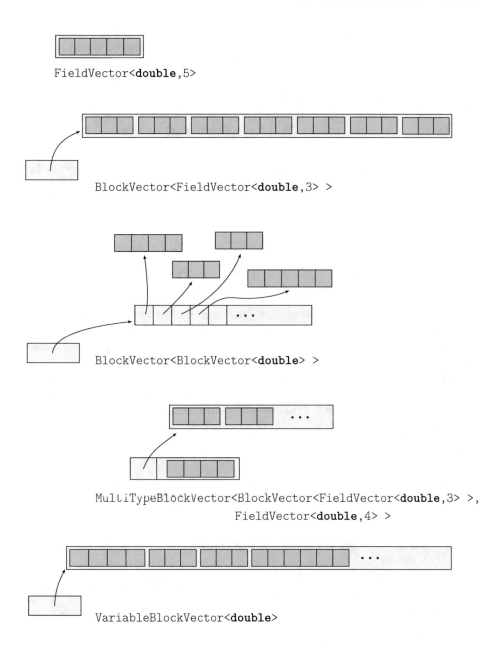

Figure 7.3.: Memory layout of different types of nested vectors. `VariableBlockVector`
provides the same functionality as `BlockVector<BlockVector>`, but
avoids the memory fragmentation.

A `CreateIterator` will forward-iterate over all entries of the outer vector. For each such entry, the size of the corresponding inner vector can be set by writing the desired size to the dereferenced iterator. So, for example, to construct a `VariableBlockVector` object with 10 entries where the i-th entry is a vector of length i, write

```
VariableBlockVector<double> v(10);
std::size_t i=0;
for (auto cIt = v.createbegin(); cIt!=v.createend(); ++cIt)
  *cIt = i++;
```

The vector is not properly initialized, and no memory is allocated, unless the iterator has stepped through all vector blocks. Using such an uninitialized vector results in an error. A `CreateIterator` is a valid standard library output iterator, and therefore standard algorithms like `std::fill` or `std::copy` can also be used to set the block sizes.

Corresponding to the two constructors, there are two methods

```
void resize(size_type nblocks, size_type m)
void resize(size_type nblocks)
```

These set the number of blocks and (the first one) a uniform block size for an existing container. Unlike when resizing a `BlockVector` object, all previous content is invalidated by this. When the second `resize` method has been called, block sizes have to be set anew using the iterator mechanism described above.

Random access and iteration over the entries works just as for a `BlockVector` of `BlockVector`s. Dereferencing an iterator (or calling **operator**`[]`) yields a proxy type that behaves just like a `BlockVector`. Individual entries can be accessed using **operator**`[]` of this proxy:

```
double a = v[i][j];
v[i][j] = 42.0 + a;
```

In all other regards, `VariableBlockVector` behaves just like an object of type `BlockVector<BlockVector>`, and no separate description is necessary.

7.3. Data Structures for Matrices

The `dune-istl` interface for matrices follows the same ideas as the interface for vectors, but since matrices have more algebraic structure the interface consists of more methods.

7.3.1. Abstract Interface

Conceptually, a `dune-istl` matrix is a two-dimensional container with additional methods for linear algebra. It is to be interpreted as a container of vectors, even though the implementation may be different. The organization is row-wise, i.e., the outer container is interpreted as a container of matrix rows.

	`Matrix()`
	Default constructor
	`Matrix(`**`const`** `Matrix& other)`
	Copy constructor
`Matrix&`	**`operator=`**`(`**`const`** `Matrix& other)`
	Assignment operator
`Matrix&`	**`operator=`**`(`**`const`** `field_type& s)`
	Assignment from scalar
`row_type&`	**`operator`**`[](size_type i)`
	Access to row
`const` `row_type&`	**`operator`**`[](size_type i)` **`const`**
	`const` access to row
`RowIterator (also: iterator)`	`begin()`, `end()`
	Iterator over rows
`size_type`	`N()` **`const`**
	Number of block rows
`size_type`	`M()` **`const`**
	Number of block columns
`bool`	`exists(size_type i, size_type j)` **`const`**
	Whether a given entry (i, j) exists in the matrix data structure

Table 7.4.: Matrix container methods

Container Interface

Construction of matrix objects depends on the individual implementations, but all must be at least default- and copy-constructible. Once a matrix object is properly set up, its number of rows and columns can be queried by the member methods[12]

```
size_type N() const    // Number of rows
size_type M() const    // Number of columns
```

respectively. Each matrix entry may be a matrix in its own right, and the methods N and M count the number of block rows and columns.

Access to (block) matrix rows is possible by **operator**`[]`. Each row is expected to implement the container part of the dune-istl vector interface; hence it also implements **operator**`[]`, and the entry a_{ij} of a matrix called 'a' can be accessed by a`[i]``[j]`. This is the natural way to index dense matrices.

Things are more complicated if the matrix data structure is sparse. Complete rows may be omitted from the data structure if they contain only zeros. The behavior of **operator**`[]` is undefined for such cases: it may return an empty row, but it may as

---

[12]As in the case of the corresponding method in the vector interface, there is no good reason for these two methods to start with capital letters.

class	block_type (also: value_type)
	Type used for matrix entries
class	field_type
	Number type used for scalars
class	row_type
	Data structure type used to implement matrix rows
class	RowIterator (also: iterator)
	Iterator over matrix rows
class	ColIterator
	Iterator over the entries of a row (same as row_type::iterator)
class	ConstRowIterator (also: const_iterator)
	const iterator over matrix rows
class	ConstColIterator
	const iterator over entries of a row (same as row_type::const_iterator)
class	allocator_type
	Standard-library-compliant allocator used for memory management (optional)

Table 7.5.: Exported static data of dune-istl matrix types

well do something else. The same problem can happen within each row: in the i-th
row, the entry in column j may not exist in the data structure. Calling

```
auto value = a[i][j];
```

then leads to undefined behavior of the second invocation of **operator**[]. To check
whether a given entry (i, j) exists in the matrix data structure, the interface offers
the method

```
bool exists(size_type i, size_type j) const
```

Only if this method returns **true** for given i and j it is safe to access a[i][j]. In
addition, each matrix row has a method

```
size_type size() const
```

which returns the number of block entries in that row that are actually stored in the
data structure.

To efficiently access only the nonzero entries of a sparse matrix, the iterator interface
has to be used. Matrices offer a standard-library-compliant forward iterator over the
rows of the matrix. This iterator may skip rows if they are completely empty. So, the
following code iterates over all nonzero rows of a matrix:

```
for (auto row = matrix.begin(); row != matrix.end(); ++row)
  std::cout << "row: " << row.index()
            << " has " << row->size() << " entries." << std::endl;
```

Observe how the non-standard iterator method index allows to obtain the number
of the current row. Unfortunately, this method prevents the use of the more readable
range-based **for** syntax in most relevant cases. It is possible to write

```
for (auto&& row : matrix)
  std::cout << "row has " << row->size() << " entries." << std::endl;
```

but the row number is not available in such a loop, whose usefulness is therefore limited.[13]

In turn, each row supports random access iteration over its nonzero entries by providing methods

```
row_type::iterator begin()
row_type::iterator end()
```

For each such entry, the column index is available by the method

```
size_type index () const
```

of the column iterator. The following code shows how to loop over the nonzero entries of a sparse matrix

```
for (auto row = matrix.begin(); row != matrix.end(); ++row)
  for (auto col = row.begin(); col != row.end(); ++col)
    std::cout << "entry: " << row.index() << ", " << col.index() << std::endl;
```

Note that depending on the type of matrix, each matrix entry may again be a sparse matrix. To really get to the scalar entries in an arbitrarily nested dune-istl matrix requires template metaprogramming.

The previous discussion has implicitly assumed that all matrix entries are represented by the same C++ type. There are heterogeneous matrix containers like MultiTypeBlockMatrix, which is the matrix cousin of MultiTypeBlockVector (see Chapter 7.3.2). In that case static loops are necessary to iterate over all entries.

Linear Algebra Operations

The list of linear algebra methods in the dune-istl matrix interface consists mainly of matrix–vector multiplications and matrix norms. Some implementations also allow matrix inversion and the solving of linear systems of equations.

First, though, as elements of a vector space, matrices can be added, and multiplied with scalars. Similar to the case of dune-istl vectors, only the class FieldMatrix has a binary **operator+**, and generally writing

```
SomeISTLMatrix a,b,c;
c = a+b;
```

will not compile. Instead, there are

```
Matrix& operator+=(const Matrix& other)
```

and

---

[13]A proxy mechanism that properly supports this kind of loops is part of DUNE 2.8.

Matrix&	**operator**+=(**const** Matrix& other) $A \longleftarrow A + B$
Matrix&	**operator**-=(**const** Matrix& other) $A \longleftarrow A - B$
Matrix&	**operator***=(**const** field_type& alpha) $A \longleftarrow \alpha A$
Matrix&	**operator**/=(**const** field_type& alpha) $A \longleftarrow \alpha^{-1} A$
void	mv(**const** X& x, Y& y) $y \longleftarrow Ax$
void	umv(**const** X& x, Y& y) $y \longleftarrow y + Ax$
void	mmv(**const** X& x, Y& y) $y \longleftarrow y - Ax$
void	usmv(**const** field_type& alpha, **const** X& x, Y& y) $y \longleftarrow y + \alpha Ax$
void	umtv(**const** X& x, Y& y) $y \longleftarrow y + A^T x$
void	mmtv(**const** X& x, Y& y) $y \longleftarrow y - A^T x$
void	usmtv(**const** field_type& alpha, **const** X& x, Y& y) $y \longleftarrow y + \alpha A^T x$
void	umhv(**const** X& x, Y& y) $y \longleftarrow y + A^H x$
void	mmhv(**const** X& x, Y& y) $y \longleftarrow y - A^H x$
void	usmhv(**const** field_type& alpha, **const** X& x, Y& y) $y \longleftarrow y + \alpha A^H x$
real_type	frobenius_norm() The Frobenius norm of the matrix
real_type	frobenius_norm2() The squared Frobenius norm of the matrix
real_type	infinity_norm() The $\|\cdot\|_\infty$-norm, if the matrix has scalar entries (see text)
real_type	infinity_norm_real() The real variant of the $\|\cdot\|_\infty$-norm, if the matrix has scalar entries (see text)
void	solve(X& x, **const** Y& b) $x \longleftarrow A^{-1} b$
void	invert() $A \longleftarrow A^{-1}$
Matrix&	leftmultiply(**const** Matrix& B) $A \longleftarrow BA$

Table 7.6.: Matrix linear algebra methods. The symbol A^H denotes the conjugate transpose of the matrix A.

```
Matrix& operator-=(const Matrix& other)
```

and matrices are added by writing

```
c = a;
c += b;
```

This will not run faster than a naive implementation of a=b+c;, but at least it makes the fact that non-optimizing compilers will emit *two* loops obvious to the reader of the source code.

Adding a sparse matrix b to a sparse matrix c may require inserting additional entries in the pattern of c. For important sparse matrix data structures this can be quite expensive [133]. Therefore—and because such non-matching additions happen only rarely in finite element applications—the dune-istl matrix interface does not specify the behavior of **operator+=** and **operator-=** for the case that the pattern of b is not a subset of the pattern of c. For multiplication with scalars, there are two methods

```
Matrix& operator*=(const field_type& s)
Matrix& operator/=(const field_type& s)
```

This concludes the description of the vector space interface for dune-istl matrices. Next, there is a number of methods for matrix–vector multiplication. Suppose that A is a matrix, and x and y are vectors of appropriate length. The method

```
void mv(const X& x, Y& y) const
```

replaces the content of the vector y by the product Ax. All other multiplication methods have a similar signature, but add their results to y rather than replacing the content of y. For example, the method

```
void umv(const X& x, Y& y) const
```

adds the product Ax to the content of y. Similarly,

```
void mmv(const X& x, Y& y) const
```

subtracts the product Ax from y, and

```
void umtv(const X& x, Y& y) const
```

adds the result of $A^T x$ to y. See Table 7.6 for a complete list of these methods.

Two types of norms are provided by the dune-istl matrix interface. Let $A \in \mathbb{R}^{n \times m}$ be a matrix with scalar entries a_{ij}. The Frobenius norm of A is

$$\|A\|_F := \sqrt{\sum_{i=0}^{n-1} \sum_{j=0}^{m-1} a_{ij}^2},$$

and is available through the method

```
real_type frobenius_norm() const
```

To avoid the costly square root, which is not always needed, there is also the method

```
real_type frobenius_norm2() const
```

which returns $\|A\|_F^2$.

The definition of the Frobenius norm extends in a straightforward way to the case of nested block matrices. Suppose that the entries of A are not scalars, but matrices A_{ij} in their own right. Then, the Frobenius norm can be written in a recursive way as

$$\|A\|_F := \sqrt{\sum_{i=0}^{n-1} \sum_{j=0}^{m-1} \|A_{ij}\|_F^2}.$$

Therefore, the frobenius_norm method will return the correct value even for matrix types with arbitrarily complex nesting structure.

Things are more complicated for the second type of norm. Let A be a matrix with scalar entries again. The row-sum norm is

$$\|A\|_\infty := \max_{i=0,\ldots,n-1} |a_i|_1 = \max_{i=0,\ldots,n-1} \sum_{j=0}^{m-1} |a_{ij}|.$$

It is denoted with the symbol $\|\cdot\|_\infty$ because it is the induced norm of the vector infinity norm:

$$\|A\|_\infty = \max_{x \neq 0} \frac{\|Ax\|_\infty}{\|x\|_\infty}, \qquad \text{with} \qquad \|x\|_\infty := \max_{i=0,\ldots,m-1} |x_i|, \qquad x \in \mathbb{R}^m.$$

For matrices with scalar entries, this norm is available through the method

```
real_type infinity_norm() const
```

Unfortunately, this method is difficult to implement correctly for nested matrices. Therefore, most dune-istl matrices implement the row-sum norm correctly for scalar matrices, but provide only the following recursive approximation for nested block matrices:

1 Method infinity_norm(A)
2 **if** A *is scalar* **then**
3 $\quad$ **return** $|A|$
4 **else**
5 $\quad$ **foreach** *row* i **do**
6 $\quad\quad$ $s_i \longleftarrow \sum_j$ infinity_norm(A_{ij})
7 $\quad$ **end**
8 $\quad$ **return** $\max_i s_i$
9 **end**

Be warned, however, that this is not the induced norm of the vector ∞-norm.

For complex-valued matrices, computing the absolute values of matrix entries involves the square-root function. The method

```
real_type infinity_norm_real() const
```

replaces the one-norm $|a_i|_1 := \sum_{j=0}^{m-1} |a_{ij}|$ in the definition of $\|A\|_\infty$ by the method `one_norm_real` from the dune-istl vector interface. That is, for complex entries $|z| := \sqrt{(\operatorname{Re} z)^2 + (\operatorname{Im} z)^2}$ is replaced by $|z|_r := |\operatorname{Re} z| + |\operatorname{Im} z|$. This approximation saves calls to the costly square root method.

As an optional extension of the interface, some matrices provide methods for matrix inversion and solving of linear systems of equations. Typically, this is the case for matrix implementations for which cheap canonical algorithms for these tasks exist. For these matrices,

```
void invert()
```

will invert the matrix in situ. The method

```
void solve(X& x, const Y& b)
```

will solve the system $Ax = b$. Finally, several matrix implementations provide the method

```
Matrix& leftmultiply(const Matrix& B)
```

which replaces the matrix A by the product BA.

7.3.2. Matrix Implementations

Currently, the DUNE core modules provide eight different implementations of the dune-istl matrix interface. Not all of them reside in dune-istl itself. The FieldMatrix and DiagonalMatrix classes are contained in the dune-common module, because they are used by grid implementations.

FieldMatrix

The class

```
template<class T, int n, int m>
class FieldMatrix
```

implements a dense matrix of static size. The data is allocated within the object, hence stack-allocated FieldMatrix objects should only be used for small matrices. FieldMatrix is the canonical matrix type to use together with FieldVector objects. In a typical situation, 3×3 FieldMatrix objects serve as the entries of a matrix for linear elasticity (as in Figure 7.2(a)).

The entry type T should be a real or complex number type such as **double**, **float**, complex<**double**>, etc. With the same reasoning as for the FieldVector type, the

FieldMatrix class ends the nesting recursion, i.e., it is not possible to use other dune-istl matrices as entries of a FieldMatrix.

Since a small dense matrix type is useful beyond linear algebra, the class is not implemented in the dune-istl module, but rather in dune-common, in the file dune/common/fmatrix.hh. For example, many grid implementations use it for the derivative of the mappings F from the reference element to the grid elements.

The FieldMatrix class implements the usual default- and copy constructors. Following the dune-istl matrix interface it can also be constructed from scalars. The line

```
FieldMatrix<double,3,3> m(0);
```

constructs a 3×3 matrix filled with zeros. Note the following pitfall: the line

```
FieldMatrix<double,3,3> m(1);
```

will *not* set up an identity matrix! The proper way to set up such a matrix is to assign from a ScaledIdentityMatrix (see below):

```
FieldMatrix<double,3,3> m = ScaledIdentityMatrix<double,3>(1);
```

Finally, a FieldMatrix can be constructed from a std::initializer_list. Another way to really set up a 3×3 identity matrix is

```
FieldMatrix<double,3,3> matrix = {{1,0,0},
                                  {0,1,0},
                                  {0,0,1}};
```

Since FieldMatrix implements a dense matrix, access using **operator**[] is straightforward. The return value of **operator**[] is a reference to a FieldVector, which is used to implement the matrix rows. Hence it is possible to write

```
FieldMatrix<double,3,3> matrix;
[...]
FieldVector<double,3>& firstRow = matrix[0];
```

to handle single rows, and

```
double b = matrix[i][j];
matrix[i][j] = 1.0;
```

for individual entries. Iterator access as described in Chapter 7.3.1 works just as well, but cannot be expected to be faster than direct access using square brackets.

As FieldMatrix objects are expected to be small, the FieldMatrix class implements the optional methods solve and invert. Internally, hand-coded formulas are used for small matrices, and an LU-decomposition is used for larger ones. Note that these algorithms cannot exploit symmetry if the matrix is symmetric. If it is known that a given FieldMatrix is symmetric, one may want to use a different solver rather than calling solve.

In addition to the vector space methods of the dune-istl matrix interface, the FieldMatrix class implements the corresponding binary (i.e., two-argument) operations

```
FieldMatrix operator+(const FieldMatrix& other) const
FieldMatrix operator-(const FieldMatrix& other) const
FieldMatrix operator*(const FieldMatrix& other) const
```

Hence it is possible to write code like

```
FieldMatrix<double,3,3> A, B, C;
auto D = A + 0.5 * B * C;
```

which is much more readable than code that uses only the interface methods.

DiagonalMatrix and ScaledIdentityMatrix

There are two implementations of sparse matrices with particular static patterns and static sizes:

```
template<class T, int n>
class DiagonalMatrix
template<class T, int n>
class ScaledIdentityMatrix
```

The first one is contained in dune/common/diagonalmatrix.hh (in the dune-common module), and the second one in dune/istl/scaledidmatrix.hh (in dune-istl). The DiagonalMatrix class implements a diagonal matrix, which internally only stores the diagonal entries. More radically, ScaledIdentityMatrix represents multiples of the identity matrix. Such matrix implementations can lead to vast reductions in run-time and memory footprint. Certain phase-field models, for example, involve the vector Laplacian, which is simply the regular Laplacian acting separately on all components of a vector field [111]. The stiffness matrix of such a system can be written as a normal Laplace matrix, each entry however being replaced by a scaled multiple of the $m \times m$ identity (m being the number of vector coefficients). Such a stiffness matrix can be implemented as

```
BCRSMatrix<ScaledIdentityMatrix<double,m> >
```

which uses a lot less memory than the naive

```
BCRSMatrix<FieldMatrix<double,m,m> >
```

which stores the identity matrices explicitly. As a consequence, matrix–vector products become much faster, because a dedicated algorithm for multiplication with diagonal matrices is used. Finally, smaller matrices save memory bandwidth, and hence further increase execution speed.

As a second application, DiagonalMatrix objects are used to speed up the implementations of structured grids. In the YaspGrid grid manager (Chapter 5.10.2), all elements are axis-aligned hypercubes. Hence, the derivative of the map F from the reference hypercube to any element is always diagonal. YaspGrid therefore uses DiagonalMatrix objects for the return value of the jacobianTransposed and jacobianInverseTransposed methods.

Both matrix types can be default- and copy constructed. Construction from a scalar is also supported, and fills all matrix elements on the diagonal with that scalar. Hence, in contrast to the `FieldVector` implementation, both

```
DiagonalMatrix<double,3> m(1);
```

and

```
ScaledIdentityMatrix<double,3> m(1);
```

really do produce identity matrices. Finally, `DiagonalMatrix` objects can be set up from a `std::initializer_list`. The code

```
DiagonalMatrix<double,3> matrix = {1,2,3};
```

creates a 3×3 matrix with the given numbers on the diagonal.

`DiagonalMatrix` and `ScaledIdentityMatrix` implement all matrix–vector multiplications and norms mandated by the dune-istl interface. These are all special-purpose implementations, and therefore much faster than the corresponding methods of `FieldMatrix`. Access to individual matrix elements works as described in the section on the general matrix container interface. There are iterators over rows and over entries of rows, and a row-entry iterator will only stop at the diagonal entries. Random-access requires more care: Elements can be accessed directly using two pairs of square brackets:

```
matrix[0][0] = 3.14;
double pi = matrix[0][0];
```

However, this is only well-defined for diagonal elements. Accessing off-diagonal entries this way leads to undefined results, because they do not exist in memory. In particular, reading off-diagonals may not return 0. To really get random-access to the off-diagonal entries requires to copy either matrix into a `FieldMatrix`:

```
DiagonalMatrix<double,3> a = {1,1,1};
FieldMatrix<double,3,3> b = a;
double c = b[2][0];  // well-defined: c equals zero
```

Finally, due to their special structure, diagonal matrices are trivial to invert. Therefore, both `DiagonalMatrix` and `ScaledIdentityMatrix` implement the optional interface methods

```
template<class Vector>
void solve(Vector& x, const Vector& b) const
```

and

```
void invert()
```

Both use the obvious optimal algorithms for inverting diagonal matrices. So, for example, given a `DiagonalMatrix` object D and `FieldVector` objects x and b of the correct length,

```
D.solve(x,b);
```

will solve the linear system $Dx = b$. Such systems occur, e.g., in the inner loops of multigrid smoothers for the vector Laplace problem [111].

Matrix

The following two sections again describe dense matrix implementations. The class

```
template<class B, class A=std::allocator<B> >
class Matrix
```

is a dense matrix with run-time number of rows and columns. It is the default choice for element stiffness matrices. A matrix object can be constructed by giving the number of rows and columns

```
Matrix<double> matrix(10,10);
```

which yields a 10×10 scalar matrix with uninitialized values. Alternatively, the Matrix object can be default-constructed, and the size can be set separately:

```
Matrix<double> matrix;
matrix.setSize(10,10);
```

The size of the matrix can be changed at any time; however, changing the size invalidates all matrix data. The Matrix class implements the full dune-istl matrix interface, with the exception of the invert and solve methods. It also implements the row and column iterators. However, as the matrix is dense, accessing individual block entries by matrix[i][j] is just as efficient.

MultiTypeBlockMatrix

In contrast to the Matrix class , the MultiTypeBlockMatrix is a dense *heterogeneous* matrix with a static number of rows and columns. It is the matrix analogon to MultiTypeBlockVector—each matrix entry has to be present, but each can be a different C++ type. This makes it the natural matrix data structure to use for mixed finite element and multi-physics problems. It is provided in the file dune/istl/multitypeblockmatrix.hh.

The class declaration is

```
template<typename... Args>
class MultiTypeBlockMatrix
```

The template parameter is *variadic*, i.e., the class accepts a variable number of template arguments. Each argument is expected to implement a matrix row. More specifically, each template argument must be a MultiTypeBlockVector class, for which each entry is a dune-istl matrix. Needless to say, all these vectors have to have the same number of entries. For example, the following code sets up a C++ type for the 2×2 block matrix of Figure 7.2(c), where the upper left block is a sparse matrix of small 2×2 matrix blocks, the lower right block is a scalar matrix, and the two off-diagonal matrices are sparse matrices consisting of 2×1 and 1×2 blocks, respectively:

```
using UpperLeft  = BCRSMatrix<FieldMatrix<double,2,2> >;
using UpperRight = BCRSMatrix<FieldMatrix<double,2,1> >;
using LowerLeft  = BCRSMatrix<FieldMatrix<double,1,2> >;
using LowerRight = BCRSMatrix<double>;

using MyMultiTypeMatrix
    = MultiTypeBlockMatrix<MultiTypeBlockVector<UpperLeft,UpperRight>,
                           MultiTypeBlockVector<LowerLeft,LowerRight> >;
```

Various constructors exist for such a type. Objects can be default-constructed and filled with data later, or they can be constructed with references to existing matrix rows. See the online documentation for a full list.

For the same reason as for the `MultiTypeBlockVector` class, there are no iterators over rows and columns of a `MultiTypeBlockMatrix` class. However, entries of a `MultiTypeBlockMatrix` can be accessed using **operator**[] and the static natural numbers `_0`, `_1`, `_2`, etc. from `dune/common/indices.hh` (see the section on `MultiTypeBlockVector`). To demonstrate this, the following code sets the size of the lower-right matrix block for an object of the example type given above:

```
myMultiTypeMatrix[_1][_1].setSize(10,10);
```

Similarly, the following code sets the upper left scalar matrix entry to 3.14:

```
myMultiTypeMatrix[_0][_0][0][0][0][0] = 3.14;
```

See Chapter 10.8 for a real-world example of the use of the `MultiTypeBlockMatrix` class.

In contrast to the entry-wise access, there is no difference at all between a `MultiTypeBlockMatrix` and other `dune-istl` matrices as far as matrix–vector products and matrix norms are concerned. `MultiTypeBlockMatrix` implements all operations listed in Table 7.6 (with the exception of the `solve`, `inverse`, and `leftmultiply` methods), and they behave as expected. For example, the following code constructs vectors x and b that match the matrix type defined above, and computes the residual $r(x) := b - Ax$:

```
using Vector = MultiTypeBlockVector<BlockVector<FieldVector<double,2> >,
                                    BlockVector<double> >;
Vector x = ...;
Vector b = ...;
auto r = b;
myMultiTypeMatrix.mmv(x,r);    // r -= Ax
```

Since these matrix–vector products are typically all that is needed to implement Krylov-type iterative solvers, the solvers presented below in Chapter 7.4.1 work flawlessly with matrices of type `MultiTypeBlockMatrix`.

$$\text{Matrix:} \qquad A = \begin{pmatrix} 0 & 3 & 0 & 0 \\ 4 & 1 & 0 & 0 \\ 0 & 5 & 9 & 2 \\ 6 & 0 & 0 & 5 \end{pmatrix}, \qquad \text{nonzeroes}(A) = 8,$$

Representation:

val	3	4	1	5	9	2	6	5
col	1	0	1	1	2	3	0	3

row	0	1	3	6	8

Figure 7.4.: Example of a matrix stored in compressed-row-storage (CRS) format

BCRSMatrix

The class

```
template<class B, A=std::allocator<B> >
class BCRSMatrix
```

implements a general sparse matrix in compressed-row-storage (CRS) format (the prefix BCRS abbreviates Block-CRS). This is the format most frequently used for finite element implementations, and therefore BCRSMatrix is the standard choice for finite element stiffness matrices. It is defined in the header dune/istl/bcrsmatrix.hh.

To understand how BCRSMatrix objects are constructed and used, we need to briefly review the CRS format. For more details read, e.g., [133], where the format is called *sparse row-wise format*. Sparse matrices in the CRS format consist of three arrays, which we will call val, col, and row. The first one stores all nonzero values in the matrix, ordered from top-left to bottom-right.[14] The second array col, of the same size as the first, contains the corresponding column index of each value. Finally, the third array row contains one integer entry for each row, which is the index of the first entry in this row in the other two arrays. Figure 7.4 shows how the data structure looks like for an example 4×4 matrix. To access the entry in, say, the third row and second column, the first step is to look up the third value in row, which is '3'. Additionally, one notes the following value, which is '6'. One then searches the entries of col between 3 and 6 for the desired column number, which in this case is '1'. If, as in this example, the column index is found, the corresponding entry of val is the desired matrix entry. Otherwise, the matrix entry is zero.

The CRS data structure is very compact, and very little information is stored for the pattern. Iterating over all nonzero entries of a row is cheap, and with the column index available, multiplying a vector from the right can be implemented efficiently. This makes CRS matrices attractive to use with Krylov solvers, which

---

[14] Variations of the format do not order within each row, but the BCRSMatrix implementation does.

use only matrix–vector products. A bit surprisingly, even though iterating over the nonzero entries of a column is expensive, multiplying a vector from the left is also cheap [133, Chapter 7.14]. Direct access to a matrix element (i, j) on the other hand, involves a binary search on all elements of the i-th row, and is therefore relatively expensive. Nevertheless, Gauss–Seidel and Jacobi smoothers for multigrid methods can be implemented cheaply, too.

As a further consequence, once a CRS matrix is set up, it is very expensive to add or remove further matrix elements, because the contents of the (large) val and col arrays would need to be shifted. Some implementations allow a little extra memory at the end of each row for this purpose, but of course this increases the overall memory consumption and decreases the cache performance.

The BCRSMatrix class implements a sparse matrix using the CRS format. True to the dune-istl nesting strategy, each entry of a BCRSMatrix can be a dune-istl matrix or a number type, set by the first template parameter B. For example,

BCRSMatrix<**double**>

is the canonical choice for sparse matrices with scalar entries. For vector-valued equations with, say, 3 vector coefficients, then

BCRSMatrix<FieldMatrix<**double**,3,3> >

is more appropriate, because it lowers the amount of data needed to store the matrix pattern (in comparison to a matrix with scalar entries), and gives the solution and assembly algorithms more information about the matrix structure.

The BCRSMatrix class implements the entire matrix interface described in Chapter 7.3.1, including all norms and matrix–vector products. Note that for the reasons given on Page 260, the method infinity_norm will *not* compute the matrix norm induced by the vector ∞-norm unless the entries of the BCRSMatrix are scalar. As there are no canonical ways to invert a general sparse matrix, the BCRSMatrix class does not implement the invert and solve methods.

Infrastructure for printing BCRSMatrix objects to the screen or writing them to files is available in the files dune/istl/io.hh and dune/istl/matrixmarket.hh.

How to Construct a BCRSMatrix Object

The main challenge concerning the implementation of a CRS matrix for finite element simulations is how to set it up. The most important use for a CRS data structure is to hold the stiffness matrix. The values of this matrix are computed during a loop over the grid elements (Chapter 2.1.3), and unless the grid entities are ordered in very particular ways, the entries of the matrix are accessed in essentially random order. However, the values can only be written into the matrix if the pattern (i.e., the col and row fields) are already set up. A naive implementation would iterate twice over all elements: once to determine the matrix pattern, and a second time for the actual values. Other approaches can set up the matrix with a single loop, but have other drawbacks.

The BCRSMatrix class provides three different ways to set up the pattern. They are called *build modes*, and they are different compromises between ease of use, memory consumption, and speed. A fourth way uses the MatrixIndexSet class, which makes matrix setup even easier at the price of a larger memory footprint.

To construct a BCRSMatrix object, call one of the constructors

```
BCRSMatrix(size_type n,
           size_type m,
           BuildMode bm)
BCRSMatrix(size_type n,     // This constructor only for implicit build mode
           size_type m,
           size_type avg,
           double compressionBufferSize,
           BuildMode bm)
```

Both create an empty matrix with uninitialized pattern. Each has a parameter of the enumeration type BuildMode, which can take one of the three values BCRSMatrix::random, BCRSMatrix::implicit, and BCRSMatrix::row_wise (see below for an explanation of the avg and compressionBufferSize parameters). The second constructor can only be used with the implicit build mode. In addition, BCRSMatrix objects can be default-constructed. In this case, the matrix build mode and dimensions are set by the methods

```
void setSize(size_type rows, size_type columns)
void setBuildMode(BuildMode bm)
void setImplicitBuildModeParameters(size_type avg,
                                    double compressionBufferSize)
```

We discuss each of the four approaches in turn, and give complete example programs for each build mode. The example programs are part of a single file which is attached to the electronic form of this document. It is not reprinted in the appendix, because what is shown here is pretty much the entire file.

The random Build Mode The random build mode constructs the entire pattern before any actual matrix entry is accepted by the BCRSMatrix object. It requires a good upper bound for the number of nonzero entries in each row. Ideally, these numbers are known exactly.

To construct a matrix pattern in random mode, first set the number of nonzero entries for each row. Loop over the row numbers $i = 0, \ldots, n-1$, and for each row number i call the member method

```
void setrowsize(size_type i, size_type numberOfNonzeroEntriesInIthRow)
```

Ideally, the argument numberOfNonzeroEntriesInIthRow should be the exact number of nonzero entries in the i-th row. It is also possible to provide a larger number here, in which case the row and val arrays will contain chunks of unused memory. This will not lead to errors or undefined behavior, but it increases the memory consumption and may hamper execution speed.

After all row sizes have been set, call the method

```
void endrowsizes()
```

Then, the actual pattern has to be set. For each nonzero entry a_{ij} that will appear in the matrix, call

```
void addindex(size_type row, size_type col)
```

Entering more indices in a row than previously announced is an error. Once all entries have been entered, call

```
void endindices()
```

At this point, the matrix pattern is completely initialized, and numerical values can be written to it. To show a complete example, the following code sets up the 4×4 matrix from Figure 7.4:

```
15    BCRSMatrix<double> matrix(4, 4, BCRSMatrix<double>::random);
16
17    matrix.setrowsize(0,1);
18    matrix.setrowsize(1,2);
19    matrix.setrowsize(2,3);
20    matrix.setrowsize(3,2);
21    matrix.endrowsizes();
22
23    matrix.addindex(0,1);
24    matrix.addindex(1,0);
25    matrix.addindex(1,1);
26    matrix.addindex(2,1);
27    matrix.addindex(2,2);
28    matrix.addindex(2,3);
29    matrix.addindex(3,0);
30    matrix.addindex(3,3);
31    matrix.endindices();
32
33    matrix[0][1] = 3;
34    matrix[1][0] = 4;
35    matrix[1][1] = 1;
36    matrix[2][1] = 5;
37    matrix[2][2] = 9;
38    matrix[2][3] = 2;
39    matrix[3][0] = 6;
40    matrix[3][3] = 5;
41
42    // From the file dune/istl/io.hh
43    printmatrix(std::cout, matrix, "random-built matrix", "--");
```

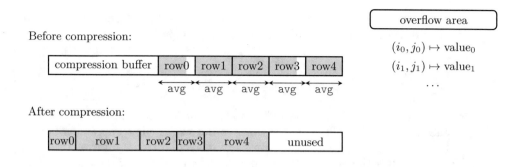

Before compression:

After compression:

Figure 7.5.: Memory layout used by the `implicit` build mode. The compression buffer is needed by the in situ compression algorithm. If it is too small, the algorithm fails.

Accessing a matrix entry that is not part of the pattern results in an error message.

The `implicit` Build Mode   Using the `random` scheme to set up a finite element stiffness matrix requires two passes over the grid: one to set up the matrix pattern, and one to assemble the actual values. This can be costly both in terms of time and memory: A full pass over the grid elements and their degrees of freedom takes time, and determining the row sizes and patterns during such a pass requires the allocation of additional bookkeeping memory. In contrast, the `implicit` build mode allows to set up and assemble a complete matrix in a single pass, provided there is an educated guess of the *average* number of matrix entries per row. It keeps its own memory footprint as low as possible, and does not require the calling code to keep additional memory at all.

When a `BCRSMatrix` object is constructed with the `implicit` build mode, it first allocates a chunk of memory with enough space for the expected number of entries, preceded by a free area called the *compression buffer*. Then, the user can start providing matrix entries, including both positions and numerical values. If the anticipated number of entries of a row is exceeded, further entries are written to a dynamically allocated second buffer called *overflow area*, which stores entries as coordinate–value pairs. After all matrix entries have been inserted, a compression step optimizes the matrix data layout, and integrates any entries from the overflow area into the standard CRS scheme. This algorithm works in-place, but it needs the compression buffer, which is overwritten in the process. The end result is a CRS data layout that is completely contiguous in a single stretch of memory (Figure 7.5).

To use the mode, construct a matrix object with the dedicated constructor

```
BCRSMatrix(size_type n, size_type m,
           size_type avg,
           double compressionBufferSize,
           BuildMode buildMode)
```

Alternatively, use one of the other constructors (setting the buildMode parameter to implicit), and provide the extra information using the methods

```
void setSize(size_type rows, size_type columns)
void setImplicitBuildModeParameters(size_type avg,
                                    double compressionBufferSize)
```

The parameter avg denotes the expected average number of matrix entries per row. The parameter compressionBufferSize denotes the size of the compression buffer, expressed as a fraction of the expected number of matrix entries of all rows together. In other words, if N is the number of matrix rows, the total size of the compression buffer will be

$$S_{\mathrm{compression}} = \mathtt{N} \times \mathtt{avg} \times \mathtt{compressionBufferSize}.$$

The matrix can then be filled with entries. However, instead of using **operator**[] for matrix rows and columns, the method

```
B& entry(size_type row, size_type col)
```

must be used. It returns a reference to the value, and implicitly creates the entry if it does not exist yet. Note, however, that newly created entries are not automatically set to zero, and hence the first access needs to be an assignment. New entries are placed in the correct matrix rows until the number of average entries is exhausted for that row. Further entries are the placed in the overflow area, which is slower, but does not have a size limit.

After the entry method has been called for each nonzero matrix entry, a call to

```
CompressionStatistics compress()
```

reorganizes the data into the standard CRS data layout (Figure 7.5). No matrix entries may be added after this step. The compression algorithm has linear time complexity, and it does not allocate further memory. It uses and overwrites the compression buffer, and if that buffer is too small the compression algorithm aborts by throwing an exception. The precise definition of "too small"—and hence a good choice for the compressionBufferSize parameter—does not only depend on the overall number of entries in the final matrix, but also on how they are distributed across the matrix rows. For example, simulations on adaptively refined grids can be expected to have more variance in the number of nonzero entries per row, and may need a larger compression buffer. See the online documentation of the BCRSMatrix class for details.

The compress method returns an object of type CompressionStatistics, which contains information on the compression step. It can be used to tune the parameters avg and compressionBufferSize.

The following example shows how to build the example matrix from Figure 7.4 using the `implicit` build mode:

```
50    BCRSMatrix<double> matrix(4, 4,
51                          2,    // Expected average number of nonzeros
52                                // per row
53                          0.2,  // Size of the compression buffer,
54                                // as a fraction of the expected
55                                // total number of nonzeros
56                          BCRSMatrix<double>::implicit);
57
58    matrix.entry(0,1) = 3;
59    matrix.entry(1,0) = 4;
60    matrix.entry(1,1) = 1;
61    matrix.entry(2,1) = 5;
62    matrix.entry(2,2) = 9;
63    matrix.entry(2,3) = 2;
64    matrix.entry(3,0) = 6;
65    matrix.entry(3,3) = 5;
66
67    // Compress matrix data structure
68    auto compressionStatistics = matrix.compress();
69
70    // From the file dune/istl/io.hh
71    printmatrix(std::cout, matrix, "implicitly-built matrix", "--");
72
73    // Show the actual average number of nonzeros per row for this matrix
74    std::cout << "Average number of nonzeros per row: "
75              << compressionStatistics.avg << std::endl;
```

The code is shorter than the previous one, and it is faster because pattern and numerical values of the matrix are set together.[15] Still, using the matrix is as efficient as using matrices built with the random mode, because they end up having the same memory layout. The price for these advantages is that the algorithm will fail if the compression buffer is too small, and that it will occupy unused memory if the buffer is too large (Figure 7.5). Whether this price is acceptable depends on the needs of the user.

The `row_wise` Build Mode The third build mode requires providing the pattern one row after the other. This is possible, e.g., for stiffness matrices for structured grids, or when reading matrices from files. Rows for which the pattern has been set can be filled with values right away. As the price for this flexibility, the values and column indices of the entries are not stored in a single contiguous chunk of memory, but rather each row is allocated separately. This may affect the cache performance.

---

[15]This does not make much of a difference in this toy example, but it does matter for the assembly of finite elements, as loops over the degrees of freedom of a finite element space can be costly.

To insert patterns row-wise, the BCRSMatrix class provides a separate iterator called CreateIterator. This iterator forward-iterates over the matrix rows. The entire range of matrix rows is obtained by calling the methods

```
CreateIterator BCRSMatrix::createbegin()
CreateIterator BCRSMatrix::createend()
```

A CreateIterator does not have all the features of a regular iterator—in particular, it cannot be dereferenced. However, matrix pattern entries can be inserted by calling

```
void CreateIterator::insert(size_type j)
```

Calling this method will add the j-th entry of the current row to the pattern. For each matrix row, the method can be called as many times as desired, and the column indices do not have to be given in any particular order. Advancing the iterator to the next row definitely commits the row pattern. The previous row can then be filled with values immediately. Once the CreateIterator has iterated over all rows, the matrix pattern is completely set, and cannot be changed anymore.

Here is how the example matrix from Figure 7.4 is set up using the row-wise build mode:

```
82    BCRSMatrix<double> matrix(4, 4, BCRSMatrix<double>::row_wise);
83
84    auto ci = matrix.createbegin();
85
86    ci.insert(1);
87    ++ci;
88    matrix[0][1] = 3;
89
90    ci.insert(0);
91    ci.insert(1);
92    ++ci;
93    matrix[1][0] = 4;
94    matrix[1][1] = 1;
95
96    ci.insert(1);
97    ci.insert(2);
98    ci.insert(3);
99    ++ci;
100   matrix[2][1] = 5;
101   matrix[2][2] = 9;
102   matrix[2][3] = 2;
103
104   ci.insert(0);
105   ci.insert(3);
106   ++ci;    // Important: advance to one-past-the-last-row,
107            // to commit the last row pattern
108   matrix[3][0] = 6;
```

```
109    matrix[3][3] = 5;
110
111    // From the file dune/istl/io.hh
112    printmatrix(std::cout, matrix, "row_wise-built matrix", "--");
```

The row_wise build mode is the most convenient mode if the pattern is known in row-wise order. However, because of the row-wise storage allocation, the matrix data is fragmented in memory, and operations like matrix–vector products may not be as fast as for matrices set up with the other build modes.

The MatrixIndexSet Approach The previous build modes tried to avoid having to do two passes over the grid in order to set up a sparse stiffness matrix efficiently. This had its cost—some sort of initial guess of the matrix row sizes was required for the implicit build mode, and a bad guess could lead to suboptimal linear algebra performance or even program termination.

In some situations, however, matrix setup cost is not an issue at all. In relation to the time spent actually using the matrix, the time needed to do several grid iterations may be irrelevant. Likewise, memory consumption is not always an issue, either. In such situations, matrix setup should simply be convenient. The dune-istl module provides the MatrixIndexSet class for this (in the file dune/istl/matrixindexset.hh). It uses the random build mode internally, but for practical purposes it constitutes a fourth matrix build mode. To use it, construct a MatrixIndexSet object, and add all pattern entries with the method

```
void add(size_type i, size_type j)
```

The entries can be added in arbitrary order. Once done, the entire pattern is copied into a BCRSMatrix object with the method

```
template<class MatrixType>
void exportIdx(MatrixType& matrix) const
```

Here is how the example matrix of Figure 7.4 is set up using the MatrixIndexSet:

```
119    MatrixIndexSet pattern(4,4);
120
121    pattern.add(0,1);
122    pattern.add(1,0);
123    pattern.add(1,1);
124    pattern.add(2,1);
125    pattern.add(2,2);
126    pattern.add(2,3);
127    pattern.add(3,0);
128    pattern.add(3,3);
129
130    BCRSMatrix<double> matrix;
131
132    pattern.exportIdx(matrix);
```

```
133
134    matrix[0][1] = 3;
135    matrix[1][0] = 4;
136    matrix[1][1] = 1;
137    matrix[2][1] = 5;
138    matrix[2][2] = 9;
139    matrix[2][3] = 2;
140    matrix[3][0] = 6;
141    matrix[3][3] = 5;
142
143    // From the file dune/istl/io.hh
144    printmatrix(std::cout, matrix, "matrix built with a MatrixIndexSet", "--");
```

The code is straightforward. The catch is that a `MatrixIndexSet` object can get large—for scalar matrices it consumes almost as much memory as the final matrix itself. Also, it does a lot of allocations and deallocations, and hence it may not be very fast. This is the price for the additional convenience.

BDMatrix and BTDMatrix

The classes `BDMatrix` and `BTDMatrix` implement (block-)diagonal and (block-)tridiagonal matrices with dynamic size, respectively. Their declarations are

```
template<class B, class A=std::allocator<B> >
class BDMatrix
template<class B, class A=std::allocator<B> >
class BTDMatrix
```

A given initialized object of either type can be used just like a general sparse matrix. However, the matrix construction and setup is easier, and the implementations are a little bit more time- and space-efficient. To construct such matrices, call the constructor with a single size (diagonal and tridiagonal matrices are quadratic):

```
BDMatrix<double> a(10);
BTDMatrix<FieldMatrix<double,2,2> > b(10);
```

and set the entries using **operator**[] for rows and columns:

```
// Make an identity matrix
BDMatrix<double> identity(10);
for (std::size_t i=0; i<identity.N(); i++)
  identity[i][i] = 1.0;
```

Even though both matrices are sparse, for every row **operator**[] is very fast because the patterns are very regular and known a priori. Accessing an entry that is not in the diagonal or tridiagonal pattern leads to undefined behavior, and must be avoided.

BDMatrix objects can also be constructed from initializer lists

```
BDMatrix<double> m = {1.0, 1.0, 1.0};
```

As for `Matrix` objects, sizes of `BDMatrix` and `BTDMatrix` objects can be changed at any moment by a call to

```
void setSize(size_type n)
```

Finally, both `BDMatrix` and `BTDMatrix` implement the `solve` method, and `BDMatrix` even implements the `invert` method. This works even for nested matrices, provided that the entry type implements `invert`, too. Mutatis mutandis, the same holds for solving linear systems of equations. The `solve` method of `BTDMatrix` implements a dedicated algorithm for tridiagonal matrices [75], with linear time and space complexity.

7.4. Solvers and Preconditioners

The `dune-istl` module implements a range of solvers for algebraic problems using the matrix and vector interfaces described above. Most of the solvers are iterative ones, and need proper preconditioning. Therefore, `dune-istl` provides an interface for preconditioners, and a set of preconditioner implementations as well. All these solvers and preconditioners run on sequential as well as on MPI-parallel machines. Of particular importance is the algebraic multigrid (AMG) preconditioner, which has been shown to scale to several hundred thousand cores [31, 100]. While this section describes the solvers and preconditioners, the AMG preconditioner has its own dedicated Section 7.5.

7.4.1. Solvers

The solver interface in `dune-istl` is based on the concept of operator equations. Let X, Y be two abstract spaces, and A an operator

$$A : X \to Y.$$

Then, for a given $y \in Y$, the task is to find an $x \in X$ such that

$$A(x) = y. \tag{7.1}$$

If such an x exists, it can be formally written using the inverse of the operator A:

$$x = A^{-1}(y).$$

For `dune-istl`, the spaces X and Y will always be Euclidean spaces. In principle, the operator A may be linear or nonlinear. However, only solvers for linear operators are currently implemented. If the operator A is linear, it can be identified with a matrix which, by abuse of notation, we also denote by A. We recover the standard linear system of equations

$$Ax = y.$$

In this case, we frequently denote the vector y as b.

The abstract operator view motivates how the solver interface is constructed. The foundation is the abstract base class

```
template<class X, class Y>
class InverseOperator
```

(Chapter 4.4.1). The template parameters X and Y must be dune-istl vector data types, and they represent the two spaces X and Y. Classes derived from InverseOperator represent inverse operators A^{-1}, and applying such inverse operators to a vector y is equivalent to solving the operator system $A(x) = y$. Consequently, the two main (and only) methods of the InverseOperator class are

```
virtual void apply(X& x, Y& y, InverseOperatorResult& res) = 0
virtual void apply(X& x, Y& y,
                   double reduction,
                   InverseOperatorResult& res) = 0
```

Both apply the inverse operator to the vector given in y, and return the result in x (i.e., they solve the system $A(x) = y$). If the implementation uses an iterative algorithm, then the content of the parameter x will be used as the initial iterate x^0. Note that the second parameter y is a non-**const** reference: The solver implementation may modify this variable, and indeed, many dune-istl solvers leave the residual in y after termination. Be aware of this: it is the source of subtle bugs.

While the first apply method lets the implementation decide how precise the evaluation of $A^{-1}y$ should be, the second one allows to request a residual reduction factor in the reduction parameter. The implementation will try to produce an $x^k \in X$ such that the residual reduction

$$\text{red}(x^k) := \frac{\|y - Ax^k\|}{\|y - Ax^0\|}$$

is less than the number given in reduction.

Both apply methods expect an object of type InverseOperatorResult as their last arguments. Upon termination of the apply method, this object is filled with information regarding the solution procedure. This information is available directly from public data members of the class. These are

```
int iterations       // Number of iterations
double reduction     // Reduction achieved: red(x^k) = ||b − A(x^k)||/||b − A(x^0)||
bool converged       // True if convergence criterion has been met
double conv_rate     // Convergence rate (average reduction per step)
double elapsed       // Elapsed time in seconds
```

In addition, the InverseOperatorResult class has a method

```
void clear()
```

which resets all values to reasonable defaults.

Actual implementations of the solver interface derive from the InverseOperator class and implement the apply methods. Several such implementations can be found in the file dune/istl/solvers.hh. For example, the class

```
template<class X>
class CGSolver : public InverseOperator<X,X>
```

implements a standard preconditioned conjugate gradient solver [146]. Observe that the only template parameter is the vector type X used for iterates and the right-hand-side vector. The matrix type is given to the solver when calling the constructor

```
template<class LinearOperator, class Preconditioner>
CGSolver(LinearOperator& linearOperator,
         Preconditioner& preconditioner,
         real_type reduction, int maxIterations, int verbose)
```

The first parameter of this represents the operator of Equation 7.1. In the linear case, Operator is not directly a matrix type, but rather a dune-istl matrix wrapped in the interface class MatrixAdapter. This extra redirection buys some additional flexibility: Nonlinear operators can replace the linear ones, and it is possible to implement matrix-free linear operators. Finally, several distributed solver algorithms need to modify the subdomain matrices, which is more easily done with the Operator layer. The second template parameter Preconditioner represents the preconditioner, to be chosen from the list in Chapter 7.4.2, or custom-implemented.

As an example, code to solve a linear system with this implementation can be seen in the following excerpt from the complete example program of Chapter 3.3:

```
376    // Choose an initial iterate that fulfills the Dirichlet conditions
377    Vector x(basis.size());
378    x = b;
379
380    // Turn the matrix into a linear operator
381    MatrixAdapter<Matrix,Vector,Vector> linearOperator(stiffnessMatrix);
382
383    // Sequential incomplete LU decomposition as the preconditioner
384    SeqILU<Matrix,Vector,Vector> preconditioner(stiffnessMatrix,
385                                                1.0);  // Relaxation factor
386
387    // Preconditioned conjugate gradient solver
388    CGSolver<Vector> cg(linearOperator,
389                        preconditioner,
390                        1e-5, // Desired residual reduction factor
391                        50,   // Maximum number of iterations
392                        2);   // Verbosity of the solver
393
394    // Object storing some statistics about the solving process
395    InverseOperatorResult statistics;
396
397    // Solve!
398    cg.apply(x, b, statistics);
```

Line 381 constructs a `MatrixAdapter` from the matrix given in the `stiffnessMatrix` object. Line 384 sets up a preconditioner; in this case an ILU preconditioner [138]. Line 388 constructs the actual CG solver, and the last line calls it with a right-hand-side vector b. After this call, x will contain the approximate solution, b the corresponding residual, and the `statistics` object will contain the required wall-time, number of iterations, etc.

There is a number of solver implementations in the file `dune/istl/solvers.hh`. Some of the more important ones are:

- `CGSolver`, `BiCGStabSolver`: These are implementations of the standard conjugate gradient and BiCGStab algorithms for symmetric positive definite and nonsymmetric matrices, respectively [138]. Both algorithms require a preconditioner.

- `LoopSolver`: Implements the standard preconditioned Richardson iteration

$$x^{x+1} = x^k + P(b - Ax^k)$$

for a given preconditioner $P \approx A^{-1}$.

- `GradientSolver`: This class solves the system $Ax = b$ by minimizing the scalar functional $J(x) = \frac{1}{2}x^T Ax - b^T x$ by a gradient method (provided that A is symmetric). In other words, iterates are computed by

$$x^{x+1} = x^k + \lambda^k P(-\nabla J(x^k)) = x^k + \lambda^k P(b - Ax^k),$$

where $P : Y \to X$ is a preconditioner, and $\lambda^k > 0$ is the optimal damping parameter

$$\lambda^k := \frac{\langle Pr^k, r^k \rangle}{\langle APr^k, Pr^k \rangle}, \qquad r^k := b - Ax^k.$$

- `MINRESSolver`, `RestartedGMResSolver`: These implement the well-known MIN-RES [84] and GMRES [139] solvers for indefinite symmetric and unsymmetric linear systems, respectively.

The `dune-istl` module does not implement direct sparse solvers itself, but it allows to call several standard open source implementations through its interface. The classes `UMFPack`, `Cholmod` and `SuperLU` from the files `dune/istl/umfpack.hh`, `dune/istl/cholmod.hh`, and `dune/istl/superlu.hh`, respectively, provide access to the UMFPACK[16] [46], CHOLMOD[17] [41] and SUPERLU[18] [116] solvers. Such solvers are very fast for small and medium-sized problems, and their speed does not depend on the condition number of the matrix. Even though the underlying libraries only support matrices with scalar entries, the dune-istl layer around them allows to use

---

[16]http://faculty.cse.tamu.edu/davis/suitesparse.html
[17]See footnote 16
[18]https://portal.nersc.gov/project/sparse/superlu

them with block matrices as well. Depending on the precise matrix type this may involve an internal copy of the matrix, which increases the storage requirements, but has negligible impact on run-time. Please see the online class documentation for details on the direct solver implementations.

7.4.2. Preconditioners

All iterative solvers of the previous section need to be preconditioned to achieve reasonable speed on any non-trivial finite element problem. For this, dune-istl offers an interface for preconditioners, together with various standard implementations. All of these run in sequential and distributed environments.

Recall that, mathematically, a preconditioner P for the equation $Ax = b$ is an approximation of the inverse of A,

$$P : Y \to X, \qquad P \approx A^{-1}.$$

The system $Ax = b$ is preconditioned by multiplying with P from the left

$$PAx = Py.$$

If done right, the matrix PA of this system is much better conditioned than the original matrix A.

Preconditioners are hardly ever computed and stored as actual matrices. Rather, they exist in the form of algorithms that implement the application of P to a residual vector $r \in \mathbb{R}^n$. A lot of standard preconditioners can be interpreted as iterative solution algorithms for the equation $Av = r$ doing one step with an initial iterate $v = 0 \in \mathbb{R}^n$.

Since preconditioners are certain forms of inverse operators, their interface in dune-istl is close to the InverseOperator interface. All preconditioners in dune-istl must inherit from the abstract base class

```
template<class X, class Y>
class Preconditioner
```

The template parameters X and Y must be implementations of the dune-istl vector interface, and are used for iterate and residual vector objects, respectively. The central method of the Preconditioner interface is

```
virtual void apply(X& v, const Y& r)
```

It applies the preconditioner P to the residual r, i.e., it computes $v = Pr$, and returns the result in the first argument v. If the preconditioner is implemented as an iterative method for the system $Av = r$, then the value of v when calling apply is the initial iterate. For standard preconditioner applications it should be equal to $0 \in \mathbb{R}^n$.

In addition, the `Preconditioner` interface mandates a setup- and a finalization method

```
virtual void pre(X& x, Y& b)
virtual void post(X& x)
```

The solver must call these before the first and after the last call to `apply`, respectively. The methods may, for example, take care of memory management tasks of the preconditioner. The parameter x of both methods is the iterate vector, and b is the right-hand side of the linear system. Many standard preconditioners disregard these parameters, but they may be useful, e.g., for certain scaling algorithms.

The `dune-istl` module provides several standard preconditioners in the file `dune/istl/preconditioners.hh`:

- The `Richardson` preconditioner approximates A^{-1} by a scaled identity matrix λI. The scalar λ can be selected in the constructor. Choosing $\lambda = 1$ here is the easiest way to have no preconditioner at all.

- The preconditioners `SeqJac`, `SeqGS`, `SeqSOR`, and `SeqSSOR` implement Jacobi, Gauss–Seidel, SOR and SSOR preconditioning, respectively. The preconditioners carry the `Seq` prefix to distinguish them from their parallel cousins, which implement the same algorithms for distributed linear systems of equations.

- `SeqILU` implements preconditioning by incomplete LU decomposition, either of zero-th or higher order [138]. The `SeqILDL` preconditioner does the same using the incomplete LDL (Cholesky) decomposition for symmetric matrices.

- `AMG` is an algebraic multigrid preconditioner. It will be discussed in detail in Chapter 7.5.

- The class `InverseOperator2Preconditioner` allows to wrap any solver (i.e., any class derived from `InverseOperator`) as a preconditioner.

In addition to the matrix and vector types, some of these preconditioners (e.g., `SeqJac`, `SeqSOR`, `SeqSSOR`) have an additional integer template parameter, the block recursion level k. These preconditioners operate on the entries of the outermost matrix type of the nesting hierarchy. The block recursion level parameter determines how inversion of blocks on the matrix diagonal is done. If these blocks are matrices and $k = 1$ (the default), then the methods `invert` and `solve` of that diagonal entry are used, whatever is more appropriate. If $k > 1$ then the preconditioner is called recursively for that diagonal block, with a block recursion level of $k - 1$. This allows for various trade-offs between execution speed and approximation power of the preconditioner. The best choice for k depends on the application and the matrix type.

7.4.3. Distributed Solvers

So far, this text has only covered `dune-istl` in situations with a single process only. However, the solvers and preconditioners are fully parallelized, and can be used on distributed machines of almost arbitrary size [31, 100]. There are two different conceptual approaches to distributed computing in dune-istl. We describe here the one that relates more directly to the previous discussions. A different one, based on abstract distributed index sets and a data consistency model described in [14], is disregarded for lack of space. In both approaches, however, the machine model is expected to be distributed; shared memory parallelism is not currently exploited in dune-istl.

Unlike alternative implementations of distributed linear algebra like EPETRA[19] or PETSc [6, 7], dune-istl does not use distributed matrix and vector data types. Matrices and vectors only ever exist on individual processes, and their relationships to other matrices and vectors on other processes must be managed by a separate agent. Conceptually, this is in contrast also to what happens in the dune-grid module, where grid objects are distributed objects that represent both the different parts of the grid and the information about the mutual relationships together.

The dune-istl parallelization concept that will be discussed here focuses on preconditioned Krylov solvers. It argues that such solvers are constructed from certain building blocks, and only some of these blocks need to know about the data distribution. The main solver implementation can then function both in sequential and in distributed situations, as long as properly parallelized building blocks are used.

The dune-istl interface identifies three building blocks used by Krylov solvers that need to know the data distribution, namely matrix–vector multiplication, preconditioner evaluation, and scalar products. These three blocks are formalized as interfaces in dune-istl. Matrix–vector multiplication is done by LinearOperator classes, preconditioners are covered by Preconditioner implementations, and there is an interface for scalar products.

The dune-istl module provides three implementations of linear operators. The first is the MatrixAdapter class, which has already been encountered in Chapter 7.4.1 on the dune-istl solver interface, and which appears in many code examples of this book. It simply wraps the local matrix without modifying it. This is in contrast to the other two, OverlappingSchwarzOperator and NonoverlappingSchwarzOperator, which do perform certain modifications to the matrix near subdomain boundaries. These are motivated by Schwarz-type domain decomposition algorithms, and will not be discussed here. Details are given in [14]. There is rarely the need to write a custom linear operator implementation, and we therefore omit the interface description here.

Sequential preconditioners have been discussed in Chapter 7.4.2. The interface does not change for distributed preconditioners, but implementations may need to keep additional information for parallel preconditioning. For example, a parallel Jacobi preconditioner may keep a consistent copy of the matrix diagonal. An example implementation is shown in the next section.

The interface for scalar products, finally, is fairly simple. Implementations need to derive from the abstract base class ScalarProduct, and (mainly) implement the two pure virtual methods

---

[19]https://trilinos.github.io/capability-areas.html

```
field_type dot(const X& x, const X& y)
```

and

```
real_type norm(const X& x)
```

Given a `LinearOperator` implementation, a parallel preconditioner, and a matching
scalar product implementation, these can be plugged into a Krylov solver to obtain a
solver that works on a distributed machine. For example, the `CGSolver` class has a
dedicated constructor that accepts a special scalar-product object, in addition to the
linear operator and preconditioner that are required in any situation:

```
template<class LinearOperator, class ScalarProduct, class Preconditioner>
CGSolver(LinearOperator& linearOperator,
         ScalarProduct& scalarProduct,
         Preconditioner& preconditioner,
         real_type reduction, int maxIterations, int verbosity)
```

A CG solver constructed with this mechanism will work as expected if the linear
system of equations is distributed across several processes. The solver itself does not
make any assumptions on the data distribution, but of course the assumptions of the
preconditioner, the linear operator, and the scalar product must match.

7.4.4. Example: Solving the Poisson Equation with a Distributed Preconditioned CG Method

This section demonstrates the use of the distributed `dune-istl` solver interfaces in a
complete example. To allow a simple comparison, the problem setting and algorithm
are precisely as in Chapter 6.7: The program solves the Poisson equation on the
L-shaped domain of Figure 6.7 with a CG solver preconditioned by a Jacobi method.
The code of the new example is therefore largely identical to the code in Chapter 6.7,
and only the parts that differ are discussed here. The complete example code is
printed in Appendix B.7; readers of an electronic version of the book can also find it
by clicking on the icon in the margin. For the grid file `l-grid-refined.msh`, which
is used by the program, click on the second icon.

The example here focuses on how to write distributed linear algebra algorithms
in the context of the `dune-istl` interfaces. This is not the same as showing how to
conveniently solve distributed linear algebra problems using `dune-istl`. Readers more
interested in that should look towards higher levels modules such as `dune-pdelab` or
`dune-fem`.

The program file starts off with the problem assembler. As in the example of
Chapter 6.7, the assembler is implemented in form of four separate methods. These
assemble the weak form of the Poisson equation in a Lagrangian finite element space
with a given volume term. The methods are `assembleElementStiffnessMatrix` and
`assembleElementVolumeTerm` for the element problems, and `getOccupationPattern`

and `assemblePoissonProblem` for the global assembly. Next follow the two data handle classes for communication between subdomains. The class `LBVertexDataHandle` is used for transmitting vertex data while distributing the grid, whereas `VertexData-Update` is used to make vectors in additive representation consistent.

As explained in the previous section, distributed computing in `dune-istl` works by combining parallel linear operators, preconditioners, and scalar products with an otherwise sequential solver implementation. For the algorithm implemented here, the `MatrixAdapter` class already used in the sequential codes in this book can be used as the linear operator. The data model simply requires linear operators to implement a matrix–vector product with a given matrix, and that is what `MatrixAdapter` does.

In contrast, preconditioner and scalar product have to be written from scratch. The first new piece of code is therefore the implementation of a preconditioner. It consists of the same code as in Chapter 6.7, but now using the preconditioner interface described in Chapter 7.4.2. The preconditioner is implemented in a class `JacobiPreconditioner` that inherits from the interface class `Preconditioner`. In the constructor it receives the grid view and a reference to the complete matrix of the *interior* element of the local process in additive representation. The constructor then uses the grid view communication methods and the `VertexUpdate` data handle to compute and store a consistent copy of the matrix diagonal. No preparation that depends on x or b is necessary—the methods `pre` and `post` exist, but remain empty:

```
341  template<class GridView, class Matrix, class Vector>
342  class JacobiPreconditioner : public Preconditioner<Vector,Vector>
343  {
344  public:
345    // Constructor
346    JacobiPreconditioner (const GridView& gridView, const Matrix& matrix)
347    : gridView_(gridView), matrix_(matrix)
348    {
349      Vector diagonal(matrix_.N());
350      for (std::size_t i=0; i<diagonal.size(); ++i)
351        diagonal[i] = matrix_[i][i];
352
353      consistentDiagonal_ = diagonal;
354      VertexDataUpdate<GridView,Vector> matrixDataHandle(gridView,
355                                                         diagonal,
356                                                         consistentDiagonal_);
357
358      gridView_.communicate(matrixDataHandle,
359                            All_All_Interface,
360                            ForwardCommunication);
361    }
362
363    // Prepare the preconditioner
364    virtual void pre(Vector& x, Vector& b) override
365    {}
366
367    // Apply the preconditioner
```

```
368    virtual void apply(Vector& v, const Vector& r) override
369    {
370      auto rConsistent = r;
371
372      VertexDataUpdate<GridView,Vector> vertexUpdateHandle(gridView_,
373                                                           r,
374                                                           rConsistent);
375
376      gridView_.communicate(vertexUpdateHandle,
377                            InteriorBorder_InteriorBorder_Interface,
378                            ForwardCommunication);
379
380      for (std::size_t i=0; i<matrix_.N(); i++)
381        v[i] = rConsistent[i] / consistentDiagonal_[i];
382    }
383
384    // Clean up
385    virtual void post(Vector& x) override
386    {}
387
388    // Category of the preconditioner
389    virtual SolverCategory::Category category() const override
390    {
391      return SolverCategory::sequential;
392    }
393
394 private:
395    const GridView gridView_;
396    const Matrix& matrix_;
397    Vector consistentDiagonal_;
398 };
```

Application of the preconditioner to a vector r in additive representation is implemented in the method `apply`. The Jacobi preconditioner simply divides r component-wise by the consistent matrix diagonal. The result is a consistent preconditioned residual, which is just what the `CGSolver` expects.

The last method `category` is required. It is part of a mechanism to ensure that `LinearOperator`, preconditioner, and scalar product match. This particular implementation has to return the value `sequential`, even though the code is for distributed computing. This is because the `MatrixAdapter` category is hard-wired to `sequential`, and the two new implementations of this example have to comply with this choice.

After the preconditioner follows the implementation of the custom scalar product. Unlike the parallel scalar products available in dune-istl, the one here expects pairs of vectors with one being in additive and one in consistent representation:

```
404 class AdditiveScalarProduct : public ScalarProduct<Vector>
405 {
```

```
406    using typename ScalarProduct<Vector>::field_type;
407    using typename ScalarProduct<Vector>::real_type;
408
409  public:
410    // Constructor
411    AdditiveScalarProduct (const GridView& gridView)
412      : gridView_(gridView)
413    {}
414
415    // Dot product of a consistent vector x and an additive vector y,
416    // or vice versa
417    virtual field_type dot (const Vector& x, const Vector& y) const override
418    {
419      return gridView_.comm().sum(x.dot(y));
420    }
421
422    // Norm of a vector given in additive representation
423    virtual real_type norm (const Vector& x) const override
424    {
425      // Construct consistent representation of x
426      auto xConsistent = x;
427
428      VertexDataUpdate<GridView,Vector> vertexUpdateHandle(gridView_,
429                                                            x,
430                                                            xConsistent);
431
432      gridView_.communicate(vertexUpdateHandle,
433                            InteriorBorder_InteriorBorder_Interface,
434                            ForwardCommunication);
435
436      // Local scalar product of x with itself
437      auto localNorm2 = x.dot(xConsistent);
438
439      // Sum over all subdomains
440      return std::sqrt(gridView_.comm().sum(localNorm2));
441    }
442
443    // Scalar product, linear operator, and preconditioner must be
444    // of the same category
445    virtual SolverCategory::Category category() const override
446    {
447      return SolverCategory::sequential;
448    }
449
450  private:
451    const GridView gridView_;
452  };
```

The constructor takes a grid view as its single argument, to be able to use the grid communication facilities. The `dot` method itself, in Lines 417–420, only uses the global communication facilities of `dune-grid`. As the two input vectors are expected to be one in additive and one in consistent representation, computing the global scalar product is simply a local scalar product on each subdomain followed by a global sum. Of course, this simple approach only works for non-overlapping grid partitions. What is more, the code that assembles the argument vectors for the scalar product must take care not to assemble contributions on *ghost* elements.

The implementation of the `norm` method does need local communication, on the other hand. The method expects a vector in additive representation. To compute a norm, the vector's consistent representation is needed in addition. That representation is computed just as in Chapter 6.7 using the `VertexDataUpdate` call-back class. Like the preconditioner, the scalar product has the category `sequential`—otherwise it cannot be combined with the `MatrixAdapter` class.

These are all the auxiliary classes needed for the example. What is left is the `main` method. It begins as in Chapter 6.7, by loading and distributing the grid. That part is repeated here for convenience:

```
457  int main(int argc, char *argv[])
458  {
459    // Set up MPI
460    const MPIHelper& mpiHelper = MPIHelper::instance(argc, argv);
461
462    // Set up the grid
463    constexpr int dim = 2;
464    using Grid = UGGrid<dim>;
465    using GridView = Grid::LeafGridView;
466
467    std::shared_ptr<Grid> grid = GmshReader<Grid>::read("l-shape-refined.msh");
468
469    std::vector<double> dataVector;
470    auto gridView = grid->leafGridView();
471
472    if (mpiHelper.rank()==0)
473    {
474      // The initial iterate as a function
475      auto initialIterate = [](auto p){return std::min(p[0],p[1]);};
476
477      // Sample on the grid vertices
478      dataVector.resize(gridView.size(dim));
479      for (const auto& vertex : vertices(gridView, Partitions::interiorBorder))
480      {
481        auto index = gridView.indexSet().index(vertex);
482        dataVector[index] = initialIterate(vertex.geometry().corner(0));
483      }
484    }
```

```
485
486    // Copy vertex data into associative container
487    std::map<Grid::LocalIdSet::IdType, double> persistentContainer;
488    const auto& idSet = grid->localIdSet();
489
490    for (const auto& vertex : vertices(gridView))
491      persistentContainer[idSet.id(vertex)]
492          = dataVector[gridView.indexSet().index(vertex)];
493
494    // Distribute the grid and the data
495    LBVertexDataHandle<Grid, std::map<Grid::LocalIdSet::IdType,double> >
496                                    dataHandle(grid, persistentContainer);
497    grid->loadBalance(dataHandle);
498
499    // Get gridView object again after load-balancing,
500    // to make sure it is up-to-date
501    gridView = grid->leafGridView();
502
503    // Copy data back into the array
504    dataVector.resize(gridView.size(dim));
505
506    for (const auto& vertex : vertices(gridView))
507      dataVector[gridView.indexSet().index(vertex)]
508          = persistentContainer[idSet.id(vertex)];
```

Next, the subproblems are assembled on the individual processes. This is also identical to the code of Chapter 6.7:

```
516    using Vector = BlockVector<double>;
517    using Matrix = BCRSMatrix<double>;
518
519    Vector rhs;
520    Matrix stiffnessMatrix;
521
522    // Assemble the Poisson system in a first-order Lagrange space
523    Functions::LagrangeBasis<GridView,1> basis(gridView);
524
525    auto sourceTerm = [](const FieldVector<double,dim>& x){return -5.0;};
526    assemblePoissonProblem(basis, stiffnessMatrix, rhs, sourceTerm);
527
528    // Determine Dirichlet degrees of freedom by marking all those
529    // whose Lagrange nodes comply with a given predicate.
530    auto dirichletPredicate = [](auto p)
531    {
532      return p[0]< 1e-8 || p[1] < 1e-8 || (p[0] > 0.4999 && p[1] > 0.4999);
533    };
534
535    // Interpolating the predicate will mark all Dirichlet degrees of freedom
536    std::vector<bool> dirichletNodes;
537    Functions::interpolate(basis, dirichletNodes, dirichletPredicate);
```

```
538
539      ////////////////////////////////////////
540      //    Modify Dirichlet matrix rows
541      ////////////////////////////////////////
542
543      // Loop over the matrix rows
544      for (size_t i=0; i<stiffnessMatrix.N(); i++)
545      {
546        if (dirichletNodes[i])
547        {
548          auto cIt    = stiffnessMatrix[i].begin();
549          auto cEndIt = stiffnessMatrix[i].end();
550          // Loop over nonzero matrix entries in current row
551          for (; cIt!=cEndIt; ++cIt)
552            *cIt = (i==cIt.index()) ? 1.0 : 0.0;
553        }
554      }
555
556      // Set Dirichlet values
557      for (std::size_t i=0; i<dirichletNodes.size(); i++)
558        if (dirichletNodes[i])
559          rhs[i] = dataVector[i];
```

The only difference in this part of the code is that the version in Chapter 6.7 contained the code to compute the consistent matrix diagonal, which is now done by the JacobiPreconditioner class.

After this setup, the actual parallel solving of the distributed linear system is simple:

```
567      // Set the initial iterate
568      Vector x(basis.size());
569      std::copy(dataVector.begin(), dataVector.end(), x.begin());
570
571      // Set up the preconditioned conjugate-gradient solver
572      double reduction  = 1e-3;  // Desired residual reduction factor
573      int maxIterations = 50;    // Maximum number of iterations
574
575      MatrixAdapter<Matrix,Vector,Vector> linearOperator(stiffnessMatrix);
576      JacobiPreconditioner<GridView,Matrix,Vector> preconditioner(gridView,
577                                                         stiffnessMatrix);
578      AdditiveScalarProduct<GridView,Vector> scalarProduct(gridView);
579
580      CGSolver<Vector> cg(linearOperator,
581                          scalarProduct,
582                          preconditioner,
583                          reduction,
584                          maxIterations,
585                          (mpiHelper.rank()==0) ? 2 : 0);   // Only rank 0
586                                                            // will print output
```

```
587
588    // Object storing some statistics about the solving process
589    InverseOperatorResult statistics;
590
591    // Solve!
592    cg.apply(x, rhs, statistics);
```

The code looks very similar to the sequential use of the `CGSolver` class in Chapter 7.4.1. The difference are the special preconditioners and scalar product that know about the data decomposition. These are constructed in Lines 577 and 578, and handed to the `CGSolver` constructor.

The only other difference to the sequential code is the output control. Recall that the last parameter of the `CGSolver` constructor is an integer verbosity level ranging from 0 to 2. Simply setting this to 2 as in the sequential example would result in screen output on all processes. This is typically not what is desired, and the code therefore sets 2 only on the master process, and 0 everywhere else.

When this program is run with

```
mpirun -np <number_of_processes> ./istl-distributed-poisson
```

the screen output is identical to the implementation of Chapter 6.7. The program also writes the result to a set of VTK files (the code for that is not shown here). These look identical to the ones from Figure 6.8.

7.5. The Algebraic Multigrid Preconditioner

Of all preconditioners for finite element problems known in the literature, the most efficient ones are usually multigrid methods. They are based on constructing approximations of the problem on various levels of resolution, and combining corrections on all these levels for a very efficient preconditioning.

Multigrid methods come in two flavors: *geometric multigrid* needs a hierarchy of (typically) nested grids to construct the different approximations. When they work, geometric multigrid methods are usually very fast, and a fair number of theoretical results can be shown [37, 88]. However, these methods also have their drawbacks. The standard coarse approximation spaces work poorly when the PDE has discontinuous or highly oscillatory coefficients [155]. Also, in real-world applications, the necessary hierarchies of grids can be difficult to obtain, because while it is easy to refine a given coarse grid, coarsening a given unstructured fine grid is challenging.

Algebraic multigrid (AMG) methods therefore take a different approach. In their pure form, they consider only the sparse matrix as input, and try to construct approximations of the linear algebra problem at various levels of resolution based on algebraic information alone. For this, off-diagonal entries a_{ij} of the matrix are interpreted as a measure for the connectedness between degrees of freedom i and j, and coarsening such that strongly coupled degrees of freedom are grouped together. Much less is known about AMG methods theoretically, but in practice they do frequently perform very well [155].

The dune-istl module contains the implementation of a state-of-the-art algebraic multigrid preconditioner, optimized for robustness and speed. It has been described in detail in [30]. The implementation runs on sequential and distributed machines, and it scales to large numbers of processes [31, 100]. As the direct setup of the distributed AMG is a bit technical, we only explain the sequential implementation here. The distributed AMG can be used easily, e.g., with the help of the dune-pdelab module.

7.5.1. Agglomeration AMG

The AMG implementation in dune-istl is of so-called *agglomeration type*. Consider a sparse linear system of equations

$$Ax = b,$$

with given invertible matrix $A \in \mathbb{R}^{n \times n}$, and vector $b \in \mathbb{R}^n$. In an initial setup phase, an AMG method constructs a hierarchy of matrices $A_L, \ldots, A_0$ of sizes $n_L > n_{L-1} > \ldots > n_0$ that represent the original matrix $A = A_L$ on $L + 1$ different levels of resolution. The representations are built by constructing prolongation operators $P_{l-1} : \mathbb{R}^{n_{l-1}} \to \mathbb{R}^{n_l}$ that map coarse vectors to finer ones. For this, an agglomeration AMG method partitions the n_l degrees of freedom on level l into groups (so-called *aggregates*), according to some measure of connection strength that is defined in terms of the matrix entries only [155]. Each such group then corresponds to one degree of freedom on level $l - 1$. The two levels are connected by the prolongation operator

$$P_{l-1} : \mathbb{R}^{n_{l-1}} \to \mathbb{R}^{n_l}, \qquad (P_{l-1})_{ij} = \begin{cases} 1 & \text{if } i \text{ is in group } j \text{ on level } l \\ 0 & \text{otherwise.} \end{cases}$$

which can be interpreted as piecewise constant interpolation. Using such a prolongation operator, vectors in $\mathbb{R}^{n_{l-1}}$ can be injected into $\mathbb{R}^{n_l}$ by

$$v \mapsto P_{l-1}v. \tag{7.2}$$

Conversely, linear functionals $r : \mathbb{R}^{n_l} \to \mathbb{R}$ can be restricted to $\mathbb{R}^{n_{l-1}}$ by applying the transpose of the prolongation operator

$$r \mapsto P_{l-1}^T r. \tag{7.3}$$

Finally, the matrix A^{l-1}, $l > 0$ is computed from A^l by Galerkin restriction

$$A^{l-1} := P_{l-1}^T A^l P_{l-1}.$$

Starting from $A^L = A$, this allows to construct the hierarchy $A_L, \ldots, A_0$ of matrices.

Given prolongation operators and a matrix hierarchy, an AMG iteration works just like a geometric multigrid one:[20] First, a few iterations of a simple iterative

---

[20] As usual, the iteration is turned into a preconditioner by applying it to the linear system with the residual to be preconditioned as the right-hand side, and starting from a zero vector.

method like SSOR [138] are applied to *smooth* the error. The resulting residual equation has mainly low-frequency components, and can be approximated on the next-coarser level. Using the precomputed matrix there, and restricting the residual by (7.3) leads to a linear problem on the next-coarser level. If that level is $l = 0$, then the problem is solved there exactly. Otherwise, the multigrid iteration is applied recursively. Afterwards, the resulting correction vector is prolongated onto level l using (7.2) (possibly with a scaling factor η), and another round of smoother iterations is applied.

The following pseudo-code shows the entire procedure. One iteration of the smoother on level l is written as a multiplication with a matrix S^l:

1 AMG(l, ν, γ, η, $\{A^j\}_{j=0}^L$, $\{S^j\}_{j=0}^L$, $\{x^j\}_{j=0}^L$, $\{b^j\}_{j=0}^L$, $\{P_j\}_{j=1}^L$)
2 **if** $l = 0$ **then**
3   // Solve exactly on the coarsest level
4 $x^0 \longleftarrow (A^0)^{-1}b^0$
5 **else**
6   // Presmoothing
7   **foreach** $i = 0, \ldots, \nu - 1$ **do**
8 $\mid$ $x^l \longleftarrow x^l + S^l(b^l - A^l x^l)$
9   **end**
10   // Restrict residual
11 $b^{l-1} \longleftarrow P_{l-1}^T(b^l - A^l x^l)$
12 $x^{l-1} \longleftarrow 0$
13   // Do multigrid on coarser level
14   **foreach** $i = 0, \ldots, \gamma - 1$ **do**
15 $\mid$ AMG($l - 1$, ν, γ, η, $\{A^j\}_{j=0}^L$, $\{S^j\}_{j=0}^L$, $\{x^j\}_{j=0}^L$, $\{b^j\}_{j=0}^L$, $\{P_j\}_{j=1}^L$)
16   **end**
17   // Scaled application of the coarse correction
18 $x^l \longleftarrow x^l + \eta P_{l-1} x^{l-1}$
19   // Postsmoothing
20   **foreach** $i = 0, \ldots, \nu - 1$ **do**
21 $\mid$ $x^l \longleftarrow x^l + S^l(b^l - A^l x^l)$
22   **end**
23 **end**

The agglomeration AMG method is not suitable as a solver, but it can be a robust preconditioner [30, 155].

7.5.2. The `dune-istl` Implementation of AMG

The `dune-istl` module implements an agglomeration AMG method in the class

```
template<class Operator, class Vector, class Smoother>
class AMG
```

which is contained in the file dune/istl/paamg/amg.hh. Unlike most other things in dune-istl, it is kept in a dedicated namespace Amg, and hence the full name of the AMG class is Dune::Amg::AMG. The class implements the abstract preconditioner interface of Chapter 7.4.2.

Of the three template parameters, the first one, Operator, is the matrix type for A, encapsulated in a MatrixAdapter class (Chapter 7.4.1). Then, Vector is a vector type with a block structure that matches the one of the matrix, and Smoother is the type of the smoother to be used. Any iterative method of Chapter 7.4.2 can be used; popular choices are the classes

```
template<class Matrix, class Vector, class Vector>
class SeqJac
template<class Matrix, class Vector, class Vector>
class SeqSOR
template<class Matrix, class Vector, class Vector>
class SeqSSOR
```

from the file dune/istl/preconditioners.hh. Further smoothers are easily implemented.

Objects of type AMG are obtained by calling one of the constructors. The most important one is

```
template<class Criterion>
AMG(const Operator& fineOperator, const Criterion& criterion,
    const SmootherArgs& smootherArgs=SmootherArgs())
```

It constructs an AMG preconditioner for a given matrix and coarsening criterion. The AMG constructor receives three objects: The matrix, the coarsening criterion object, and a set of parameters for the smoother. The parameter fineOperator is the matrix $A \in \mathbb{R}^{n \times n}$ of the original problem. For consistency with the dune-istl solver and preconditioner interfaces described in Chapter 7.4, the matrix is not given directly, but rather it has to be wrapped in a MatrixAdapter object (which then determines the class template parameter Operator). The total number L of coarsening steps is not set by the user, but rather determined automatically from the coarsening criterion.

The criterion parameter is a policy class that implements the agglomeration strategy for grouping degrees of freedom into aggregates, which form the degrees of freedom on the next-coarser level. Usually, the class to use here is

```
template<class CoarseningPolicy>
class CoarsenCriterion
```

from the file dune/istl/paamg/matrixhierarchy.hh. It is a wrapper for its template parameter CoarseningPolicy, which does the actual work. Two choices for CoarseningPolicy are predefined in dune/istl/paamg/aggregates.hh:

```
template<class Matrix, class Norm>
class UnSymmetricCriterion
```

```
template<class Matrix, class Norm>
class SymmetricCriterion
```

While the first one coarsens by considering (block) matrix entries directly, the second one combines each off-diagonal (block) entry a_{ij} with its corresponding entry from A^T to determine the connection strength between the degrees of freedom i and j. The precise criteria are geared towards PDE problems with jumping coefficients, and are described in [30].

The template parameter `Matrix` of the two `CoarseningPolicy` implementations is the dune-istl matrix type used to store A (not wrapped). The `Norm` template parameter allows to specify the function used to measure the connection strength represented by a matrix entry. This is relevant for vector-valued problems, where each matrix entry is a small matrix itself. The name `Norm` is historical and badly chosen, because in fact not all implementations are norms in the mathematical sense. Implementations of the `Norm` interface have to provide

```
real_type operator()(const MatrixEntry& entry) const
```

where `MatrixEntry` can be a number type or a matrix type like `FieldMatrix`. The file dune/istl/paamg/aggregates.hh provides a few alternatives. For scalar problems, `FirstDiagonal` is a good choice.

No matter what the actual `Norm` and `CoarseningPolicy` classes are, the wrapper class `CoarsenCriterion` always has the same constructor:

```
CoarsenCriterion(int maxLevel=100,
                 int coarsenTarget=1000,
                 double minCoarsenRate=1.2,
                 double prolongDamp=1.6,
                 AccumulationMode accumulate=successiveAccu)
```

This constructor allows to specify a number of parameters that control the agglomeration process, like the upper limit `maxLevel` for the depth L of the multigrid hierarchy. The number `coarsenTarget` specifies a target number of degrees of freedom for the coarsest level. No coarser levels will be constructed if the currently coarsest one already has less than `coarsenTarget` degrees of freedom. Similarly, if the coarsening rate, i.e., the ratio of degrees of freedom between successive levels, falls below the threshold given in the `minCoarsenRate` parameter, the coarsening will stop. The parameter `prolongDamp` allows to set the scaling factor η for the prolonged correction $x^{l-1} \mapsto \eta P_{l-1} x^{l-1}$ in Line 18 of the AMG iteration on Page 293.

The `accumulate` parameter is for distributed situations only: it controls whether the coarser parts of the matrix hierarchy are kept on a smaller number of processes (`successiveAccu`, to save communication cost) or not (`noAccu`). Note that all constructor parameters of the `CoarsenCriterion` class have default values, and do not usually have to be set explicitly.

Member methods of the `CoarsenCriterion` class allow to control further parameters. For example, the method

```
void setGamma(std::size_t gamma)
```

controls the number of multigrid iterations performed on each level (denoted by γ in the algorithm on Page 293). In the language of multigrid theory, a value of 1 will produce a V-cycle, and a value of 2 will produce a W-cycle [155]. Most other parameters influence properties of the aggregates. The methods

```
void setAlpha(double alpha)
```

and

```
void setBeta(double beta)
```

allow to set the bounds that determine whether connections between degrees of freedom are marked as "strong" ($\geq \alpha$) or "isolated" ($\leq \beta$). Desired minimum and maximum aggregate sizes can be set by the methods

```
void setMinAggregateSize(std::size_t size)
```

and

```
void setMaxAggregateSize(std::size_t size)
```

The desired maximum diameter of an aggregate (in the graph-theoretic sense, see [30, 54]) can be set by the method

```
void setMaxDistance(std::size_t size)
```

For algebraic problems that result from discretizations of isotropic PDEs, reasonable values for the aggregate size and diameter are set by

```
void setDefaultValuesIsotropic(std::size_t dim, std::size_t diameter=2)
```

where `dim` is the dimension of the PDE domain. Finally, the method

```
void setDebugLevel(int level)
```

controls the amount of debugging information printed by the AMG method. A value of 0 will lead to no information being printed at all from the preconditioner. Higher values mean more information.

In contrast to the coarsening criterion, the smoother objects are constructed internally by the AMG class. Their setup is controlled by the `smootherArgs` object given to the AMG constructor, which holds the parameters

```
int iterations
```

for the number ν of smoother iterations, and

```
double relaxationFactor
```

for the relaxation factor of SOR and related methods [138]. The type `SmootherArgs` is exported by the `AMG` class, but can also be constructed as

```
using SmootherArgs = Amg::SmootherTraits<Smoother>::Arguments;
```

The coarsest-level solver is selected automatically. Its type depends on which of the sparse direct solvers supported by `dune-istl` is found by the build system. If a sparse direct solver is found, and the coarsest problem only lives on a single process, then that direct solver is used to solve the problem on level 0. Otherwise, a BiCGStab iteration preconditioned with the smoother is used as a fall-back.

7.5.3. Example: Sequential Algebraic Multigrid

To show how the AMG method is used in a sequential setting we now discuss a complete example problem. The following code solves a linear system of equations with a preconditioned CG method and the AMG preconditioner. Unlike in the previous examples, the code does not contain a finite element assembler to set up the linear system. Rather, the matrix and right-hand side vector are read from two files in MATRIXMARKET format.[21] Besides showing how this is done in DUNE, reading the matrix from a file also leads to a much shorter example code.

The complete C++ file of the example is printed in Appendix B.8, and electronic versions of this document also contain it as a file attachment. The matrix and vector files are also embedded, and available through the two green icons. They correspond to a first-order Lagrange discretization of the Poisson problem described in Chapter 3.3. Indeed, they were written by the example code of that chapter, with the only modification being that the number of global refinement steps in Line 288 was set to 4, to obtain a larger algebraic problem. Consult the matrix market website for a wealth of other symmetric positive definite sparse matrices.

The entire example is contained in the `main` method. That method starts by setting up the problem. First, it defines the C++ types for matrices and vectors. As the example treats a scalar problem, `BCRSMatrix<double>` and `BlockVector<double>` are suitable types. Then, one object of each type is created, and filled by reading two files in the MATRIXMARKET format:

```
15  int main(int argc, char *argv[])
16  {
17    using Matrix = BCRSMatrix<double>;
18    using Vector = BlockVector<double>;
19    Matrix A;
20    Vector b;
21    loadMatrixMarket(A, "getting-started-poisson-fem-4-refinements-matrix.mtx");
22    loadMatrixMarket(b, "getting-started-poisson-fem-4-refinements-rhs.mtx");
```

To hold the CG iterates, the code sets up a vector x of the correct size, and initializes it with the value 0:

---

[21] http://math.nist.gov/MatrixMarket

```
27    Vector x(b.size());
28    x=b;
```

The next block sets up the smoother. An SSOR smoother is chosen, which works nicely with symmetric problems stemming from scalar elliptic equations with constant coefficients:

```
33    using Smoother = SeqSSOR<Matrix,Vector,Vector>;
34
35    Amg::SmootherTraits<Smoother>::Arguments smootherArgs;
36    smootherArgs.iterations = 3;
37    smootherArgs.relaxationFactor = 1;
```

The number ν of pre- and post-smoothing iterations is set to 3, and the overrelaxation factor of the SSOR method is set to 1. Note that no smoother object is constructed—that is done by the AMG class itself.

The next block sets up the coarsening:

```
42    using Norm = Amg::FirstDiagonal;
43    using Criterion
44        = Amg::CoarsenCriterion<Amg::UnSymmetricCriterion<Matrix,Norm> >;
45
46    Criterion criterion(15,                    // Maximum number
47                                               // of multigrid levels
48                         2000);                // Create coarse levels until
49                                               // problem size gets smaller
50                                               // than this
51    criterion.setDefaultValuesIsotropic(2);    // Aggregate sizes and shapes
52    criterion.setAlpha(.67);                   // Connections above this value
53                                               // are "strong"
54    criterion.setBeta(1.0e-4);                 // Connections below this value
55                                               // are treated as zero
56    criterion.setGamma(1);                     // Number of
57                                               // coarse-level iterations
58    criterion.setDebugLevel(2);                // Print some debugging output
```

As the matrix is symmetric with the exception of the modifications done for the Dirichlet boundary constraints (Chapter 2.1.4), there is no difference between the symmetric and unsymmetric criteria, and therefore the simpler unsymmetric one is chosen. Most of the parameter values set by the code have reasonable defaults and could be omitted.

The next block of code sets up the preconditioner and the CG solver. The AMG constructor is invoked in Line 66. Before, the matrix A needs to be wrapped in a MatrixAdapter object both for the preconditioner and for the solver. This happens in Line 65. Constructing the conjugate gradient solver then happens as in all previous examples:

```
62    using LinearOperator = MatrixAdapter<Matrix,Vector,Vector>;
63    using AMG = Amg::AMG<LinearOperator,Vector,Smoother>;
```

```
64
65    LinearOperator linearOperator(A);
66    AMG amg(linearOperator, criterion, smootherArgs);
67
68    CGSolver<Vector> amgCG(linearOperator,
69                           amg,           // Preconditioner
70                           1e-5,          // Desired residual reduction
71                           50,            // Maximum number of iterations
72                           2);            // Verbosity
```

The same holds for starting the solver; there is no difference to previous examples:

```
76    InverseOperatorResult r;
77    amgCG.apply(x,b,r);
78
79    std::cout << "CG+AMG did " << r.iterations <<  " iterations." << std::endl;
```

This already ends the example program. Even though the code computes the solution to a finite element problem, the result cannot be written to a VTK file, because the corresponding grids and function space bases are not known to the program.

When run, the code produces the following output:

```
%%MatrixMarket
%%MatrixMarket
Level 0 has 8113 unknowns, 8113 unknowns per proc (procs=1)
aggregating finished.
Level 1 has 1809 unknowns, 1809 unknowns per proc (procs=1)
operator complexity: 1.21792
Using a direct coarse solver (UMFPack)
Building hierarchy of 2 levels (including coarse solver)
                                        took 0.118297 seconds.
=== Dune::IterativeSolver
 Iter          Defect           Rate
    0         6.62899
    1        0.156942        0.0236751
    2        0.023006         0.14659
    3       0.0066211        0.287799
    4      0.00102052        0.154132
    5     0.000145506         0.14258
    6     2.10809e-05        0.144879
=== rate=0.121266, T=0.622719, TIT=0.103786, IT=6
CG+AMG did 6 iterations.
```

Comparing this to the behavior of the CG method with an ILU preconditioner used in Chapter 3.3, it is obvious that the AMG preconditioner performs much better for this kind of problems. It needs less than one quarter of the number of iterations, which is even more impressive considering that the AMG example does not solve the same problem as the one in Chapter 3.3, but rather a larger one obtained by two additional steps of uniform refinement.

8. Local Finite Elements and the dune-localfunctions Module

The discrete function spaces used in the finite element method are constructed by defining spaces on the individual grid elements, and specifying inter-element continuity conditions. For example, Lagrangian finite elements on triangles use complete polynomial spaces on each element, and full continuity across element boundaries. In contrast, the vector-valued Raviart–Thomas elements use true subspaces of the full polynomial spaces, and require only continuity of the normal components at the edge midpoints [32].

The definition of finite element spaces therefore has a local and a global aspect to it, and both have to be represented in software. In DUNE, the local aspect is handled by the dune-localfunctions module. It covers the handling of function spaces on individual grid elements. In DUNE jargon, finite-dimensional function spaces on individual elements are called *local finite elements*. Combining these local function spaces to global ones defined on the entire grid involves the above-mentioned continuity conditions. These are dealt with by the dune-functions module (to be discussed in Chapter 10), which builds on top of dune-localfunctions.

The dune-localfunctions module fulfills two purposes. First, it prescribes an *interface* to local finite elements. This interface tries to be general enough to encompass the needs of most implementors of finite element codes, while at the same time keeping the run-time overhead of the interface to a minimum. Secondly, the module contains *implementations* of this interface. The set of implementations contains common elements like the Lagrange and Raviart–Thomas elements, but a few exotic ones as well. The aim is to collect an ever-growing library of local finite element implementations, and make them available through the common interface.

Conceptually, local finite element spaces are quite separate from the rest of the finite element machinery. For their definition and implementation one needs individual grid elements, but not the entire grid. For the important subcase of *affine families* (see Chapter 8.1 below) it is even sufficient to only know the reference elements. Consequently, the dune-localfunctions module has very few relations to other DUNE core modules. It only depends on dune-common and dune-geometry.[1] It is therefore easy (and not frowned upon) to use the dune-localfunctions shape function library with finite element codes other than DUNE.

---

[1] In fact, this is the main reason why grid element geometries are handled in the separate module dune-geometry, rather than in dune-grid itself.

© Springer Nature Switzerland AG 2020
O. Sander, *DUNE — The Distributed and Unified Numerics Environment*, Lecture Notes in Computational Science and Engineering 140, https://doi.org/10.1007/978-3-030-59702-3_8

8.1. Finite Elements and Affine Families

In his classic textbook [42], Ciarlet defines finite elements as triplets (T, P, Σ) with the following properties:

1. T is a closed subset of $\mathbb{R}^d$ with a nonempty interior and a Lipschitz-continuous boundary. (Each such set T corresponds to an element of the grid.)

2. P is a space of functions defined over the set T,

3. Σ is a finite set of linearly independent linear forms $\boldsymbol{\sigma}_i$, $0 \le i < N$, defined over the space P.

The linear forms are called the *degrees of freedom* of the finite element. They must be unisolvent on P, that is, for each set of numbers $\alpha_0, \ldots, \alpha_{N-1}$ there must exist exactly one function ϕ in P such that $\boldsymbol{\sigma}_i(\phi) = \alpha_i$ for all $i = 0, \ldots, N-1$. This implies that P must be N-dimensional. It further implies the existence of a unique basis $\phi_0, \ldots, \phi_{N-1}$ of P such that $\{\phi_i\}_{i=0}^{N-1}$ and $\{\boldsymbol{\sigma}_i\}_{i=0}^{N-1}$ are dual to each other, that is, $\boldsymbol{\sigma}_i(\phi_j) = 1$ if $i = j$, and zero otherwise. The functions $\phi_0, \ldots, \phi_{N-1}$ are called *shape functions*.

To give two examples, let T be a simplex element of a grid in $\mathbb{R}^d$. Then, for the p-th order Lagrange element on T, the space P is the set of all polynomials in d variables on T with order at most p. The degrees of freedom $\boldsymbol{\sigma}_i : P \to \mathbb{R}$ are the point evaluations at the N Lagrange nodes, and they are linearly independent if the Lagrange nodes are suitably positioned. The dual basis to the $\boldsymbol{\sigma}_i$ consists precisely of the Lagrange polynomials. Similarly, for the space of lowest-order Raviart–Thomas elements on T, P is the space of those functions $v : T \to \mathbb{R}^d$ that have the form

$$v(x) = \vec{a} + bx, \tag{8.1}$$

where $\vec{a} \in \mathbb{R}^d$ and $b \in \mathbb{R}$. The degrees of freedom $\boldsymbol{\sigma}_i$ are the values of the normal components at the element facet centers.

Certain finite elements are invariant under affine transformations. For example, let T_1 and T_2 be two grid elements that can be transformed into each other by an affine transformation $A : T_1 \to T_2$. Then, the corresponding Lagrange finite elements are affine invariant in the sense that if $v \in P_{T_1}$, then $v \circ A^{-1} : T_2 \to \mathbb{R}$ is in P_{T_2}. The degrees of freedom transform accordingly, meaning that $\boldsymbol{\sigma}_{i,T_2}(v \circ A^{-1}) = v(A^{-1} a_{i,T_2}) = v(a_{i,T_1}) = \boldsymbol{\sigma}_{i,T_1}(v)$ for all $i = 0, \ldots, N-1$, if a_{i,T_1}, a_{i,T_2} are the Lagrange nodes on T_1 and T_2, respectively. This does not hold for all finite elements! For example, if $v : T_1 \to \mathbb{R}^d$ is a Raviart–Thomas function, i.e., of the form (8.1), then $(v \circ A^{-1})(x) = \vec{a} + bA^{-1}x$ is typically *not* of the same form.

Finite elements that are invariant under affine transformations are said to form so-called *affine families* [42]. It is sufficient to define them on a single element, the reference element, and to construct the spaces on the actual grid elements by affine transformations. Non-affine finite elements can also be constructed by transformations from the reference element, but these transformations are not simply affine, and

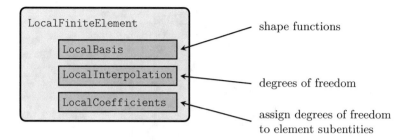

Figure 8.1.: The three parts of the local finite element interface

depend on the particular finite element (T, P, Σ). The dune-localfunctions module currently focuses on affine families of finite elements. The collection of local finite element implementations also contains a few non-affine ones, but these are given for the reference element only. The user needs to know and apply the correct transformations to the grid elements herself.

8.2. The Interface for Finite Elements Defined on the Reference Element

While the finite element interface of dune-localfunctions does not quite mirror Ciarlet's definition as a triple, it nevertheless draws quite a bit of inspiration from it. In DUNE, the corresponding concept is called a *local finite element*. The qualifier *local* is added to emphasize that such a finite element refers to a single element only. Furthermore, the word *local* is sometimes used in dune-localfunctions to emphasize that the interface currently only covers finite elements defined on reference elements.

In DUNE, a local finite element consists of three things:

1. A set of shape functions on a fixed reference element T_{ref},

2. a mechanism to evaluate the degrees of freedom $\{\sigma_i\}_{i=0}^{N-1}$ for any sufficiently smooth function v on T_{ref},

3. a map that assigns each degree of freedom to a subentity of T_{ref}, and gives it an index to distinguish it from other degrees of freedom assigned to the same subentity.

The similarities to the definition of Ciarlet are obvious. The element T does not appear explicitly in the interface, because the interface covers only finite elements on the reference element T_{ref}. However, the set of shape functions in Item 1 corresponds directly to the space P, and the degrees of freedom of Item 2 are taken directly from Ciarlet's definition.

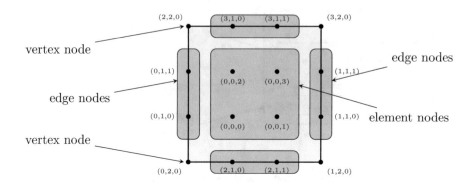

Figure 8.2.: Degrees of freedom of a third-order Lagrange element on a quadrilateral (visualized by their evaluation points), and their assignments to subentities of the reference quadrilateral. Triples of numbers (s, c, i) denote the number s of the associated subentity, the codimension c of that subentity, and the number i of the degree of freedom within that subentity.

The third item does not appear in the definition. It is not required by the mathematical theory of finite elements, but it is an important tool for their implementation. When local finite elements on individual grid elements are combined to finite element spaces on the entire grid, local degrees of freedom must be assigned to global indices, and possible continuity constraints must be incorporated. Depending on the space this can be a highly nontrivial problem. To support algorithms for this, the dune-localfunctions interface assigns each degree of freedom to a subentity of the reference element (possibly to the single subentity of codimension 0, i.e., the reference element itself), and gives it a unique index with respect to all degrees of freedom assigned to that subentity. There are natural such assignments for most finite elements that are in practical use. Figure 8.2 shows the assignment for a third-order Lagrange element. Some of the implementations in dune-localfunctions allow to apply certain symmetry transformations, e.g., edge flips, to these assignments, to facilitate the construction of global finite element spaces.

Interfaces in dune-localfunctions use the duck typing paradigm (Chapter 4.4.2). Local finite elements are represented in dune-localfunctions as classes that implement the interface class

```
class LocalFiniteElement
```

It can be interpreted as a container that holds three subobjects that implement the three parts of a local finite element, and hands them out on request. Implementations of the LocalFiniteElement interface are the only objects that typical users are expected to construct directly. In many cases, they are received from a global function space basis object, like the ones provided by the dune-functions module (Chapter 10).

The `LocalFiniteElement` interface consists primarily of the three methods

```
const LocalBasisType& localBasis() const
const LocalInterpolationType& localInterpolation() const
const LocalCoefficientsType& localCoefficients() const
```

These simply return **const** references to the three subobjects. In addition, the `LocalFiniteElement` interface provides the method

```
GeometryType type() const
```

which returns the type of the reference element on which the local finite element is defined. Finally, the method

```
unsigned int size() const
```

returns $N = \dim P = |\Sigma|$, the number of shape functions and degrees of freedom N.

The `LocalFiniteElement` class exports all types used in its public interface, but unlike the interface classes from `dune-grid`, it collects them all in a single type container called `LocalFiniteElement::Traits`. This type in turn exports:

```
class LocalBasisType          // The type returned by localBasis()
class LocalCoefficientsType   // The type returned by localCoefficients()
class LocalInterpolationType  // The type returned by localInterpolation()
```

These are occasionally useful.

8.2.1. Shape Functions and the `LocalBasis` Interface

The shape functions $\phi_i : T_{\text{ref}} \to \mathbb{R}^m$ of the finite element are implemented in classes that follow the `LocalBasis` interface. True to the duck typing paradigm there is no actual class called `LocalBasis`, and the interface only exists by convention.

The expected interface methods are listed in Table 8.1. The central ones are those that evaluate shape functions or their derivatives. It is assumed that one rarely needs to deal with single shape functions. Instead, most finite element algorithms require access to all shape functions at a given point at once. Therefore, to make the common case fast, the `LocalBasis` interface only provides methods to query *all* shape functions of the finite element together. In particular, the method

```
void evaluateFunction(const Traits::DomainType& in,
                      std::vector<Traits::RangeType>& out) const
```

evaluates all ϕ_i at a given point ξ in T_{ref}. The position ξ is provided as the first argument in an implementation-specific type (usually a `FieldVector`). The type `Traits` is exported by each `LocalBasis` implementation; see below. The output values are returned through the second argument, which is a reference to a `std::vector` of a second implementation-specific type. For scalar shape functions this second type is usually `FieldVector<double,1>`. It is an array type to accommodate vector-valued

void	evaluateFunction(**const** Traits::DomainType& in, std::vector<Traits::RangeType>& out) **const**
	Evaluate all shape functions at a given position
void	evaluateJacobian(**const** Traits::DomainType& in, std::vector<Traits::JacobianType>& out) **const**
	Evaluate Jacobians of all shape functions at a given position
void	partial(**const** std::array<**unsigned int**,Traits::dimDomain>& order, **const** Traits::DomainType& in, std::vector<Traits::RangeType>& out) **const**
	Evaluate partial derivatives of all shape functions at a given position
unsigned int	size() **const**
	The number of shape functions

Table 8.1.: Methods of the LocalBasis interface. The Traits type is exported by the LocalBasis implementation classes.

shape functions. The variable out is not expected to have the correct size upon entering the evaluateFunction method. Instead, the method itself will adjust out to the correct length before filling it.

In a similar fashion, the

```
void evaluateJacobian(const Traits::DomainType& in,
                      std::vector<Traits::JacobianType>& out) const
```

method returns first derivatives of all shape functions at a given point $\xi \in T_{\text{ref}}$ provided as the first method argument in. Note that for a shape function $\phi_i : T_{\text{ref}} \to \mathbb{R}^m$, the first derivative (i.e., the Jacobian) is an $m \times d$-matrix, where $d = \dim T_{\text{ref}}$. Therefore, the output argument of evaluateJacobian is a std::vector of an implementation-specific *matrix* type. In particular, for scalar-valued shape functions it is usually FieldMatrix<**double**,1,d>. Again, the out container is not expected to have the correct size before entering the method.

To see all this in actual code, the following code snippet constructs a second-order Lagrange finite element on a reference triangle, and prints all shape function values and gradients at the element center (see below for the Traits type):

```
const int dim = 2;
auto triangleCenter = ReferenceElements<double,dim>::simplex().position(0,0);

using LocalFEType = LagrangeSimplexLocalFiniteElement<double,double,dim,2>;
LocalFEType localFE;

// Get shape function values
using ValueType = LocalFEType::Traits::LocalBasisType::Traits::RangeType;
std::vector<ValueType> values;
localFE.localBasis().evaluateFunction(triangleCenter,values);
```

```
for (auto&& value : values)
  std::cout << value << std::endl;

// Get shape function Jacobians
using JacobianType = LocalFEType::Traits::LocalBasisType::Traits::JacobianType;
std::vector<JacobianType> jacobians;
localFE.localBasis().evaluateJacobian(triangleCenter,jacobians);

for (auto&& jacobian : jacobians)
  std::cout << jacobian << std::endl;
```

Shape function values and first derivatives are sufficient for many finite element discretizations. However, sometimes higher derivatives are needed. For this, the LocalBasis interface offers the method

```
void partial(const std::array<unsigned int,dim>& order,
             const Traits::DomainType& in,
             std::vector<Traits::RangeType>& out) const
```

which computes partial derivatives of the shape functions at a given point. In general, derivatives of arbitrary order can be requested through this interface; however, implementations of the LocalBasis interface are not required to implement them all. The behavior when an unimplemented partial derivative is requested is undefined.

The method partial has three parameters. The second and third ones, in and out, behave as for the previous methods: in receives the evaluation point $\xi \in T_{\text{ref}}$, and the result is returned in out. The array out contains one partial derivative for each shape function, and hence its type is

```
std::vector<Traits::RangeType>&
```

The desired partial derivative is specified using the first parameter, called order. It uses the multi index notation traditionally seen in analysis textbooks. Specifically, a multi-index $\alpha = (\alpha_0, \ldots, \alpha_{d-1}) \in \mathbb{N}_0^d$ designates the partial derivative

$$\partial^\alpha := \frac{\partial^{\alpha_0 + \cdots + \alpha_{d-1}}}{\partial \xi_0^{\alpha_0} \ldots \partial \xi_{d-1}^{\alpha_{d-1}}}.$$

For example, the following code snippet computes the traces of the Hessians of third-order Lagrangian shape functions on a square element at $\xi = (0.5, 0.5)$:

```
const int dim = 2;
LagrangeCubeLocalFiniteElement<double,double,dim,3> localFiniteElement;
const auto& localBasis = localFiniteElement.localBasis();
std::vector<double> traceOfHessian(localBasis.size(), 0.0);

FieldVector<double,dim> xi = {0.5, 0.5};
for (int i=0; i<dim; i++)
```

307

```
{
  std::array<unsigned int, dim> alpha;        // The multi-index
  std::fill(alpha.begin(), alpha.end(), 0);
  alpha[i] = 2;                                // Differentiate twice
                                               // in the i-th direction

  std::vector<FieldVector<double,1> > out;
  localBasis.partial(alpha, xi, out);
  for (size_t j=0; j<out.size(); j++)
    traceOfHessian[j] += out[j];
}
```

The reason why partial returns individual partial derivatives instead of full differentials is efficiency. On the one hand, the number of partial derivatives that make up a higher-order differential grows exponentially with the order. On the other hand, it is argued that many applications that need higher-order shape function derivatives will not actually need all of these components. If the full differential is needed it can still be assembled from the components.

Two auxiliary functions provide some information about the shape function set as such. The method

```
unsigned int size() const
```

returns $N = \dim P$, the number of shape functions in the set. The method

```
unsigned int order() const
```

gives some information on the polynomial order of the shape functions. If all shape functions in the set are polynomials, then order returns the order of the highest-order polynomial in the set. This information can be used to select appropriate quadrature rules. If the shape functions in the set are not polynomials, the behavior of the order method is undefined.

Like for the LocalFiniteElement interface, implementations of the LocalBasis interface must export the types used in their public interface through a type called Traits. A prototype class for this type exists in the file dune/localfunctions/ common/localbasis.hh. It has more content than the corresponding Traits type for the LocalFiniteElement interface. For example,

```
LocalBasis::Traits::RangeType
```

is the vector type used to store shape function values, and

```
LocalBasis::Traits::RangeFieldType
```

is the number type for individual components of such values. Similarly, the types

```
LocalBasis::Traits::DomainType
```

and

```
LocalBasis::Traits::DomainFieldType
```

are container and number type, respectively, for coordinates on the reference element. Finally,

```
LocalBasis::Traits::JacobianType
```

is the type used for shape function derivatives. See the online documentation of the file dune/localfunction/common/localbasis.hh for more details.

8.2.2. Degrees of Freedom and the `LocalInterpolation` Interface

The second part of the interface provides a way to evaluate the degrees of freedom of the local finite element. For sufficiently smooth functions f defined on the reference element, it computes the values $\sigma_0(f), \ldots, \sigma_{N-1}(f)$ of Ciarlet's definition. Implementations have to follow an interface class which is called LocalInterpolation (the name Interpolation is historic), and which has only a single method

```
template<class F, class C>
void interpolate(const F& f, std::vector<C>& out) const
```

Similar to the interface for shape functions, access to individual degrees of freedom is not possible; one can only evaluate all N of them at once. The first parameter f of the interpolate method is the function for which the degrees of freedom are to be evaluated. The result, i.e., the values $\sigma_0(f), \ldots, \sigma_{N-1}(f)$ are returned in the second argument, which is a std::vector of a number type C. Both the type of the function f and the type C of the values $\sigma_i(f)$ are template parameters of the method. The number type C is not used anywhere else in the dune-localfunctions interface, and therefore hardly any restrictions hold for it.

The template parameter F is the type of the function f for which the degrees of freedom are to be evaluated. The precise interface that F is expected to fulfill depends on the finite element. For example, if the degrees of freedom involve point evaluations of f, the type F has to be a function object, i.e., it has to implement

```
C operator()(const DomainType& xi)
```

where DomainType is the type used for points ξ in the reference element. Instead of returning an object of type C, the method can also return a different number type, if that type can be converted to C. Degrees of freedom that involve integral terms are typically implemented using numerical quadrature. There again f only needs to implement point evaluations.

For elements that use derivatives as degrees of freedom (see, e.g., [42, Chapter 2.2]), the derivatives of f must be available. The interface then expected of f is defined in the dune-functions module (Chapter 10.5):[2] A C++ object is a differentiable

---

[2]Even though dune-localfunctions does not technically depend on dune-functions, the duck typing idea allows to use interfaces from dune-functions in dune-localfunctions.

function if there is a free method `derivative` that returns a function that implements the derivative. In other words, the first derivative of f at a point $\xi \in T_{\text{ref}}$ must be available as

```
auto firstDer = derivative(f)(xi);
```

If necessary, the process must be nestable to allow for higher derivatives:

```
auto secondDer = derivative(derivative(f))(xi);
```

See Chapter 10.5 on `dune-functions` for more details.

It is not required that f implements a dedicated interface for directional derivatives for finite elements that use them. Rather, implementations of `LocalInterpolation` will request the full derivative of f as described above, and compute the directional derivatives from this.

8.2.3. Subentity Assignments and the `LocalCoefficients` Interface

For the finite element method, the spaces P on the individual grid elements are combined to a single space of discrete functions defined on the entire grid. This frequently involves continuity conditions between functions on neighboring elements. The space of finite element functions on the entire grid is spanned by a basis just like the element function spaces are, and it has a set of global degrees of freedom as well.

Remember from Chapter 2.1.3 that the assembly of finite element problems proceeds element by element. When assembling on an element T, the software needs to relate the local number $0 \leq p < N$ of a shape function ϕ_p or degree of freedom σ_p on T to a global basis function or degree of freedom (this is the map $p \mapsto i$, $q \mapsto j$ appearing in Chapter 2.1.3). This can be a nontrivial problem because, due to the continuity constraints, there is rarely a simple one-to-one relationship between local and global numbers. For many finite element spaces, shape functions from more than one element need to be mapped to the same global degree of freedom. This is reasonable easy for first-order Lagrange finite elements, where each degree of freedom corresponds to a grid vertex; however, for more exotic elements like higher-order Raviart–Thomas elements this identification can be quite difficult.

The construction of maps from local to global indices is one of the features of the `dune-functions` module (Chapter 10), and it is beyond the scope of `dune-localfunctions` (which deals with *local* things, after all). However, the module `dune-localfunctions` does offer some infrastructure to help implement local-to-global maps. The idea underlying this infrastructure is that while Ciarlet's formal definition of finite elements makes no reference to the geometry of the element T, virtually all degrees of freedom used in practice can be associated with faces of the (polytopal) domain of definition. For example, the shape functions of a first-order Lagrange finite element are associated to the corners (i.e., the zero-dimensional faces) of T_{ref}. For higher-order Lagrangian finite elements the shape functions are associated to edges, vertices, and even the element itself (Figure 8.2). More than one shape function may be associated to a particular face of T_{ref}. Even for very elaborate finite elements like the Argyris element [42, Chapter 2.2], each shape function or local degree of freedom can be attached to a face of the element. The same holds for Raviart–Thomas and Brezzi–Douglas–Marini elements that use edge- and facet moments as degrees of freedom [32].[3]

The `LocalCoefficients` interface describes the assignment of individual shape functions and degrees of freedom to faces (or subentities, in DUNE terminology) of the reference element. For every degree of freedom it stores

1. the codimension of the associated subentity,

2. the local number of that subentity,

3. the index in the set of all shape functions for this element associated to the same subentity.

This information is kept in objects of the class

```
class LocalKey
```

which allow access to it by the methods

```
unsigned int subEntity() const
unsigned int codim() const
unsigned int index() const
```

For the correct local numbering of the subentities see the documentation of the `ReferenceElements` in Chapter 5.5.

The `LocalCoefficient` interface class now associates the indices of the shape functions with their corresponding `LocalKey` objects. It has only two public methods

```
std::size_t size() const
```

and

```
const LocalKey& localKey(std::size_t i) const
```

The first one simply returns the number of shape functions of the local finite element. The second one takes a local shape function number i between 0 and $N - 1$, and returns the corresponding assignment to an element subentity in form of a `LocalKey` reference. For example, the following code prints these assignments for all degrees of freedom of a third-order Lagrange element on a triangle

---

[3]Exceptions are, e.g., the spaces of spline functions used in isogeometric analysis [44]. While such functions can be viewed as finite elements in some sense, the degrees of freedom cannot be reasonably assigned to element faces. However, face assignments are only a means to an end, and global basis function indices of spline spaces can be computed using alternative ways. Therefore, such spaces are still first-class citizens in `dune-functions` (Chapter 10.3.1).

```
LagrangeSimplexLocalFiniteElement<double,double,2,3> localFE;

for (size_t i=0; i<localFE.size(); i++)
  std::cout << localFE.localCoefficients().localKey(i) << std::endl;
```

The output produced by this is

```
[ subEntity: 0, codim: 2, index: 0 ]
[ subEntity: 0, codim: 1, index: 0 ]
[ subEntity: 0, codim: 1, index: 1 ]
[ subEntity: 1, codim: 2, index: 0 ]
[ subEntity: 1, codim: 1, index: 0 ]
[ subEntity: 0, codim: 0, index: 0 ]
[ subEntity: 2, codim: 1, index: 0 ]
[ subEntity: 1, codim: 1, index: 1 ]
[ subEntity: 2, codim: 1, index: 1 ]
[ subEntity: 2, codim: 2, index: 0 ]
```

From the second column it can be seen that 3 degrees of freedom are assigned to vertices, 6 are assigned to edges, and one is assigned to the element itself.

8.3. Implementations of the Local Finite Element Interface

The `dune-localfunctions` module offers a range of implementations of the finite element interface. This section briefly describes the more important ones.

8.3.1. Affine-Equivalent Finite Elements

Affine-equivalent finite elements are defined on a reference element, and mapped to grid elements by affine transformations. The following such elements are available in version 2.7 of `dune-localfunctions`.

Lagrange Elements

Lagrange elements use complete spaces of polynomials on simplex elements, and tensor products of polynomials on cubes; the degrees of freedom are point evaluations at certain fixed points $a_0, \ldots, a_{N-1} \in T_{\mathrm{ref}}$ (the Lagrange nodes). The p-th order Lagrangian shape functions $\phi_0, \ldots, \phi_{N-1}$ are those polynomials (multi-polynomials, if T_{ref} is not a simplex) of order p such that

$$\phi_i(a_j) = \delta_{ij} := \begin{cases} 1 & \text{if } i = j, \\ 0 & \text{otherwise} \end{cases}$$

for all $i, j = 0, \ldots, N - 1$.

There are two types of Lagrange element implementations in dune-localfunctions, both in the header file dune/localfunctions/lagrange.hh. The classes

```
template<class DomainField, class RangeField, int dim, int order>
class LagrangeSimplexLocalFiniteElement
```

and

```
template<class DomainField, class RangeField, int dim, int order>
class LagrangeCubeLocalFiniteElement
```

provide local Lagrange finite elements of any order for simplex and cube elements of any dimension. Here and throughout this section, the template parameters DomainField and RangeField are the number types used for domain coordinates and shape function values, respectively, and dim is the dimension of the domain T_{ref}. The number order is the polynomial or multinomial order of the space. The implementation classes have both the domain of definition and the polynomial order hard-coded into their C++ types. The compiler is therefore expected to be able to optimize code involving these classes well. Corresponding classes exist for three-dimensional prisms and pyramids.

For more run-time flexibility, dune-localfunctions also provides the class

```
template<class LagrangePoints,
         unsigned int dim,
         class DomainField, class RangeField>
class LagrangeLocalFiniteElement
```

Here, both the domain of definition (except for its dimension) and the polynomial order are run-time parameters, and have to be passed to the constructor

```
LagrangeLocalFiniteElement(const GeometryType& type, unsigned int order)
```

In contrast to the Lagrange implementation with compile-time domain type, this implementation supports all element types that can be described by the construction of Chapter 5.5.2; including four- and higher-dimensional elements that are neither simplices nor cubes. The template parameter LagrangePoints allows to control the distribution of the Lagrange points. Using the class EquidistantPointSet here produces the standard uniform spacing. See the online documentation for alternatives.

There is no implementation with compile-time reference element type and run-time order, or vice versa. For such cases, Chapter 8.4 describes a general way to switch between different implementation classes at run-time.

Incidentally, the local finite element interface provides no direct way to obtain the positions of the Lagrange nodes. These are rarely needed in finite element computations, but there is a trick to get them: For any $i = 0, \ldots, d-1$ with $d = \dim T_{\text{ref}}$, let c_i be the i-th coordinate function $c_i : (\xi_0, \ldots, \xi_{d-1}) \mapsto \xi_i$. Since $\sigma_j(c_i) = c_i(a_j) = (a_j)_i$ for the j-th Lagrange node $a_j \in T_{\text{ref}}$, one can extract the i-th coordinate of all Lagrange nodes by evaluating the degrees of freedom $\{\sigma_i\}_{i=0}^{N-1}$ on the coordinate functions c_i. The following snippet of code does this:

Figure 8.3.: Example shape functions from the RefinedP0LocalFiniteElement (left) and RefinedP1LocalFiniteElement (center and right) implementations

```
60    constexpr int dim=2;
61    LagrangeSimplexLocalFiniteElement<double,double,dim,3> localFE;
62
63    std::vector<FieldVector<double,dim> > lagrangePoints(localFE.size());
64
65    for (int i=0; i<dim; i++)
66    {
67      auto ithCoord = [&i](const FieldVector<double,dim>& x)
68      {
69        return x[i];
70      };
71
72      std::vector<double> out;
73      localFE.localInterpolation().interpolate(ithCoord,out);
74
75      for (std::size_t j=0; j<out.size(); j++)
76        lagrangePoints[j][i] = out[j];
77    }
78
79    for (const auto& p : lagrangePoints)
80      std::cout << p << std::endl;
```

The same idea also works for determining the Lagrange node positions of a global Lagrange function defined on an entire finite element grid.

Refined Finite Elements

The RefinedP1LocalFiniteElement implementation is a variation of second-order Lagrangian finite elements. Certain situations require a space that has better approximation properties than first-order Lagrange spaces, but quadratic shape functions have certain stability problems (e.g., [81]). The implementation class

```
template<class DomainField, class RangeField, int dim>
class RefinedP1LocalFiniteElement
```

achieves this by (conceptually) refining the reference element once, and using the first-order Lagrange basis for the resulting grid (Figure 8.3). The shape functions are only piecewise differentiable. Calling the evaluateJacobian method for points where the shape functions are not differentiable leads do undefined behavior, and the user is expected to know where these points are. Similarly, the class

```
template<class DomainField, class RangeField, int dim>
class RefinedP0LocalFiniteElement
```

implements a space of piecewise constant functions, where not even `evaluateFunction` can be called for every point. To use the refined local finite elements include the header `dune/localfunctions/refined.hh`.

Hierarchical Basis Finite Elements

Hierarchical finite element methods consider the finite element space V_h split into a direct sum

$$V_h = V_h^0 \oplus V_h^{\text{ext}},$$

where V_h^0 is typically a first-order Lagrange space, and V_h^{ext} is an extension. Correspondingly, the shape function sets for such spaces consist of the first-order Lagrange shape functions and additional extension functions, which are zero at the element vertices. They can still be associated to subentities of the reference element, and they are sometimes called *bubble functions*. The degrees of freedom are affine functions of point evaluations, but since the basis is not Lagrangian the resulting coefficients are not point-values of the function. Hierarchical finite element spaces have applications in error estimation and for constructing multi-level solvers [9].

The `dune-localfunctions` module implements second-order hierarchical finite elements only. They are available for simplex elements under the name of

```
template<class DomainField, class RangeField, int dim>
class HierarchicalP2LocalFiniteElement
```

from the header `dune/localfunctions/hierarchical/hierarchical.hh`. That file additionally contains a few similar finite elements; check the online documentation for details.

Crouzeix–Raviart and Rannacher–Turek Elements

The Crouzeix–Raviart element is the simple-most example of a useful finite element that does not form subspaces of the Sobolev space H^1. It is defined for simplex elements only, and uses the space of first-order polynomials [45]. In contrast to the first-order Lagrange element, however, the degrees of freedom are the point evaluations at the facet midpoints. Correspondingly, global Crouzeix–Raviart functions are not everywhere continuous, but only at the facet midpoints of adjacent elements. The `dune-localfunctions` module provides a Crouzeix–Raviart implementation in the class

```
template<class DomainField, class RangeField, int dim>
class CrouzeixRaviartLocalFiniteElement
```

from the file dune/localfunctions/crouzeixraviart.hh. As usual DomainField and RangeField are the number types for coordinates and shape function values, and dim is the dimension of the domain simplex.

The Rannacher–Turek element adapts the idea of Crouzeix and Raviart to quadrilateral and hexahedral grid elements [136]. There, a small trick is needed because the naive definition of a Crouzeix–Raviart element on a quadrilateral element is ill-defined. While some authors propose to use facet averages as degrees of freedom, the implementation of dune-localfunctions uses point evaluations at the facet midpoints just as for the Crouzeix–Raviart element. The implementation is contained in the class

```
template<class DomainField, class RangeField, unsigned int dim>
class RannacherTurekLocalFiniteElement
```

provided by the header dune/localfunctions/rannacherturek.hh.

Dual Mortar Finite Elements

Dual mortar finite elements were introduced by Wohlmuth [164] as a smart replacement of the Lagrange test functions in the mortar method [23]. Unlike the other local finite elements in this chapter, dual mortar finite elements usually live only on the boundary of a grid, rather than on the grid itself. They are affine on each element but not globally continuous, and have the interesting property of being biorthogonal to the first-order Lagrange shape functions, in the sense that

$$\int_F \phi_i \theta_j \, dx = \delta_{ij} \int_F \theta_j \, dx, \qquad F \text{ is facet of } T_{\text{ref}}, \qquad (8.2)$$

for all first-order Lagrange shape functions ϕ_i and dual mortar shape functions θ_j. As a result, the boundary mass matrices that appear in the mortar method have only diagonal entries, which makes them trivial to invert. The degrees of freedom of the dual finite elements are weighted linear combinations of point evaluations at the element vertices, and they are most naturally assigned to the element vertices.

The dune-localfunctions module provides first-order dual mortar finite elements for simplices and cubes in all dimensions. They are implemented in the classes

```
template<class DomainField, class RangeField, int dim, bool faceDual=false>
class DualP1LocalFiniteElement
```

for simplices and

```
template<class DomainField, class RangeField, int dim, bool faceDual=false>
class DualQ1LocalFiniteElement
```

for cubes, both from the file dune/localfunctions/dualmortarbasis.hh. The template parameters DomainField, RangeField, and dim have the usual meaning.

The parameter faceDual allows to control the biorthogonality property (8.2). When set to **true**, the shape functions $\theta_0, \ldots, \theta_{N-1}$ fulfill (8.2) to the letter. If faceDual is set to **false**, on the other hand, (8.2) holds with the integration domain replaced by T_{ref} itself. In most cases, **true** will need to be set here, for the following reason: The dune-grid grid interface is not set up to support finite element spaces on the grid boundary directly. Therefore, a trick is necessary if such spaces are needed, e.g., by the mortar method. The shape functions provided by the DualP1LocalFiniteElement class live on grid elements, but they are only ever evaluated at the element boundaries. This way, a separate indexing mechanism for degrees of freedom of boundary finite element spaces can be avoided, and the infrastructure of Chapter 5.6 can be used instead.

As finite element spaces on the grid boundary are useful beyond the mortar method, we demonstrate this trick by the following example, which constructs a uniform simplex grid on the unit square, computes the left integral of (8.2) for all domain boundary facets, and prints the results. Computing these integrals is one important aspect of the mortar method. A full implementation unfortunately requires more infrastructure—it is recommended to use the dune-grid-glue module[4] for that.

The example program consists of a main method only. The file is embedded into the electronic document, and available through the icon in the margin. It is also printed in Appendix B.9. The code first sets up a uniform 2×2 triangle grid:

```
19  int main(int argc, char *argv[])
20  {
21    // Set up MPI if available
22    MPIHelper::instance(argc, argv);
23
24    // Construct a 2x2 structured triangle grid
25    constexpr int dim = 2;
26    using Grid = UGGrid<dim>;
27
28    auto grid = StructuredGridFactory<Grid>::createSimplexGrid({0.0,0.0},
29                                                               {1.0,1.0},
30                                                               { 2,  2}),
31    auto gridView = grid->leafGridView();
```

Next, it constructs a first-order Lagrange local finite element for the shape functions $\phi_0, \ldots, \phi_{N-1}$ of Equation (8.2), a dual mortar finite element with faceDual set to **true** for the test functions $\theta_0, \ldots, \theta_{N-1}$, and a dense matrix object that will hold the integrals (8.2):

```
36    LagrangeSimplexLocalFiniteElement<double,double,dim,1> lagrangeFE;
37    DualP1LocalFiniteElement<double,double,dim,true> dualFE;
38
39    // Matrix that will store the integrals
40    Matrix<double> facetMassMatrix;
```

Note how both finite elements are defined on the triangle even though we are only interested in integrals over boundary edges. This makes immediate sense for the trial

---

[4]https://dune-project.org/modules/dune-grid-glue

basis functions $\phi_0, \ldots, \phi_{N-1}$, because in a standard mortar method the functions ϕ_i in (8.2) really are boundary restrictions of finite element functions defined on the entire grid. Using dual mortar finite elements defined on the triangle, however, is the above-mentioned trick: only its values on the domain boundary will be used. Nevertheless, the matrix will contain entries for all degrees of freedom of the triangle for simplicity, but only the ones pertaining to the boundary degrees of freedom will be used.

The code then loops over the facets of the domain boundary, by iterating over all intersections of all elements, and skipping those that are interior. It also acquires the current reference element (always a triangle here), which is later used to check whether given degrees of freedom are associated to the current intersection:

```
44    for (const auto& element : elements(gridView))
45    {
46      for (const auto& intersection : intersections(gridView, element))
47      {
48        if (intersection.boundary())
49        {
50          auto refElement = referenceElement<double,dim>(element.type());
```

To prepare for the integration on the current boundary facet, the mass matrix is set to the proper size (which is always 3×3 in this example), and initialized by zero:

```
54          facetMassMatrix.setSize(lagrangeFE.size(), dualFE.size());
55          facetMassMatrix = 0;
```

Then, the code fetches a second-order quadrature rule on the current intersection, and loops over the quadrature points:

```
59          constexpr auto intersectionDim
60            = std::decay_t<decltype(intersection)>::mydimension;
61          const auto& quad
62            = QuadratureRules<double,intersectionDim>::rule(intersection.type(),
63                                                             2);
64
65          for (const auto& quadPoint : quad)
66          {
```

The body of this loop is the core part of this example: First, the Lagrange shape functions $\phi_0, \ldots, \phi_{N-1}$ and the dual mortar shape functions $\theta_0, \ldots, \theta_{N-1}$ have to be evaluated at the quadrature point quadPoint. Both sets of shape functions are defined on the entire current grid element, but the position of the evaluation point (i.e., the quadrature point) is given in the coordinate system of the current intersection (which is a part of the element boundary). To obtain the necessary coordinate transformation to the element coordinate system, the code uses the intersectionInside method provided by the Intersection interface class (cf. Chapter 5.4):

```
68          auto quadPosInElement
69            = intersection.geometryInInside().global(quadPoint.position());
```

318

```
70
71        std::vector<FieldVector<double,1> > lagrangeValues, dualValues;
72        lagrangeFE.localBasis().evaluateFunction(quadPosInElement,
73                                                 lagrangeValues);
74        dualFE.localBasis().evaluateFunction(quadPosInElement, dualValues);
```

The actual loop to compute the integrals (8.2) then needs to skip those degrees of freedom that are not associated to the current facet. This is done by asking the current degree of freedom for the vertex it is associated with using the `LocalKey::subEntity` method, and then having the reference element check whether the facet associated to the current intersection contains this vertex. Note also that, as we are integrating over a part of the element boundary, the Jacobian determinant *of the intersection* is used:

```
78        for (std::size_t i=0; i<lagrangeFE.size(); i++)
79        {
80          LocalKey lagrangeLocalKey
81            = lagrangeFE.localCoefficients().localKey(i);
82
83          auto lagrangeSubEntities
84            = refElement.subEntities(intersection.indexInInside(),
85                                     1,   // The codimension of facets
86                                     lagrangeLocalKey.codim());
87
88          if (lagrangeSubEntities.contains(lagrangeLocalKey.subEntity()))
89          {
90            for (std::size_t j=0; j<dualFE.size(); j++)
91            {
92              LocalKey dualLocalKey = dualFE.localCoefficients().localKey(j);
93
94              auto dualSubEntities
95                = refElement.subEntities(intersection.indexInInside(),
96                                         1,   // The codimension of facets
97                                         dualLocalKey.codim());
98
99              if (dualSubEntities.contains(dualLocalKey.subEntity()))
100             {
101               auto integrationElement
102                 = intersection.geometry()
103                     .integrationElement(quadPoint.position());
104               facetMassMatrix[i][j] += quadPoint.weight()
105                                      * integrationElement
106                                      * lagrangeValues[i]
107                                      * dualValues[j];
108             }
109           }
110         }
111       }
112     } // End of quadrature loop
```

This ends the quadrature loop. All that is left to do is to print the current mass matrix, and to proceed to the next intersection:

```
117          printmatrix(std::cout, facetMassMatrix, "facetMassMatrix", "--");
118
119        } // End of 'if boundary'
120      } // End of intersection loop
121    } //End of element loop
```

When run, the code prints eight 3×3 matrices for the eight boundary intersections of the grid:

```
facetMassMatrix [n=3,m=3,rowdim=3,coldim=3]
--    0   2.50e-01   1.39e-17   0.00e+00
--    1   1.39e-17   2.50e-01   0.00e+00
--    2   0.00e+00   0.00e+00   0.00e+00
facetMassMatrix [n=3,m=3,rowdim=3,coldim=3]
--    0   2.50e-01   1.39e-17   0.00e+00
--    1   1.39e-17   2.50e-01   0.00e+00
--    2   0.00e+00   0.00e+00   0.00e+00
facetMassMatrix [n=3,m=3,rowdim=3,coldim=3]
--    0   2.50e-01   1.39e-17   0.00e+00
--    1   1.39e-17   2.50e-01   0.00e+00
--    2   0.00e+00   0.00e+00   0.00e+00
facetMassMatrix [n=3,m=3,rowdim=3,coldim=3]
--    0   0.00e+00   0.00e+00   0.00e+00
--    1   0.00e+00   2.50e-01   0.00e+00
--    2   0.00e+00   1.39e-17   2.50e-01
facetMassMatrix [n=3,m=3,rowdim=3,coldim=3]
--    0   2.50e-01   1.39e-17   0.00e+00
--    1   1.39e-17   2.50e-01   0.00e+00
--    2   0.00e+00   0.00e+00   0.00e+00
facetMassMatrix [n=3,m=3,rowdim=3,coldim=3]
--    0   2.50e-01   0.00e+00   2.78e-17
--    1   0.00e+00   0.00e+00   0.00e+00
--    2   2.78e-17   0.00e+00   2.50e-01
facetMassMatrix [n=3,m=3,rowdim=3,coldim=3]
--    0   0.00e+00   0.00e+00   0.00e+00
--    1   0.00e+00   2.50e-01   0.00e+00
--    2   0.00e+00   1.39e-17   2.50e-01
facetMassMatrix [n=3,m=3,rowdim=3,coldim=3]
--    0   2.50e-01   0.00e+00   2.78e-17
--    1   0.00e+00   0.00e+00   0.00e+00
--    2   2.78e-17   0.00e+00   2.50e-01
```

Note that they are indeed diagonal. Of the three diagonal entries, one is always zero. This entry corresponds to the triangle vertex that is not on the current intersection.

320

8.3.2. Finite Elements that are not Affine-Equivalent

The dune-localfunctions module also offers a few finite elements that do not form affine families. All these elements are provided for the reference elements only. Users have to provide the necessary non-standard transformations to the actual grid elements themselves.

Raviart–Thomas and Brezzi–Douglas–Marini Elements

The Raviart–Thomas and Brezzi–Douglas–Marini (BDM) finite elements are both conforming discretizations of the space $H(\text{div})$ of vector fields with divergence in L^2. These spaces occur primarily in mixed formulations of elliptic equations like the Poisson problem or the equations of linear elasticity [32, 37].

Both Raviart–Thomas and BDM elements are polynomial on each element. However, the requirement to have divergence in L^2 is weaker than weak differentiability, and does not require the functions to be globally continuous. Rather, it only requires continuity of the normal components of the vector fields at the element boundaries. Consequently, the degrees of freedom include moments of the vector field normal components on the element facets.

Raviart–Thomas and BDM elements do not form affine families in the sense of Chapter 8.1, because boundary normals do not stay normal under affine transformations. Therefore it is necessary to use the Piola transform to transform gradients from the reference element T_{ref} to an element T [32]. Originally defined for simplex elements only, Raviart–Thomas and BDM elements have also been constructed for cube grids. However, these constructions are more difficult [32, Chap. 2.4].

The dune-localfunctions module provides implementations of Raviart–Thomas elements in the file dune/localfunctions/raviartthomas.hh, and of the BDM elements in dune/localfunctions/brezzidouglasmarini.hh. For simplex domains, Raviart–Thomas elements of any order and dimension are implemented by the class

```
template<unsigned int dim, class DomainField, class RangeField>
class RaviartThomasSimplexLocalFiniteElement
```

A short description on how some of these elements are constructed is given in [49]. For cube elements, only a handful of dimensions and orders are available, and both order and dimension are compile time parameters:

```
template<class DomainField, class RangeField,
         unsigned int dim, unsigned int order>
class RaviartThomasCubeLocalFiniteElement
```

Check the online documentation for details.

Implementations for BDM elements exist for simplex and cube domains, with domain dimension and order to be selected at compile-time:

```
template<class DomainField, class RangeField,
         unsigned int dim, unsigned int order>
```

```
class BDMSimplexLocalFiniteElement
template<class DomainField, class RangeField,
        unsigned int dim, unsigned int order>
class BDMCubeLocalFiniteElement
```

Again, only a few low orders and dimensions are available at the time of writing. The current list is available in the online documentation.

Monomial Finite Elements

For use in Discontinuous Galerkin (DG) methods [55, 94], dune-localfunctions offers monomial finite elements, which have shape functions of the form

$$\phi_i = \xi_0^{\alpha_0} \xi_1^{\alpha_1} \dots \xi_{d-1}^{\alpha_{d-1}}, \qquad (\alpha_0, \dots, \alpha_{d-1}) \in \mathbb{N}_0^d. \tag{8.3}$$

The corresponding local finite element is implemented in the class

```
template<class DomainField, class RangeField, int dim, int order>
class MonomialLocalFiniteElement
```

in the file dune/localfunctions/monomial.hh. The parameters DomainField and RangeField are the usual number types, dim is the domain dimension, and the parameter order specifies the highest polynomial order of the space. Naturally, this construction works for all types of reference elements. The degrees of freedom are computed by L^2-projection. That is, for a given function $f : T_{\text{ref}} \to \mathbb{R}$, a call to interpolate will return coefficients $v_i = \boldsymbol{\sigma}_i(f)$, $i = 0, \dots, N-1$ such that

$$\mathcal{L}_f^2(v_0, \dots, v_{N-1}) := \left\| f - \sum_{i=0}^{N-1} v_i \phi_i \right\|_{L^2(T_{\text{ref}})}^2$$

is minimal in $\mathbb{R}^N$, where the ϕ_i are the shape functions given by (8.3). These coefficients are difficult to interpret directly. As DG methods do not involve continuity conditions, the LocalCoefficients class assigns them all to the element itself.

Orthonormal Finite Elements

Also for DG methods, dune-localfunctions has finite elements with pairwise L^2-orthonormal shape functions, i.e.,

$$\int_{T_{\text{ref}}} \phi_i \phi_j \, d\xi = \delta_{ij}$$

for all $\phi_0, \dots, \phi_{N-1}$. This property can speed up certain calculations, because element mass matrices are now cheap to set up and invert [94]. The local finite element is implemented in the class

```
template<unsigned int dim, class DomainField, class RangeField>
class OrthonormalLocalFiniteElement
```

in the file `dune/localfunctions/orthonormal.hh`. Reference element type and polynomial order are set at run-time via the constructor

```
OrthonormalLocalFiniteElement(const GeometryType& type, unsigned int order)
```

The degrees of freedom are computed in the same way as for the monomial finite elements, and they all get assigned to the element itself. To retain orthonormality on non-reference elements, the shape functions require suitable scaling.

8.4. Run-Time Selection of Local Finite Elements

Some of the finite element implementations available in `dune-localfunctions` determine parameters like the reference element type or the polynomial order at compile-time. This is not required by the interface—both could be run-time parameters as well. However, fixing such features at compile time allows the compiler to optimize more aggressively, and in many cases it is indeed known a priori that only a single fixed finite element order will be used, or that the grid will contain, e.g., only simplex elements. Nevertheless, it is also frequently required to use several different local finite element implementations in a single simulation. The prime example are grids that mix simplex and cube elements: a finite element assembler for element stiffness matrices must switch between different local finite elements at run-time, depending on the type of the current element. For some implementations like the `LagrangeLocalFiniteElement` of Chapter 8.3.1, this simply involves calling the constructor with different `GeometryType` arguments. However, it is frequently also convenient to switch between completely unrelated `LocalFiniteElement` implementations at run-time. One way to implement this would be a second programming interface using abstract base classes and virtual inheritance (Chapter 4.4.1), [5] or a type erasure mechanism similar to what is implemented in `dune-functions` (Chapter 10.5.2).

At the time of writing, `dune-localfunctions` does not implement full type erasure for local finite elements. Instead, it offers a container that can store one local finite element object from a given list of implementation types. This container is called

```
template<class... Implementations>
class LocalFiniteElementVariant
```

and it is implemented in the file `localfiniteelementvariant.hh` in the directory `dune/localfunctions/common/`. It is called Variant because it is very similar to `std::variant` from the C++17 standard library [99]: A type-safe union that can hold an object of any one of the types given to it in the list of template arguments.

The resulting object implements the full `LocalFiniteElement` interface presented in Chapter 8.2. It behaves like a local finite element in all respects, and can be used as a drop-in replacement for a `LocalFiniteElement` implementation. In addition, it has a method

---

[5] ...which does indeed exist, but its future is unclear...

```
const auto& variant() const
```

that provides access to the `std::variant` object that is used internally to keep the data.

Objects of type `LocalFiniteElementVariant` are constructed by calling the constructor with the object that is supposed to be stored:

```
template<class Implementation>
LocalFiniteElementVariant(Implementation&& impl)
```

By template trickery, the parameter type `Implementation` is restricted to those that appear in the variadic list `Implementations` given to the class. Unlike objects of type `std::variant`, objects of type `LocalFiniteElementVariant` are always default constructible, and explicitly listing `std::monostate` is not required.

As an example, we show code that sets up a `LocalFiniteElementVariant` that switches between Crouzeix–Raviart and Rannacher–Turek elements (Chapter 8.3.1), to be used with grids that contain triangle and quadrilateral elements:

```
constexpr int dim = 2;
using TriangleLFEType = CrouzeixRaviartLocalFiniteElement<double,double,dim>;
using QuadLFEType     = RannacherTurekLocalFiniteElement<double,double,dim>;
LocalFiniteElementVariant<TriangleLFEType,QuadLFEType> lfeVariant;

// Init with a triangle element
lfeVariant = TriangleLFEType();
std::cout << "Finite element on a triangle has "
          << lfeVariant.size() << " degrees of freedom." << std::endl;

// Init with a quadrilateral element
lfeVariant = QuadLFEType();
std::cout << "Finite element on a quadrilateral has "
          << lfeVariant.size() << " degrees of freedom." << std::endl;
```

The approach using a type-safe union is limited in the sense that all possible types must be known at compile-time. However, in many practical situations the list of possible local finite element classes really is known and small. The advantage is that the implementation is much simpler than full type erasure, and it is usually more efficient, too.

9. Quadrature Rules

Both finite element and finite volume methods are based on integral expressions, and these are usually computed using numerical quadrature. Computer codes implementing such methods need to provide quadrature rules, and DUNE is no exception. Although forming a feature set of its own, the numerical quadrature of DUNE is not given in a separate module. Instead, it is part of dune-geometry, because it conceptually depends on the reference elements defined there.

9.1. Numerical Integration

Integration domains in finite element and finite volume computations are typically the reference elements, i.e., certain fixed polytopes in $\mathbb{R}^d$ (Chapter 2.1.3). Cubes and simplices are by far the most common, but a few others can occur as well. Methods that use general polyhedral grids typically do not use integration over the elements. Most integrands in finite element computations are polynomials or at least differentiable on each element. Singular integrands occur in certain situations, like the XFEM method for fracture mechanics [71].

This section briefly reviews a few ways to construct quadrature rules. Besides dedicated simplex rules, DUNE provides an algorithm that constructs quadrature rules for all DUNE reference elements of Chapter 5.5.2.

9.1.1. One-Dimensional Integration

Most research in numerical quadrature has focused on one-dimensional integrals. Let $f : [a,b] \to \mathbb{R}$ be a given integrable function. Numerical integration tries to approximate the integral $\int_a^b f(s)\,ds$ by the sum

$$I_h(f) := \sum_{i=0}^{n} w_i f(s_i),$$

where the $s_i \in [a,b]$, $i = 0, \dots, n$ are called *quadrature points*, and the $w_i \in \mathbb{R}$ are corresponding *weights* that sum up to $b - a$. The collection of points and weights together is called a *quadrature rule*. To reproduce the desirable property that $\int_a^b f(s)\,ds \geq 0$ for all f that are pointwise nonnegative, only rules with positive weights are usually considered. Most textbooks define quadrature rules on a fixed interval like $[-1,1]$; transformation to other domains works as in Chapter 2.1.3.

© Springer Nature Switzerland AG 2020
O. Sander, *DUNE — The Distributed and Unified Numerics Environment*, Lecture Notes in Computational Science and Engineering 140, https://doi.org/10.1007/978-3-030-59702-3_9

The most common quadrature rules are based on polynomial interpolation of the function f at the points s_i, and the weights w_i are given as the integrals over the corresponding Lagrange polynomials [52]. Quadrature rules differ by how the points are distributed in $[a, b]$. For example, Newton–Cotes rules use uniformly spaced points. Better approximation properties can be obtained by the Gauss quadrature rules, where the $n + 1$ points are given as the roots of certain orthogonal polynomials. For integrals without a weight function, these polynomials are the Legendre polynomials, and the corresponding rules are called Gauss–Legendre rules. The integration error is bounded by the $(2n + 2)$-nd derivative of f; in particular they can integrate polynomials of order up to $2n + 1$ exactly [52, Satz 9.12]. Such rules are said to have *order* $2n + 1$.

A generalization of this approach allows to integrate functions with certain singularities. Suppose that the integrand f can be written as a product $f(s) = \omega(s)g(s)$, where ω is a known positive weight function. Then it is possible to construct rules

$$\int_a^b f(s)\, ds = \int_a^b \omega(s)g(s) \approx \sum_{i=0}^n w_i g(s_i),$$

whose error is bounded in terms of the $(2n + 2)$-nd derivative of g rather than that of f. In particular, if g is a polynomial of degree not greater than $2n + 1$, then the integral will be evaluated exactly. The quadrature points of such rules are the roots of the polynomials orthogonal with respect to the scalar product $\langle g, h \rangle_\omega := \int_a^b \omega g h\, ds$. For the domain $[-1, 1]$ one obtains the Gauss–Chebyshev quadrature rules (for $\omega(s) = (1 - s^2)^{-1/2}$), the Gauss–Jacobi rules (for $(\omega(s) = (1 - s)^\alpha (1 + s)^\beta$, $\alpha, \beta > -1$), and various others.

In some instances of the finite element method, the integrands on a single element are only piecewise polynomial. Examples are the Hsieh–Clough–Tocher or Powell–Sabin elements [37], or the refined elements of Chapter 8.3.1. To properly integrate such functions, the quadrature rules need to respect the piecewise structure. Appropriate rules are constructed by combining multiple scaled copies of Gauss rules. Such rules are called *composite rules*. Particularly useful as building blocks for composite rules are the Gauss–Lobatto rules. These are the Gauss–Legendre rules, but with their two outer points modified such that $s_0 = a$ and $s_n = b$. Since these boundary points coincide with boundary points of neighboring subdomains, the number of function evaluations for the composite rule is reduced.

9.1.2. Multi-Dimensional Integrals

Most integrals appearing in finite element or finite volume simulations are on two- or three-dimensional domains. For such domains the approach of the previous section can be reused: Select a set of points $\{s_i\}_i$ in the domain such that the corresponding polynomial interpolation problem is well posed, then compute the corresponding weights $\{w_i\}_i$ as the integrals over the Lagrange polynomials. A wealth of rules for cubes, simplices, and other domains have been constructed using this and other techniques [149]. A lot of them are available in DUNE, but their detailed description is beyond the scope of this book.

Rules for multi-dimensional domains can also be constructed by multiplication of one-dimensional rules. This is particularly easy for cubes, which are among the most import domains in finite element and finite volume methods. The construction is defined recursively. Let D^{d-1} be a domain in $\mathbb{R}^{d-1}$, and $\{\tilde{s}_j, \tilde{w}_j\}_{j=0}^{\tilde{n}}$ a p-th order quadrature rule for D^{d-1}. Suppose there is also a rule $\{s_j, w_j\}_{j=0}^m$ of the same order for the one-dimensional domain $[0,1]$. Then a rule of order p for $D^d := D^{d-1} \times [0,1]$ can be constructed by setting

$$\int_{D^d} f(x)\, dx \approx \sum_{i=0}^{\tilde{n}} \sum_{j=0}^m \tilde{w}_i w_j f(\underbrace{\tilde{s}_i, s_j}_{\in D^d})$$

(Stroud [149, Chap. 2.2]). Repeatedly applying this construction leads to p-th order quadrature rules for hypercubes of all dimensions. If $n+1$-point Gauss–Legendre rules are used to integrate over $[0,1]$, a tensor product rule of order $2n+1$ for the d-dimensional hypercube consists of $(n+1)^d$ points.

A similar approach allows the construction of rules for simplices and pyramids. Let again D^{d-1} be a region in $\mathbb{R}^{d-1}$, and let D^d be the cone over D^{d-1} with vertex $(0, \ldots, 0, 1) \in \mathbb{R}^d$, defined as

$$D^d := \Big\{ x \in \mathbb{R}^d \ : \ x_i = \tilde{x}_i(1-\lambda),\ i = 0, \ldots, n-2,\ x_{d-1} = \lambda,$$
$$0 \le \lambda \le 1, (\tilde{x}_0, \ldots, \tilde{x}_{d-2}) \in D^{d-1} \Big\}.$$

This is called the conical product. Examples of such product domains are the simplices (each simplex being a cone over a lower-dimensional simplex), and the three-dimensional pyramid, where the base D^{d-1} is a square.

Construction of tensor product quadrature rules for cubes works because the integral of a monomial over $[0,1]^d$ can be written as the product of a one-dimensional integral with an integral over $[0,1]^{d-1}$. The same is possible for integrals over cones. Let $D^{d-1}(\lambda)$, $0 \le \lambda \le 1$ be the intersection of D^d and the plane $x_{d-1} = \lambda$. An integral over $D^{d-1}(\lambda)$ can be evaluated by affinely transforming $D^{d-1}(\lambda)$ onto D^{d-1}, and integrating there. For each monomial $x_0^{\alpha_0} \ldots x_{d-1}^{\alpha_{d-1}}$ of degree $p = |\alpha| := \alpha_0 + \cdots + \alpha_{d-1}$ we get

$$\int_{D^d} x_0^{\alpha_0} \ldots x_{d-1}^{\alpha_{d-1}}\, dx_0 \ldots dx_{d-1}$$
$$= \int_0^1 x_{d-1}^{\alpha_{d-1}} \left[\int_{D^{d-1}} (1-x_{d-1})^{d-1} \prod_{j=0}^{d-2} [x_j(1-x_{d-1})]^{\alpha_j}\, dx_0 \ldots dx_{d-2} \right] dx_{d-1}$$
$$= \int_0^1 (1-x_{d-1})^{d-1+\beta} x_{d-1}^{\alpha_{d-1}}\, dx_{d-1} \int_{D^{d-1}} x_0^{\alpha_0} \ldots x_{d-2}^{\alpha_{d-2}}\, dx_0 \ldots dx_{d-2}$$

with $\beta := \alpha_0 + \cdots + \alpha_{d-2}$. The integral with respect to x_{d-1} in the third line can be written as

$$\int_0^1 (1 - x_{d-1})^{d-1} \left[(1 - x_{d-1})^\beta x_{d-1}^{\alpha_{d-1}} \right] dx_{d-1}, \qquad (9.1)$$

and the term in brackets is a polynomial of degree $p = |\alpha|$. The trick is now to see that (9.1) is a weighted integral that can be integrated exactly with a p-th order Gauss–Jacobi rule.

Theorem 9.1 (Stroud [149, Thm. 2.5-1]) *Suppose that the domain D^d can be written as a cone over D^{d-1} with apex $(0, \ldots, 0, 1) \in \mathbb{R}^d$. Given a quadrature rule $\{\tilde{s}_i, \tilde{w}_i\}_{i=0}^{\tilde{n}}$ of degree p for D^{d-1}, and a quadrature rule $\{s_i, w_i\}_{i=0}^{m}$ of degree p for the weighted one-dimensional integral $\int_0^1 (1 - t)^{d-1} g(t) \, dt$, then a rule of order p for the cone D^d is given by*

$$\int_{D^d} f(x) \, dx \approx \sum_{i=0}^{\tilde{n}} \sum_{j=0}^{m} \tilde{w}_i w_j \underbrace{f(\tilde{s}_i, s_j)}_{\in D^d}.$$

While the repeated application of tensor multiplication leads to formulas for hypercubes, repeated application of the conical product leads to simplex rules. Combining the two allows to deal with domains like the three-dimensional pyramid (the conical product of a square with an interval) and the three-dimensional prism (the tensor product of a triangle with an interval). This construction corresponds directly to the recursive construction of the reference elements themselves given in Chapter 5.5.2. Therefore, quadrature rules for all reference elements of DUNE can be defined by tensor- and conical multiplication.

Unfortunately, quadrature rules constructed by multiplication are not optimal. First of all, these rules are asymmetric: By the nature of the conical product, the quadrature points cluster near one simplex vertex. For finite element computations this means that the approximation error for integrals over an element may depend on the map from the element onto the reference element. Also, rules constructed by multiplication of lower-dimensional rules typically contains more points than necessary for a given accuracy. DUNE therefore only uses those rules for the case that no other rules are available.

9.2. The Dune Quadrature Rule Interface

By definition, quadrature rules are sets of points in $\mathbb{R}^d$ with associated scalar weights. So while the derivation and analysis of quadrature rules may sometimes be difficult, the corresponding data structures needed to store them are not.

Most functionality for numerical integration in DUNE is available from the single header file `dune/geometry/quadraturerules.hh`. The quadrature points are represented by a class

```
template<class ctype, int dim>
class QuadraturePoint
```

with an interface that exports only three types:

```
constexpr int dimension = dim;            // Domain dimension
using Field = ctype;                      // Number type for coordinates
                                          // and weights
using Vector = FieldVector<ctype,dim>;    // Type for point position
```

and two methods

```
const Vector& position() const
const ctype& weight() const
```

One can access a quadrature point's position $s \in \mathbb{R}^d$ and weight $w \in \mathbb{R}$, and that is pretty much all there is to it. The QuadraturePoint class is statically parametrized by the dimension dim of the integration domain, and by a number type ctype. That type is used for the weight w, and for the coordinates of the point s.

All quadrature rules in dune-geometry are implemented by dedicated classes. These inherit from an abstract base class

```
template<class ctype, int dim>
class QuadratureRule
```

which in turn inherits from std::vector<QuadraturePoint<ctype,dim> >. As a consequence, any quadrature rule object will have the full interface of std::vector. Integrating a given function f that implements

```
double operator()(const FieldVector<double,dim>& x)
```

with a quadrature rule called quad is therefore as easy as

```
double result = 0;
for (const auto& quadPoint : quad)
  result += quadPoint.weight() * f(quadPoint.position());
```

In addition, QuadratureRule objects yield the type of the integration domain (which must be one of the DUNE reference elements), and the polynomial order of the quadrature rule by the methods

```
virtual GeometryType type() const
```

and

```
virtual int order() const
```

respectively.

A quadrature rule of a given type for a given reference element and order is something that is known and fixed. Once it is constructed it will not change while the program

329

is executed. This, together with the fact that the total number of quadrature rules used by a program is relatively small, makes them good candidates for singleton implementations [72]. That means that only one quadrature rule object exists for each given type, reference element, and order. The user never constructs quadrature rules directly (the constructors are private), but instead uses a dedicated factory class. This factory keeps the existing objects, constructs new QuadratureRule objects if necessary, and hands out references to them on request.

In dune-geometry, the factory class for quadrature rules is

```
template<class ctype, int dim>
class QuadratureRules
```

As its only public member it has the static method

```
static const QuadratureRule& rule(const GeometryType& gt,
                                  int order,
                                  QuadratureType::Enum qt
                                     = QuadratureType::GaussLegendre)
```

It returns a reference to a singleton object of a quadrature rule for the domain specified by gt and the order given in order. The dimension of gt has to match the dimension given as the second template argument to the QuadratureRules class.

The type of the quadrature rule that is returned depends on the reference element type, and on the requested order. If gt is a simplex and order is not too large, then a dedicated symmetric simplex rule is returned. Otherwise, an appropriate rule is created by multiplication of one-dimensional rules. The last parameter qt allows to select the type of the one-dimensional quadrature rule. The returned rule is a product rule obtained by tensor-multiplying the one-dimensional rule as described in Section 9.1.2.[1] Currently, the following one-dimensional quadrature rules can be selected:

- GaussLegendre: Quadrature rules of Gauss–Legendre type for the domain $[0, 1]$

- GaussLobatto: Gauss–Lobatto rules for $[0, 1]$

- GaussJacobi_1_0, GaussJacobi_2_0: These are rules of Gauss–Jacobi type, for integrals with the weight function $\omega(s) = (1 - s)^\alpha (1 + s)^\beta$. The two rules correspond to the cases $\alpha = 1, \beta = 0$ and $\alpha = 2, \beta = 0$.

Point positions and weights for these rules are taken from tables that are created semi-automatically using the symbolic computation program MAXIMA.[2] Hence if higher orders are needed, the corresponding tables can be updated fairly easily.

Composite rules can be constructed from other rules in a modular way. There is a separate class

---

[1] As of dune-geometry 2.7, for simplices, the special structure of (9.1) as a weighted integral is not actually exploited, yet.

[2] http://maxima.sourceforge.net

```
template<class ctype, int dim>
class CompositeQuadratureRule
```

(in dune/geometry/quadraturerules/compositequadraturerule.hh) which transforms any given quadrature rule into a composite one. Suppose a quadrature rule is
given as an object quad for a reference element gt. Then a composite rule can be
constructed by

```
CompositeQuadratureRule<ctype,dim> compositeQuad(quad,refinement);
```

This will create a rule by doing refinement steps of uniform refinement of the reference
element, and transforming the original rule to act on each of the subelements. So to
integrate the function f from the example above on a tetrahedron with a second-order
composite rule on $(2^3)^m$ subelements for some $m \in \mathbb{N}$, the code is

```
#include <dune/geometry/quadraturerules.hh>
[...]
const int dim = 3;
int quadOrder = 2;
GeometryType type = GeometryTypes::simplex(dim);   // The reference element
                                                   // to integrate over
const auto& quad = QuadratureRules<double, dim>::rule(type, quadOrder);

CompositeQuadratureRule<double,dim> compositeQuad(quad,
                                          m);      // Refine domain
                                                   // m times

double result = 0;
for (const auto& quadPoint : compositeQuad)
  result += quadPoint.weight() * f(quadPoint.position());
```

Use this, e.g., for the refined finite elements of Chapter 8.3.1.

Part III.

Solving Partial Differential Equations

10. Function Space Bases and Discrete Functions

In previous chapters we have described the DUNE interface for grids, linear algebra, and finite element shape functions. Now we move one level up in the abstraction hierarchy and consider global function spaces and functions. By *global* we mean function spaces and functions that are defined not on a single element, but on entire grid views (i.e., non-hierarchical grids). This functionality is covered by the dune-functions module.

The main purpose of dune-functions is the description of discrete function spaces defined on a grid view. Still, the central concept in the code is not the function space itself, but rather a *basis* of such a space. This is because even though finite element spaces play a central role in theoretical considerations of the finite element method, actual computations use coefficient representations of discrete functions, given with respect to a particular basis. Also, for many different finite element spaces, more than one basis is used in practice. For example, the space of second-order Lagrangian finite elements is used both with the nodal (Lagrange) basis [37], and with the hierarchical basis [9]. Discontinuous Galerkin spaces can be described in terms of Lagrange bases, monomial bases, Legendre bases and more [94]. It is therefore important to be able to distinguish these different representations of the same space in the application code. For these reasons, the main dune-functions interface represents a basis of a discrete function space, and not the space itself.

Finite element function spaces frequently exhibit a fair amount of structure. In particular, vector-valued and mixed finite element spaces can be written as products of simpler spaces. Even more, such spaces have a natural structure as a tree, with scalar-valued or otherwise irreducible spaces forming the leaves, and products forming the inner nodes. The dune-functions module allows to systematically construct new bases by multiplication of existing bases. The resulting tree structures are reproduced as type information in the code. These ideas originally appeared in [123].

For the basis functions in such a non-trivial tree structure, there is no single canonical way to index them. Keeping all degrees of freedom in a single standard array would require indexing by a contiguous, zero-starting set of integer numbers. On the other hand, from the tree structure of the basis follows a natural indexing by multi-indices, which can be used to address nested vector and matrix data types like the ones provided by dune-istl. Closer inspection reveals that these two possibilities are just

---

[0]A lot of text in this chapter is directly taken from the two articles [61] and [62], published jointly with Christian Engwer, Carsten Gräser, and Steffen Müthing. I would like to thank them for the permission to reuse these documents here.

© Springer Nature Switzerland AG 2020

O. Sander, *DUNE — The Distributed and Unified Numerics Environment*, Lecture Notes in Computational Science and Engineering 140, https://doi.org/10.1007/978-3-030-59702-3_10

the extreme cases of a wider scale of indexing rules. While some of them are somewhat contrived, many others really are useful in applications. The dune-functions module provides a systematic way to construct many of these rules.

The first four sections of this chapter deal with function space bases. The abstract construction is described in Chapter 10.1, and Chapter 10.2 explains the corresponding programming interface. Chapters 10.3 and 10.4 show how to obtain basis trees from smaller and larger trees, respectively.

The second part of this chapter covers general functions $f : \mathcal{D} \to \mathcal{R}$ between sets $\mathcal{D}$ and $\mathcal{R}$. An interface for such functions is given as part of the C++11 standard, and dune-functions simply reuses this interface.

However, functions in a finite element context commonly have additional features. First of all, the derivative of a function is frequently needed. The first derivative is the most common, but higher derivatives also occur. Secondly, functions are frequently defined in a piecewise way with respect to a finite element grid. In that case, evaluation using a global coordinate is usually very expensive. The natural way to evaluate such a function uses an element and local coordinates on that element. The dune-functions module uses *local functions* that can be bound to individual elements, and can then be evaluated in local coordinates of that element. This mechanism makes the common case of multiple evaluations within one element fast.

Chapter 10.5 explains the dune-functions programming interface for functions. Chapter 10.6 then shows how discrete functions can be obtained by combining bases and coefficient vectors. Chapter 10.7 describes VTK output for functions. Finally, in Chapter 10.8 we give a complete example showing how dune-functions can be used to solve the Stokes equation using Taylor–Hood elements.

10.1. Function Space Bases

Before we can explain the programming interface for bases of discrete function spaces in Chapter 10.2, we need to say a few words about how these bases can be endowed with an abstract tree structure. Readers who are only interested in finite element spaces of scalar-valued functions may try to proceed directly to Chapter 10.2. They should only know that whenever a local finite element tree is mentioned there, this tree consists of a single node only, which is the local finite element basis. Similarly, for a scalar finite element space the tree of multi-indices used to index the basis functions simply represents a contiguous, zero-starting set of natural numbers.

10.1.1. Trees of Function Spaces

All function spaces that are considered in this chapter are defined on a single fixed domain Ω. The focus is on spaces of functions that are piecewise polynomial with respect to a grid, but at this point that is not actually required yet.

For a set R we denote by $R^\Omega := \{f : \Omega \to R\}$ the set of all functions mapping from Ω to R. We write $P_k(\Omega) \subset \mathbb{R}^\Omega$ for the space of all scalar-valued continuous piecewise polynomials of degree at most k on Ω with respect to some given triangulation. We will omit the domain if it can be inferred from the context.

Considering the different finite element spaces that appear in the literature, there are some that we will call *irreducible*. By this term we mean all spaces of scalar-valued functions, but also others like the Raviart–Thomas space that cannot easily be written as a combination of simpler spaces. Many other finite element spaces arise naturally as a combination of simpler ones. There are primarily two ways how two vector spaces V and W can be combined to form a new one: sums and products (also known as internal and external sums, respectively).

For sums, both spaces need to have the same range space R. Then the vector space sum

$$V + W := \{v + w \ : \ v \in V, \ w \in W\}$$

in R^Ω will have that same range space. For example, a P_2-space can be viewed as a P_1-space plus a hierarchical extension spanned by bubble functions [9]. XFEM spaces [122] are constructed by adding particular weighted Heaviside functions to a basic space to capture certain discontinuities. The dune-functions module does not currently support constructing sums of finite element bases, but this may be added in later versions.

The second way to construct finite element spaces from simpler ones uses Cartesian products. Let $V \subset (\mathbb{R}^{r_1})^\Omega$ and $W \subset (\mathbb{R}^{r_2})^\Omega$ be two function spaces. Then we define the product of V and W as

$$V \times W := \big\{(v, w) \ : \ v \in V, \ w \in W\big\}.$$

Functions from this space take values in $\mathbb{R}^{r_1} \times \mathbb{R}^{r_2} = \mathbb{R}^{r_1 + r_2}$.

The product operation allows to build vector-valued and mixed finite element spaces of arbitrary complexity. For example, the space of first-order Lagrangian finite elements with values in $\mathbb{R}^3$ can be seen as the product $P_1 \times P_1 \times P_1$. The lowest-order Taylor–Hood element is the product $P_2 \times P_2 \times P_2 \times P_1$ of $P_2 \times P_2 \times P_2$ for the velocities with P_1 for the pressure. More factor spaces can be included easily, if necessary. We call such products of spaces *composite spaces*.

In the Taylor–Hood space, the triple $P_2 \times P_2 \times P_2$ forms a semantic unit—it contains the components of a velocity field. The associativity of the product allows to write the Taylor–Hood space as $(P_2 \times P_2 \times P_2) \times P_1$, which makes the semantic relationship clearer. Grouped expressions of this type are conveniently visualized as trees. This suggests to interpret composite finite element spaces as tree structures. In these structures, leaf nodes represent scalar or otherwise irreducible spaces, and inner nodes represent products of their children. Subtrees then represent composite finite element spaces. Figure 10.1 shows the Taylor–Hood finite element space in such a tree representation. All these trees are *rooted* and *ordered*, i.e., they have a dedicated root node, and the children of each node have a fixed given ordering. Based on this child ordering we associate to each child the corresponding zero-based index.

While the inner tree nodes may initially appear like useless artifacts of the tree representation, they are often extremely useful because we can treat the subtrees

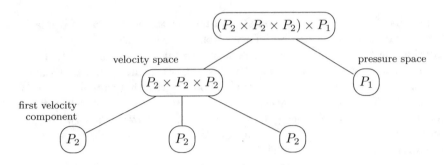

Figure 10.1.: Function space tree of the Taylor–Hood space $(P_2 \times P_2 \times P_2) \times P_1$

rooted in those nodes as individual trees in their own right. This often allows to reuse existing algorithms that expect to operate on those subtrees in more complex settings.

10.1.2. Trees of Function Space bases

The multiplication of finite-dimensional spaces naturally induces a corresponding operation on bases of such spaces. We introduce a generalized tensor product notation: Consider linear ranges $R_0, \ldots, R_{m-1}$ of function spaces $R_0^\Omega, \ldots, R_{m-1}^\Omega$, and the i-th canonical basis vector $\mathbf{e}_i$ in $\mathbb{R}^m$. Then

$$\mathbf{e}_i \otimes f := (0, \ldots, 0, \underbrace{f}_{i\text{-th entry}}, 0, \ldots, 0) \in \prod_{j=0}^{m-1}\left(R_j^\Omega\right) = \left(\prod_{j=0}^{m-1} R_j\right)^\Omega,$$

where 0 in the j-th position denotes the zero-function in R_j^Ω. Let Λ_i be a function space basis of the space $V_i = \operatorname{span} \Lambda_i$ for $i = 0, \ldots, m-1$. Then a natural basis Λ of the product space

$$V_0 \times \cdots \times V_{m-1} = \prod_{i=0}^{m-1} V_i = \prod_{i=0}^{m-1} \operatorname{span} \Lambda_i$$

is given by

$$\Lambda = \Lambda_0 \sqcup \cdots \sqcup \Lambda_{m-1} = \bigsqcup_{i=0}^{m-1} \Lambda_i := \bigcup_{i=0}^{m-1} \mathbf{e}_i \otimes \Lambda_i. \tag{10.1}$$

The product $\mathbf{e}_i \otimes \Lambda_i$ is to be understood element-wise, and the "disjoint union" symbol $\sqcup$ is used here as a simple short-hand notation for (10.1) and not to be understood as an associative binary operation. Using this new notation we have

$$\operatorname{span} \Lambda = \operatorname{span}\left(\Lambda_0 \sqcup \cdots \sqcup \Lambda_{m-1}\right) = (\operatorname{span} \Lambda_0) \times \cdots \times (\operatorname{span} \Lambda_{m-1}).$$

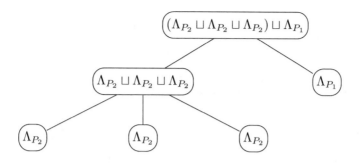

Figure 10.2.: Function space basis tree of the Taylor–Hood space $(P_2 \times P_2 \times P_2) \times P_1$

Similarly to the case of function spaces, bases can be interpreted as trees. If we associate a basis Λ_V to each space V in the function space tree, then the induced natural function space basis tree is obtained by simply replacing V by Λ_V in each node. For the Taylor–Hood basis this leads to the tree depicted in Figure 10.2.

10.1.3. Indexing Basis Functions by Multi-Indices

To work with the basis of a finite element space, the basis functions need to be indexed. Indexing the basis functions is what allows to address the corresponding vector and matrix coefficients in suitable vector and matrix data structures. In simple cases, indexing means simply enumerating the basis functions with natural numbers, but for many applications hierarchically structured vector and matrix data structures are more natural or efficient. This leads to the idea of hierarchically structured multi-indices.

Definition 10.1 (Multi-indices) *A tuple $I \in \mathbb{N}_0^k$ for some $k \in \mathbb{N}_0$ is called a multi-index of length k, and we write $|I| := k$. The set of all multi-indices is denoted by $\mathcal{N} = \bigcup_{k \in \mathbb{N}_0} \mathbb{N}_0^k$.*

To establish some structure in a set of multi-indices it is convenient to consider prefixes.

Definition 10.2 (Multi-index prefixes)

1. *If $I \in \mathcal{N}$ takes the form $I = (I^0, I^1)$ for $I^0, I^1 \in \mathcal{N}$, then we call I^0 a prefix of I. If additionally $|I^1| > 0$, then we call I^0 a strict prefix of I.*

2. *For $I, I^0 \in \mathcal{N}$ and a set $\mathcal{M} \subset \mathcal{N}$:*

   a) *We write $I = (I^0, \dots)$, if I^0 is a prefix of I,*

   b) *we write $I = (I^0, \bullet, \dots)$, if I^0 is a strict prefix of I,*

   c) *we write $(I^0, \dots) \in \mathcal{M}$, if I^0 is a prefix of some $I \in \mathcal{M}$,*

   d) *we write $(I^0, \bullet, \dots) \in \mathcal{M}$, if I^0 is a strict prefix of some $I \in \mathcal{M}$.*

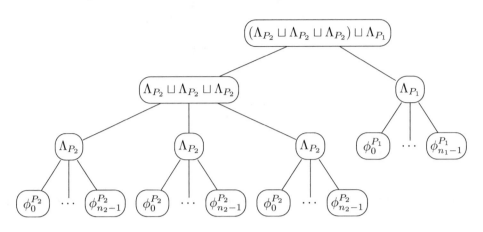

Figure 10.3.: Tree of basis vectors for the Taylor–Hood basis

It is important to note that the multi-indices from a given set do not necessarily all have the same length. For an example, Figure 10.3 illustrates the set of all basis functions of a Taylor–Hood basis by extending the basis tree of Figure 10.2 by leaf nodes for individual basis functions. A possible indexing of the basis functions of the Taylor–Hood basis Λ_{TH} then uses multi-indices of the form $(0, i, j)$ for velocity components, and $(1, k)$ for pressure components. For the velocity multi-indices $(0, i, j)$, the $i = 0, \dots, 2$ determines the component of the velocity vector field, and the $j = 0, \dots, n_2 - 1 := |\Lambda_{P_2}| - 1$ determines the number of the scalar P_2 basis function that determines this component. For the pressure multi-indices $(0, k)$ the $k = 0, \dots, n_1 - 1 := |\Lambda_{P_1}| - 1$ determines the number of the P_1 basis function for the scalar P_1 function that determines the pressure.

It is evident that the complete set of these multi-indices can again be associated to a rooted tree. In this tree, the multi-indices correspond to the leaf nodes, their strict prefixes correspond to interior nodes, and the multi-index digits labeling the edges are the indices of the children within the ordered tree. Prefixes can be interpreted as paths from the root to a given node.

This latter fact can be seen as the defining property of index trees. Indeed, a set of multi-indices (together with all its strict prefixes) forms a tree as long as it is consistent in the sense that the multi-indices can be viewed as the paths to the leafs in an ordered tree. That is, the children of each node are enumerated using consecutive zero-based indices and paths to the leafs (i.e., the multi-indices) are built by concatenating those indices starting from the root and ending in a leaf. Since the full structure of this tree is encoded in the multi-indices associated to the leafs we will—by a slight abuse of notation—call the set of multi-indices itself a tree from now on.

Definition 10.3 *A set $\mathcal{I} \subset \mathcal{N}$ is called an* index tree *if for any $(I, i, \dots) \in \mathcal{I}$ there are also $(I, 0, \dots), (I, 1, \dots), \dots, (I, i - 1, \dots) \in \mathcal{I}$, but $I \notin \mathcal{I}$.*

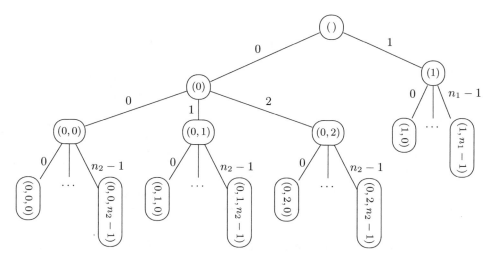

Figure 10.4.: Index tree for the Taylor–Hood basis inherited from the basis tree

The index tree for the example indexing of the Taylor–Hood basis given above is shown in Figure 10.4.

Definition 10.4 *Let $(I, \dots) \in \mathcal{I}$, i.e., I is a prefix of multi-indices in $\mathcal{I}$. Then the size of $\mathcal{I}$ relative to I is given by*

$$\deg^{+}_{\mathcal{I}}[I] := \max \left\{ k \; : \; \exists (I, k, \dots) \in \mathcal{I} \right\} + 1. \tag{10.2}$$

In terms of the ordered tree associated with $\mathcal{I}$ this corresponds to the out-degree of I, i.e., the number of direct children of the node indexed by I.

Using the idea of multi-index trees, an indexing of a function space basis is an injective map from the leaf nodes of a tree of basis functions to the leafs of an index tree.

Definition 10.5 *Let M be a finite set and $\iota : M \to \mathcal{N}$ an injective map whose range $\iota(M)$ forms an index tree. Then ι is called an* index map *for M. The index map is called* uniform *if additionally $\iota(M) \subset \mathbb{N}_0^k$ for some $k \in \mathbb{N}$, and* flat *if $\iota(M) \subset \mathbb{N}_0$.*

Continuing the Taylor–Hood example, if all basis functions $\Lambda_{\mathrm{TH}} = \{\phi_I\}$ of the whole finite element tree are indexed by multi-indices of the above given form, and if $\bar{x}$ is a coefficient vector that has a compatible hierarchical structure, then a finite element function (v_h, p_h) with velocity v_h and pressure p_h defined by the coefficient vector $\bar{x}$ is given by

$$(v_h, p_h) = \sum_{i=0}^{2} \sum_{j=0}^{n_2-1} \bar{x}_{(0,i,j)} \phi_{(0,i,j)} + \sum_{k=0}^{n_1-1} \bar{x}_{(1,k)} \phi_{(1,k)}, \tag{10.3}$$

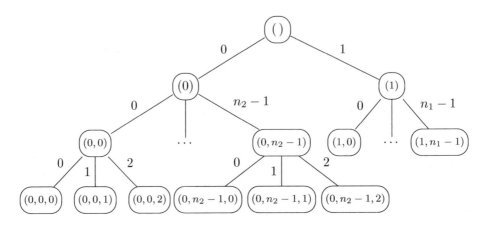

Figure 10.5.: Index tree for Taylor–Hood with blocking of velocity components

with basis functions

$$\phi_{(0,i,j)} = \mathbf{e}_0 \otimes (\mathbf{e}_i \otimes \phi_j^{P_2}), \qquad i = 0,1,2, \qquad \text{and} \qquad \phi_{(1,k)} = \mathbf{e}_1 \otimes \phi_k^{P_1}.$$

Introducing the corresponding index map $\iota : \Lambda_{\mathrm{TH}} \to \mathcal{N}$ with $\iota(\phi_I) = I$ on the set Λ_{TH} of all basis functions, we can write this in compact form as

$$(v_h, p_h) = \sum_{\phi \in \Lambda_{\mathrm{TH}}} \bar{x}_{\iota(\phi)} \phi = \sum_{I \in \iota(\Lambda_{\mathrm{TH}})} \bar{x}_I \phi_I.$$

Alternatively, the individual velocity and pressure fields v_h and p_h are given separately by

$$v_h = \sum_{i=0}^{2} \sum_{j=0}^{n_2-1} \bar{x}_{(0,i,j)} (\mathbf{e}_i \otimes \phi_j^{P_2}), \qquad\qquad p_h = \sum_{k=0}^{n_1-1} \bar{x}_{(1,k)} \phi_k^{P_1}.$$

In the previous example, the index tree was isomorphic to the basis function tree depicted in Figure 10.3. However, one may also be interested in constructing multi-indices that do not mimic the structure of the basis function tree: For example, to increase data locality in assembled matrices for the Taylor–Hood basis it may be preferable to group all velocity degrees of freedom corresponding to a single P_2 basis function together, i.e., to use the index $(0, j, i)$ for the j-th P_2 basis function for the i-th component. The corresponding alternative index tree is shown in Figure 10.5. Figure 10.6 shows the corresponding layouts of a hierarchical stiffness matrix.

Yet another alternative is to index all basis functions from the Taylor–Hood basis with a single natural number. This can be represented by an index tree with $3n_2 + n_1$ leaf nodes all directly attached to a single root. Different variations of such a tree differ by how the degrees of freedom are ordered.

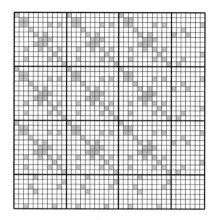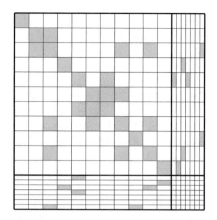

Figure 10.6.: Two matrix occupation patterns for different indexings of the Taylor–Hood bases. Left: Corresponding to the index tree of Figure 10.4. Right: Corresponding to the index tree of Figure 10.5.

10.1.4. Strategy-Based Construction of Multi-Indices

Let Λ be the set of basis functions of a finite element basis tree. In principle, dune-functions allows any indexing scheme that is given by an index map, i.e., any map $\iota : \Lambda \to \mathcal{N}$ that is injective and whose range $\iota(\Lambda)$ is an index tree. In practice, out of this large set of maps, dune-functions allows to construct the most important ones systematically using certain transformation rules.

Consider a tree of function space bases in the sense of Section 10.1.2. We want to construct an indexing for this tree, that is an index tree $\mathcal{I}$ and a bijection ι from the set of all basis functions Λ to the multi-indices in $\mathcal{I}$. The construction proceeds recursively. To describe it, we assume in the following that Λ is a node in the function space basis tree, i.e., it is the set of all basis functions corresponding to a node $V := \operatorname{span} \Lambda$ in the function space tree.

To end the recursion, we assume that an index map $\iota : \Lambda \to \mathcal{N}$ is given if $V = \operatorname{span} \Lambda$ is a leaf node of the function space tree. The most obvious choice would be a flat zero-based index of the basis functions of Λ. However, other choices are possible. For example, in case of a discontinuous finite element space, each basis function $\phi \in \Lambda$ could also be associated to a two-digit multi-index $\iota(\phi) = (i, k)$ that combines the index i of the grid element that forms the support of ϕ and the index k of ϕ within this element.

For the actual recursion, if Λ is any non-leaf node in the function space basis tree, then it takes the form

$$\Lambda = \Lambda_0 \sqcup \cdots \sqcup \Lambda_{m-1} = \bigcup_{i=0}^{m-1} \mathbf{e}_i \otimes \Lambda_i,$$

343

where $\Lambda_0, \ldots, \Lambda_{m-1}$ are the direct children of Λ, i.e., the sets of basis functions of the child spaces $\{\text{span}\,\Lambda_i\}_{i=0,\ldots,m-1}$ of the product space

$$\text{span}\,\Lambda = \text{span}\big(\Lambda_0 \sqcup \cdots \sqcup \Lambda_{m-1}\big) = (\text{span}\,\Lambda_0) \times \cdots \times (\text{span}\,\Lambda_{m-1}).$$

For the recursive construction we assume that an index map $\iota_i : \Lambda_i \to \mathcal{N}$ on Λ_i is given for any $i = 0, \ldots, m-1$. The task is to construct an index map $\iota : \Lambda \to \mathcal{N}$ from the maps ι_i. In the following we describe four strategies to achieve this; all have been implemented in `dune-functions`. When reading about these strategies, remember that any $\phi \in \Lambda$ has a unique representation $\phi = \mathbf{e}_i \otimes \hat{\phi}$ for $i \in \{0, \ldots, m-1\}$ and some $\hat{\phi} \in \Lambda_i$. It will be necessary to distinguish the special case that all children Λ_i are identical.

Definition 10.6 *An inner node Λ will be called* power node *if all of its children Λ_i are identical and equipped with identical index maps ι_i. An inner node that is not a power node is called* composite node.

This definition is needed because some of the following strategies can only be applied to power nodes.

- **BlockedLexicographic**: This strategy prepends the child index to the multi-index within the child basis. That is, the index map $\iota : \Lambda \to \mathcal{N}$ is given by

$$\iota(\mathbf{e}_i \otimes \hat{\phi}) = (i, \iota_i(\hat{\phi})).$$

  It is straightforward to show that ι is always an index map for Λ. To demonstrate the strategy the following table shows the multi-indices at inner nodes when the basis functions of the subtrees $\Lambda_0, \Lambda_1, \ldots$ are labeled by multi-indices $(I^0), (I^1), \ldots$ for Λ_0, $(K^0), (K^1), \ldots$ for Λ_1, and so on.

indices for Λ_0	indices for Λ_1	$\ldots$	indices for Λ
$\iota_0(\hat{\phi}_{0,0}) = (I^0)$			$\iota(\mathbf{e}_0 \otimes \hat{\phi}_{0,0}) = (0, I^0)$
$\iota_0(\hat{\phi}_{0,1}) = (I^1)$			$\iota(\mathbf{e}_0 \otimes \hat{\phi}_{0,1}) = (0, I^1)$
	$\iota_1(\hat{\phi}_{1,0}) = (K^0)$		$\iota(\mathbf{e}_1 \otimes \hat{\phi}_{1,0}) = (1, K^0)$
	$\iota_1(\hat{\phi}_{1,1}) = (K^1)$		$\iota(\mathbf{e}_1 \otimes \hat{\phi}_{1,1}) = (1, K^1)$
	$\iota_1(\hat{\phi}_{1,2}) = (K^2)$		$\iota(\mathbf{e}_1 \otimes \hat{\phi}_{1,2}) = (1, K^2)$
		$\ldots$	$\ldots$

- **BlockedInterleaved**: This strategy is only well-defined for power nodes. It appends the child index to the multi-index within the child basis. That is, the index map $\iota : \Lambda \to \mathcal{N}$ is given by

$$\iota(\mathbf{e}_i \otimes \hat{\phi}) = (\iota_i(\hat{\phi}), i).$$

An example is given in the following table:

indices for Λ_0	indices for Λ_1	...	indices for Λ
$\iota_0(\hat{\phi}_{0,0}) = (I^0)$			$\iota(\mathbf{e}_0 \otimes \hat{\phi}_{0,0}) = (I^0, 0)$
	$\iota_1(\hat{\phi}_{1,0}) = (I^0)$		$\iota(\mathbf{e}_1 \otimes \hat{\phi}_{1,0}) = (I^0, 1)$
		...	...
$\iota_0(\hat{\phi}_{0,1}) = (I^1)$			$\iota(\mathbf{e}_0 \otimes \hat{\phi}_{0,1}) = (I^1, 0)$
	$\iota_1(\hat{\phi}_{1,1}) = (I^1)$		$\iota(\mathbf{e}_1 \otimes \hat{\phi}_{1,1}) = (I^1, 1)$
		...	...
$\iota_0(\hat{\phi}_{0,2}) = (I^2)$			$\iota(\mathbf{e}_0 \otimes \hat{\phi}_{0,2}) = (I^2, 0)$
	$\iota_1(\hat{\phi}_{1,2}) = (I^2)$		$\iota(\mathbf{e}_1 \otimes \hat{\phi}_{1,2}) = (I^2, 1)$
		...	...

To see that this strategy does not work for general composite nodes, consider $\iota_0(\Lambda_0) = \{0\}$ and $\iota_1(\Lambda_1) = \{(0,0)\}$. Then $\iota(\Lambda) = \{(0,0), (0,0,1)\}$ which is not an index tree.

Unlike the previous two strategies, the following two do not introduce new multi-index digits. Such strategies are called *flat*.

- **FlatLexicographic**: This strategy merges the roots of all index trees $\iota_i(\Lambda_i)$ into a single new one. Assume that we split the multi-index $\iota_i(\hat{\phi})$ according to

$$\iota_i(\hat{\phi}) = (i_0, I), \tag{10.4}$$

where $i_0 \in \mathbb{N}_0$ is the first digit. The index map $\iota : \Lambda \to \mathcal{N}$ is then given by

$$\iota(\mathbf{e}_i \otimes \hat{\phi}) = (L_i + i_0, I),$$

where the offset L_i for the first digit is computed by

$$L_i = \sum_{j=0}^{i-1} \deg^+_{\iota_j(\Lambda_j)}[()].$$

In this expression, () is the empty index prefix, and therefore the sum is over the out-degrees of the roots of the index trees $\iota_j(\Lambda_j)$, $j = 0, \ldots i-1$ (see Definition 10.4). This construction offsets the first digits of the multi-indices of all basis functions from Λ_j with $j > 0$ such that they form a consecutive sequence. This guarantees that ι is always an index map for Λ. An example is given in the following table:

indices for Λ_0	indices for Λ_1	...	indices for Λ
$\iota_0(\hat{\phi}_{0,0}) = (0, I^0)$			$\iota(\mathbf{e}_0 \otimes \hat{\phi}_{0,0}) = (0, I^0)$
$\iota_0(\hat{\phi}_{0,1}) = (1, I^1)$			$\iota(\mathbf{e}_0 \otimes \hat{\phi}_{0,1}) = (1, I^1)$
	$\iota_1(\hat{\phi}_{1,0}) = (0, K^0)$		$\iota(\mathbf{e}_1 \otimes \hat{\phi}_{1,0}) = (2, K^0)$
	$\iota_1(\hat{\phi}_{1,1}) = (0, K^1)$		$\iota(\mathbf{e}_1 \otimes \hat{\phi}_{1,1}) = (2, K^1)$
	$\iota_1(\hat{\phi}_{1,2}) = (1, K^2)$		$\iota(\mathbf{e}_1 \otimes \hat{\phi}_{1,2}) = (3, K^2)$
		...	...

The digit "0" deliberately appears twice in the column for Λ_1, to demonstrate that a consecutive first digit is not required.

- **FlatInterleaved**: This strategy again only works for power nodes. It also merges the roots of all child index trees $\iota_i(\Lambda_i)$ into a single one, but it interleaves the children. Again using the splitting $\iota_i(\hat{\phi}) = (i_0, I)$ introduced in (10.4), the index map $\iota : \Lambda \to \mathcal{N}$ is given by

$$\iota(\mathbf{e}_i \otimes \hat{\phi}) = (i_0 m + i, I),$$

where the fixed stride m is given by the number of children of Λ. The following table shows an example:

indices for Λ_0	indices for Λ_1	$\ldots$	indices for Λ
$\iota_0(\hat{\phi}_{0,0}) = (0, I^0)$			$\iota(\mathbf{e}_0 \otimes \hat{\phi}_{0,0}) = (0, I^0)$
	$\iota_1(\hat{\phi}_{1,0}) = (0, I^0)$		$\iota(\mathbf{e}_1 \otimes \hat{\phi}_{1,0}) = (1, I^0)$
		$\ldots$	$\ldots$
$\iota_0(\hat{\phi}_{0,1}) = (1, I^1)$			$\iota(\mathbf{e}_0 \otimes \hat{\phi}_{0,1}) = (m + 0, I^1)$
	$\iota_1(\hat{\phi}_{1,1}) = (1, I^1)$		$\iota(\mathbf{e}_1 \otimes \hat{\phi}_{1,1}) = (m + 1, I^1)$
		$\ldots$	$\ldots$
$\iota_0(\hat{\phi}_{0,2}) = (2, I^2)$			$\iota(\mathbf{e}_0 \otimes \hat{\phi}_{0,2}) = (2m + 0, I^2)$
	$\iota_1(\hat{\phi}_{1,2}) = (2, I^2)$		$\iota(\mathbf{e}_1 \otimes \hat{\phi}_{1,2}) = (2m + 1, I^2)$
		$\ldots$	$\ldots$

Again, for this interleaved strategy, ι may not be an index map for general composite nodes.

These four strategies are offered by `dune-functions`, but there are others that are sometimes useful. Experimentally, `dune-functions` therefore also provides a way to use self-implemented custom rules.

To further illustrate the four index transformation strategies, we return to the Taylor–Hood example. While the indexing schemes proposed for this example so far were introduced in an ad-hoc way, we will now systematically apply the four strategies. Recall that the Taylor–Hood basis is denoted by

$$\Lambda_{\mathrm{TH}} = (\Lambda_{P_2} \sqcup \Lambda_{P_2} \sqcup \Lambda_{P_2}) \sqcup \Lambda_{P_1}.$$

For the bases $\Lambda_{P_1}, \Lambda_{P_2}$ of the elementary spaces P_1, P_2 we consider fixed given flat index maps

$$\iota_{P_1}(\Lambda_{P_1}) \to \mathbb{N}_0, \qquad\qquad \iota_{P_2}(\Lambda_{P_2}) \to \mathbb{N}_0.$$

These are typically constructed by enumerating the grid entities the basis functions are associated to. Then the interior product space basis

$$\Lambda_V = \Lambda_{P_2} \sqcup \Lambda_{P_2} \sqcup \Lambda_{P_2}$$

	BL(BL)	BL(BI)	BL(FL)	BL(FI)	FL(BL)	FL(BI)	FL(FL)	FL(FI)
$v_{x_0,0}$	$(0,0,0)$	$(0,0,0)$	$(0,0)$	$(0,0+0)$	$(0,0)$	$(0,0)$	(0)	$(0+0)$
$v_{x_0,1}$	$(0,0,1)$	$(0,1,0)$	$(0,1)$	$(0,3+0)$	$(0,1)$	$(1,0)$	(1)	$(3+0)$
$v_{x_0,2}$	$(0,0,2)$	$(0,2,0)$	$(0,2)$	$(0,6+0)$	$(0,2)$	$(2,0)$	(2)	$(6+0)$
$v_{x_0,3}$	$(0,0,3)$	$(0,3,0)$	$(0,3)$	$(0,9+0)$	$(0,3)$	$(3,0)$	(3)	$(9+0)$
$\vdots$	$\vdots$	$\vdots$	$\vdots$	$\vdots$	$\vdots$	$\vdots$	$\vdots$	$\vdots$
$v_{x_1,0}$	$(0,1,0)$	$(0,0,1)$	$(0,n_2+0)$	$(0,0+1)$	$(1,0)$	$(0,1)$	(n_2+0)	$(0+1)$
$v_{x_1,1}$	$(0,1,1)$	$(0,1,1)$	$(0,n_2+1)$	$(0,3+1)$	$(1,1)$	$(1,1)$	(n_2+1)	$(3+1)$
$v_{x_1,2}$	$(0,1,2)$	$(0,2,1)$	$(0,n_2+2)$	$(0,6+1)$	$(1,2)$	$(2,1)$	(n_2+2)	$(6+1)$
$v_{x_1,3}$	$(0,1,3)$	$(0,3,1)$	$(0,n_2+3)$	$(0,9+1)$	$(1,3)$	$(3,1)$	(n_2+3)	$(9+1)$
$\vdots$	$\vdots$	$\vdots$	$\vdots$	$\vdots$	$\vdots$	$\vdots$	$\vdots$	$\vdots$
$v_{x_2,0}$	$(0,2,0)$	$(0,0,2)$	$(0,2n_2+0)$	$(0,0+2)$	$(2,0)$	$(0,2)$	$(2n_2+0)$	$(0+2)$
$v_{x_2,1}$	$(0,2,1)$	$(0,1,2)$	$(0,2n_2+1)$	$(0,3+2)$	$(2,1)$	$(1,2)$	$(2n_2+1)$	$(3+2)$
$v_{x_2,2}$	$(0,2,2)$	$(0,2,2)$	$(0,2n_2+2)$	$(0,6+2)$	$(2,2)$	$(2,2)$	$(2n_2+2)$	$(6+2)$
$v_{x_2,3}$	$(0,2,3)$	$(0,3,2)$	$(0,2n_2+3)$	$(0,9+2)$	$(2,3)$	$(3,2)$	$(2n_2+3)$	$(9+2)$
$\vdots$	$\vdots$	$\vdots$	$\vdots$	$\vdots$	$\vdots$	$\vdots$	$\vdots$	$\vdots$
p_0	$(1,0)$	$(1,0)$	$(1,0)$	$(1,0)$	$(3+0)$	(n_2+0)	$(3n_2+0)$	$(3n_2+0)$
p_1	$(1,1)$	$(1,1)$	$(1,1)$	$(1,1)$	$(3+1)$	(n_2+1)	$(3n_2+1)$	$(3n_2+1)$
p_2	$(1,2)$	$(1,2)$	$(1,2)$	$(1,2)$	$(3+2)$	(n_2+2)	$(3n_2+2)$	$(3n_2+2)$
$\vdots$	$\vdots$	$\vdots$	$\vdots$	$\vdots$	$\vdots$	$\vdots$	$\vdots$	$\vdots$

Table 10.1.: Different indexing strategies for the Taylor–Hood basis functions. The abbreviations signify BlockedLexicographic (BL), BlockedInterleaved (BI), FlatLexicographic (FL), and FlatInterleaved (FI).

together with the index map ι_{P_2} is a power node in the sense of Definition 10.6, while the tree root

$$\Lambda_{\mathrm{TH}} = \Lambda_V \sqcup \Lambda_{P_1}$$

is a composite node. The basis functions for the k-th component of the velocity are denoted by

$$v_{x_k,i} = \mathbf{e}_0 \otimes (\mathbf{e}_k \otimes \phi_i^{P_2})$$

where $i = 0, \ldots, n_2 - 1$ for $n_2 = |\Lambda_{P_2}| = \dim P_2$, whereas the basis functions for the pressure are denoted by

$$p_j = \mathbf{e}_1 \otimes \phi_j^{P_1},$$

where $j = 0, \ldots, n_1 - 1$ for $n_1 = |\Lambda_{P_1}| = \dim P_1$.

As two of the above given strategies can be used for composite nodes, while all four can be applied to power nodes we obtain eight different index maps for the Taylor–Hood basis Λ_{TH}. They are listed in Table 10.1, where the label $X(Y)$ means that strategy X is used for the outer product and strategy Y for the inner product. For X and Y we use the abbreviations BL (BlockedLexicographic), BI (BlockedInterleaved), FL (FlatLexicographic), and FI (FlatInterleaved). The index maps depicted in Figure 10.4 and Figure 10.5 are reproduced for the strategies BL(BL) and BL(BI), respectively.

10.1.5. Localization to Single Grid Elements

For the most part, access to finite element bases happens element by element (cf. Chapter 2.1.3). It is therefore important to consider the restrictions of bases to single grid elements. In contrast to the previous sections we now require that there is a finite element grid for the domain Ω. For simplicity we will assume that all bases consist of functions that are defined piecewise with respect to this grid, but it is actually sufficient to require that the restrictions of all basis functions to elements of the grid can be constructed cheaply.

Consider the restrictions of all basis functions $\phi \in \Lambda$ of a given tree to a single fixed grid element T. Of these restricted functions, we discard all those that are constant zero functions on T. All others form the *local basis* on T

$$\Lambda|_T := \big\{ \phi|_T \ : \ \phi \in \Lambda, \quad \mathrm{int}(\mathrm{supp}\,\phi) \cap T \neq \emptyset \big\}.$$

The local basis forms a tree that is isomorphic to the original function space basis tree, with each global function space basis Λ replaced by its local counterpart $\Lambda|_T$.

For a given index map ι of Λ, this natural isomorphism from the global to the local tree naturally induces a localized version of ι given by

$$\iota|_T : \Lambda|_T \to \mathcal{I}, \qquad\qquad \iota|_T(\phi_T) := \iota(\phi).$$

This is the map that associates shape functions on a given grid element T to the multi-indices of the corresponding global basis functions. Note that the map $\iota|_T$ itself is not an index map in the sense of Definition 10.5 since $\iota|_T(\Lambda|_T)$ is only a subset of the index tree $\iota(\Lambda)$, and not always an index tree itself.

In order to index the basis functions in $\Lambda|_T$ efficiently we introduce an additional local index map

$$\iota^{\mathrm{local}}_{\Lambda|_T} : \Lambda|_T \to \mathcal{N},$$

such that $\iota^{\mathrm{local}}_{\Lambda|_T}(\Lambda|_T)$ is an index tree. The index $\iota^{\mathrm{local}}_{\Lambda|_T}(\phi|_T)$ is called the *local index* of ϕ (with respect to T). To distinguish it from the indices generated by ι we call $\iota(\phi)$ the *global index* of ϕ. The local index is typically used to address the element stiffness matrix. In principle, this indexing can use another non-flat index tree, which does not have to coincide with the index tree for the global basis. This means that the local index of a shape function can again be a multi-index, but the types, lengths and orderings can be completely unrelated to the corresponding global indices. This would allow to use nested types for element stiffness matrices and load vectors. As explained in Chapter 10.2, the dune-functions *implementation* is fairly restrictive here, and only allows flat local indices, i.e., $\iota^{\mathrm{local}}_{\Lambda|_T}(\Lambda|_T) \subset \mathbb{N}_0$.

In addition, for each leaf local basis $\hat{\Lambda}|_T$ of the full local basis tree we introduce another local index map

$$\iota^{\mathrm{leaf\text{-}local}}_{\hat{\Lambda}|_T} : \hat{\Lambda}|_T \to \mathbb{N}_0.$$

As there is no hierarchical structure involved, this index is simply a natural number. The index $\iota_{\hat{\Lambda}|_T}^{\text{leaf-local}}(\phi|_T)$ is called the *leaf-local index* of ϕ (with respect to T).

In an actual programming interface one typically accesses basis functions by indices directly. We will later see that in dune-functions the leaf-local index is the shape function index of the dune-localfunctions module. Hence the dune-functions API needs to implement the map

$$\iota_T^{\text{leaf}\to\text{local}} := \iota_{\Lambda|_T}^{\text{local}} \circ (\iota_{\hat{\Lambda}|_T}^{\text{leaf-local}})^{-1}$$

mapping leaf-local indices to local indices, and

$$\iota_T^{\text{local}\to\text{global}} := \iota|_T \circ (\iota_{\Lambda|_T}^{\text{local}})^{-1}$$

mapping local indices to global multi-indices.

10.2. Programming Interface for Function Space Bases

The design of the dune-functions interface for bases of function spaces follows the ideas of the previous section. The main interface concept are global basis objects that represent trees of function space bases. These trees can be localized to individual elements of the grid. Such a localization provides access to the (tree of) shape functions there, together with the two shape-function index maps $\iota_T^{\text{leaf}\to\text{local}}$ and $\iota_T^{\text{local}\to\text{global}}$. The structure of the interface is visualized in Figure 10.7.

The global basis interface is not enforced by deriving from specific base classes. Instead, dune-functions is based on C++-style duck-typing, i.e., any C++ type providing the required interface is a valid implementation of that interface (Chapter 4.4.2). Internally, dune-functions depends on the dune-typetree module, which implements abstract compile-time tree data structures. Unlike all other DUNE modules mentioned so far, dune-functions keeps all of its code in a dedicated namespace Dune::Functions.

10.2.1. The Interface for Global Function Space Bases

We start by describing the user interface for global bases. Since we are discussing duck-typing interfaces, all class names used below are generic. A tree of global bases is implemented by one class which, in the following, we will call GlobalBasis, and which can have an arbitrary number of template parameters. All types and methods listed in the following interface declaration shall be public members of the generic implementation class GlobalBasis.

As each basis implementation may require its own specific data for construction, we do not enforce a precise set of constructors. Each GlobalBasis may have one or several constructors with implementation-dependent lists of arguments.

The main purpose of a GlobalBasis is to give access to basis functions and their indices. Most of this access happens through the localization of the basis to single

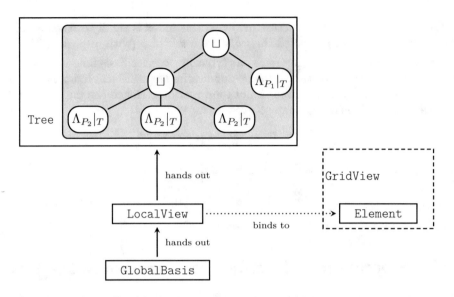

Figure 10.7.: Overview of the classes making up the interface to finite element bases

grid elements (Section 10.1.5). In the programming interface, this localization is called `LocalView`. Objects of type `LocalView` are obtained from `GlobalBasis` objects through the method

```
LocalView localView() const
```

The precise return type of the `localView` method is implementation-dependent. Objects created by the method have undefined state, and need to be attached to individual grid elements in a process called *binding*. The details are explained in Section 10.2.2.

Several methods of a `GlobalBasis` provide information on the sizes of the bases contained in the tree. The total number of basis functions of the global basis is exported via the method

```
size_type dimension() const
```

This method can be used to allocate vector containers if flat multi-indices are used. However, the information provided by the `dimension` method is generally not sufficient to allocate hierarchical containers to be accessed by more general multi-indices. Therefore, the basis provides additional structural information of those multi-indices via the method

```
size_type size(const SizePrefix& prefix) const
```

The parameter prefix is a multi-index itself. If $\mathcal{I}$ is the set of all global multi-indices of the basis and prefix is a prefix for this set, then size(prefix) returns the size $\deg_{\mathcal{I}}^{+}[\text{prefix}]$ of $\mathcal{I}$ relative to prefix defined in (10.2), i.e., the number of direct children of the node prefix in the index tree. If prefix is not a prefix for $\mathcal{I}$ the result is undefined. If prefix $\in \mathcal{I}$, i.e., the prefix is itself one of the multi-indices then the result is zero. The type SizePrefix is always a container of type ReservedVector (from the dune-common module; more details are given in Chapter 10.2.4). Like all other types used in the GlobalBasis interface, the implementation class has to export it. For convenience there is also the method

```
size_type size() const
```

returning the same value as size({}), i.e., the number of children of the root of the index tree. For a scalar basis, this is again the overall number of basis functions.

Finally, each GlobalBasis provides access to the grid view it is defined on by the method

```
const GridView& gridView() const
```

The corresponding type is exported as GridView. If the grid view was modified (e.g., by local grid refinement), the result of calling any method of the basis is undefined until the basis has been explicitly updated. For this, there is the method

```
void update(const GridView& gridView)
```

which tells the basis to adapt its local state to the new grid view.

10.2.2. The Interface for the Localized Basis

The localization of a function space basis to a single grid element is represented by an interface called LocalView. Objects of type LocalView are returned by the GlobalBasis method localView, and there is no way to construct such objects directly. All types and methods listed in the following interface declaration are public members of the generic class LocalView.

A freshly constructed LocalView object is not completely initialized yet. To truly have the object represent the basis localization on a particular element, it must be *bound* to that element. This is achieved by calling

```
void bind(const Element& element)
```

where Element is defined as GridView::Codim<0>::Entity. Once this method has been called, the LocalView object is fully set up and can be used. The call may incorporate expensive computations needed to precompute the local basis functions and their global indices. The local view can be bound to another element at any time by calling the bind method again. To set the local view back to the unbound state again, call the method

```
void unbind()
```

The local view will store a copy of the element it is bound to, which is accessible via

```
const Element& element() const
```

A bound `LocalView` object provides information about the size of the local basis at the current element. The total number of basis functions associated to the local view at the current element is returned by

```
size_type size() const
```

In the language of Chapter 10.1, this method computes the number $|\Lambda|_T|$.

To allow preallocation of buffers for local functions, the method

```
size_type maxSize() const
```

returns the maximum return value of the `size` method for all elements in the grid view associated to the global basis, i.e., it computes $\max_T |\Lambda|_T|$. As this information does not depend on a particular element, the method `maxSize` can even be called in unbound state.

As an example, suppose that `basis` is an object of type `TaylorHoodBasis`, which implements the Taylor–Hood basis that has been used for examples in the previous section. The following code loops over all elements of the grid view and prints the numbers of degrees of freedom per element:

```
auto localView = basis.localView();

for (auto&& element : elements(basis.gridView()))
{
  localView.bind(element);
  std::cout << "Element with " << localView.size()
    << " degrees of freedom" << std::endl;
}
```

Access to the actual local basis functions is provided by the method

```
const Tree& tree() const
```

This encapsulates the set $\Lambda|_T$ of basis functions localized to the element T, organized in the tree of function space bases. While the tree itself can be queried in unbound state, the local view must be bound in order to use most of the tree's methods. A detailed discussion of the interface of the tree object is given below.

For any of the local basis functions in the local tree accessible by `tree`, the global multi-index is provided by the method

```
MultiIndex index(size_type i) const
```

The argument for this method is the local index of the basis function within the tree as returned by node.localIndex(k) (see Page 355 for the localIndex method); here node is a leaf node of the tree provided by tree(), and k is the number of the shape function within the corresponding local finite element (see below). Hence the index method implements the map $\iota_T^{\mathrm{local}\to\mathrm{global}}$ introduced in Section 10.1.5, which maps local indices to global multi-indices. Accessing the same global index multiple times is expected to be cheap, because implementations are supposed to pre-compute and cache indices during binding. The result of calling index(size_type) in unbound state is undefined.

Extending the previous example, the following loop prints the global indices for each degree of freedom of each element:

```
auto localView = basis.localView();

for (auto&& element : elements(basis.gridView()))
{
  localView.bind(element);
  for (std::size_t i=0; i<localView.size(); i++)
    std::cout << localView.index(i) << std::endl;
}
```

When this code is run for a Taylor–Hood basis on a two-dimensional triangle grid, it will print 15 multi-indices per element, because a Taylor–Hood element has 12 velocity degrees of freedom and 3 pressure degrees of freedom per triangle.

Finally, the global basis of type GlobalBasis is known by the LocalView object, and exported by the globalBasis method.

```
const GlobalBasis& globalBasis() const
```

Therefore, code that is given only a LocalView object can retrieve the global basis from it, and the grid view from there.

10.2.3. The Interface for the Tree of Local Bases

The local view provides access to local basis functions of an element by exporting a Tree object, which keeps the local basis functions in its leaves. The tree structure is encoded in the type of the Tree object, using the infrastructure of the dune-typetree module.

The object returned by the LocalView::tree method is not actually a tree, but rather (a **const** reference to) the root node of the tree. To navigate within the tree, any non-leaf node allows to access its children using the two methods

```
template<class... ChildIndices>
auto child(ChildIndices... childIndices)
```

and

```
template<class ChildTreePath>
auto child(ChildTreePath childTreePath)
```

The parameters of these methods are the paths from the current node to the desired descendants.

For the first method, the path is passed as a sequence of indices. Indices referring to children of a power node can be passed as run-time integers, typically of type `std::size_t`. Indices referring to children of a composite node have to be passed statically as objects of type `Dune::index_constant<i>`. [1] For convenience global constants `_0,_1, ...` of this type are implemented in the `Dune::Indices` namespace, in the header `dune/common/indices.hh`. Continuing the example of the previous section, if `localView` is a local view of the `TaylorHoodBasis` localized to a particular grid element, then the leaf node for the second velocity component can be obtained by

```
using namespace Dune::Indices;   // Namespace containing
                                 // the index constants _0, _1, _2, ...
const auto& node = localView.tree().child(_0, 1);
```

Note how the index constant `Dune::Indices::_0` is used to address the velocity node, because the tree root is a composite node whose child nodes are of different types. Within the velocity subtree, all three children are identical, and the second one can be accessed by the run-time integer 1.

The second `child` method allows to pass the tree path in a dedicated container. Such a container needs to handle sequences of static and run-time values. The `dune-functions` module uses `Dune::TypeTree::HybridTreePath` for this, which we describe in detail in Chapter 10.2.4. Using a `HybridTreePath` object, the example looks as follows:

```
auto treePath = TypeTree::treePath(_0, 1);
const auto& node = localView.tree().child(treePath);
```

At each of its leaf nodes, the localized basis function tree provides the set of all corresponding shape functions. The method for this is

```
const FiniteElement& finiteElement() const
```

The object returned by this method is a `LocalFiniteElement` as specified in the dune-localfunctions module (Chapter 8). As such, it provides access to shape function values and derivatives, to the evaluation of degrees of freedom in the sense of [42], and to the assignment of local degrees of freedom to element faces. The numbering used by `dune-localfunctions` for the shape functions are the leaf-local indices defined in Section 10.1.5. For example, if `node` is a leaf node in the localized Taylor–Hood tree, the following code prints all shape function values of the leaf shape function set at the point $(0, 0, 0)$ in local coordinates of the appropriate reference element:

---

[1] . . . which is a shortcut for `std::integral_constant<std::size_t, i>`, cf. Chapter 7.2.2.

```
const auto& localBasis = node.finiteElement().localBasis();
std::vector<double> values;
localBasis.evaluate({0,0,0}, values);
for (auto v : values)
  std::cout << v << std::endl;
```

To obtain the entries of the element stiffness matrix that correspond to a given shape function from a given leaf node, the local index needs to be computed from the leaf-local index of that shape function. For each local basis function the method

```
size_type localIndex(size_type i) const
```

returns the local index within all local basis functions of the current element associated to the full local tree. The argument to this method is the index of the local basis functions within the leaf. In other words, the localIndex method implements the map $\iota_T^{\text{leaf}\to\text{local}}$ introduced in Section 10.1.5. The return value is *not* a multi-index. While in principle all basis functions of the local subtree could be indexed using general multi-indices, the dune-functions module only supports flat indices here to keep the implementation simple.

While LocalFiniteElement objects are only available at leaf nodes, the following methods work at every node in the tree again. Calling

```
size_type size() const
```

returns the total number of local basis functions within the subtree rooted at the present node. In particular, calling this method for the tree root yields the number of rows and/or columns of the element stiffness matrix.

Finally, all nodes provide access to the element which they are bound to via the method

```
const Element& element() const
```

10.2.4. Multi-Indices

Multi-indices appear in several places in dune-functions. They are used as global indices to identify individual basis functions of a function space basis, and for indexing inner nodes of basis and index trees as well. From an implementation point of view, basis and index trees differ considerably. Only the localized basis tree explicitly appears in the programming interface, whereas index trees appear only implicitly in the form of sets of indices with the appropriate structure (Definition 10.3). These differences require separate multi-index implementations for the different types of trees. We discuss implementations for both types of trees in turn.

Multi-Index Implementations for Basis Trees

The tree of localized basis functions is the only tree that explicitly appears in the dune-functions programming interface. The tree structure is encoded as C++ type information using the tools from the `dune-typetree` module. Navigation in this tree requires to manipulate paths from the root to particular nodes. In principle, such a path is a sequence of integers.

To understand the implementation, remember that non-leaf tree nodes can be of two types, *power* and *composite* (Section 10.1.4). Since composite nodes have children of different types, it is not possible to access those children using a run-time index. Instead, the child index in a composite node has to be encoded in a static way, using the `index_constant` type already seen in Chapter 7.2.2. On the other hand, all children of a power node have the same C++ type, and can be accessed using a dynamic index of type `std::size_t`.

In a typical tree, composite and power nodes appear together. It is therefore necessary to have a container that can store both compile-time and run-time integers. This is achieved by the class

```
template<class... I>
class HybridTreePath
```

from the `dune-typetree` module. Conceptually, a `HybridTreePath` is a fixed-size container, where each entry can be of different type. The types of the individual entries are passed as template parameters. If the type used for an entry is `std::size_t`, then this entry will have a dynamic value. If, on the other hand, the type is `Dune::index_constant<i>`, then its value is static, and can be used for compile-time decisions.

An object of type `HybridTreePath` can be used to access the nodes of a localized basis tree if dynamic tree path entries only appear as child indices for power nodes in the tree while all other entries are static. For example, to access the leaf nodes corresponding to the velocity components in the Taylor–Hood ansatz tree depicted in Figure 10.2 one would use multi-indices of the type

```
HybridTreePath<Dune::index_constant<0>, std::size_t>;
```

whereas the multi-index for the pressure leaf node would use the type

```
HybridTreePath<Dune::index_constant<1> >;
```

To construct objects of these types, call

```
using namespace Dune::Indices;
HybridTreePath<Dune::index_constant<0>, std::size_t> i00(_0,0);
HybridTreePath<Dune::index_constant<0>, std::size_t> i01(_0,1);
HybridTreePath<Dune::index_constant<0>, std::size_t> i02(_0,2);
```

for the velocity leaf nodes, and

```
HybridTreePath<Dune::index_constant<1> > i1(_1);
```

for the pressure node. As mentioned above, the constants `_0` and `_1` are predefined in the namespace `Dune::Indices`.

This way of construction is overly verbose, because static indices have to be provided both as template and as constructor arguments. To simplify the construction of such objects, the dune-typetree module provides the helper function

```
template<class... I>
auto TypeTree::treePath(I... i)
```

which creates a HybridTreePath object with the given entries. As this is a free method rather than a constructor, the entries have to be given only once, and their types are inferred. The multi-indices of the previous example can be constructed using

```
auto i00 = TypeTree::treePath(_0, 0);
auto i01 = TypeTree::treePath(_0, 1);
auto i02 = TypeTree::treePath(_0, 2);
auto i1  = TypeTree::treePath(_1);
```

which is much shorter.

To access entries of a HybridTreePath object, the object has the method

```
template<std::size_t i>
constexpr decltype(auto)
   operator[](const Dune::index_constant<i>& indexVariable) const
```

Depending on the template parameter i, the return type is either std::size_t or Dune::index_constant<i>. Note that this construction is necessary although the function is already **constexpr**: Since HybridTreePath objects are intended to select children of different type in run-time contexts, they have to encode compile-time index values into the run-time index objects. The latter is only possible by making the types of those index objects dependent on their compile-time values.

However, as objects of type Dune::index_constant can be implicitly converted to std::size_t, there is also

```
auto operator[](std::size_t i) const
```

Hence, to get the first digit of a tree path it is possible to write

```
std::size_t d0 = myHybridTreePath[0];
```

For a tree path in the Taylor–Hood tree this will return 0 or 1 as expected. However, this return value is not usable in compile-time situations anymore.

These are just the more important methods of the HybridTreePath class. For a complete description see the online documentation of the dune-typetree module.

Multi-Index Implementations for Index Trees

Index trees are formed by the multi-indices that are used to label basis functions. Conceptually, there are two such trees in the dune-functions interface: the tree of global indices, and the tree of local indices. To keep the implementation simple, dune-functions only allows flat (i.e., single-digit) multi-indices for the local index tree. Therefore, only data types for global indices need to be discussed.

Unlike the tree paths of the previous section, global indices are run-time constructs. A single C++ type represents all such indices for a given basis, even if that basis has a non-trivial tree structure. The exact type is selected by the basis implementation, and can differ from basis to basis. It mainly depends on whether the index is uniform, i.e., whether all indices from the set have the same number of digits. Having purely dynamic multi-indices can be inconvenient when accessing containers such as `std::tuple` or `MultiTypeBlockVector` (from the dune-istl module, Chapter 7). However, it has the advantage that standard run-time loops can be used to iterate over the indices.

Dynamic multi-indices are random-access containers holding entries of a fixed integer type. All implement a common interface, consisting of the two member functions

```
std::size_t size() const
```

and

```
auto operator[](std::size_t i) const
```

The `size` method returns the number of digits of the multi-index, and `operator[]` allows to access each entry by its position. Since multi-indices are typically not changed by user code, both methods are `const`.

In the following we will give an overview of the types used to represent multi-indices in dune-functions. In the most general case, not all multi-indices for a given basis have the same number of digits. As examples, consider columns 1, 2, 5, and 6 of Table 10.1, which give such numberings for the Taylor–Hood basis. In these cases, multi-indices are typically represented by the class

```
template<class T, int k>
class ReservedVector
```

from the dune-common module, which is parameterized by the entry type T and a capacity k. It implements a standard-library-compatible random-access container with a dynamic size, which may not exceed k entries. In contrast to a fully dynamic vector implementation like `std::vector<T>`, the class `ReservedVector` stores its entries on the stack. This avoids dynamic memory management, and makes the implementation much more efficient. The global multi-indices typically have a small number of digits only with a known upper bound. Hence the overhead of always using a buffer of size k even for indices with less than k digits will typically be small.

However, many bases can be indexed by uniform index trees, i.e., sets of indices where all indices have the same number of digits. In that case, the capacity of a `ReservedVector` can be set to the correct length, and no buffer space is wasted. However, in addition to the buffer, each `ReservedVector` object has to store the container length, which is not needed when the index set is known to be uniform. `GlobalBasis` objects that implement uniform index sets can therefore opt to use a fixed-size container type like `std::array` instead of `ReservedVector`.

Finally, if the basis is indexed with a flat index, i.e., a multi-index with only a single digit, then using an array can be cumbersome. Morally, flat multi-indices are simply non-negative integer numbers. However, if i is a std::array of length 1, using it to access the corresponding entry of a std::vector called vec has to be written as

```
auto value = vec[i[0]];
```

To allow the more intuitive syntax

```
auto value = vec[i];
```

dune-functions implements the FlatMultiIndex class for the case that the index of a basis tree is flat. Objects of type FlatMultiIndex behave like objects of type std::array<T,1>, but additionally, they allow to cast their content to T&. Therefore, objects of type FlatMultiIndex can be directly used like number types, and like multi-index types as well.

10.3. Constructing Trees of Function Space Bases

There are various ways to construct finite element bases in dune-functions. A set of standard bases is provided directly. These can then be combined to form trees. Conversely, subtrees can be extracted, and they (almost) act like complete bases in their own right.

10.3.1. Basis Implementations Provided by dune-functions

The dune-functions module contains a collection of standard finite element bases. These can be directly used in finite element simulation codes. At the time of writing there are:

- LagrangeBasis: Lagrange basis of order k, where k is a run-time or compile-time parameter. This implementation works on all kinds of conforming grids, including grids with more than one element type. At the time of writing, higher-order spaces are implemented only partially. Check the online class documentation for the current status.

- LagrangeDGBasis: A k-th order Discontinuous-Galerkin (DG) basis with Lagrange shape functions. As a DG basis, it also works well on non-conforming grids. The polynomial order k is a compile-time parameter.

- RannacherTurekBasis: An H^1-nonconforming scalar basis, which adapts the idea of the Crouzeix–Raviart basis to cube grids [136].

- BSplineBasis: A B-spline basis on a structured, axis-aligned grid as described, e.g., in [44]. Arbitrary orders, dimensions, and knot vectors are supported, allowing, e.g., to work with C^1 elements for fourth-order differential equations.

A `BSplineBasis` object implements a basis on a single patch, and the grid must correspond to this patch. For this to work, several restrictions apply for the grid. It must be structured and axis-aligned, and consist of (hyper-)cube elements only. Further, the element indices must be ordered lexicographically and increase from the lower left to the upper right domain corner. The element spacing must match the knot spans. Unfortunately, not all these requirements can be checked for by the basis, so users have to be a bit careful. Using `YaspGrid` objects works well.

Unlike in standard finite element bases, in a B-spline basis the basis functions cannot be associated to grid entities such as vertices, edges, or elements. The `dune-localfunctions` programming interface nevertheless mandates that a `LocalCoefficient` object, which assigns shape functions to faces of the reference element, must be available even for a B-spline basis. For the `BSplineBasis` implementation, the behavior of this object is undefined.

- `TaylorHoodBasis`: An implementation of a first-order Taylor–Hood basis. It exists mainly to serve as an example of how to directly implement a basis with a non-trivial tree. Generally, non-trivial product bases can be easily constructed in a generic way. This approach is described in Chapter 10.3.2 and it is the preferred way to construct a Taylor–Hood basis.

For all bases listed above, the shape functions provided by `tree.finiteElement()` are implemented in terms of coordinates of the reference element T_{ref}. That is, if a grid element T is obtained by the transformation $F : T_{\text{ref}} \to T$, then the implemented localized shape function representing the restriction of the basis function ϕ to the element T is given by $\hat{\phi}|_T = \phi \circ F$. Finite elements that form non-affine families [42] may require additional transformations. This is the case for the following global bases implementations:

- `RaviartThomasBasis`: The standard Raviart–Thomas basis [32] for problems in $H(\text{div})$. Available for different orders and element types.

- `BrezziDouglasMariniBasis`: The standard Brezzi–Douglas–Marini basis, which is an alternative basis for $H(\text{div})$-conforming problems [32].

Both of these bases require the Piola transformation to properly pull back the basis functions onto the reference element. As of `dune-functions` 2.7, this transformation is *not* performed by the `dune-functions` implementation, and is expected to happen in user code.[2] For a detailed discussion of the template parameters and constructor arguments of the basis implementations listed above we refer to the online documentation.

---

[2]Dune 2.8 changed this, and the Piola transformation is now done by the bases themselves.

10.3.2. Combining Bases into Trees

The basis implementations of the previous section can be combined by multiplication to form new bases. This produces the tree structures described in Section 10.1. The multiplication code resides in the `BasisFactory` namespace, which is a nested namespace within `Dune::Functions`. Therefore, the examples in this section need a

using namespace Dune::Functions::BasisFactory;

to compile.

The methods to combine bases into trees do not operate on the basis classes of the previous section directly. Rather, they combine so-called *pre-bases*, of which there is one for each basis. The reason for this is that it is technically challenging to combine the actual user-visible basis types in a tree hierarchy that itself again implements the interface of a function space basis. Therefore, the multiplication operators are applied to pre-basis objects, and return pre-basis objects of the resulting tree. The pre-basis of the final basis tree can then be turned into an actual basis.

Since all pre-bases in the product pre-basis have to know some common information like, e.g., the grid view, doing this product construction manually is verbose and error prone. As a more user-friendly and safer solution a global basis can be constructed by a call to

```
template<class GridView, class PreBasis>
auto makeBasis(const GridView& gridView, PreBasis&& preBasis)
```

The pre-basis argument encodes the product. The actual basis is constructed automatically by the `makeBasis` function from the pre-bases in a consistent way. This also determines a suitable multi-index type automatically, which otherwise would have to be done by the user.[3]

In the simple-most case, the basis tree consists of a single leaf. This leaf is then, e.g., one of the basis implementations of the previous section. As a convention, for each global basis `FooBarBasis` there is a function `BasisFactory::fooBar()` (defined in the same header file as `FooBarBasis`), creating a suitable pre-basis object that stores all basis-specific information. This means that in particular one can write

auto raviartThomasBasis = makeBasis(gridView, raviartThomas<k>());

to obtain a Raviart–Thomas basis for the given grid view. This call to `makeBasis` is equivalent to constructing the basis directly:

RaviartThomasBasis<GridView,k> raviartThomasBasis(gridView);

---

[3]This interface description is in fact slightly simplified: The user-provided arguments of `makeBasis` are not pre-bases themselves but pre-basis-factory objects that can construct the corresponding pre-bases. This mechanism allows to delay passing the shared information (e.g. the grid view) to the construction of the real pre-bases which is triggered by `makeBasis`. However, to simplify the presentation we will ignore the technical difference of a pre-basis and its pre-basis-factory in the following.

Note that the raviartThomas function, just like the corresponding functions for other bases, does not need the grid view as argument.

If FooBarBasis has template and/or constructor parameters, then by convention they are given in the same order as the template and method parameters of the BasisFactory::fooBar() function. As the only difference, the former has the grid view type and object prepended.

The pre-basis combining several bases in a product is called CompositePreBasis, and it is defined in the header file compositebasis.hh in the folder dune/functions/functionspacebases/. It implements a composite tree node as introduced in Definition 10.6. Analogously to the above description, a pre-basis for a tree with a composite root can be constructed using the global function

```
template<class... ChildPreBasis>
auto composite(ChildPreBasis&&... childPreBasis)
```

contained in the namespace BasisFactory. The method has an unspecified number of parameters, of unspecified type. The arguments are expected to be pre-basis objects themselves. They can either be plain pre-bases constructed by, e.g., lagrange<1>() or raviartThomas<k>(), or composite- or power pre-bases constructed by the composite or power function (see below), respectively.

As an example, to combine a Raviart–Thomas basis with a zero-order Lagrange basis (let's say for solving the mixed formulation of the Poisson equation [37]), the appropriate call is

```
auto mixedBasis = makeBasis(
  gridView,
  composite(
    raviartThomas<0>(),
    lagrange<0>()
));
```

Combining three copies of a first-order Lagrange basis for a displacement field in elasticity theory is done by

```
auto displacementBasis = makeBasis(
  gridView,
  composite(
    lagrange<1>(),
    lagrange<1>(),
    lagrange<1>()
));
```

The examples produce the trees shown in Figure 10.8.

The second example is not as elegant as it could be. First of all, it is inconvenient and unnecessarily wordy to list the same scalar Lagrange basis three times. Secondly, that number may depend on a parameter. Finally, the implementation can benefit from the explicit knowledge that all children are equal. For these reasons, dune-functions

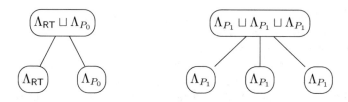

Figure 10.8.: Example composite bases

offers a second way to combine bases: The `PowerPreBasis`. The interface is again a single method

```
template<std::size_t k, class ChildPreBasis>
auto power(ChildPreBasis&& childPreBasis)
```

provided in the file dune/functions/functionspacebases/powerbasis.hh. It combines k copies of a subtree of type `ChildPreBasis` in a new tree. Therefore, the displacement vector field basis from above is more easily written as

```
auto displacementBasis = makeBasis(
  gridView,
  power<3>(
    lagrange<1>()
  ));
```

Since `composite` and `power` create pre-bases themselves, all these techniques can be combined. To obtain the p-th order Taylor–Hood basis, write

```
auto taylorHoodBasis = makeBasis(
  gridView,
  composite(
    power<dim>(
      lagrange<p+1>()),
    lagrange<p>()
  ));
```

The call to `power` produces the `dim`-component Lagrange basis of order p+1 for the velocity, and the call to `composite` combines this with a p-th order Lagrange basis for the pressure. Note that this is the preferred way to construct a Taylor–Hood basis in contrast to

```
auto taylorHoodBasis = makeBasis(gridView, taylorHood());
```

and

```
auto taylorHoodBasis = TaylorHoodBasis<GridView>(gridView);
```

These variants mainly exist as an implementation example.

The previous discussion has left out the question of how the degrees of freedom in the combined tree are numbered. In Section 10.1.3 it was explained how the indices of the degrees of freedom form a separate tree by their multi-index structure, and how this tree is constructed from the basis tree by a set of strategies. These ideas are reflected in the design of the dune-functions programming interface. First of all, each of the bases of Section 10.3.1 implements a numbering of its degrees of freedom, and generally these numberings cannot be changed. To select a numbering for a non-trivial basis, each call to composite or power can be augmented by an additional flag indicating an IndexMergingStrategy. The four implemented strategies are

- BlockedLexicographic

- BlockedInterleaved

- FlatLexicographic

- FlatInterleaved

and have been described in Section 10.1.4. For each strategy FooBar there is a function BasisFactory::fooBar() creating the flag (in the header functionspacebases/ basistags.hh). For example, a Taylor–Hood basis with the indexing listed in the second column (labeled BL(BI)) of Table 10.1 can be created using

```
auto taylorHoodBasis = makeBasis(
  gridView,
  composite(
    power<dim>(
      lagrange<p+1>(),
      blockedInterleaved()),
    lagrange<p>(),
    blockedLexicographic()
  ));
```

This will lead to multi-indices of length three and two for velocity and pressure degrees of freedom, respectively. The same ordering of basis functions with a uniform indexing scheme with multi-index length two (Column 4 labeled BL(FI) in Table 10.1) is obtained by

```
auto taylorHoodBasis = makeBasis(
  gridView,
  composite(
    power<dim>(
      lagrange<p+1>(),
      flatInterleaved()),
    lagrange<p>(),
    blockedLexicographic()
  ));
```

Finally, a flat indexing scheme still preserving the same ordering (Column 8 labeled FL(FI) in Table 10.1) is obtained by

```
auto taylorHoodBasis = makeBasis(
  gridView,
  composite(
    power<dim>(
      lagrange<p+1>(),
      flatInterleaved()),
    lagrange<p>(),
    flatLexicographic()
));
```

If no strategy is given, composite will use the BlockedLexicographic strategy, whereas power will use BlockedInterleaved.

10.4. Treating Subtrees as Separate Bases

The previous section has shown how trees of bases can be combined to form bigger trees. It is also possible to extract subtrees from other trees and treat these subtrees as basis trees in their own right. The programming interface for such subtree bases is called SubspaceBasis. It mostly coincides with the interface of a global basis, but additionally to the GlobalBasis interface the SubspaceBasis provides information about how the subtree is embedded into the global basis (which in this context is called the *root basis*). More specifically, the method

```
const GlobalBasis& rootBasis() const
```

provides access to the root basis, and the method

```
const TypeTree::HybridTreePath& prefixPath() const
```

returns the path of the subtree associated to the SubspaceBasis within the full tree. For convenience a global basis behaves like a trivial SubspaceBasis, i.e., it has the method rootBasis returning the basis itself, and prefixPath returning an empty tree-path. Note that a SubspaceBasis differs from a full global basis because the global multi-indices are a subset of those of the root basis, and thus they are in general neither consecutive nor zero-based. Instead, those multi-indices allow to access containers storing coefficients for the full root basis.

SubspaceBasis objects are created using a global factory function, which is given the root basis and the path to the desired subtree. The path can either be passed as a single HybridTreePath object (see Section 10.2.4), or as a sequence of individual indices:

```
template<class GlobalBasis, class... PathIndices>
auto subspaceBasis(const GlobalBasis& rootBasis,
                   const TypeTree::HybridTreePath<PathIndices...>& prefixPath)

template<class GlobalBasis, class... PathIndices>
auto subspaceBasis(const GlobalBasis& rootBasis,
                   const PathIndices&... indices)
```

For example, suppose that `taylorHoodBasis` is any one of the implementations of the Taylor–Hood basis defined in Section 10.3.2. Then

```
auto velocityBasis = subspaceBasis(taylorHoodBasis, _0);
```

will extract the subtree of velocity degrees of freedom, and

```
auto pressureBasis = subspaceBasis(taylorHoodBasis, _1);
```

will extract the (trivial) subtree of pressure degrees of freedom. The possibly non-consecutive multi-indices of a `SubspaceBasis` are best illustrated by extracting a single velocity component

```
auto velocityX2Basis = subspaceBasis(taylorHoodBasis, _0, 2);
```

For this example the following table shows the multi-indices of the `SubspaceBasis` extracted from the full basis, with columns representing the different index merging strategies also used in Table 10.1:

	BL(BL)	BL(BI)	BL(FL)	BL(FI)	FL(BL)	FL(BI)	FL(FL)	FL(FI)
$v_{x_2,0}$	$(0,2,0)$	$(0,0,2)$	$(0,2n_2+0)$	$(0,0+2)$	$(2,0)$	$(0,2)$	$(2n_2+0)$	$(0+2)$
$v_{x_2,1}$	$(0,2,1)$	$(0,1,2)$	$(0,2n_2+1)$	$(0,3+2)$	$(2,1)$	$(1,2)$	$(2n_2+1)$	$(3+2)$
$v_{x_2,2}$	$(0,2,2)$	$(0,2,2)$	$(0,2n_2+2)$	$(0,6+2)$	$(2,2)$	$(2,2)$	$(2n_2+2)$	$(6+2)$
$v_{x_2,3}$	$(0,2,3)$	$(0,3,2)$	$(0,2n_2+3)$	$(0,9+2)$	$(2,3)$	$(3,2)$	$(2n_2+3)$	$(9+2)$
$\vdots$	$\vdots$	$\vdots$	$\vdots$	$\vdots$	$\vdots$	$\vdots$	$\vdots$	$\vdots$

Most of the resulting multi-index sets do not form trees.

10.5. Functions

The second important feature of `dune-functions` is a programming interface for general functions

$$f : \mathcal{D} \to \mathcal{R} \tag{10.5}$$

between sets $\mathcal{D}$ and $\mathcal{R}$. In many important cases $\mathcal{D}$ will be the domain Ω, and $\mathcal{R}$ a Euclidean space, but the `dune-functions` interface handles the general case.

Like in the rest of `dune-functions`, duck typing is used for the interface (Chapter 4.4.2). The model classes form a conceptual hierarchy (shown in Figure 10.9), even though no inheritance is involved at all. Implementors of the interface need to write classes that have all the methods and behavior specified by the model classes. Concept checking is used to produce readable error messages. The following sections explain the individual ideas in detail.

10.5.1. Function Objects and Functions

The C++ language has a standard way to implement functions given mathematically as (10.5). A language construct is called a *function object* if it is an ordinary function, a function pointer, or an object (or reference to an object) of a class providing an

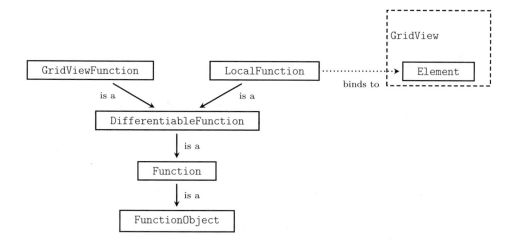

Figure 10.9.: Relationship of the various model classes in the function interface

operator(). We will adopt the common convention to call the latter a *functor*. In the following we will denote a function object as *function* if **operator**() does not return **void**. In other words, a function f is anything that can appear in an expression of the form

```
auto y = f(x);
```

for an argument x of suitable type. Examples of such constructs are:

- free functions
  ```
  double sinSquared(double x)
  {
    return std::sin(x) * std::sin(x);
  }
  ```

- lambda expressions
  ```
  auto sinSquaredLambda = [](double x){return std::sin(x) * std::sin(x); };
  ```

- functors
  ```
  struct SinSquared
  {
    double operator()(double x)
    {
      return std::sin(x) * std::sin(x);
    }
  };
  ```

This list is not exhaustive.

All three examples are called in the same way:

```
double a = sinSquared(3.14);          // Free function
double b = sinSquaredLambda(3.14);    // Lambda expression
SinSquared sinSquaredObject;
double c = sinSquaredObject(3.14);    // Functor
```

Argument and return value do not have to be **double** at all; any type is possible. They can be scalar or vector types, floating point or integer, and even more exotic data like matrices, tensors, and strings.

To pass a function as an argument to a C++ method, the type of that argument must be explicitly stated in the signature of the called method, for example

```
void evaluateAt42(double (*f)(double))   // Accepts a function pointer
void evaluateAt42(SinSquared f)          // Accepts SinSquared function objects
```

To allow *any* function type as method argument, it must be a template parameter:

```
template<class F>
void evaluateAt42(F&& f)
{
    std::cout << "Value of f(42): " << f(42) << std::endl;
}
```

Any of the example functions from above can then be used as an argument of the method evaluateAt42:

```
evaluateAt42(sinSquared);         // Call with a free function
evaluateAt42(sinSquaredLambda);   // Call with a lambda expression
evaluateAt42(sinSquaredObject);   // Call with a functor
```

10.5.2. Type Erasure and `std::function`

For the previous example, three different instatiations of the evaluateAt42 method are created by the compiler, for the three different function types used here. Sometimes, however, the precise type of a function is not known at compile-time but selected depending on run-time information. This behavior is commonly referred to as *dynamic dispatch*. The classic way to implement it uses inheritance and the **virtual** keyword: All classes implementing functions must inherit from a common base class, and a pointer to this class is then passed around instead of the function itself.

The disadvantages of this approach have been discussed in Chapter 4.4.1. Particularly relevant for function interfaces is the fact that in a derived class, the return value of **operator**() must match the return value used in the base class (unless it is a pointer or reference type, which does not occur in a functions interface, though). However, it is frequently convenient to also allow return values that are only *convertible* to the return value of the base class. This is not possible in C++.

The C++ standard library therefore uses *type erasure* rather than inheritance to implement run-time polymorphism. Starting with C++11, the standard library contains a class [98, Sect. 20.8.11]

```
template<class Range, class... Args>
class function<Range(Args...)>
```

that wraps all functions that map a set of type arguments Args... to a type (convertible to) Range behind a single C++ type std::function. In the context of dune-functions, we are only interested in the case that Args... is a single type Domain.

Given two types Domain and Range, any function object that accepts a Domain as argument and returns something convertible to Range can be stored in an object of type std::function<Range(Domain)>. For example, reusing the three implementations of $\sin^2(x)$ from Section 10.5.1, one can write

```
std::function<double(double)> polymorphicF;
polymorphicF = sinSquared;            // Assign a free function
polymorphicF = sinSquaredLambda;      // Assign a lambda expression
polymorphicF = SinSquared();          // Assign a functor
double a = polymorphicF(3.14);        // Evaluate
```

Note how different C++ constructs are all assigned to the same object. One can even write

```
// Okay: Return type int can be converted to double
polymorphicF = [](double x) -> int
{
  return floor(x);
};
```

but not

```
// Error: Return type std::complex<double> cannot be converted to double
polymorphicF = [](double x) -> std::complex<double>
{
  return std::complex<double>(x,0);
};
```

Looking at implementations of std::function, one can see that virtual functions and heap allocations are used *internally*, but they are hidden to the outside. Considering this, one may not expect any run-time gains for type erasure over dynamic polymorphism using virtual methods. While std::function itself does not have any virtual methods, each call to **operator**() does get routed through a virtual method. Additionally, each call to the copy constructor or assignment operator invokes a heap allocation. The virtual function call is the price for run-time polymorphism. It can only be avoided by smart compiler devirtualization [96]. To alleviate the cost of the heap allocation, std::function typically implements a technique called small object optimization [104], which is a trade-off between run-time and space requirements. Objects of type std::function need more memory with it, but the heap allocation is avoided if the function object is small. Measurements of the performance penalty caused type erasure in dune-functions can be found in [62].

10.5.3. Differentiable Functions

The extension of the concept for a function (10.5) to a differentiable function requires to also provide access to its derivative

$$Df : \mathcal{D} \to L(\mathcal{D}, \mathcal{R})$$

where, in the simplest case, $L(\mathcal{D}, \mathcal{R})$ is the set of linear maps from the affine hull of $\mathcal{D}$ to $\mathcal{R}$. For example, for a function $f : \mathbb{R}^n \to \mathbb{R}^m$ the derivative Df maps each vector $x \in \mathcal{D} = \mathbb{R}^n$ to a linear map $Df(x)$ from $\mathcal{D} = \mathbb{R}^n$ to $\mathcal{R} = \mathbb{R}^m$. Since we can identify any linear map $M : \mathbb{R}^n \to \mathbb{R}^m$ with a matrix $M \in \mathbb{R}^{m \times n}$, we can identify $L(\mathcal{D}, \mathcal{R})$ with $\mathbb{R}^{m \times n}$. Hence Df is itself a function mapping vectors from $\mathbb{R}^n$ to matrices from $\mathbb{R}^{m \times n}$.

To provide access to derivatives, the dune-functions module extends the ideas from the previous section in a natural way. A C++ construct F is a differentiable function if, in addition to having **operator**() as described above, there is a free method

```
Function derivative(const F& f)
```

that returns a function that implements the derivative. The typical way to do this will be a **friend** function as illustrated in the following prototype implementation of a polynomial:

```cpp
template<class T>    // T is the number type
class Polynomial
{
public:
  Polynomial(const std::vector<T>& coefficients)
  : coefficients_(coefficients) {}

  T operator() (const T& x) const
  {
    // The Horner scheme
    auto y = coefficients_.back();
    for (int i=coefficients_.size()-2; i>=0, --i)
      y = x * y + coefficients_[i];
    return y;
  }

  friend Polynomial derivative(const Polynomial& p)
  {
    std::vector<T> dpCoefficients(p.coefficients_.size()-1);
    for (size_t i=1; i<p.coefficients_.size(); ++i)
      dpCoefficients[i-1] = p.coefficients_[i]*i;
    return Polynomial(dpCoefficients);
  }
```

```
private:
  std::vector<T> coefficients_;
};
```

To use this class, write

```
auto f = Polynomial<double>({1, 2, 3});
double a = f(3.14);
double b = derivative(f)(3.14);
```

Note, however, that the derivative method may be expensive, because it needs to compute the entire derivative function, not just a single value. It is therefore usually preferable to call it only once and to store the derivative function separately:

```
auto df = derivative(f);
double b = df(3.14);
```

Higher derivatives are obtained by calling derivative repeatedly:

```
auto df  = derivative(f);
auto ddf = derivative(df);
double c = ddf(3.14);
```

Functions supporting these operations are implementations of an interface called DifferentiableFunction.

When combining differentiable functions and dynamic polymorphism, the class std::function cannot be used as is, because it does not provide access to the derivative method. However, it can serve as inspiration for more general type erasure wrappers. The dune-functions module provides the class

```
template<class Signature,
         template<class> class DerivativeTraits=DefaultDerivativeTraits,
         size_t bufferSize=56>   // Buffer size in bytes
                                 // for small object optimization
class DifferentiableFunction
```

in the file dune/functions/common/differentiablefunction.hh. Partially, it is a reimplementation of std::function. The first template parameter Signature is equivalent to the template parameter Range(Args...) of std::function (with Args... being a single type Domain), and DifferentiableFunction implements a method

```
Range operator()(const Domain& x) const
```

This method works essentially like the one in std::function, except for the fact that its argument type is const Domain& instead of Domain, because arguments of a mathematical function (10.5) are always immutable. In addition, the type erasure wrapper DifferentiableFunction implements a free method

```
friend DerivativeInterface derivative(const DifferentiableFunction& f)
```

that wraps the corresponding method of the function implementation f. It allows to call the `derivative` method for objects whose precise type is determined only at run-time:

```
DifferentiableFunction<double(double)> polymorphicF;
polymorphicF = Polynomial<double>({1, 2, 3});
double a = polymorphicF(3.14);
auto polymorphicDF = derivative(polymorphicF);
double b = polymorphicDF(3.14);
```

While the domain of a derivative function is $\mathcal{D}$, the same as the one of the original function, its range is $L(\mathcal{D}, \mathcal{R})$. Unfortunately, it is not possible to always infer the best C++ type for objects from $L(\mathcal{D}, \mathcal{R})$ from the types for $\mathcal{D}$ and $\mathcal{R}$. To deal with this problem, dune-functions offers the `DerivativeTraits` class that maps the signature of a function to the range type of its derivative. The line

```
using DerivativeRange = DerivativeTraits<Range(Domain)>::Range;
```

is used to access the type that should be used to represent elements of $L(\mathcal{D}, \mathcal{R})$. The template `DefaultDerivativeTraits` is specialized for common combinations of DUNE matrix and vector types, and provides reasonable defaults for the derivative ranges. However, it is also possible to use other types by passing a custom `DerivativeTraits` template to the `DifferentiableFunction` class. This allows, e.g., to use optimized application-specific matrix and vector types or suitable representations for generalized derivative concepts.

10.5.4. `GridView` Functions and Local Functions

An important class of functions in any finite element application is the class of discrete functions, i.e., functions that are defined piecewise with respect to a grid. For a grid $\mathcal{T}$ on Ω a function $f : \Omega \to \mathcal{R}$ is called a *discrete function* if it has a natural piecewise definition with respect to $\mathcal{T}$. Such functions are typically too expensive to evaluate in global coordinates, i.e., at points $x \in \Omega$ directly. Luckily this is hardly ever necessary. Instead, one often knows an element T and local coordinates $\xi \in T_{\text{ref}}$ of x with respect to T that allow to evaluate $f(x) = f(F_T(\xi))$ cheaply. Formally, this means that we have localized versions

$$f_T := f \circ F_T : T_{\text{ref}} \to \mathcal{R}$$

of f for each grid element T, where F_T is the usual map from the reference element T_{ref} to T.

To support this kind of function evaluation, DUNE uses a *binding* mechanism very similar to the one used in Chapter 10.2 to localize basis trees. To evaluate a discrete function $f : \Omega \to \mathcal{R}$ at a local position ξ within a given element, the following code is used:

```
auto localF = localFunction(f);
localF.bind(element);
auto y = localF(xi);
```

Here we first obtain a *local function* localF, which represents the restriction of f to a single element. This function is then bound to a specific element using the method

```
void bind(const Codim<0>::Entity& element)
```

This explicit binding allows the function to perform any required setup procedures before being evaluated on that element. Afterwards the local function can be evaluated using the standard function interface of Chapter 10.5.1, but now using local coordinates $\xi = F_T^{-1}(x)$ with respect to the element that the local function is bound to. The same localized function object can be used for other elements by calling bind with a different element. Functions supporting these operations are called *grid view functions*, and are said to implement the GridViewFunction model. The model implemented by local functions is called LocalFunction.

Since functions in a finite element context are usually at least piecewise differentiable, grid view functions as well as local functions provide the full interface of differentiable functions as outlined in Section 10.5.3. To completely grasp the semantics of the interface, observe that strictly speaking localization does not commute with taking the derivative. Formally, a localized version of the derivative is given by

$$(Df)_T : T_{\text{ref}} \to L(\mathcal{D}, \mathcal{R}), \qquad (Df)_T = (Df) \circ F_T. \tag{10.6}$$

In contrast, the derivative of a localized function is given by

$$D(f_T) : T_{\text{ref}} \to L(T_{\text{ref}}, \mathcal{R}), \qquad D(f_T) = ((Df) \circ F_T) \cdot DF_T.$$

However, in the dune-functions implementation, the derivative of a local function does by convention always return values *in global coordinates*. Hence, the functions dfe1 and dfe2 obtained by

```
auto df = derivative(f);
auto dfT1 = localFunction(df);
dfT1.bind(element);

auto fT = localFunction(f);
fT.bind(element);
auto dfT2 = derivative(fT);
```

both *behave the same*, implementing $(Df)_T$ as in (10.6). This is motivated by the fact that $D(f_T)$ is hardly ever used in applications, whereas $(Df)_T$ is needed frequently. To express this mild inconsistency in the interface, a local function uses a special DerivativeTraits implementation that forwards the derivative range to the one of the corresponding global function.

Again, type erasure classes allow to use grid view functions and local functions in a polymorphic way. The class

```
template<class Signature,
         class GridView,
         template<class> class DerivativeTraits=DefaultDerivativeTraits,
         size_t bufferSize=56>   // Buffer size in bytes
                                 // for small object optimization
class GridViewFunction
```

stores any function that models the `GridViewFunction` concept with given signature and grid view type. Similarly, functions modeling the `LocalFunction` concept can be stored in the class

```
template<class Signature,
         class Element,
         template<class> class DerivativeTraits=DefaultDerivativeTraits,
         size_t bufferSize=56>   // Buffer size in bytes
                                 // for small object optimization
class LocalFunction
```

These type erasure classes can be used in combination:

```
GridViewFunction<double(GlobalCoordinate), GridView> polymorphicF;
polymorphicF = f;
auto polymorphicLocalF = localFunction(polymorphicF);
polymorphicLocalF.bind(element);
LocalCoordinate xLocal = ... ;
auto y = polymorphicLocalF(xLocal);
```

Note that, as described above, the `DerivativeTraits` used in `polymorphicLocalF` are not the same as the ones used by `polymorphicF`. Instead, they are a special implementation forwarding to the global derivative range even for the domain type `LocalCoordinate`.

10.6. Combining Global Bases and Coefficient Vectors

We now bring the interfaces for bases and functions together. Combining a function space bases and a coefficient vector yields a discrete function, by the linear combination shown exemplarily in (10.3). Conversely, discrete and non-discrete functions can be projected onto the span of a basis, which yields a corresponding coefficient vector. In dune-functions, this process is called *interpolation*, although it is not always an interpolation in the strict sense of the word.

10.6.1. Vector Backends

In both cases, individual basis functions need to be associated with corresponding entries of a container data type that holds vector coefficients. While trivial in theory, in practice there is a gap here because the multi-index types used by dune-functions to label basis functions (Section 10.2.4) cannot be used to access entries of standard random-access containers.

The gap is bridged by a concept called *vector backends*. These are shim classes that abstract away implementation details of particular container classes, and make them addressable by multi-indices. The dune-functions module currently offers such a backend for containers from dune-istl and the C++ standard library, but others can be added easily. This makes it possible to combine dune-functions function space bases with basically any linear algebra implementation.[4]

There are two parts to the vector backend concept: When interpreting a vector of given coefficients with respect to a basis, access is only required in a non-mutable way. In dune-functions this functionality is encoded in the ConstVectorBackend concept, which solely requires direct access by **operator**[] using the multi-indices provided by the function space basis:

```
const auto& operator[](Basis::MultiIndex index) const
```

For interpolation of given functions a corresponding mutable access is needed as well. Furthermore, it must be possible to resize the vector to match the index tree generated by the basis. These two additional methods make up the VectorBackend concept:

```
auto& operator[](Basis::MultiIndex index)
void resize(const Basis& basis)
```

Note that the argument of the resize member function is not a number, but the basis itself. This is necessary because resizing nested containers requires information about the whole index tree.

For the vector types implemented in the dune-istl and dune-common modules, such a backend can be obtained using

```
template<class SomeDuneISTLVector>
auto istlVectorBackend(SomeDuneISTLVector& x)
```

or

```
template<class SomeDuneISTLVector>
auto istlVectorBackend(const SomeDuneISTLVector& x)
```

(from the header dune/functions/backends/istlvectorbackend.hh). Depending on the **const**-ness of the argument, the resulting object implements the VectorBackend or only the ConstVectorBackend interface. Even though these methods have the istl prefix in their names, they actually also work well for containers from the C++ standard library like std::array and std::vector.

---

[4]Similar shim classes are needed to access matrix entries by pairs of multi-indices. At the time of writing there are no matrix backend implementations in dune-functions, but they may appear in future versions.

Since a non-trivial product function space basis corresponds to functions with a non-scalar range, we additionally have to map the components of the spanned product function space to components of a function range type. If, for example, the functions from the power function space generated by the basis

```
auto basis = makeBasis(
  gridView,
  power<dim>(
    lagrange<1>()
));
```

should be interpreted as vector fields, one would map the dim leaf nodes of this function space tree to the dim entries of a FieldVector<**double**,dim>. Whenever combining bases and range types dune-function uses a default mapping generalizing this idea to more complex nested bases: Assume that y is an object of the function range type. Then a leaf node with tree path i0, ..., in is associated to the entry y[i0]...[in]. In the exceptional case that the range type does not provide an **operator**[] it is directly used for all leaf nodes in the ansatz space. This last rule allows to interpolate a scalar function into all components of a basis at once. For additional flexibility, users can also provide custom mappings to be used instead of this one. However, we will not discuss the corresponding interface and rely on the implicitly used default implementation in the following.

10.6.2. Interpreting Coefficient Vectors as Finite Element Functions

To combine a basis and a coefficient vector to a discrete function, dune-functions provides the method

```
template<class Range, class Basis, class Coefficients>
auto makeDiscreteGlobalBasisFunction(const Basis& basis,
                                     const Coefficients& coefficients)
```

For given basis basis and coefficient vector coefficients this returns an object representing the corresponding finite element function. This object implements the GridViewFunction concept for the grid view the basis is defined on, with the range type Range. The basis can either be a global basis or a SubspaceBasis. In the latter case the coefficient vector has to correspond to the full basis nevertheless, but only the coefficients associated with the subspace basis functions will be used.

For the type Coefficients used to represent the coefficient vector there are two choices. Either it implements the ConstVectorBackend concept, as for example all objects returned by the istlVectorBackend method do. If Coefficients does not implement this concept, then the code assumes that it is a dune-istl-style container and tries to wrap it with an ISTLVectorBackend. That way, the method makeDiscreteGlobalBasisFunction can be called with dune-istl or standard library containers directly, but all others have to be wrapped in an appropriate backend explicitly.

It has to be noted that the range type of the resulting function can in general not be determined automatically from the basis and coefficient type, because there may be more than one plausible choice. For example, a scalar function could return **double** or FieldVector<**double**,1>. Hence the range type Range has to be given explicitly by the user. The mapping from the different leaf nodes of the basis to the entries of Range follows the procedure described at the end of Chapter 10.6.1.

To give an example of how makeDiscreteGlobalBasisFunction is used, we construct yet another instance of the Taylor–Hood basis

```
auto taylorHoodBasis = makeBasis(
  gridView,
  composite(
    power<dim>(
      lagrange<p+1>()),
    lagrange<p>()
));
```

By default, the merging strategies are BlockedLexicographic for the composite node, and BlockedInterleaved for the power node. The resulting indices are the ones from Column 2 of Table 10.1. An appropriate vector container type for this is

```
MultiTypeBlockVector<BlockVector<FieldVector<double,dim> >,
                     BlockVector<double>
               >;
```

Let x be an object of this type. To obtain the corresponding velocity field as a discrete function, write

```
// Create SubspaceBasis for the velocity field
auto velocityBasis = subspaceBasis(taylorHoodBasis, _0);

// Fix a range type for the velocity field
using VelocityRange = FieldVector<double,dim>;

// Create a function for the velocity field only
// but using the vector x for the full taylorHoodBasis.
auto velocityFunction
    = Functions::makeDiscreteGlobalBasisFunction<VelocityRange>(velocityBasis,
                                                                x);
```

Note that the dim leaf nodes of the function space tree spanned by velocityBasis are automatically mapped to the dim components of the VelocityRange type. The resulting function created in the last line implements the full GridViewFunction interface described in Chapter 10.5.4.

10.6.3. Interpolation

In various parts of a finite element or finite volume simulation code, given functions need to be approximated in spaces spanned by a global basis. For example, initial value functions may be given in closed form, but need to be transferred to a finite element representation to be usable. Similarly, Dirichlet values given in closed form may need to be evaluated on the set of Dirichlet degrees of freedom. Depending on the finite element space, the approximation may take different forms. Nodal interpolation is the natural choice for Lagrange elements, but for other spaces L^2-projections or Hermite-type interpolation may be more appropriate. For historical reasons, dune-functions always speaks of *interpolation*.

The dune-functions module provides a set of methods for interpolation in the file dune/functions/functionspacebases/interpolation.hh. These methods are canonical in the sense that they use the LocalInterpolation functionality of the dune-localfunctions interface (Chapter 8 on each element for the interpolation. This is appropriate for many, but not for all finite element spaces. For example, no reasonable local interpolation can be defined for B-spline bases, and therefore the standard interpolation functionality cannot be used with the BSplineBasis class. Prior to DUNE 2.8 this approach would also fail for non-affine finite elements, because LocalInterpolation did not apply non-standard transformations to the reference element.

The interpolation functionality is implemented in two global functions. The first deals with the simple case of a given function and basis, where the function is to be projected onto the span of the basis, yielding a coefficient vector describing the result:

```
template<class Basis, class Coefficients, class Function>
void interpolate(const Basis& basis,
                 Coefficients&& coefficients,
                 const Function& f)
```

Note that this will only work if the range type of f and the global basis basis are compatible. The dune-functions module implements a compatibility layer that allows to use different vector (or matrix) types from the DUNE core modules and scalar types like, e.g., **double** for the range of f, as long as the number of scalar entries of this range type is the same as the dimension of the range space of the function space spanned by the basis. This also implies the assumption that the coefficients for individual basis functions are scalar. The type of the coefficient vector coefficients either has to implement the VectorBackend concept, or it has to be wrappable by the istlVectorBackend. For example, consider the function

$$f_1 : \mathbb{R}^2 \to \mathbb{R} \qquad f_1 = \exp(-\|x\|^2)$$

implemented as

```
auto f1 = [](const FieldVector<double,2>& x)
{
  return exp(-1.0*x.two_norm2());
};
```

Additionally, consider a scalar second-order Lagrange space

```
Functions::LagrangeBasis<GridView,2> p2basis(gridView);
```

and an empty coefficient vector x1, not necessarily of correct size:

```
std::vector<double> x1;
```

Then, the single line

```
interpolate(p2basis, x1, f1);
```

will fill x1 with the nodal values of the function f1.

This interpolation works equally well for non-trivial basis trees and subtrees obtained by the subspaceBasis function, provided that the coefficient vector matches the basis and that the function range can be mapped to the product space associated to the basis. Consider the following Taylor–Hood basis for a two-dimensional grid that uses flat multi-indices to label its degrees of freedom:

```
using namespace Functions::BasisBuilder;

auto taylorHoodBasis = makeBasis(
  gridView,
  composite(
    power<GridView::dimension>(
      lagrange<2>(),
      flatLexicographic()),
    lagrange<1>(),
    flatLexicographic()
));
```

A suitable coefficient vector for such a basis is, for example,

```
BlockVector<double> x2;
```

In this situation, interpolation of f1 into the pressure components of the Taylor–Hood basis can be achieved by

```
using namespace Indices;
interpolate(subspaceBasis(taylorHoodBasis, _1), x2, f1);
```

Similarly we can interpolate a given vector field f2 into the non-trivial subtree representing the velocity using

```
auto f2 = [](const FieldVector<double,2>& x) {
  return x;    // Simple example function
};

interpolate(subspaceBasis(taylorHoodBasis, _0), x2, f2);
```

It is even possible to interpolate into the full taylorHoodBasis if the range type of the provided function has the same nesting structure as the basis.

In some situations it is also desirable to interpolate only on a part of the domain. Algebraically, the interpolation is then performed as before, but only a subset of all coefficients are written to the container. The most frequent use-case is the interpolation of Dirichlet data onto the algebraic degrees of freedom on the Dirichlet boundary. All others degrees of freedom must not be touched, as they may contain, e.g., a suitable initial iterate obtained by some other means.

To support this kind of interpolation, a variant of the `interpolate` method allows to explicitly mark a subset of coefficient vector entries to be written:

```
template<class Basis, class Coefficients, class Function, class BitVector>
void interpolate(const Basis& basis,
                 Coefficients&& coefficients,
                 const Function& f,
                 const BitVector& bitVector)
```

Conceptually, the additional `bitVector` argument must be a container of booleans having the same nesting structure as `coefficients`. Its entries are treated as boolean values indicating if the corresponding entry of `coefficients` should be written. For example, for flat global indices `std::vector<bool>` and `std::vector<char>` work nicely. The class `BitSetVector<N>` (from the `dune-common` module) can be used as a space-optimized alternative to `std::vector<std::bitset<N> >`. For example, to interpolate the `f2` function defined above only into the boundary velocity degrees of freedom, first set up a suitable bit-vector:

```
std::vector<char> isBoundary;
auto isBoundaryBackend = Functions::istlVectorBackend(isBoundary);
isBoundaryBackend.resize(taylorHoodBasis);
std::fill(isBoundary.begin(), isBoundary.end(), false);
forEachBoundaryDOF(subspaceBasis(taylorHoodBasis, _0),
  [&] (auto&& index) {
    isBoundaryBackend[index] = true;
  });
```

This uses the convenience method

```
template<class Basis, class FunctionObject>
void forEachBoundaryDOF(const Basis& basis, FunctionObject&& predicate);
```

(from the file `dune/functions/functionspacebases/boundarydofs.hh`), which implements a loop over all degrees of freedom associated to entities located on the domain boundary. The algorithm will invoke the call-back function `predicate` for each such degree of freedom, passing its global index as the call-back argument.

The actual interpolation is then a single line:

```
interpolate(subspaceBasis(taylorHoodBasis, _1), x2, f2, isBoundary);
```

See Chapter 10.8.3 for a more involved example.

10.7. Writing Functions to a VTK File

Functions in the sense of Chapter 10.5 can be written to VTK files for visualization and post-processing [107]. The interface for this extends the VTKWriter class of the dune-grid module, which was presented in Chapter 5.8.1. It is advised to read that chapter before proceeding.

The VTKWriter class allows to write one grid view to a VTK file, and to attach cell (i.e., element) and vertex data to it. As explained in Chapter 5.8.1, such data can be handed to the VTKWriter object using the member methods

```
template<class Container>
void addCellData(const Container& v, const std::string& name, int ncomps=1)
```

and

```
template<class Container>
void addVertexData(const Container& v, const std::string& name, int ncomps=1)
```

The Container arguments to these methods are expected to hold numerical data associated to grid elements or vertices, respectively, and VTK interprets such data as piecewise constant or linear functions.

However, VTKWriter also directly accepts functions defined on the grid domain Ω as data. There are two methods of the VTKWriter class for this:

```
template<class F>
void addCellData(F&& f, VTK::FieldInfo vtkFieldInfo)

template<class F>
void addVertexData(F&& f, VTK::FieldInfo vtkFieldInfo)
```

Both methods take a function f as their first parameter. The VTKWriter object evaluates this function at either the element centers (addCellData), the vertices (addVertexData, if the data mode is VTK::conforming), or at the element corners (addVertexData, if the data mode is VTK::nonconforming), and writes the resulting coefficients to the file.

The VTK::FieldInfo object is used to pass additional information about the function to the writer. Most users will only need its constructor

```
FieldInfo(std::string name,
          FieldInfo::Type type,
          std::size_t size,
          VTK::Precision prec = VTK::Precision::float32)
```

The first parameter is the name to be used by VTK to denote the function. The second one describes the range $\mathcal{R}$ of the function. Possible values are:

- VTK::FieldInfo::Type::scalar: The function components are written as separate scalar fields,

- VTK::FieldInfo::Type::vector: they are written as a vector-valued field (will always be zero-padded to three components).

The third parameter size gives the number of components of the range, which is not extracted automatically from the range type of the function type F. Finally, the prec parameter allows to select the data type used to store numerical values of the function in the VTK file, overriding the choice taken in the constructor. Of interest here are mainly VTK::Precision::float32 and VTK::Precision::float64.

To see the function-writing features in action, the following examples shows how to write a scalar function as element data of a two-dimensional grid in single (4-byte) precision:

```
auto f1 = [](const FieldVector<double,2>& x)
  {
    return std::sin(x.two_norm());
  };


vtkWriter.addCellData(f1,
                 VTK::FieldInfo("first example function",
                              VTK::FieldInfo::Type::scalar, 1));
```

Alternatively, the following code writes a 2-valued function with double precision:

```
auto f2 = [](const FieldVector<double,2>& x)
  {
    return FieldVector<double,2>({x[0], -x[1]})
  };


vtkWriter.addVertexData(f2,
                 VTK::FieldInfo("second example function",
                              VTK::FieldInfo::Type::vector, 2,
                              VTK::Precision::float64));
```

Functions written this way end up in VTK as piecewise constant (addCellData) or piecewise linear (addVertexData) functions. The VTK file format does support higher-order geometries and corresponding data files [107], but writing those from dune-grid is not currently implemented. Instead, the module allows to subsample functions on virtual refinements of a given grid. That way, functions that are not piecewise constant or piecewise linear can be visualized.[5]

Writing a subsampled function to a VTK file is implemented in a separate class SubsamplingVTKWriter, from the header subsamplingvtkwriter.hh in dune/grid/io/file/vtk/. This class inherits from VTKWriter, and it has the same template parameters and the same public member methods as the VTKWriter class. The only difference is the constructor

```
SubsamplingVTKWriter(const GridView& gridView,
                     RefinementIntervals intervals,
                     bool coerceToSimplex = false)
```

---

[5]More advanced VTK writing features are provided by the dune-vtk module; see www.dune-project.org/modules/dune-vtk.

Its first parameter is the grid view that is to be written to the file. The second one specifies the virtual refinement, encoded by objects of the dedicated type `RefinementIntervals`. Objects of this type can be obtained from two global methods. First,

```
RefinementIntervals refinementIntervals(int intervals)
```

produces a uniform refinement where each element edge is subdivided into `intervals` new ones. The method

```
RefinementIntervals refinementLevels(int levels)
```

on the other hand, will perform `levels` steps of standard uniform refinement (leading to 2^{levels} new edges per edge in total). If the optional `coerceToSimplex` parameter is set to **true**, then all elements are split into simplices before applying refinement. An example showing the use of the `SubsamplingVTKWriter` class is given in Chapter 10.8.

Note that in order to simplify the internal implementation, `SubsamplingVTKWriter` objects always write VTK files with the data mode `VTK::nonconforming`. If the input data functions are continuous, this is not visible in the output, but the resulting VTK grid will contain more vertices than expected. If the input function is discontinuous, on the other hand, the `SubsamplingVTKWriter` will automatically reproduce the discontinuities without the user having to set any further options.

As `SubsamplingVTKWriter` inherits from `VTKWriter`, it can be used together with a `VTKSequenceWriter` (see Chapter 5.8.2), which delegates the actual writing to a `VTKWriter` object. Writing time-dependent data with spatial subsampling is therefore straightforward.

10.8. Example: Solving the Stokes Equation with `dune-functions`

We close the chapter by showing a complete example that demonstrates the features of `dune-functions`. The program will solve the stationary Stokes equation using Taylor–Hood finite elements. It is contained in a single file, which is printed in Appendix B.10. If you read this document in electronic form, the file can also be accessed by clicking on the icon in the margin.

10.8.1. The Stokes Equation

The Stokes equation models a viscous incompressible fluid in a d-dimensional domain Ω. There are two unknowns in this problem: a stationary fluid velocity field $\mathbf{u} : \Omega \to \mathbb{R}^d$, and the fluid pressure $p : \Omega \to \mathbb{R}$. Together, they have to solve the boundary value problem

$$-\Delta\mathbf{u} - \nabla p = 0 \qquad \text{in } \Omega,$$
$$\text{div}\,\mathbf{u} = 0 \qquad \text{in } \Omega,$$
$$\mathbf{u} = \mathbf{g} \qquad \text{on } \partial\Omega,$$

where we have omitted the physical parameters. The boundary value problem only determines the pressure p up to a constant function. The pressure is therefore usually normalized such that $\int_\Omega p\,dx = 0$.

Due to the constraint $\text{div}\,\mathbf{u} = 0$, the corresponding weak form of the equation is a saddle-point problem. Introducing the spaces

$$\mathbf{H}_\mathbf{g}^1(\Omega) := \left\{\mathbf{v} \in H^1(\Omega, \mathbb{R}^d) \;:\; \text{tr}\,\mathbf{v} = \mathbf{g}\right\},$$
$$L_{2,0}(\Omega) := \left\{q \in L_2(\Omega) \;:\; \int_\Omega q\,dx = 0\right\},$$

and the bilinear forms

$$a(\mathbf{u}, \mathbf{v}) := \int_\Omega \nabla\mathbf{u}\nabla\mathbf{v}\,dx, \qquad \text{and} \qquad b(\mathbf{v}, q) := \int_\Omega \text{div}\,\mathbf{v} \cdot q\,dx,$$

the weak form of the Stokes equation is: Find $(\mathbf{u}, p) \in \mathbf{H}_\mathbf{g}^1(\Omega) \times L_{2,0}(\Omega)$ such that

$$a(\mathbf{u}, \mathbf{v}) + b(\mathbf{v}, p) = 0 \qquad \text{for all } \mathbf{v} \in \mathbf{H}_0^1(\Omega)$$
$$b(\mathbf{u}, q) \qquad = 0 \qquad \text{for all } q \in L_{2,0}(\Omega).$$

If $\mathbf{g}$ is sufficiently smooth, this variational problem has a unique solution. The Taylor–Hood element is the standard way to discretize this saddle point problem [37], and will be used in the following implementation.

10.8.2. The Driven-Cavity Benchmark

For our example we choose to simulate a two-dimensional driven cavity. This is a standard benchmark for the Stokes problem in the literature [142]. Let Ω be the unit square $[0, 1]^2$, and set the Dirichlet boundary conditions for the velocity $\mathbf{u}$ to

$$\mathbf{u}(x) = \mathbf{g}(x) := \begin{cases} (0, 1)^T & \text{if } x \in \{0\} \times [0, 1] \\ (0, 0)^T & \text{elsewhere on } \partial\Omega. \end{cases}$$

The interpretation of this is a fluid container that is closed on all but one side. While the fluid remains motionless on the closed sides, an external agent drives a constant upward motion on the left vertical side. The domain and boundary conditions are depicted in Figure 10.10, left. The corresponding solution is shown on the right side of the same figure. The velocity forms a vortex, while the pressure forms extrema in the two left corners.

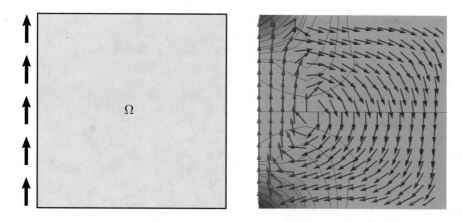

Figure 10.10.: Driven cavity. Left: domain and boundary conditions; right: simulation result. The arrows show the *normalized* velocity.

10.8.3. Implementation

The implementation consists of an assembler for the Stokes equation and a `main` method. Both will be discussed in the following.

The `main` Method

The `main` method sets up the algebraic Stokes problem, calls a linear solver, and writes the result to a VTK file. It begins by setting up MPI and the grid. We choose to discretize the domain using a structured 4×4 quadrilateral grid, which we get by using the `YaspGrid` grid implementation from the `dune-grid` module:

```
269  int main(int argc, char *argv[])
270  {
271    // Set up MPI, if available
272    MPIHelper::instance(argc, argv);

280    constexpr int dim = 2;
281    using Grid = YaspGrid<dim>;
282    FieldVector<double,dim> upperRight = {1, 1};
283    std::array<int,dim> nElements = {4, 4};
284    Grid grid(upperRight,nElements);
285
286    using GridView = typename Grid::LeafGridView;
287    GridView gridView = grid.leafGridView();
```

On the grid view, we then set up the function space basis for the Taylor–Hood element. This is as simple as

```
295     using namespace Functions::BasisFactory;
296
297     constexpr std::size_t p = 1; // Pressure order for Taylor-Hood
298
299     auto taylorHoodBasis = makeBasis(
300           gridView,
301           composite(
302             power<dim>(
303               lagrange<p+1>(),
304               blockedInterleaved()),
305             lagrange<p>()
306           ));
```

This way of constructing a Taylor–Hood basis from instances of Lagrange bases has been discussed in Section 10.3.2. The indexing strategies used here are BlockedInterleaved for the velocity subtree and BlockedLexicographic (the default) for the root. This results in the indexing scheme spelled out in Column 2 of Table 10.1, i.e., velocity degrees of freedom have indices $(0, i, j)$ where i is the Lagrange node and j is the component in $\mathbb{R}^d$, and pressure degrees of freedom have indices $(1, i)$.

Before being able to assemble the stiffness matrix of the Stokes system we need to pick suitable data structures for the linear algebra. These data structures should have a blocking structure that matches the multi-indices used by the Taylor–Hood basis we just constructed. More concretely, the appropriate vector container will be a pair of vectors, where the first one has entries in $\mathbb{R}^d$ for the velocity and the second one has entries in $\mathbb{R}$ for the pressure degrees of freedom. Analogously, the matrix must consist of 2×2 sparse matrices where the $(0,0)$-block has entries in $\mathbb{R}^{d \times d}$, the $(0, 1)$-block has entries in $\mathbb{R}^{d \times 1}$, the $(1, 0)$-block has entries in $\mathbb{R}^{1 \times d}$, and the $(1, 1)$-block has entries in $\mathbb{R}$ (as on the right side of Figure 10.6). The following code sets up such vector and matrix types. It uses the nesting machinery from dune-istl, but data types from other linear algebra libraries could be used as well:

```
314     using VelocityVector = BlockVector<FieldVector<double,dim>>;
315     using PressureVector = BlockVector<double>;
316     using Vector = MultiTypeBlockVector<VelocityVector, PressureVector>;
317
318     using Matrix00 = BCRSMatrix<FieldMatrix<double,dim,dim>>;
319     using Matrix01 = BCRSMatrix<FieldMatrix<double,dim,1>>;
320     using Matrix10 = BCRSMatrix<FieldMatrix<double,1,dim>>;
321     using Matrix11 = BCRSMatrix<double>;
322     using MatrixRow0 = MultiTypeBlockVector<Matrix00, Matrix01>;
323     using MatrixRow1 = MultiTypeBlockVector<Matrix10, Matrix11>;
324     using Matrix = MultiTypeBlockMatrix<MatrixRow0,MatrixRow1>;
```

The heterogeneous containers MultiTypeBlockVector and MultiTypeBlockMatrix are discussed in detail in Chapters 7.2.2 and 7.3.2, respectively. However, it must be emphasized that the use of such advanced data structures is by no means mandatory. As detailed in Sections 10.1.4 and 10.3.2 it is trivial to make the Taylor–Hood basis use flat global indices, which work directly with standard container types like `std::vector`.

Now that we have chosen the C++ types for the matrix and vector data structures we can actually assemble the system. Assembling the right-hand-side vector rhs is easy, because, apart from the Dirichlet boundary data (which we will insert later), all its entries are zero. An all-zero vector of the correct type and size is set up by the following lines:

```
332    Vector rhs;
333
334    auto rhsBackend = Functions::istlVectorBackend(rhs);
335
336    rhsBackend.resize(taylorHoodBasis);
337    rhs = 0;
```

The backend object returned by `istlVectorBackend` connects the dune-functions basis with dune-istl linear algebra containers (Chapter 7.2). In particular, it offers convenient resizing of an entire hierarchy of nested vectors from given function space basis trees. Line 337 fills the entire vector with zeros in one go, but observe that this is actually a dune-istl feature.

To obtain the stiffness matrix we first create an empty matrix object of the correct type. The actual assembly is factored out into a separate method:

```
341    Matrix stiffnessMatrix;
342    assembleStokesMatrix(taylorHoodBasis, stiffnessMatrix);
```

As the matrix assembly is a central part of this example we explain it in detail below, after having covered the `main` method.

Suppose now that we have the correct stiffness matrix assembled in the object `stiffnessMatrix`. We still need to modify the linear system to include the Dirichlet boundary information. In a first step we need to determine all degrees of freedom with Dirichlet boundary conditions. To store this information we use a vector of flags which has the same structure as `VectorType` and is again initialized using the ISTLVectorBackend:

```
351    using VelocityBitVector = std::vector<std::array<char,dim> >;
352    using PressureBitVector = std::vector<char>;
353    using BitVector = TupleVector<VelocityBitVector, PressureBitVector>;
354
355    BitVector isBoundary;
356
357    auto isBoundaryBackend = Functions::istlVectorBackend(isBoundary);
358    isBoundaryBackend.resize(taylorHoodBasis);
359
360    using namespace Indices;
361    for (auto&& b0i : isBoundary[_0])
362      for (std::size_t j=0; j<b0i.size(); ++j)
363        b0i[j] = false;
364    std::fill(isBoundary[_1].begin(), isBoundary[_1].end(), false);
```

We now want to mark all the velocity degrees of freedom on the Dirichlet boundary. In the driven-cavity example, the entire boundary is Dirichlet boundary. To mark the boundary degrees of freedom for the velocity subtree write:

```
368    Functions::forEachBoundaryDOF(
369            Functions::subspaceBasis(taylorHoodBasis, _0),
370            [&] (auto&& index) {
371              isBoundaryBackend[index] = true;
372            });
```

The convenience method `forEachBoundaryDOF` (from the file `boundarydofs.hh` in `dune/functions/functionspacebases/`) implements a loop over all degrees of freedom associated to entities located on the domain boundary. Here, it only considers velocity degrees of freedom because we called it with the corresponding subspace basis. Nevertheless, the global indices handed out by `forEachBoundaryDOF` correspond to the full tree, and can therefore by used to access the `isBoundary` container via the `ISTLVectorBackend`.

Now that we have determined the set of Dirichlet degrees of freedom, we define a method implementing the actual Dirichlet values function **g**, and write its point values into the right-hand-side vector `rhs`:

```
376    using Coordinate = GridView::Codim<0> ::Geometry::GlobalCoordinate;
377    using VelocityRange = FieldVector<double,dim>;
378    auto&& g = [](Coordinate x)
379    {
380      return VelocityRange{0.0, (x[0] < 1e-8) ? 1.0 : 0.0};
381    };
382
383    Functions::interpolate(Functions::subspaceBasis(taylorHoodBasis, _0),
384                           rhs,
385                           g,
386                           isBoundary);
```

Observe how the `dune-functions` interface allows to interpolate lambda objects, which makes the code short and readable. Again the operation is constrained to the velocity degrees of freedom by passing the corresponding subspace basis only. The `isBoundary` vector given as the last argument restricts the interpolation to only the boundary degrees of freedom.

The stiffness matrix is modified in a more manual fashion. For each Dirichlet degree of freedom we need to fill the corresponding matrix row with zeros, and write a 1 on the diagonal (Chapter 2.1.4). The following algorithm does this by looping over all grid elements, and for each element looping over all Dirichlet degrees of freedom. This is less efficient than simply looping over all matrix rows, but it allows to avoid implementing iterators for the nested sparse matrix data type `MatrixType`:

```
395    auto localView = taylorHoodBasis.localView();
396    for(const auto& element : elements(gridView))
397    {
```

```
398        localView.bind(element);
399        for (size_t i=0; i<localView.size(); ++i)
400        {
401          auto row = localView.index(i);
402          // If row corresponds to a boundary entry,
403          // modify it to be an identity matrix row.
404          if (isBoundaryBackend[row])
405            for (size_t j=0; j<localView.size(); ++j)
406            {
407              auto col = localView.index(j);
408              matrixEntry(stiffnessMatrix, row, col) = (i==j) ? 1 : 0;
409            }
410        }
411    }
```

Access to the matrix entries needs a matrix analogue to the vector backends—a translation layer that converts multi-indices to the correct sequence of instructions required to access the matrix data structure. Such a backend is more challenging to write, as it requires handling row and column indices at the same time. At the time of writing dune-functions does not provide matrix backends. Instead, in Line 408, the example code uses a small helper method `matrixEntry` that is defined in the example file itself. It is much simpler than a generic matrix backend, because it is written directly for the matrix data type of the Stokes problem:

```
196    template<class Matrix, class MultiIndex>
197    decltype(auto) matrixEntry(
198            Matrix& matrix, const MultiIndex& row, const MultiIndex& col)
199    {
200      using namespace Indices;
201      if ((row[0]==0) && (col[0]==0))
202        return matrix[_0][_0][row[1]][col[1]][row[2]][col[2]];
203      if ((row[0]==0) && (col[0]==1))
204        return matrix[_0][_1][row[1]][col[1]][row[2]][0];
205      if ((row[0]==1) && (col[0]==0))
206        return matrix[_1][_0][row[1]][col[1]][0][col[2]];
207      return matrix[_1][_1][row[1]][col[1]];
208    }
```

Note how the outer indices for the `MultiTypeBlockMatrix` have to be encoded statically, because different matrix entries have different types. The inner matrices can be addressed directly using run-time integers.

Finally, we can solve the linear system. Dedicated Stokes solvers frequently operate on some sort of Schur complement, and hence they need direct access to the submatrices [101]. This can be elegantly done using the nested matrix type used in this example. However, efficiently solving the Stokes system is an art, which we do not want to get into here. Instead, we use a GMRes solver [139], without any preconditioner at all. This is known to converge, albeit slowly. The advantage is that it can be written down in only a few lines. The following code shows the typical way of using `dune-istl` to solve a linear system of equations, and is not particular to `dune-functions` at all:

```
418    // Initial iterate: Start from the rhs vector,
419    // that way the Dirichlet entries are already correct.
420    Vector x = rhs;
421
422    // Turn the matrix into a linear operator
423    MatrixAdapter<Matrix,Vector,Vector> stiffnessOperator(stiffnessMatrix);
424
425    // Fancy (but only) way to not have a preconditioner at all
426    Richardson<Vector,Vector> preconditioner(1.0);
427
428    // Construct the iterative solver
429    RestartedGMResSolver<Vector> solver(
430            stiffnessOperator,   // Operator to invert
431            preconditioner,      // Preconditioner
432            1e-10,               // Desired residual reduction factor
433            500,                 // Number of iterations between restarts,
434                                 // here: no restarting
435            500,                 // Maximum number of iterations
436            2);                  // Verbosity of the solver
437
438    // Object storing some statistics about the solving process
439    InverseOperatorResult statistics;
440
441    // Solve!
442    solver.apply(x, rhs, statistics);
```

Observe how the `RestartedGMResSolver` object is completely oblivious to the fact that the matrix has a nesting structure.

Once the iterative solver has terminated, the result is written to a VTK file. We construct velocity and pressure discrete functions by combining the appropriate coefficient vectors and basis subtrees:

```
450    using VelocityRange = FieldVector<double,dim>;
451    using PressureRange = double;
452
453    auto velocityFunction
454            = Functions::makeDiscreteGlobalBasisFunction<VelocityRange>(
455                    Functions::subspaceBasis(taylorHoodBasis, _0), x);
456    auto pressureFunction
457            = Functions::makeDiscreteGlobalBasisFunction<PressureRange>(
458                    Functions::subspaceBasis(taylorHoodBasis, _1), x);
```

Then, we write the resulting velocity as a vector field, and the resulting pressure as a scalar field. We subsample the grid twice to get an approximation of the second-order velocity functions—currently the VTKWriter class of dune-grid natively only writes piecewise linear functions:

```
467    SubsamplingVTKWriter<GridView> vtkWriter(
468            gridView,
469            refinementLevels(2));
470    vtkWriter.addVertexData(
471            velocityFunction,
472            VTK::FieldInfo("velocity", VTK::FieldInfo::Type::vector, dim));
473    vtkWriter.addVertexData(
474            pressureFunction,
475            VTK::FieldInfo("pressure", VTK::FieldInfo::Type::scalar, 1));
476    vtkWriter.write("stokes-taylorhood-result");
```

When run, this program produces a file called `stokes-taylorhood-result.vtu`. The file can be opened in PARAVIEW, and the outcome looks like the image on the right in Figure 10.10.

The Global Assembler

Now that we have covered the `main` method, we can turn to the assembler for the Stokes stiffness matrix. We begin with the global assembler, which is implemented in the method `assembleStokesMatrix` called in Line 342 of the `main` method. The global assembler sets up the matrix pattern, loops over all elements, and accumulates the element stiffness matrices in the global matrix. The signature of the method is

```
214    template<class Basis, class Matrix>
215    void assembleStokesMatrix(const Basis& basis, Matrix& matrix)
```

The only arguments it gets are the finite element basis and the matrix to fill. Observe that the Taylor–Hood basis is not hard-wired here, so we could call the method with a different basis. However, not surprisingly the local assembler for the Stokes problem makes relatively tight assumptions on the basis tree structure, so there is little practical freedom here. Ideally, a global assembler should be fully generic, and all knowledge about the current spaces and differential operators should be confined to the local assembler. Real discretization frameworks like `dune-pdelab` do achieve this separation, but for the example here we are less strict, to avoid technicalities.

The first few lines of the `assembleStokesMatrix` method set up the matrix occupation pattern, and initialize all matrix entries with zero:

```
219    // Set matrix size and occupation pattern
220    setOccupationPattern(basis, matrix);
221
222    // Set all entries to zero
223    matrix = 0;
```

The method `setOccupationPattern` that constructs the matrix pattern is included in the example file itself. It is easy to understand for everyone who understands the rest of the assembly code, and we therefore omit a detailed description.

Next comes the actual element loop. We first request a `localView` object from the finite element basis:

```
228    auto localView = basis.localView();
```

After that starts the loop over the grid elements. For each element, we bind the `localView` object to the element. From then on all enquiries to the local view will implicitly refer to this element:

```
233    for (const auto& element : elements(basis.gridView()))
234    {
235      // Bind the local FE basis view to the current element
236      localView.bind(element);
```

We then create the element stiffness matrix, and call the separate method assemble-ElementStiffnessMatrix to fill it. For simplicity the code uses a dense matrix type even though it is known a priori that the stationary Stokes matrix does not contain entries in the pressure diagonal block. As local shape function indices are always flat (Chapter 10.2.3), a matrix data type without nesting is used for the element stiffness matrix:

```
242      Dune::Matrix<double> elementMatrix;
243      assembleElementStiffnessMatrix(localView, elementMatrix);
```

The `assembleElementStiffnessMatrix` method is discussed in detail below. It gets only the `localView` object in addition to the `elementMatrix`. The former object contains all necessary information required to assemble the stiffness matrix on the element it is bound to. After the call to `assembleElementStiffnessMatrix` the `elementMatrix` object contains the element stiffness matrix for the current element. The code loops over the entries of the element stiffness matrix and adds them onto the global matrix:

```
248      for (size_t i=0; i<elementMatrix.N(); i++)
249      {
250        // The global index of the i-th local degree of freedom
251        // of the current element
252        auto row = localView.index(i);
253
254        for (size_t j=0; j<elementMatrix.M(); j++ )
255        {
256          // The global index of the j-th local degree of freedom
257          // of the current element
258          auto col = localView.index(j);
259          matrixEntry(matrix, row, col) += elementMatrix[i][j];
260        }
261      }
```

The type returned in Lines 252 and 258 for the global row and column indices is a multi-index. It has length 3 for velocity degrees of freedom and length 2 for pressure degrees of freedom. Line 259 uses the helper function `matrixEntry` again to access the nested global stiffness matrix using those multi-indices.

The preceding loops write into all four of the matrix blocks, even though it is known that for the Stokes system the lower right block contains only zeros. A more optimized version of the code would leave out the lower right submatrix altogether.

The Local Assembler

It remains to investigate the method that assembles the element stiffness matrices. Its signature is

```
37  template<class LocalView>
38  void assembleElementStiffnessMatrix(const LocalView& localView,
39                                       Matrix<double>& elementMatrix)
```

It only receives the local view of the Taylor–Hood basis, expected to be bound to an element, and the empty element matrix. There is no explicit requirement that the `LocalView` object be a local view of a Taylor–Hood basis, but the assumption is made implicitly in various parts of the local assembler. The first few lines of the `assembleElementStiffnessMatrix` method gather some information about the element the method is to work on. In particular, from the `localView` object it extracts the element itself, and the element's dimension and geometry:

```
44    using Element = typename LocalView::Element;
45    const Element element = localView.element();
46
47    constexpr int dim = Element::dimension;
48    auto geometry = element.geometry();
```

Next, the element stiffness matrix is initialized. The `localView` object knows the total number of degrees of freedom of the element it is bound to, and since the matrix has only scalar entries, this is the correct number of matrix rows and columns:

```
53    elementMatrix.setSize(localView.size(), localView.size());
54    elementMatrix = 0;        // Fills the entire matrix with zeros
```

Finally, we ask for the sets of velocity and pressure shape functions:

```
59    using namespace Indices;
60    const auto& velocityLocalFiniteElement
61        = localView.tree().child(_0,0).finiteElement();
62    const auto& pressureLocalFiniteElement
63        = localView.tree().child(_1).finiteElement();
```

The two objects returned in Lines 60–63 are `LocalFiniteElements` in the dune-localfunctions sense of the word. These lines also show the tree structure of the Taylor–Hood basis in action: The expression

```
localView.tree().child(_0,0)
```

returns the first child of the first child of the root, i.e., the basis for the first component of the velocity field, and

```
localView.tree().child(_1)
```

returns the basis for the pressure space. As the root of the tree combines two different bases, the static identifiers _0 and _1 from the Dune::Indices namespace are needed to specify its children. The inner node for the velocities combines d times the same basis, and hence the normal integer 0 can be used to address its first child. This particular implementation of the local Stokes assembler is actually "cheating", because it exploits the knowledge that the same basis is used for all velocity components. Therefore, only the first leaf of the velocity subtree is acquired in Line 60, and then used for all components. Using separate local finite elements for each component is wasteful because the same shape function values and gradients would have to be computed d times.

Next, the code constructs a suitable quadrature rule and loops over the quadrature points. The formula for the quadrature order combines information about the element type, the shape functions, and the differential operator. It computes the lowest order that will integrate the weak form of the Stokes equation exactly on a cube grid:

```
68    int order = 2*(dim*velocityLocalFiniteElement.localBasis().order()-1);
69    const auto& quad = QuadratureRules<double,dim>::rule(element.type(), order);
70
71    // Loop over all quadrature points
72    for (const auto& quadPoint : quad)
73    {
```

The quadrature loop starts like the simpler local assembler codes seen previously. First, we get the inverse transposed Jacobian of the map F from the reference element to the grid element, and the Jacobian determinant for the integral transformation formula:

```
76    // The transposed inverse Jacobian of the map from the
77    // reference element to the element
78    const auto jacobianInverseTransposed
79            = geometry.jacobianInverseTransposed(quadPoint.position());
80
81    // The multiplicative factor in the integral transformation formula
82    const auto integrationElement
83            = geometry.integrationElement(quadPoint.position());
```

With these preparations done, we can assemble the first part of the stiffness matrix, corresponding to the velocity–velocity coupling. For two d-valued velocity basis functions $\boldsymbol{\varphi}_i^k = \mathbf{e}_k \varphi_i$ and $\boldsymbol{\varphi}_j^l = \mathbf{e}_l \varphi_j$ we need to compute

$$a_T(\boldsymbol{\varphi}_i^k, \boldsymbol{\varphi}_j^l) := \int_T \nabla \boldsymbol{\varphi}_i^k \nabla \boldsymbol{\varphi}_j^l \, dx = \delta_{kl} \int_T \nabla \varphi_i \nabla \varphi_j \, dx$$

on the current element T, where φ_i and φ_j are the corresponding scalar basis functions, and $\mathbf{e}_k$, $k = 0, \ldots, d-1$ are the canonical basis vectors in $\mathbb{R}^d$. The code first computes the derivatives of the velocity shape functions at the current quadrature point, and then uses the matrix in jacobianInverseTransposed to transform the shape functions gradients to gradients of the actual basis functions defined on the grid element:

```
92      std::vector<FieldMatrix<double,1,dim> > referenceGradients;
93      velocityLocalFiniteElement.localBasis().evaluateJacobian(
94                                              quadPoint.position(),
95                                              referenceGradients);
96
97      // Compute the shape function gradients on the grid element
98      std::vector<FieldVector<double,dim> > gradients(referenceGradients.size());
99      for (size_t i=0; i<gradients.size(); i++)
100         jacobianInverseTransposed.mv(referenceGradients[i][0], gradients[i]);
```

With the velocity basis function gradients at hand we can assemble the velocity contribution to the stiffness matrix:

```
105     for (size_t i=0; i<velocityLocalFiniteElement.size(); i++)
106       for (size_t j=0; j<velocityLocalFiniteElement.size(); j++ )
107         for (size_t k=0; k<dim; k++)
108         {
109           size_t row = localView.tree().child(_0,k).localIndex(i);
110           size_t col = localView.tree().child(_0,k).localIndex(j);
111           elementMatrix[row][col] += ( gradients[i] * gradients[j] )
112                             * quadPoint.weight() * integrationElement;
113         }
```

Noteworthy here are the Lines 109–110 which, for two given shape functions from the finite element basis tree, obtain the flat numbering used to index the element stiffness matrix. The expression child(_0,k) singles out the tree leaf for the k-th component of the velocity basis. The loop variables i and j run over the shape functions in this set, and

```
localView.tree().child(_0,k).localIndex(i);
```

returns the corresponding scalar index for the i-th shape function in the set of *all* shape functions of the Taylor–Hood basis on this element. This is the *local* index $\iota_{\Lambda|T}^{\text{local}}(\cdot)$ of Section 10.1.5. Line 112 then updates the corresponding (scalar) element matrix entry with the correctly weighted product of the two gradients $\nabla\varphi_i$ and $\nabla\varphi_j$.

Once this part is understood, computing the velocity–pressure coupling terms is easy. For a given velocity basis function φ_i^k and pressure basis function θ_j we need to compute

$$b_T(\varphi_i^k, \theta_j) := \int_T \left(\operatorname{div} \varphi_i^k \right) \theta_j \, dx$$

$$= \int_T \sum_{l=0}^{d-1} \frac{\partial(\varphi_i^k)_l}{\partial x_l} \, \theta_j \, dx = \int_T \frac{\partial \varphi_i}{\partial x_k} \, \theta_j \, dx = \int_T (\nabla \varphi_i)_k \, \theta_j \, dx.$$

At this point in the code the value of $\nabla\varphi_i$ at the current quadrature point has been computed already, but value of θ_i is still unknown. The values for all i are evaluated by the following two lines:

```
122     std::vector<FieldVector<double,1> > pressureValues;
123     pressureLocalFiniteElement
124         .localBasis().evaluateFunction(quadPoint.position(),pressureValues);
```

Then, the actual matrix assembly of the bilinear form $b_T(\cdot,\cdot)$ is:

```
129     for (size_t i=0; i<velocityLocalFiniteElement.size(); i++)
130       for (size_t j=0; j<pressureLocalFiniteElement.size(); j++ )
131         for (size_t k=0; k<dim; k++)
132         {
133           size_t vIndex = localView.tree().child(_0,k).localIndex(i);
134           size_t pIndex = localView.tree().child(_1).localIndex(j);
135
136           auto value = gradients[i][k] * pressureValues[j]
137                      * quadPoint.weight() * integrationElement;
138           elementMatrix[vIndex][pIndex] += value;
139           elementMatrix[pIndex][vIndex] += value;
140         }
```

Line 133 computes the flat local index of φ_i^k again, and Line 134 computes the index for θ_j (remember that _1 denotes the pressure basis). Finally, Lines 136–139 then compute the integrand value $(\nabla\varphi_i)_k\theta_j$, and add the resulting terms to the matrix. This concludes the implementation of the local assembler for the Stokes problem.

11. Discretizing Partial Differential Equations with dune-pdelab

All parts of DUNE that we have described so far have been rather low-level in nature. The features of the DUNE core modules described in Part II of this book are of great help to people writing their own finite element or finite volume codes; however, they are not sufficient to actually solve even a simple PDE without a substantial amount of additional coding. This can be witnessed by looking at the code example in Chapter 3.3, which solves the Poisson equation using only the core modules and dune-functions. Even though this is one of the easiest PDE examples to implement, it still takes about 200 lines of source code.

The missing part is a library to assemble stiffness matrices and load vectors using the function spaces provided by the dune-functions module. Such functionality is of a higher abstraction level than what has been presented previously. A general assembly framework will use the grids, shape functions, and function spaces described in earlier chapters, and it will allow people to solve simple PDE problems with only little additional coding.

The dune-pdelab module is one such framework.[1] It is extremely powerful and flexible, being able to handle a large range of discrete spaces and boundary conditions. At the same time it is very efficient, the overhead introduced by the flexibility being kept to a minimum [20]. Primarily, dune-pdelab implements assemblers for finite element problems. These can be linear or nonlinear, and they can involve contributions from the grid elements as well as from the element facets. A number of assemblers for common partial differential equations are included in the module, but it is also easy to write new ones. For the solution of the resulting algebraic problems, several nonlinear solvers are provided.

As part of this, dune-pdelab implements an interface for spaces of finite element functions, which adds a few additional features on top of what the dune-functions bases offer. For example, it automatically handles the data exchange and communication when a function space is defined on a grid that is distributed across several processes. The same holds for grid changes due to local grid adaptivity. Data preserva-

---

[1] Since assembly and discretization of PDE problems is a very diverse subject, it is much more difficult to agree on a single set of abstractions and corresponding implementations than for grids and shape functions. That is why there are other DUNE modules also dedicated to the discretization of PDEs. The dune-fem module [48] (www.dune-project.org/modules/dune-fem) is aimed at adaptive and parallel schemes. In contrast, dune-fufem (www.dune-project.org/modules/dune-fufem) deliberately sacrifices a lot of the power of dune-pdelab and dune-fem for a much simpler interface and a gentler learning curve.

© Springer Nature Switzerland AG 2020
O. Sander, *DUNE — The Distributed and Unified Numerics Environment*, Lecture Notes in Computational Science and Engineering 140, https://doi.org/10.1007/978-3-030-59702-3_11

tion during grid refinement and coarsening, which was a bit tedious when implemented by hand in Chapter 5.9.2, becomes fully transparent for the user. The dune-pdelab module also provides a flexible mechanism to apply linear constraints to the degrees of freedom. The most obvious use for this are Dirichlet boundary values, as explained in Chapter 11.3, but it also allows more challenging things like constructing conforming Lagrange finite element spaces on nonconforming grids. Finally, dune-pdelab introduces a set of different linear algebra backends, which allows to select different implementations at compile-time. The default backend uses the linear algebra from dune-istl (Chapter 7), but others like EIGEN [87] can be used as well.

The dune-pdelab module is a large and powerful project. For size reasons we have to omit important aspects of it in this book. For example, we will not discuss distributed grids, instationary problems, or constrained spaces beyond mere Dirichlet constraints. Readers interested in these topics will need to consult the online resources.

11.1. Example: Linear Reaction–Diffusion Equation

We begin the chapter by showing how a simple boundary value problem can be solved using dune-pdelab. The example program solves a linear reaction–diffusion equation using three different conforming and nonconforming finite element spaces. The algebraic problem is solved using a CG solver with an SSOR preconditioner. The complete source code for the example is contained in a single file, which is printed in Appendix B.11. Readers of an electronic version of this text can also access the file by clicking on the icon in the margin.

All code in dune-pdelab is contained in the nested namespace Dune::PDELab. The example programs in this chapter contain the line **using namespace** Dune; at the top, and only give the prefix PDELab:: explicitly. Unlike other DUNE modules, dune-pdelab allows to include the entire module by means of a single header file dune/pdelab.hh.

We briefly state the example problem. Let Ω be a domain in $\mathbb{R}^d$. The Neumann problem for the linear reaction–diffusion equation is to find a function $u : \Omega \to \mathbb{R}$ such that

$$
\begin{aligned}
-\Delta u + cu &= f && \text{in } \Omega, \\
\langle \nabla u, \mathbf{n} \rangle &= 0 && \text{on } \partial\Omega,
\end{aligned}
\tag{11.1}
$$

for given functions $f, c : \Omega \to \mathbb{R}$. It is a special case of the general elliptic problem (2.1), with $\mathcal{A}(x, u, \nabla u) = \nabla u$ and $h(u) = cu$. If f is smooth enough and $c(x) \geq c_0$ a.e. for some $c_0 > 0$, this problem has a unique solution in $H^1(\Omega)$ by the Lax–Milgram theorem [37].

For the example implementation we pick Ω to be the unit square $(0,1) \times (0,1)$, $c \equiv 10$, and the volume term $f : \Omega \to \mathbb{R}$ equal to -10 on a circle of radius $1/4$ around the center of Ω, and 10 elsewhere. The domain is discretized with a structured grid consisting of 8×8 quadrilateral elements. We solve the equation three times: Twice using Lagrangian finite elements of different orders, and once using nonconforming Rannacher–Turek elements (Chapter 8.3.1). This demonstrates how easy it is to go from one finite element discretization to another.

The program consists of three parts: a parameter class ReactionDiffusionProblem that implements the functions f and c, a method solveReactionDiffusionProblem that receives a function space and does the simulation, and a main method that calls it three times for the three different spaces. The main method starts off as usual by initializing a potential MPI installation, and by setting up the 8×8 structured grid using the YaspGrid implementation:

```
137  int main(int argc, char *argv[])
138  {
139    // Set up MPI, if available
140    MPIHelper::instance(argc,argv);
141
142    // Create a structured grid
143    constexpr int dim = 2;
144    using Grid = YaspGrid<dim>;
145    FieldVector<double,dim> bbox = {1.0, 1.0};
146    std::array<int,dim> elements = {8,8};
147    Grid grid(bbox,elements);
148
149    using GridView = Grid::LeafGridView;
150    GridView gridView = grid.leafGridView();
```

The YaspGrid grid manager has been discussed in detail in Chapter 5.10.2.

Following this, the code constructs three different basis objects from the dune-functions module, and calls the method solveReactionDiffusionProblem with each one:

```
154    // Solve boundary value problem using first-order Lagrange elements
155    using Q1Basis = Functions::LagrangeBasis<GridView,1>;
156    auto q1Basis = std::make_shared<Q1Basis>(gridView);
157
158    solveReactionDiffusionProblem(q1Basis,
159                         "pdelab-linear-reaction-diffusion-result-q1",
160                         9);    // Average number of matrix entries
161                                // per row
162
163    // Solve boundary value problem using second-order Lagrange elements
164    using Q2Basis = Functions::LagrangeBasis<GridView,2>;
165    auto q2Basis = std::make_shared<Q2Basis>(gridView);
166
167    solveReactionDiffusionProblem(q2Basis,
168                         "pdelab-linear-reaction-diffusion-result-q2",
169                         25);    // Average number of matrix entries
170                                 // per row
```

399

```
171
172     // Solve boundary value problem using Rannacher-Turek elements
173     using RannacherTurekBasis = Functions::RannacherTurekBasis<GridView>;
174     auto rannacherTurekBasis = std::make_shared<RannacherTurekBasis>(gridView);
175
176     solveReactionDiffusionProblem(rannacherTurekBasis,
177                             "pdelab-linear-reaction-diffusion-result-rt",
178                             7);     // Average number of matrix entries
179                                     // per row
180   }
```

The method `solveReactionDiffusionProblem` does the actual computations. It gets
two parameters besides the finite element basis. One is the filename used for the
result file. The other one is the expected average number of entries per row of the
stiffness matrix. In this example, the `dune-pdelab` assembler uses the `BCRSMatrix`
implementation of a sparse matrix together with its *implicit* matrix build mode
(explained in Chapter 7.3.2), which needs an a priori guess of the number of nonzero
matrix entries per row. Computing these automatically is fairly expensive, and
`dune-pdelab` therefore offloads this task to the user. After these few lines, we have
already reached the end of the `main` method.

We now turn to the method `solveReactionDiffusionProblem`, which does the
main work. It will be noticed that the method is quite short, comprising less than 40
lines of code. The method signature is

```
40   template<class Basis>
41   void solveReactionDiffusionProblem(std::shared_ptr<Basis> basis,
42                                      std::string filename,
43                                      int averageNumberOfEntriesPerMatrixRow)
44   {
```

The meaning of the method parameters has been explained above.

The first thing done by the method is constructing a `dune-pdelab` style discrete
function space from the basis given as the `basis` parameter. As mentioned in the
introduction of this chapter, `dune-pdelab` has its own set of abstractions for finite
element bases and spaces, and while they share many ideas, their implementations
are mostly separate. The central concept in `dune-pdelab` is the *grid function space*,
which is roughly equivalent to the function space basis of `dune-functions`. For the
examples in this book we use the adaptor class `Experimental::GridFunctionSpace`
that allows to use `dune-functions` bases as implementations of grid function spaces:

```
48   using GridView = typename Basis::GridView;
49   using VectorBackend = PDELab::ISTL::VectorBackend<>;
50
51   using GridFunctionSpace = PDELab::Experimental::GridFunctionSpace<Basis,
52                                                     VectorBackend,
53                                                     PDELab::NoConstraints>;
54
```

```
55    GridFunctionSpace gridFunctionSpace(basis);
```

Compared to a dune-functions basis, the GridFunctionSpace of dune-pdelab offers several additional features, which require two extra template parameters. The VectorBackend type selects the container data type used for vector coefficients. The particular choice in Line 49 selects the BlockVector class from dune-istl (see Chapter 11.4.1 for the details). The third template parameter PDELab::NoConstraints in Line 53 specifies that there are no additional algebraic constraints (such as Dirichlet boundary conditions) acting on the space.

Now that we have a function space we can set up the finite element assembler. It consists of three classes. The first is a small helper class called ReactionDiffusionProblem that implements the coefficient functions f and c. It is defined near the top of the file, and very short:

```
18    template<class GridView>
19    struct ReactionDiffusionProblem
20    : public PDELab::ConvectionDiffusionModelProblem<GridView,double>
21    {
22      template<typename Element, typename Coord>
23      auto f(const Element& element, const Coord& xi) const
24      {
25        const Coord center = Coord(0.5);
26        auto distanceToCenter = (element.geometry().global(xi)-center).two_norm();
27        return (distanceToCenter <= 0.25) ? -10.0 : 10.0;
28      }
29
30      template<typename Element, typename Coord>
31      auto c(const Element& element, const Coord& xi) const
32      {
33        return 10.0;
34      }
35    };
```

Each method in this class takes an element and a position in local coordinates of this element, and returns the corresponding function value there. The code constructs such a parameter class, and then uses it to construct an object of type ConvectionDiffusionFEM, which is the actual element assembler:

```
60    using Problem = ReactionDiffusionProblem<GridView>;
61    Problem reactionDiffusionProblem;
62    using LocalOperator
63        = PDELab::ConvectionDiffusionFEM<Problem,
64                                        typename GridFunctionSpace::Traits::
65                                                            FiniteElementMap>;
66    LocalOperator localOperator(reactionDiffusionProblem);
```

In dune-pdelab terminology, such an element assembler is called a *local operator*, which explains the naming used in Lines 62–66. The operator, called PDELab::ConvectionDiffusionFEM (from the file convectiondiffusionfem.hh in the folder dune/pdelab/localoperator/) assembles the element stiffness matrices of a more general linear problem that involves convection terms together with the reaction and diffusion. The particular choice of parameter functions here turns it into an element assembler for the test problem (11.1).

The third and final step is to construct a global assembler, which loops over the grid elements, calls the local operator to obtain the element matrices, and accumulates them in the global stiffness matrix:

```
70    using MatrixBackend = PDELab::ISTL::BCRSMatrixBackend<>;
71    MatrixBackend matrixBackend(averageNumberOfEntriesPerMatrixRow);
72    using GridOperator
73        = PDELab::GridOperator<GridFunctionSpace,     // Trial function space
74                               GridFunctionSpace,     // Test function space
75                               LocalOperator,         // Element assembler
76                               MatrixBackend,         // Data structure
77                                                      // for the stiffness matrix
78                               double,                // Number type for
79                                                      // solution vector entries
80                               double,                // Number type for
81                                                      // residual vector entries
82                               double>;               // Number type for
83                                                      // stiffness matrix entries
84
85    GridOperator gridOperator(gridFunctionSpace,
86                              gridFunctionSpace,
87                              localOperator,
88                              matrixBackend);
```

The first two lines select the matrix backend, i.e., the matrix data structure used for the stiffness matrix. The choice ISTL::BCRSMatrixBackend selects the BCRSMatrix class from the dune-istl module; see Chapter 11.4 for alternative choices.

Then, the code constructs an object of type PDELab::GridOperator. In the dune-pdelab module, the word *grid operator* is a synonym for the global assembler, and the GridOperator class is the standard implementation of such an assembler. It is configured with a fairly long list of template- and run-time arguments. The first two arguments are the finite element spaces for the trial and test functions. Note that both have to be specified separately. In most cases they will be the same two spaces, but for advanced applications like Petrov–Galerkin methods (see, e.g., [63]), or error estimators (Chapter 11.5.2), it is possible to specify two separate spaces.

The next two arguments are the element assembler and the matrix backend, both of which we have just set up. Finally, we get to select three number types. These are for the entries of the solution vector (the *domain*), the residual vector (the *range*), and the entries of the stiffness matrix, respectively. In most cases **double** is the appropriate choice here.

Finally, we can solve the algebraic system. We will use a conjugate gradient solver preconditioned with an SSOR method [138]. First of all, we need a vector to hold the iterates. Since dune-pdelab hides the linear algebra implementation from the user, we do not construct a vector data type directly. Rather, we query the vector backend of the GridFunctionSpace for the appropriate vector type:

```
93    using VectorContainer = PDELab::Backend::Vector<GridFunctionSpace,double>;
94    VectorContainer u(gridFunctionSpace,0.0); // Initial value
```

Since we selected the dune-istl backend in Line 49, u will be an object of (a wrapper type around) the type BlockVector discussed in Chapter 7.2. The second constructor argument initializes all entries to zero.

The code then sets up and calls the preconditioned CG solver. This is a three-step procedure:

```
99    using LinearSolverBackend = PDELab::ISTLBackend_SEQ_CG_SSOR;
100   LinearSolverBackend linearSolverBackend(5000,2);
101
102   using LinearProblemSolver
103       = PDELab::StationaryLinearProblemSolver<GridOperator,
104                                               LinearSolverBackend,
105                                               VectorContainer>;
106
107   LinearProblemSolver linearProblemSolver(gridOperator,
108                                           linearSolverBackend,
109                                           u,
110                                           1e-10);
111   linearProblemSolver.apply();
```

First, we construct the actual solver from the dune-istl module. It is again accessed only through the linear algebra backend. Then, this solver is handed to an object of type StationaryLinearProblemSolver. This class connects the solver to the assembler: It first uses the gridOperator object to assemble the algebraic problem, and then the linearSolverBackend object to solve it, starting from the initial iterate in u. This whole process is triggered by the call to the method apply in Line 111. After this call, u contains the final iterate. All that is left to do is to write it to a file for later post-processing and visualization. The following lines write the content of u to a VTK file, which can then be looked at, e.g., using PARAVIEW:[2]

```
116   // Make a discrete function from the FE basis and the coefficient vector
117   auto uFunction
118       = Functions::makeDiscreteGlobalBasisFunction<double>(
119                                        *basis,
120                                        PDELab::Backend::native(u));
121
122   // We need to subsample, because the dune-grid VTKWriter
123   // cannot natively display second-order functions.
124   SubsamplingVTKWriter<GridView> vtkWriter(basis->gridView(),
125                                            refinementLevels(4));
```

---

[2]www.paraview.org

Figure 11.1.: Results for the linear reaction–diffusion example computed with three different finite element spaces. From left: first- and second-order Lagrangian elements, and Rannacher–Turek elements on an 8×8 grid.

```
126    vtkWriter.addVertexData(uFunction,
127                       VTK::FieldInfo("u",
128                                   VTK::FieldInfo::Type::scalar,
129                                   1));
130    vtkWriter.write(filename);
```

The code first constructs a discrete function from the dune-functions basis and the vector of coefficients. This function can then be handed to the SubsamplingVTKWriter as described in Chapter 10.7. We need to subsample the result functions, because the VTKWriter class from dune-grid currently only supports writing piecewise (multi-) linear functions.

The VTK writing code mostly uses functionality from the dune-grid and dune-functions modules. The only dune-pdelab-specific part is the call to PDELab::Backend::native(u), which provides a reference to the underlying vector data structure of the linear algebra backend (in this case a BlockVector).

When started, the program will produce the following output:

```
=== matrix setup (max) 0.00127015 s
=== matrix assembly (max) 0.00287509 s
=== residual assembly (max) 0.0014412 s
=== solving (reduction: 1e-10) === Dune::IterativeSolver
 Iter        Defect            Rate
    0        1.03754
    1        0.498995        0.48094
    2        0.0309844       0.0620935
    3        0.00108314      0.0349576
    4        9.55854e-05     0.0882486
    5        5.97994e-06     0.0625612
    6        9.76342e-07     0.16327
    7        1.16108e-08     0.0118921
    8        1.281e-09       0.110329
    9        4.06776e-11     0.0317545
```

```
=== rate=0.0697759, T=0.00437401, TIT=0.000486001, IT=9
0.00441099 s

[...]
```

The part that is printed here reports on the simulation using first-order finite elements, and it is followed by two similar blocks of output for the other two spaces. The output is quite self-explanatory: We see the list of iterations of the CG solver, giving the iteration number, the current defect (residual) norm, and the defect reduction rate. Observe that the iteration stops as soon as the defect norm drops below 10^{-10} times the initial defect norm, as requested. Interspersed are wall-time measurements of different parts of the simulation process.

The program will write three files called pdelab-linear-reaction-diffusion-result-q1.vtu and pdelab-linear-reaction-diffusion-result-q2.vtu for the Lagrange element results, and pdelab-linear-reaction-diffusion-result-rt.vtu for the simulation using the Rannacher–Turek element. Figure 11.1 shows their visualizations in PARAVIEW. It can be seen that all three approximate the same solution function, but differences between the three trial spaces are clearly noticeable. The second-order Lagrange method gives a much smoother-looking approximation than the first-order one. In the third picture, the discontinuities at the element boundaries, which are characteristic of Rannacher–Turek spaces, are clearly visible.

11.2. Implementing Element Assemblers

The previous example has used a local assembler that came ready-to-use as part of dune-pdelab. Even though dune-pdelab provides a large collection of such assemblers in the dune/pdelab/localoperator directory, sooner or later many users will want to write their own. This is the subject of this chapter.

We begin by explaining the residual formulation, which is dune-pdelab's way of looking at partial differential equations, in Chapters 11.2.1 and 11.2.2. Then, Chapter 11.2.3 will cover the implementation by explaining the LocalOperator interface class. A first complete example in Chapter 11.2.4 will show how to implement and use a local assembler for the p-Laplace problem with a reaction term. The remaining two sections will explain how to assemble skeleton and boundary terms, and demonstrate this with a second complete example implementing a Discontinuous Galerkin (DG) method.

11.2.1. The Residual Formulation

The dune-pdelab module has a particular way of looking at partial differential equations: It focuses on the residuals of the equations. To explain this we briefly revisit the generic elliptic model problem from Chapter 2.1. We will focus a bit more on nonlinear problems, because, at least as far as assembly is concerned, dune-pdelab treats all problems as nonlinear ones.

Let Ω be an open bounded subset of $\mathbb{R}^d$, and $f : \Omega \to \mathbb{R}$ a given function of sufficient smoothness. Further, suppose that $\mathcal{A} : \Omega \times \mathbb{R} \times \mathbb{R}^d \to \mathbb{R}^d$ is a function acting on positions in Ω, point values, and first derivatives, and $h : \mathbb{R} \to \mathbb{R}$ is a given function called the *reaction term*. We look for a function $u : \Omega \to \mathbb{R}$ solving the equation

$$- \operatorname{div} \mathcal{A}(x, u, \nabla u) + h(u) = f \qquad \text{in } \Omega, \tag{11.2a}$$

subject to the homogeneous Neumann boundary conditions

$$\langle \mathcal{A}(x, u, \nabla u), \mathbf{n} \rangle = 0 \qquad \text{on } \partial\Omega. \tag{11.2b}$$

The treatment of Dirichlet boundary conditions is deliberately postponed to Chapter 11.3.

The weak formulation for this problem is derived in the same way as in Section 2.1. Multiply (11.2a) with a test function v from $H^1 = H^1(\Omega)$, integrate over Ω and use Green's identity to obtain

$$\int_\Omega \mathcal{A}(x, u, \nabla u) \nabla v \, dx + \int_\Omega h(u) v \, dx = \int_\Omega f v \, dx \qquad \text{for all } v \in H^1. \tag{11.3}$$

We abbreviate this as

$$a(u, v) = l(v) \qquad \forall v \in H^1, \tag{11.4}$$

where

$$a(u, v) := \int_\Omega \mathcal{A}(x, u, \nabla u) \nabla v \, dx + \int_\Omega h(u) v \, dx$$

and

$$l(v) := \int_\Omega f v \, dx.$$

Both forms are linear in v, but $a(\cdot, \cdot)$ may be nonlinear in its first argument. For simplicity, we suppose that $\mathcal{A}$, h, f are such that the weak problem (11.4) has at least one solution in H^1.

While (11.4) is the text book version of the weak elliptic model problem, dune-pdelab focuses on the *residual formulation*. Introduce the residual form

$$r(v, w) := a(v, w) - l(w),$$

defined for functions $v, w \in H^1$. It is nonlinear in its first argument, but linear in its second one. Using the residual form, the weak problem (11.3) can evidently be rewritten as

$$\text{Find } u \in H^1 : \quad r(u, v) = 0 \qquad \text{for all } v \in H^1. \tag{11.5}$$

This is the *residual formulation* of the original boundary value problem.

We now want to compute a finite element approximation of (one of the) solutions of (11.5). For this we first introduce two finite element spaces U_h and V_h that both approximate the Sobolev space H^1. We will look for a solution in U_h, but use test functions in V_h. In many applications U_h and V_h will be the same space; however, dune-pdelab also caters to methods for which they differ, such as Petrov–Galerkin methods [63]. For simplicity we assume in the following that both spaces have the same dimension n.

If the finite element spaces are conforming, i.e., $U_h \subset H^1$ and $V_h \subset H^1$, then the discrete residual problem results from replacing H^1 by U_h and V_h in the residual formulation (11.5):

$$\text{Find } u_h \in U_h \ : \quad r(u_h, v_h) = 0 \qquad \text{for all } v_h \in V_h.$$

However, dune-pdelab more generally allows nonconforming spaces, i.e., spaces U_h, V_h that approximate a Sobolev space without being a subset of it. For this case we need to introduce a new residual form

$$r_h(\cdot, \cdot) : U_h \times V_h \to \mathbb{R}$$

that approximates $r(\cdot, \cdot)$ in a suitable way. In the following we will always write $r_h(\cdot, \cdot)$ for the discrete residual form, even if the spaces U_h, V_h are conforming.

Assumption 11.1 *All relevant boundary value problems can be written in the form*

$$\text{Find } u_h \in U_h : \quad r_h(u_h, v_h) = 0 \qquad \forall v_h \in V_h \qquad (11.6)$$

for suitable spaces U_h, V_h, and forms $r_h(\cdot, \cdot)$.

This assumption certainly excludes a number of interesting problems, but it is at the heart of the design of dune-pdelab. It is always assumed that $r_h(\cdot, \cdot)$ is linear in its second argument, but it may be nonlinear in the first one. Finally, it is assumed that the residual problem (11.6) has at least one solution.

Using bases $\{\phi_i\}_{i=0}^{n-1}$ of U_h and $\{\theta_i\}_{i=0}^{n-1}$ of V_h we can derive the algebraic formulation of the discrete residual equation (11.6). Replacing v_h in (11.6) by its basis representation $\sum_{i=0}^{n-1} v_i \theta_i$, we obtain

$$r_h\left(u_h, \sum_{i=0}^{n-1} v_i \theta_i\right) = \sum_{i=0}^{n-1} v_i r_h(u_h, \theta_i) = 0 \qquad \forall v_h \in V_h,$$

which is equivalent to

$$r_h(u_h, \theta_i) = 0 \qquad \forall i = 0, \ldots, n-1.$$

To formulate the algebraic residual, let $\bar{u} \in \mathbb{R}^n$ be the vector of coefficients of u_h with respect to the basis $\{\phi_i\}_{i=0}^{n-1}$. We define a map $\mathcal{R} : \mathbb{R}^n \to \mathbb{R}^n$ by

$$(\mathcal{R}(\bar{u}))_i := r_h(u_h, \theta_i) \qquad \text{for all } i = 0, \ldots, n-1.$$

The algebraic residual problem is then

$$\text{Find } \bar{u} \in \mathbb{R}^n : \quad \mathcal{R}(\bar{u}) = 0 \in \mathbb{R}^n. \tag{11.7}$$

Solvability of this nonlinear algebraic system of equations follows directly from the assumed solvability of (11.6).

By default, `dune-pdelab` additionally assumes that the algebraic residual $\mathcal{R}$ depends differentiably on its argument, and offers a damped Newton method [51, 130] as the standard way to solve the residual equation (11.7). For such a Newton method, the Jacobian matrix $\nabla \mathcal{R}(\bar{u})$ defined by

$$(\nabla \mathcal{R}(\bar{u}))_{i,j} := \frac{\partial (\mathcal{R}(\bar{u}))_i}{\partial \bar{u}_j}$$

needs to be computable for each Newton iterate $\bar{u}$.

If both the differential operator $\mathcal{A}$ and the reaction term h are linear, then the residual takes the simple form

$$\mathcal{R}(\bar{u}) = A\bar{u} - b, \tag{11.8}$$

where A and b are the stiffness matrix and load vector as defined in Chapter 2.1. Note that this is the negative of the residual of a linear equation as it is typically defined in text books. From such a residual, the stiffness matrix can be retrieved as the Jacobian of $\mathcal{R}$, whereas b is the *negative* value of $\mathcal{R}$ at $0 \in \mathbb{R}^n$.

11.2.2. Assembling Element Residuals and Their Derivatives

In Chapter 2.1.3 we have explained how algebraic formulations of finite element problems are assembled from their discrete counterparts. It was shown that in practice this is done by regarding the restriction of the weak equation to each grid element, and by appropriately summing up the results. The terms involved integrals over elements and element boundaries. This motivates the following assumption.

Assumption 11.2 *The residual form $r_h(\cdot, \cdot)$ can be additively split into element, skeleton, and boundary terms*

$$r_h(u_h, v_h) = \sum_{T \in \mathcal{T}} r_{h,T}^{\text{vol}}(u_h, v_h) + \sum_{\gamma^i \in \mathcal{F}} r_{h,\gamma^i}^{\text{skel}}(u_h, v_h) + \sum_{\gamma^b \in \mathcal{B}} r_{h,\gamma^b}^{\text{bnd}}(u_h, v_h), \tag{11.9}$$

where the skeleton $\mathcal{F}$ is the set of facets (i.e., $d-1$-dimensional element faces) not on the domain boundary, and $\mathcal{B}$ is the set of facets on $\partial \Omega$.[3]

---

[3] While we speak of skeleton facets here, the assumption does encompass nonconforming grids as well, in which case the elements of the skeleton $\mathcal{F}$ are the *intersections* between adjacent elements (Chapter 5.4).

Consequently, algebraic problems in `dune-pdelab` are assembled by adding element, skeleton, and boundary terms. Not all discretizations have all three terms, of course—the example residual form of Chapter 11.1 has only the element terms. For a start we focus on these element terms and postpone the others to Chapter 11.2.5.

By construction, the element, skeleton, and boundary contributions of $r_h(\cdot, \cdot)$ must be linear in the second argument v_h. They will be linear in the first argument u_h only if the original PDE problem is linear. The `dune-pdelab` module assumes these forms to be *local* in the sense that $r_{h,T}^{\text{vol}}(u_h, v_h)$ will only depend on values of u_h and v_h on the element T, and inner facet terms $r_{h,\gamma^{\text{i}}}^{\text{skel}}(u_h, v_h)$, and boundary facet terms $r_{h,\gamma^{\text{b}}}^{\text{bnd}}(u_h, v_h)$ will only depend on values of u_h and v_h on the adjacent elements of γ^{i} and γ^{b}, respectively. This is fulfilled by virtually all finite element problems, because the residual terms consist of integrals over subsets of the domains.[4] For example, the residual form

$$r_h(u_h, v_h) := \int_\Omega \Big[\mathcal{A}(x, u_h, \nabla u_h) \nabla v_h + h(u_h) v_h \Big] dx - \int_\Omega f v_h \, dx$$

corresponding to (11.4) can be split into a sum of element integrals

$$r_h(u_h, v_h) = \sum_{T \in \mathcal{T}} r_{h,T}^{\text{vol}}(u_h, v_h)$$

with

$$r_{h,T}^{\text{vol}}(u_h, v_h) := \int_T \Big(\mathcal{A}(x, u_h, \nabla u_h) \nabla v_h + h(u_h) v_h \Big) dx - \int_T f v_h \, dx,$$

which indeed only depends on the restrictions of u_h and v_h on T. An example involving skeleton and boundary terms will be given in Chapter 11.2.6.

The localization feature of the residual form motivates the construction of corresponding algebraic restrictions. Let T be a grid element, and call $\bar{u}_T \in \mathbb{R}^{n_T}$ the vector of those degrees of freedom that pertain to element T, i.e., those that $r_{h,T}^{\text{vol}}$ depends on. Then we can define the canonical restriction operator

$$\mathbf{R}_T : \mathbb{R}^n \to \mathbb{R}^{n_T}, \qquad \mathbf{R}_T : \bar{u} \mapsto \bar{u}_T.$$

Its transpose $\mathbf{R}_T^T$ extends a vector of element degrees of freedom to a global vector by padding with zeros. Likewise, for any skeleton facet γ^{i} we introduce the operator $\mathbf{R}_{T_{\text{in}}(\gamma^{\text{i}}), T_{\text{out}}(\gamma^{\text{i}})}$ that restricts to the degrees of freedom on the two neighboring elements of γ^{i}, and the operator $\mathbf{R}_{T(\gamma^{\text{b}})}$ restricting to the degrees of freedom on the element supporting a boundary facet γ^{b}. With these operators, the assembly of the overall

---

[4]It is not sufficient for non-trivial finite volume schemes; see, e.g., the multidimensional Lax–Wendroff method described in [115].

residual $\mathcal{R}$ can be written as three loops

$$\mathcal{R}(\bar{u}) = \sum_{T \in \mathcal{T}} (\mathbf{R}_T)^T \mathcal{R}_T^{\text{vol}}(\mathbf{R}_T \bar{u})$$
$$+ \sum_{\gamma^i \in \mathcal{F}} (\mathbf{R}_{T_{\text{in}}(\gamma^i), T_{\text{out}}(\gamma^i)})^T \mathcal{R}_{\gamma^i}^{\text{skel}}(\mathbf{R}_{T_{\text{in}}(\gamma^i), T_{\text{out}}(\gamma^i)})$$
$$+ \sum_{\gamma^b \in \mathcal{B}} (\mathbf{R}_{T(\gamma^b)})^T \mathcal{R}_{\gamma^b}^{\text{bnd}}(\mathbf{R}_{T(\gamma^b)}), \tag{11.10}$$

where $\mathcal{R}_T^{\text{vol}}$, $\mathcal{R}_{\gamma^i}^{\text{skel}}$, $\mathcal{R}_{\gamma^b}^{\text{bnd}}$ are the natural restrictions of the algebraic residual operators to single elements or pairs of elements. Note that this approach is in minor contrast to Chapter 2.1, where assembly used only a single loop over the elements.

Likewise, the Jacobian of $\mathcal{R}$ at $\bar{u}$ can be computed from local contributions of three loops

$$\nabla \mathcal{R}(\bar{u}) = \sum_{T \in \mathcal{T}} (\mathbf{R}_T)^T \nabla \mathcal{R}_T^{\text{vol}}(\mathbf{R}_T \bar{u}) \mathbf{R}_T$$
$$+ \sum_{\gamma^i \in \mathcal{F}} (\mathbf{R}_{T_{\text{in}}(\gamma^i), T_{\text{out}}(\gamma^i)})^T \nabla \mathcal{R}_{\gamma^i}^{\text{skel}}(\mathbf{R}_{T_{\text{in}}(\gamma^i), T_{\text{out}}(\gamma^i)} \bar{u}) \mathbf{R}_{T_{\text{in}}(\gamma^i), T_{\text{out}}(\gamma^i)}$$
$$+ \sum_{\gamma^b \in \mathcal{B}} (\mathbf{R}_{T(\gamma^b)})^T \nabla \mathcal{R}_{\gamma^b}^{\text{bnd}}(\mathbf{R}_{T(\gamma^b)} \bar{u}) \mathbf{R}_{T(\gamma^b)}. \tag{11.11}$$

As the restriction operators are independent of the PDE, users only need to provide the local residuals $\mathcal{R}_T^{\text{vol}}$, $\mathcal{R}_{\gamma^i}^{\text{skel}}$, $\mathcal{R}_{\gamma^b}^{\text{bnd}}$ together with their Jacobians $\nabla \mathcal{R}_T^{\text{vol}}$, $\nabla \mathcal{R}_{\gamma^i}^{\text{skel}}$, $\nabla \mathcal{R}_{\gamma^b}^{\text{bnd}}$ to implement a new `dune-pdelab` assembler.

11.2.3. Implementing Element Assemblers: The `LocalOperator` Interface

The local residuals $\mathcal{R}_T$, $\mathcal{R}_{\gamma^i}$, and $\mathcal{R}_{\gamma^b}$ and their Jacobians encode all information required to solve a particular discretized PDE with a Newton or Newton-like method. In `dune-pdelab`, these terms are assembled by implementations of the interface class `LocalOperator`. Given a grid element T, a skeleton intersection γ^i, or a boundary intersection γ^b, together with the degrees of freedom related to T or the elements adjacent to the intersections γ^i and γ^b, respectively, the `LocalOperator` object implements the local residual terms $\mathcal{R}_T^{\text{vol}}$, $\mathcal{R}_{\gamma^i}^{\text{skel}}$, and $\mathcal{R}_{\gamma^b}^{\text{bnd}}$, and their Jacobians $\nabla \mathcal{R}_T^{\text{vol}}$, $\nabla \mathcal{R}_{\gamma^i}^{\text{skel}}$, and $\nabla \mathcal{R}_{\gamma^b}^{\text{bnd}}$. It is used by the generic `dune-pdelab` implementation of a global assembler, which is called `GridOperator`. The interface class is

```
template<class GridFunctionSpaceU, class GridFunctionSpaceV,
         class LocalOperator,
         class MatrixBackend,
         class DomainNumberType,
         class RangeNumberType,
         class JacobianNumberType>
class GridOperator
```

where GridFunctionSpaceU and GridFunctionSpaceV are the implementations of the discrete function spaces U_h and V_h, respectively, LocalOperator is the element assembler, and MatrixBackend is a sparse matrix implementation (Chapter 11.4). The remaining three parameters DomainNumberType, RangeNumberType, and JacobianNumberType are the number types used for coefficients of primal and dual vectors (i.e., iterates and residuals), respectively, and for stiffness matrix entries. Objects of GridOperator type are constructed directly calling the constructor

```
GridOperator(const GridFunctionSpaceU & gfsu, const GridFunctionSpaceV & gfsv,
             LocalOperator& localOperator,
             const MatrixBackend& matrixBackend = MatrixBackend())
```

The GridOperator object loops over all elements, skeleton, and boundary intersections, and assembles the *global* residual $\mathcal{R}(\bar{u})$ and Jacobian $\nabla\mathcal{R}(\bar{u})$, $\bar{u} \in \mathbb{R}^n$ by calling the local operator on each element, and piecing together the results by using (11.10) and (11.11), respectively. This separation mirrors the mathematical formulation of the finite element assembly process as explained in Chapter 2.1.3, and extends it to include skeleton terms.

While the GridOperator class is a fixed implementation which the user never has to modify, writing new LocalOperator implementations occurs regularly. Such implementations must use the *duck typing* technique (Chapter 4.4.2), i.e., they must provide the methods and types required by the LocalOperator class, but compliance is not directly checked by the compiler. The LocalOperator interface uses a more flexible variant of duck typing: Some of the methods are optional, and are only required if certain flags are set.

Any implementation of the LocalOperator interface must inherit from the PDELab::LocalOperatorDefaultFlags class, which injects certain infrastructure. To make the global operator set up the matrix pattern, it must also inherit from PDELab::FullVolumePattern. If the LocalOperator assembles skeleton or boundary contributions as well, then the class may have to inherit from PDELab::FullSkeletonPattern and PDELab::FullBoundary pattern, too. In addition, the implementation has to set the flags

```
static const bool doPatternVolume
static const bool doPatternSkeleton
static const bool doPatternBoundary
```

if the corresponding Jacobian matrix pattern entries are used.

void	alpha_volume(**const** EntityGeometry& entityGeometry, 　**const** LocalFunctionSpaceU& localFunctionSpaceU, 　**const** Vector& x, 　**const** LocalFunctionSpaceV& localFunctionSpaceV, 　Residual& r) **const** Compute the element part $\mathcal{R}_T^{\mathrm{vol}}$ of the residual
void	alpha_skeleton(**const** IntersectionGeometry& intersectionGeometry, 　**const** LocalFunctionSpaceU& localFunctionSpaceUIn, 　**const** Vector& xIn, 　**const** LocalFunctionSpaceV& localFunctionSpaceVIn, 　**const** LocalFunctionSpaceU& localFunctionSpaceUOut, 　**const** Vector& xOut, 　**const** LocalFunctionSpaceV& localFunctionSpaceVOut, 　Residual& rIn, Residual& rOut) **const** Compute the skeleton part $\mathcal{R}_{\gamma_i\mathrm{i}}^{\mathrm{skel}}$ of the residual
void	alpha_boundary(**const** IntersectionGeometry& intersectionGeometry, 　**const** LocalFunctionSpaceU& localFunctionSpaceUIn, 　**const** Vector& xIn, 　**const** LocalFunctionSpaceV& localFunctionSpaceVIn, 　Residual& rIn) **const** Compute the boundary part $\mathcal{R}_{\gamma_i\mathrm{b}}^{\mathrm{bnd}}$ of the residual
void	jacobian_volume(**const** EntityGeometry& entityGeometry, 　**const** LocalFunctionSpaceU& localFunctionSpaceU, 　**const** Vector& x, 　**const** LocalFunctionSpaceV& localFunctionSpaceV, 　Jacobian& jacobian) **const** Compute the element part $\nabla\mathcal{R}_T^{\mathrm{vol}}$ of the residual Jacobian
void	jacobian_skeleton(**const** IntersectionGeometry& intersectionGeometry, 　**const** LocalFunctionSpaceU& localFunctionSpaceUIn, 　**const** Vector& xIn, 　**const** LocalFunctionSpaceV& localFunctionSpaceVIn, 　**const** LocalFunctionSpaceU& localFunctionSpaceUOut, 　**const** Vector& xOut, 　**const** LocalFunctionSpaceV& localFunctionSpaceVOut, 　Jacobian& jacobianInIn, Jacobian& jacobianInOut, 　Jacobian& jacobianOutIn, Jacobian& jacobianOutOut) **const** Compute the skeleton part $\nabla\mathcal{R}_{\gamma_i\mathrm{i}}^{\mathrm{skel}}$ of the residual Jacobian
void	jacobian_boundary(**const** IntersectionGeometry& intersectionGeometry, 　**const** LocalFunctionSpaceU& localFunctionSpaceUIn, 　**const** Vector& xIn, 　**const** LocalFunctionSpaceV& localFunctionSpaceVIn, 　Jacobian& jacobian) **const** Compute the boundary part $\nabla\mathcal{R}_{\gamma_i\mathrm{b}}^{\mathrm{bnd}}$ of the residual Jacobian

Table 11.1.: Methods of the LocalOperator interface class

The main interface consists of three methods for local element, skeleton, and boundary contributions, and three methods for the corresponding Jacobians. There are further optional methods, but those will not be discussed in this book. The list of methods is given in Table 11.1. For historical reasons, the methods to compute the residual are called `alpha` methods.[5]

To control whether a method is implemented, the `LocalOperator` has to export a list of **bool** flags:

```
static const bool doAlphaVolume
static const bool doAlphaSkeleton
static const bool doAlphaBoundary
```

Each flag represents an `alpha`/`jacobian` pair of methods. If the flag corresponding to a particular pair of methods is **false**, these methods are never called, and may even be omitted entirely from the local operator implementation. If both the `doAlphaSkeleton` and `doAlphaBoundary` flags are false, the iteration through the intersections is skipped.

To assemble the global residual or Jacobian, the `GridOperator` object iterates over the elements of the grid. For each element, it will call the appropriate `volume` method. Then it will iterate through the elements intersections and call the appropriate `skeleton` or `boundary` methods on the intersection. In this chapter and the subsequent example we focus on the volume terms. Boundary and skeleton terms, which are needed, e.g., for Neumann boundary conditions and DG methods, will be treated in Chapter 11.2.5.

The first element-related method is `alpha_volume`, which computes $\mathcal{R}_T^{\mathrm{vol}}$, the algebraic form of the term $r_{h,T}^{\mathrm{vol}}$ from the splitting (11.9). The signature of the `alpha_volume` method is

```
template<class EntityGeometry, class LocalFunctionSpaceU, class Vector,
        class LocalFunctionSpaceV, class Residual>
void alpha_volume(const EntityGeometry& entityGeometry,
                  const LocalFunctionSpaceU& localFunctionSpaceU,
                  const Vector& x,
                  const LocalFunctionSpaceV& localFunctionSpaceV,
                  Residual& r) const
```

The method has five parameters, the types of which are all determined by corresponding member template parameters. The first, `entityGeometry`, represents the grid element, and its type is a wrapper around the class `Entity` from the `dune-grid` interface (Chapter 5.3).[6] The parameters `localFunctionSpaceU` and `localFunctionSpaceV` are the finite elements to be used for the trial and test spaces, respectively. When `alpha_volume` is called, the types `LocalFunctionSpaceU` and `LocalFunctionSpaceV`

---

[5]Experienced DUNE users would expect camel-case naming here [159], as used almost everywhere else in DUNE. Unfortunately, standard-library-style naming was used originally, and it is very difficult to change it now.

[6]The wrapper gives access to the element by the `entity` method, and to the element geometry by the `geometry` method. The rationale for this construction is beyond the scope of this text.

will be instances of a class called PDELab::LocalFunctionSpace. Without going into further details, objects of this type have a method

```
const Traits::FiniteElementType& finiteElement() const
```

which returns a LocalFiniteElement class in the sense of Chapter 8. This is how to get the actual shape functions—see Chapter 8 for the details on how to use them.

The third method parameter x stores the coefficients of the local function $\mathbf{R}_T\bar{u}$. The fifth parameter r receives the residual. When the method alpha_volume is called, the container r is already of the correct length, and the values are initialized. The alpha_volume method is expected to *add* the local contribution to the content of r. That way it is easier to combine several local operator implementations to a sum operator.

The precise types of x and r are dedicated vector classes provided by dune-pdelab to hold the degrees of freedom of a finite element space for a single element. The data layout of these containers is determined by the set of shape functions on the element. Therefore, accessing individual entries requires passing the localFunctionSpaceU and localFunctionSpaceV objects along. The object x is of type PDELab::LocalVector. Access to the entry for the *i*-th shape function is provided by the method

```
template<class LocalFunctionSpace>
LocalVector::value_type& operator()(
                         const LocalFunctionSpace& localFunctionSpace,
                         size_type i)
```

On the other hand, the type Residual is a WeightedVectorAccumulationView. It is intended to be write-only. The main method is

```
template<class LocalFunctionSpace>
void accumulate(const LocalFunctionSpace& localFunctionSpace,
                size_type i,
                value_type v)
```

which adds the value v to the i-th degree of freedom of the local function space localFunctionSpace.

To obtain the Jacobian $\nabla\mathcal{R}_T^{\mathrm{vol}}$, the LocalOperator class has a method called jacobian_volume. It has the signature

```
template<class EntityGeometry, class LocalFunctionSpaceU, class Vector,
         class LocalFunctionSpaceV, class Jacobian>
void jacobian_volume(const EntityGeometry& entityGeometry,
                     const LocalFunctionSpaceU& localFunctionSpaceU,
                     const Vector& x,
                     const LocalFunctionSpaceV& localFunctionSpaceV,
                     Jacobian& jacobian) const
```

The signature is very similar to the alpha_volume method; the only difference being the last argument, which is now a matrix instead of a vector. Again, before calling jacobian_volume, the GridOperator object initializes this matrix object properly, and values have to be *added* to it. Like the residual vector, it has a dedicated method for this:

```
template<class LocalFunctionSpaceU, class LocalFunctionSpaceV>
void accumulate(const LocalFunctionSpaceV& localFunctionSpaceV, size_type i,
                const LocalFunctionSpaceU& localFunctionSpaceU, size_type j,
                value_type v)
```

Similar to its residual-vector counterpart, calling this methods adds the content of v to the matrix entry specified by the indices i and j. As the row and column numbering is determined by the local test and trial spaces, respectively, the corresponding localFunctionSpaceV and localFunctionSpaceU objects have to be passed as well.

11.2.4. Example: The p-Laplace Equation with a Reaction Term

We now give a complete example that demonstrates the use of the volume methods of a LocalOperator implementation. It implements a local assembler for the p-Laplace equation with a reaction term, and solves the discrete equation with Newton's method. The complete example code is printed in Appendix B.12, and it is available through the icon in the margin.

Let Ω be a domain, and $f : \Omega \to \mathbb{R}$ suitably smooth. The p-Laplace equation for the scalar function u is [91, 118]

$$- \operatorname{div}(|\nabla u|^{p-2} \nabla u) = f.$$

The exponent p is a fixed real number strictly larger than 1. Evidently, the p-Laplace equation is a special case of the elliptic model problem (11.2a), with

$$\mathcal{A}(x, u, \nabla u) = |\nabla u|^{p-2} \nabla u,$$

and $h(u) = 0$. Its weak form is therefore

$$\text{Find } u \in W^{1,p} \ : \quad \int_\Omega |\nabla u|^{p-2} \langle \nabla u, \nabla v \rangle \, dx = \int_\Omega f v \, dx \qquad \forall v \in W^{1,p},$$

where $W^{1,p}$ is the first-order Sobolev space based on p-integrable functions [37, 162].

The p-Laplace equation is not well posed, unless Dirichlet boundary conditions are imposed on at least a part of the boundary. Since we want to avoid such boundary conditions until Chapter 11.3, we instead enforce well-posedness by adding a linear reaction term. We obtain the weak example problem

$$\int_\Omega |\nabla u|^{p-2} \langle \nabla u, \nabla v \rangle \, dx + \int_\Omega c u v \, dx = \int_\Omega f v \, dx \qquad \forall v \in W^{1,p},$$

with an additional coefficient function $c : \Omega \to \mathbb{R}$.

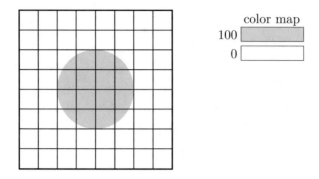

Figure 11.2.: Grid and volume term of the p-Laplace example

This is the example problem to be solved in this chapter. For Ω we choose the unit square $(0, 1)^2$. The reaction function c is set to a constant 100, and for the volume term f we pick

$$f(x) = \begin{cases} 100 & \text{if } \|x - (0.5, 0.5)\| \leq 0.25, \\ 0 & \text{otherwise,} \end{cases}$$

as illustrated in Figure 11.2.

We then discretize Ω with the structured 8×8 grid also shown in Figure 11.2. As finite element spaces U_h and V_h we select second-order Lagrange spaces. The discrete residual for the p-Laplace equation with a reaction term is

$$r_h(u_h, v_h) = \int_\Omega |\nabla u_h|^{p-2} \langle \nabla u_h, \nabla v_h \rangle \, dx + \int_\Omega c u_h v_h \, dx - \int_\Omega f v_h \, dx,$$

for any $u_h \in U_h$ and $v_h \in V_h$. It can be split in the sense of Assumption 11.2 as

$$r_h(u_h, v_h) = \sum_{T \in \mathcal{T}} r_{h,T}^{\text{vol}}(u_h, v_h), \tag{11.12}$$

with

$$r_{h,T}^{\text{vol}}(u_h, v_h) = \int_T |\nabla u_h|^{p-2} \langle \nabla u_h, \nabla v_h \rangle \, dx + \int_T c u_h v_h \, dx - \int_T f v_h \, dx.$$

The localized form (11.12) contains volume terms, but no skeleton or boundary ones. Correspondingly, we will see the methods `alpha_volume` and `jacobian_volume` in the code, but no others.

The `main` Method

After these preparations, let us start to discuss the actual code. The program file contains a class `PLaplaceLocalOperator` and a `main` method. As usual, the `main` method starts by setting up the `MPIHelper` instance and creating a grid:

```
265  int main(int argc, char *argv[])
266  {
267    // Initialize MPI
268    MPIHelper::instance(argc,argv);
269
270    constexpr int dim = 2;
271    using Grid = YaspGrid<dim>;
272    FieldVector<double,dim> upper = {1.0, 1.0};
273    std::array<int,dim> elements = {8, 8};
274    Grid grid(upper,elements);
275
276    using GridView = Grid::LeafGridView;
277    GridView gridView = grid.leafGridView();
```

The Yaspgrid implementation is used for a structured grid with 8×8 quadrilateral elements.

Next is the setup of the finite element space. As announced, the code uses the space of second-order Lagrange finite elements for both trial and test functions, but other orders would work equally well:

```
282    using Basis = Functions::LagrangeBasis<GridView,2>;
283    auto basis = std::make_shared<Basis>(gridView);
284
285    using VectorBackend = PDELab::ISTL::VectorBackend<>;
286
287    using GridFunctionSpace
288        = PDELab::Experimental::GridFunctionSpace<Basis,
289                                                  VectorBackend,
290                                                  PDELab::NoConstraints>;
291
292    GridFunctionSpace gridFunctionSpace(basis);
```

None of this is any different from the linear example that we have seen in Chapter 11.1.

Now we create the finite element assembler. First, we implement the two coefficient functions f and c. Since they are simple, we store them in lambda expressions. As we assumed the order p to be fixed, we simply store it as a number:

```
296    auto f = [](const FieldVector<double,dim>& x)
297    {
298      FieldVector<double,dim> center(0.5);
299      auto distanceToCenter = (x-center).two_norm();
300      return (distanceToCenter <= 0.25) ? 100 : 0;
301    };
302
303    auto c = [](const FieldVector<double,dim>& x)
304    {
305      return 100.0;
306    };
307
308    double p = 5.0;
```

Note that unlike in the example in Chapter 11.1, we are not required to follow a dune-pdelab interface for the coefficient functions, and we can therefore implement them in a simpler way that takes global point coordinates as arguments.

Then we construct an object of type PLaplaceLocalOperator and hand it the exponent p and the two coefficient functions:

```
313    using LocalOperator = PLaplaceLocalOperator<dim>;
314    LocalOperator localOperator(p,f,c);
```

The PLaplaceLocalOperator assembles the residual form of the p-Laplace equation on one element. It is implemented at the top of the example file, and will be explained below. The LocalOperator object in turn is handed to an object of type GridOperator:

```
318    using MatrixBackend = PDELab::ISTL::BCRSMatrixBackend<>;
319    MatrixBackend matrixBackend(25);  // 25: Expected average number of entries
320                                      //     per matrix row
321
322    using GridOperator
323        = PDELab::GridOperator<GridFunctionSpace,  // Trial function space
324                               GridFunctionSpace,  // Test function space
325                               LocalOperator,      // Element assembler
326                               MatrixBackend,      // Data structure
327                                                   // for the stiffness matrix
328                               double,             // Number type for
329                                                   // solution vector entries
330                               double,             // Number type for
331                                                   // residual vector entries
332                               double>;            // Number type for
333                                                   // stiffness matrix entries
334
335    GridOperator gridOperator(gridFunctionSpace,
336                              gridFunctionSpace,
337                              localOperator,
338                              matrixBackend);
```

Below, the GridOperator object will loop over all elements, and it will assemble the element contributions of the global residual. All template and constructor parameters here are again just as in the linear case in Chapter 11.1.

Now the global assembler has been set up, and we can construct the algebraic solver. We use a Newton method with a line search globalization, which comes ready to use in dune-pdelab. The Newton solver needs a linear solver and an initial iterate, which we set up first:

```
343    using LinearSolverBackend = PDELab::ISTLBackend_SEQ_CG_SSOR;
344    LinearSolverBackend linearSolverBackend(5000,  // Max. number of iterations
345                                            2);    // Verbosity level
346
347    // Make vector of coefficients
```

```
348    using U = PDELab::Backend::Vector<GridFunctionSpace,double>;
349    U u(gridFunctionSpace,0.0);    // Initial iterate
```

This is again as in the linear example in Chapter 11.1. Then we can construct and call the actual Newton solver:

```
354    PDELab::NewtonMethod<GridOperator,LinearSolverBackend>
355                                    newtonMethod(gridOperator,
356                                        linearSolverBackend);
357
358    newtonMethod.setReduction(1e-10);         // Requested reduction
359                                              // of the residual
360    newtonMethod.setMinLinearReduction(1e-4); // Minimal reduction of the
361                                              // inner linear solver
362    newtonMethod.setVerbosityLevel(2);        // Solver verbosity
363    newtonMethod.apply(u);
```

Lines 358–362 set parameters that control the termination criteria of the Newton solver and the inner linear iterative solver. The last line does the actual solving. In each iteration, the newtonMethod object will call the gridOperator to obtain the residual and its Jacobian. The gridOperator in turn will call the localOperator to obtain the element contributions. Then the linearSolver object will be used to solve the resulting system. Finally, the computed correction is applied using a line search strategy described in [89].

At this point, the nonlinear algebraic problem has been solved, and all we need to do is to write the result to a file. The code for this is taken directly from the linear reaction–diffusion example of Chapter 11.1:

```
368    // Make a discrete function from the finite element basis
369    // and the coefficient vector
370    auto uFunction
371        = Functions::makeDiscreteGlobalBasisFunction<double>(
372                                        *basis,
373                                        PDELab::Backend::native(u));
374
375    // We need to subsample, because the dune-grid VTKWriter
376    // cannot natively display second-order functions.
377    SubsamplingVTKWriter<GridView> vtkwriter(gridView, refinementLevels(2));
378    vtkwriter.addVertexData(uFunction,
379                    VTK::FieldInfo("u",
380                        VTK::FieldInfo::Type::scalar,
381                        1));
382    vtkwriter.write("pdelab-p-laplace-result");
```

We again use the subsampling VTK writer, because as of Version 2.7, dune-grid does not natively support writing second-order finite elements. Observe again that dune-pdelab is not involved in writing VTK files at all.

The LocalOperator Implementation

We now discuss the main part of this example: the implementation of the local operator. The class signature with inheritances and boolean flags is:

```
30  template<int dim>
31  class PLaplaceLocalOperator :
32    public PDELab::LocalOperatorDefaultFlags,
33    public PDELab::FullVolumePattern
34  {
35  public:
36    // Pattern assembly flags
37    static const bool doPatternVolume = true;
38
39    // Residual assembly flags
40    static const bool doAlphaVolume = true;
```

The class is templated with the grid dimension dim, and inherits from the standard base class PDELab::LocalOperatorDefaultFlags. Additionally inheriting from PDELab::FullVolumePattern imports code to set up the correct matrix sparsity pattern. The two **bool** flags signal to the GridOperator that this local operator implements only the alpha_volume and jacobian_volume methods. No further methods are needed for this simple example.

Next are the data members for the order p and the coefficient functions f and c, and the single constructor:

```
44    // The order of the p-Laplace term
45    const double p_;
46
47    // Source term
48    std::function<double(const FieldVector<double,dim>&)> f_;
49
50    // Reaction term
51    std::function<double(const FieldVector<double,dim>&)> c_;
52
53    PLaplaceLocalOperator(double p,
54              const std::function<double(const FieldVector<double,dim>&)>& f,
55              const std::function<double(const FieldVector<double,dim>&)>& c)
56    : p_(p),
57      f_(f),
58      c_(c)
59    {}
```

Note that p is a constant, but that f and c can be general function objects. Those function objects will be evaluated in global coordinates given as objects of type FieldVector<**double**,dim>—for coefficient functions that are defined per element the DiscreteGlobalBasisFunction class from the dune-functions module can be used (Chapter 10.6).

The heart of the `LocalOperator` implementation are the methods `alpha_volume` and `jacobian_volume`. For a given grid element T, the method `alpha_volume` computes

$$\left(\mathcal{R}_T(u_h)\right)_i = \int_T |\nabla u_h|^{p-2} \langle \nabla u_h, \nabla \theta_i \rangle \, dx + \int_T c u_h \theta_i \, dx - \int_T f \theta_i \, dx \qquad (11.13)$$

for all shape functions $\theta_0, \ldots, \theta_{n_T - 1}$ of the test space on this element. The method `alpha_volume` has the signature discussed in Chapter 11.2.3, and starts with a few initial type declarations:

```
64    template<class EntityGeometry,
65             class LocalFunctionSpaceU, class Vector,
66             class LocalFunctionSpaceV, class Residual>
67    void alpha_volume(const EntityGeometry& entityGeometry,
68                      const LocalFunctionSpaceU& localFunctionSpaceU,
69                      const Vector& x,
70                      const LocalFunctionSpaceV& localFunctionSpaceV,
71                      Residual& residual) const
72    {
73        // Extract types of shape function values and gradients
74        using TrialFE = typename LocalFunctionSpaceU::Traits::FiniteElementType;
75        using TestFE  = typename LocalFunctionSpaceV::Traits::FiniteElementType;
76        using LocalBasisU = typename TrialFE::Traits::LocalBasisType;
77        using LocalBasisV = typename TestFE::Traits::LocalBasisType;
78        using RangeU = typename LocalBasisU::Traits::RangeType;
79        using RangeV = typename LocalBasisV::Traits::RangeType;
80        using GradientU = typename LocalBasisU::Traits::JacobianType;
81        using GradientV = typename LocalBasisV::Traits::JacobianType;
```

It then constructs arrays for the trial and test function values, and the trial and test function gradients both locally (i.e., on the reference element) and transformed to the grid element T:

```
85        std::vector<RangeU> phi(localFunctionSpaceU.size());
86        std::vector<RangeV> theta(localFunctionSpaceV.size());
87
88        std::vector<GradientU> localGradPhi(localFunctionSpaceU.size());
89        std::vector<GradientV> localGradTheta(localFunctionSpaceV.size());
90
91        std::vector<GradientU> gradPhi(localFunctionSpaceU.size());
92        std::vector<GradientV> gradTheta(localFunctionSpaceV.size());
```

This example is meticulous about properly keeping trial functions $\{\phi_i\}$ and test functions $\{\theta_i\}$ separate. The assembler will therefore assemble the correct matrices for the case that the trial and test bases differ. For the common case that the two bases are known to be identical, this makes the code longer and slower than necessary. We nevertheless write all examples of this chapter in this way, because it better matches the mathematics behind the implementations and the dune-pdelab interfaces. Real implementation will want to check for identical bases, and handle that case separately.

Next is the beginning of the quadrature loop. The quadrature rule is provided by the dune-geometry module as explained in Chapter 9:

```
97    int intOrder
98       = (localFunctionSpaceU.finiteElement().localBasis().order()-1) * (p_-1)
99       + (localFunctionSpaceV.finiteElement().localBasis().order()-1);
100   const auto& quadRule
101      = QuadratureRules<double,dim>::rule(entityGeometry.entity().type(),
102                                           intOrder);
103
104   for (const auto& quadPoint : quadRule)
105     {
```

The integration order is computed by combining the orders of the trial and test shape functions, in such a way as to obtain a suitable order for residual expression (11.13).

The code then evaluates the values of the ansatz and test shape functions ϕ_i and θ_i, $i = 0, \ldots n_T - 1$ at the current quadrature point. It also evaluates the corresponding gradients. These are then transformed from reference element coordinates to world coordinates, by left multiplication with ∇F^{-T}, where F is the Jacobian of the map $F : T_{\mathrm{ref}} \to T$ that maps the reference element T_{ref} to the current grid element T. See Chapter 2.1.3 for details:

```
109       // Evaluate basis functions on reference element
110       localFunctionSpaceU.finiteElement().localBasis()
111                         .evaluateFunction(quadPoint.position(),phi);
112       localFunctionSpaceV.finiteElement().localBasis()
113                         .evaluateFunction(quadPoint.position(),theta);
114
115       // Evaluate gradient of basis functions on reference element
116       localFunctionSpaceU.finiteElement().localBasis()
117                         .evaluateJacobian(quadPoint.position(),localGradPhi);
118       localFunctionSpaceV.finiteElement().localBasis()
119                         .evaluateJacobian(quadPoint.position(),localGradTheta);
120
121       // Transform gradients from reference element to grid element
122       const auto jacInvTransp
123          = entityGeometry.geometry()
124                         .jacobianInverseTransposed(quadPoint.position());
125
126       for (std::size_t i=0; i<localFunctionSpaceU.size(); i++)
127         jacInvTransp.mv(localGradPhi[i][0], gradPhi[i][0]);
128
129       for (std::size_t i=0; i<localFunctionSpaceV.size(); i++)
130         jacInvTransp.mv(localGradTheta[i][0], gradTheta[i][0]);
```

Recall from Chapter 8.2.1 that Jacobians of shape functions are typically matrices, with only one row if the shape functions are scalar-valued. In the transformation code above, these rows have to be treated as vectors, which explains the extra [0] indexing.

The following lines then compute the values of u_h and ∇u_h at the current quadrature point, by linear combination of the coefficients of u_h given in the method parameter x and the shape function values and derivatives, respectively:

```
134    // Compute u at integration point
135    RangeU u = 0;
136    for (std::size_t i=0; i<localFunctionSpaceU.size(); ++i)
137      u += x(localFunctionSpaceU,i)*phi[i];
138
139    // Compute gradient of u
140    GradientU gradU = 0;
141    for (std::size_t i=0; i<localFunctionSpaceU.size(); ++i)
142      gradU.axpy(x(localFunctionSpaceU,i),gradPhi[i]);
```

Here it can be seen that x is not a standard container. Querying its i-th entry requires handing over the local shape function set together with the number i. This additional layer between the LocalOperator code and the container implementation is needed for finite element bases that form a tree structure, as introduced in Chapter 10.1. Note that the variable gradU is again of a matrix type, with only a single row. Code that uses gradU as a vector will access that row using an extra [0] index.

Finally, a short loop computes the actual residual entry contribution from the current quadrature point. The code actually implements $(|\nabla u_h|^2)^{\frac{1}{2}(p-2)}$ rather than $|\nabla u_h|^{p-2}$, to avoid the costly square root in the expression for $|\nabla u_h|$:

```
146    auto globalPos = entityGeometry.geometry().global(quadPoint.position());
147
148    // Integrate |∇u_h|^(p-2)⟨∇u_h, ∇θ_i⟩ + cu_hθ_i − fθ_i
149    auto factor = quadPoint.weight()
150                    * entityGeometry.geometry()
151                              .integrationElement(quadPoint.position());
152
153    for (std::size_t i=0; i<localFunctionSpaceV.size(); ++i)
154    {
155      auto value = (std::pow(gradU[0].two_norm2(), 0.5*(p_-2))
156                                    * (gradU[0]*gradTheta[i][0])
157                + c_(globalPos)*u*theta[i]
158                - f_(globalPos)*theta[i]) * factor;
159      residual.accumulate(localFunctionSpaceV, i, value);
160    }
```

Similar to the input coefficient container x, the output container residual is not a plain array. To add a number called value to the i-th entry, the call to accumulate in Line 159 is used. This extra indirection allows to support finite element bases that form trees (Chapter 10.1). This concludes the discussion of the alpha_volume method.

The `jacobian_volume` method implements the tangent matrix $\nabla \mathcal{R}_T(\mathbf{R}_T \bar{u})$, with entries

$$
\begin{aligned}
(\nabla \mathcal{R}_T(\mathbf{R}_T \bar{u}))_{i,j} &= \frac{\partial(\mathcal{R}_T(\mathbf{R}_T \bar{u}))_i}{\partial(\mathbf{R}_T \bar{u})_j} \\
&= \int_T \left[\frac{p-2}{2} \left(|\nabla u_h|^2\right)^{\frac{p-2}{2}-1} \cdot \frac{\partial}{\partial(\mathbf{R}_T \bar{u})_j} |\nabla u_h|^2 \cdot \langle \nabla u_h, \nabla \theta_i \rangle \right. \\
&\qquad\qquad\qquad\qquad \left. + |\nabla u_h|^{p-2} \cdot \langle \nabla \phi_j, \nabla \theta_i \rangle \right] dx \\
&\quad + \int_T c\theta_i \phi_j \, dx \\
&= \int_T \left[(p-2)|\nabla u_h|^{p-4} \cdot \langle \nabla u_h, \nabla \phi_j \rangle \cdot \langle \nabla u_h, \nabla \theta_i \rangle \right. \\
&\qquad\qquad\qquad\qquad \left. + |\nabla u_h|^{p-2} \cdot \langle \nabla \phi_j, \nabla \theta_i \rangle \right] dx \\
&\quad + \int_T c\theta_i \phi_j \, dx, \tag{11.14}
\end{aligned}
$$

where we have used that

$$
\frac{\partial}{\partial(\mathbf{R}_T \bar{u})_j} |\nabla u_h|^2 = 2 \sum_{l=0}^{d-1} \sum_{k=0}^{n_T-1} \bar{u}_k \frac{\partial \phi_k}{\partial x_l} \frac{\partial \phi_j}{\partial x_l} = 2 \sum_{l=0}^{d-1} \frac{\partial u_h}{\partial x_l} \frac{\partial \phi_j}{\partial x_l} = 2\langle \nabla u_h, \nabla \phi_j \rangle.
$$

The method has the signature

```
167     template<class EntityGeometry,
168             class LocalFunctionSpaceU, class Vector,
169             class LocalFunctionSpaceV, class Jacobian>
170     void jacobian_volume(const EntityGeometry& entityGeometry,
171                          const LocalFunctionSpaceU& localFunctionSpaceU,
172                          const Vector& x,
173                          const LocalFunctionSpaceV& localFunctionSpaceV,
174                          Jacobian& jacobian) const
175     {
```

Its code is very similar to `alpha_volume`. The initial part again sets up various types and objects:

```
177     using TrialFE = typename LocalFunctionSpaceU::Traits::FiniteElementType;
178     using TestFE  = typename LocalFunctionSpaceU::Traits::FiniteElementType;
179     using LocalBasisU = typename TrialFE::Traits::LocalBasisType;
180     using LocalBasisV = typename TestFE::Traits::LocalBasisType;
181     using RangeU = typename LocalBasisU::Traits::RangeType;
182     using RangeV = typename LocalBasisV::Traits::RangeType;
183     using GradientU = typename LocalBasisU::Traits::JacobianType;
184     using GradientV = typename LocalBasisV::Traits::JacobianType;
```

```
185
186        std::vector<RangeU> phi(localFunctionSpaceU.size());
187        std::vector<RangeV> theta(localFunctionSpaceV.size());
188
189        std::vector<GradientU> localGradPhi(localFunctionSpaceU.size());
190        std::vector<GradientV> localGradTheta(localFunctionSpaceV.size());
191
192        std::vector<GradientU> gradPhi(localFunctionSpaceU.size());
193        std::vector<GradientV> gradTheta(localFunctionSpaceV.size());
```

The following quadrature loop then evaluates the shape functions of the trial and test spaces. It also evaluates their gradients, and transforms them from reference element coordinates to world coordinates. For simplicity we use the same quadrature order as for the residual itself, even though this may not be optimal:

```
198        int intOrder
199            = (localFunctionSpaceU.finiteElement().localBasis().order()-1)
200            * (p_-1)
201            * (localFunctionSpaceV.finiteElement().localBasis().order()-1);
202        const auto& quadRule
203            = QuadratureRules<double,dim>::rule(entityGeometry.entity().type(),
204                                                intOrder);
205
206        for (const auto& quadPoint : quadRule)
207        {
208          // Evaluate shape functions on reference element
209          localFunctionSpaceU.finiteElement().localBasis()
210                            .evaluateFunction(quadPoint.position(),phi);
211          localFunctionSpaceV.finiteElement().localBasis()
212                            .evaluateFunction(quadPoint.position(),theta);
213
214          // Evaluate gradients of shape functions on reference element
215          localFunctionSpaceU.finiteElement().localBasis()
216                            .evaluateJacobian(quadPoint.position(),localGradPhi);
217          localFunctionSpaceV.finiteElement().localBasis()
218                            .evaluateJacobian(quadPoint.position(),localGradTheta);
219
220          // Transform gradients from reference element to grid element
221          const auto jacInvTransp
222            = entityGeometry.geometry()
223                            .jacobianInverseTransposed(quadPoint.position());
224
225          for (std::size_t i=0; i<localFunctionSpaceU.size(); i++)
226            jacInvTransp.mv(localGradPhi[i][0],gradPhi[i][0]);
227
228          for (std::size_t i=0; i<localFunctionSpaceV.size(); i++)
229            jacInvTransp.mv(localGradTheta[i][0],gradTheta[i][0]);
```

This is all very similar to the code for computing the residual itself. The code then evaluates ∇u_h at the quadrature point. Unlike for the residual itself, the point value of u_h is not required here:

```
233        // Compute gradient of u
234        GradientU gradU(0.0);
235        for (std::size_t i=0; i<localFunctionSpaceU.size(); ++i)
236          gradU.axpy(x(localFunctionSpaceU,i),gradPhi[i]);
```

The final loop then computes the expression (11.14), and adds it to the element stiffness matrix stored in the object mat:

```
240        auto factor = quadPoint.weight()
241                        * entityGeometry.geometry()
242                                    .integrationElement(quadPoint.position());
243        auto globalPos = entityGeometry.geometry().global(quadPoint.position());
244
245        // Integrate
246        for (std::size_t i=0; i<localFunctionSpaceV.size(); i++)
247          for (std::size_t j=0; j<localFunctionSpaceU.size(); j++)
248          {
249            auto value = (p_-2) * std::pow(gradU[0].two_norm2(), 0.5*p_-2)
250                * (gradU[0] * gradPhi[j][0]) * (gradU[0] * gradTheta[i][0]);
251            value += std::pow(gradU[0].two_norm2(), 0.5*p_-1)
252                            * (gradTheta[i][0] * gradPhi[j][0]);
253            value += c_(globalPos) * theta[i] * phi[j];
254            jacobian.accumulate(localFunctionSpaceV, i,
255                                localFunctionSpaceU, j,
256                                value*factor);
257          }
```

Note the use of the special `accumulate` method to update the matrix entries. Like the residual vector `residual` in the `alpha_volume` method, the matrix object `jacobian` is not an actual matrix container. The call to `accumulate` adds the value given as the last method argument to the matrix entry (i, j). The local function space objects have to be provided to control the degree-of-freedom numbering for finite element spaces that form trees.

Running the Program

When the program is run, it produces the following output:

```
  Initial defect:   2.9898e+00
=== Dune::CGSolver
  Iter        Defect            Rate
     0       2.9898e+00
     1       8.2228e-03       2.7503e-03
     2       3.7030e-04       4.5033e-02
     3       4.3728e-06       1.1809e-02
```

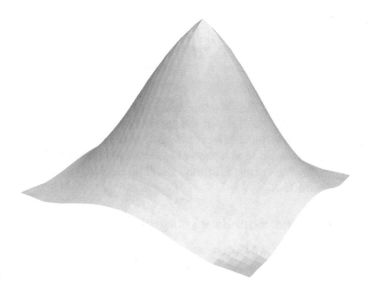

Figure 11.3.: Result of the p-Laplace example, visualized as a height field

```
=== rate=1.1351e-02, T=8.1717e-03, TIT=2.7239e-03, IT=3
  Newton iteration  1.  New defect:   2.7633e+00.  \
      Reduction (this):   9.2424e-01.  Reduction (total):   9.2424e-01
=== Dune::CGSolver
  Iter        Defect          Rate
    0       2.7633e+00
    1       1.7670e+00      6.3943e-01
    2       9.6523e-01      5.4620e-01
    3       1.3731e-01      1.4226e-01
    4       3.2171e-02      2.3429e-01
    5       1.1449e-02      3.5590e-01
    6       6.0409e-03      5.2761e-01
    7       1.1805e-03      1.9542e-01
    8       1.6940e-04      1.4349e-01
=== rate=2.9746e-01, T=2.0751e-02, TIT=2.5938e-03, IT=8
  Newton iteration  2.  New defect:   2.3743e+00.  \
      Reduction (this):   8.5923e-01.  Reduction (total):   7.9413e-01

[...]

=== Dune::CGSolver
  Iter        Defect          Rate
    0       3.1771e-10
    1       5.6390e-11      1.7749e-01
```

```
    2       3.8550e-11       6.8363e-01
    3       1.8346e-11       4.7592e-01
=== rate=3.8652e-01, T=8.0403e-03, TIT=2.6801e-03, IT=3
  Newton iteration 11.  New defect:   1.8346e-11.  \
        Reduction (this):   5.7745e-02.  Reduction (total):   6.1362e-12
```

One can see 11 iterations of the Newton solver, of which we have printed only the first and last few. Each iteration shows the current defect (i.e., the norm of $\mathcal{R}$ at the current iterate), and the reduction factor of that norm with respect to the previous iterate and the initial iterate. It also shows the iteration history of the CG solver used to solve the linear Newton correction problems. The resulting finite element function is visualized in Figure 11.3.

11.2.5. Skeleton and Boundary Integrals

The example problems that we have looked at so far in this chapter have only involved integrals over the grid elements. However, the dune-pdelab design also foresees integrals over element boundaries in its separability assumption 11.2. There, it is supposed that the residual can be written as three sums

$$r_h(u_h, v_h) = \sum_{T \in \mathcal{T}} r_{h,T}^{\mathrm{vol}}(u_h, v_h) + \sum_{\gamma^i \in \mathcal{F}} r_{h,\gamma^i}^{\mathrm{skel}}(u_h, v_h) + \sum_{\gamma^b \in \mathcal{B}} r_{h,\gamma^b}^{\mathrm{bnd}}(u_h, v_h),$$

$$u_h \in U_h, \quad v_h \in V_h,$$

where $r_{h,T}^{\mathrm{vol}}$ involves integral terms over elements T, $r_{h,\gamma^i}^{\mathrm{skel}}$ involves integrals over intersections γ^i between two adjacent elements, and $r_{h,\gamma^b}^{\mathrm{bnd}}$ represents domain boundary terms. Skeleton terms are required, for example, for nonconforming discretizations like the Discontinuous Galerkin (DG) method [4, 55, 94]. Neumann and Robin boundary conditions require integrals over the domain boundary.

While the previous sections have only discussed and used the volume terms $r_{h,T}^{\mathrm{vol}}$, it is now time for the other ones. The LocalOperator class, which defines the interface for assemblers of the local residual terms, has member methods for the skeleton and boundary contributions $r_{h,\gamma^i}^{\mathrm{skel}}$ and $r_{h,\gamma^b}^{\mathrm{bnd}}$, and their Jacobians. We begin by discussing the methods related to domain boundary terms, because their interfaces are a bit simpler than the ones related to skeleton integrals. The first method is alpha_boundary, which computes the contributions to the residual from grid boundary element facets. Its interface is

```
template<class IntersectionGeometry, class LocalFunctionSpaceU, class Vector,
        class LocalFunctionSpaceV, class Residual>
void alpha_boundary(const IntersectionGeometry& intersectionGeometry,
                    const LocalFunctionSpaceU& localFunctionSpaceUIn,
                    const Vector& xIn,
                    const LocalFunctionSpaceV& localFunctionSpaceVIn,
                    Residual& residualIn) const
```

The GridOperator will call this method once for each boundary intersection. The method must add $\mathcal{R}^{\text{bnd}}_{\gamma^{\text{b}}}$, the algebraic equivalent of $r^{\text{bnd}}_{h,\gamma^{\text{b}}}$ to the local residual vector residualIn given as the last parameter.

The method looks very similar to the alpha_volume method discussed in Chapter 11.2.3. It has five parameters, all of which have free types that are determined by corresponding member template parameters. Unlike alpha_volume, which received the element it was supposed to operate on, the alpha_boundary method receives a boundary intersection γ^{b} as its first parameter intersectionGeometry, which represents the domain of the integral term (see Chapter 5.4 for details on intersections). Much like the entityGeometry parameter of the alpha_volume method, intersectionGeometry objects are only wrappers around Intersection objects in the sense of the dune-grid interface.[7] The actual intersections can be retrieved by calling the method

```
const Intersection& intersection() const
```

of the intersectionGeometry object.

The parameters localFunctionSpaceUIn and localFunctionSpaceVIn are the local finite elements to be used for the trial and test spaces. They are defined on the element T_{in} that borders the current boundary intersection (which dune-grid calls the *inside* element). The alpha_boundary implementation has to compute the restriction of the shape functions to the actual boundary itself. LocalFiniteElement objects in the sense of the dune-localfunctions module can be retrieved from these objects by calling the finiteElement method, as could already be seen in the previous example.

The container xIn stores the local coefficients of the current configuration on the element T_{in}. Accessing individual entries requires the trial function space localFunctionSpaceUIn together with the index, as in the alpha_volume method. Likewise, residualIn is where the method writes the resulting residual. Their types are the same custom types as in the alpha_volume method. In particular, this means that values must be added to residualIn using its method accumulate.

To obtain the Jacobian of the boundary terms of the residual with respect to the coefficients of u_h, there is the method

```
template<class IntersectionGeometry,
         class LocalFunctionSpaceU, class Vector,
         class LocalFunctionSpaceV, class Jacobian>
void jacobian_boundary(const IntersectionGeometry& intersectionGeometry,
                       const LocalFunctionSpaceU& localFunctionSpaceUIn,
                       const Vector& xIn,
                       const LocalFunctionSpaceV& localFunctionSpaceVIn,
                       Jacobian& jacobianIn) const
```

It computes the matrix

$$\nabla \mathcal{R}^{\text{bnd}}_{\gamma^{\text{b}}}(u_h) \in \mathbb{R}^{n_1 \times n_2},$$

---

[7] ... for reasons beyond the scope of this text.

where n_1 is the number of degrees of freedom of the test function space on the element, and n_2 is the number of trial functions on the element. Entry (i, j) of the matrix will contain the number

$$\frac{\partial (\mathcal{R}^{\mathrm{bnd}}_{\gamma^{\mathrm{b}}})_i}{\partial u_j}.$$

The parameters of `jacobian_boundary` are the same as for the `alpha_boundary` method. The only difference is the fifth parameter, which receives the result. While the `alpha_boundary` method has a vector data type here, the `jacobian_boundary` method expects a matrix. The syntactic requirements of that matrix type are the same as for the `jacobian_volume` method discussed in Chapter 11.2.3. An example implementation of the method is given in the following section.

As for the `volume` methods, the `GridOperator` class does not automatically recognize the method `alpha_boundary` in a `LocalOperator` implementation. If the `LocalOperator` implementation has a method `alpha_boundary`, it has to set

```
constexpr static bool doAlphaBoundary = true;
```

as a static member constant of the `LocalOperator` implementation class. Otherwise the `alpha_boundary` method will not be called by the `GridOperator` object.

A second method of the `LocalOperator` interface allows to compute the residual on the intersections γ^{i} of the skeleton of the grid. If the grid is conforming, then the skeleton naturally decomposes into element facets, but nonconforming grids can be handled as well.

Residual terms on the skeleton are assembled by the method

```
template<class IntersectionGeometry,
         class LocalFunctionSpaceU, class Vector,
         class LocalFunctionSpaceV, class Residual>
void alpha_skeleton(const IntersectionGeometry& intersectionGeometry,
                    const LocalFunctionSpaceU& localFunctionSpaceUIn,
                    const Vector& xIn,
                    const LocalFunctionSpaceV& localFunctionSpaceVIn,
                    const LocalFunctionSpaceU& localFunctionSpaceUOut,
                    const Vector& xOut,
                    const LocalFunctionSpaceV& localFunctionSpaceVOut,
                    Residual& residualIn, Residual& residualOut) const
```

If `doAlphaSkeleton` is set by the `LocalOperator` implementation, it is called by the `GridOperator` for each intersection that is *not* part of the domain boundary.[8] The signature is easy to understand when having understood the `alpha_boundary` method. The current intersection is again handed to the method in form of the intersection geometry object `intersectionGeometry`. Then, there are parameters for the shape function set, the current configuration, and the resulting residual. However, unlike the `alpha_boundary` method, which involves only one element, intersections in

---

[8]The orientation in which the intersection is visited is unspecified, except that the *inside* entity will always be an ***interior*** entity.

the skeleton are shared by *two* grid elements. Therefore, alpha_skeleton has two parameters for each object. They are distinguished by the suffix In (for the *inside* element) and Out (for the *outside* element). For example, localFunctionSpaceUIn contains the ansatz local finite element space of the *inside* element of the intersection, and xOut contains the local configuration of the outside element.

To compute the derivatives of the skeleton residual terms, the LocalOperator has the jacobian_skeleton method:

```
template<class IntersectionGeometry,
        class LocalFunctionSpaceU, class Vector,
        class LocalFunctionSpaceV, class Jacobian>
void jacobian_skeleton(const IntersectionGeometry& intersectionGeometry,
                const LocalFunctionSpaceU& localFunctionSpaceUIn,
                const Vector& xIn,
                const LocalFunctionSpaceV& localFunctionSpaceVIn,
                const LocalFunctionSpaceU& localFunctionSpaceUOut,
                const Vector& xOut,
                const LocalFunctionSpaceV& localFunctionSpaceVOut,
                Jacobian& jacobianInIn, Jacobian& jacobianInOut,
                Jacobian& jacobianOutIn, Jacobian& jacobianOutOut)
                                                              const
```

Its seven input parameters are the same as for the alpha_skeleton method. However, it has *four* output matrix parameters instead of the two vector ones of alpha_skeleton. These four matrices are the blocks of a 2×2 block matrix

$$\begin{pmatrix} \texttt{jacobianInIn} & \texttt{jacobianInOut} \\ \texttt{jacobianOutIn} & \texttt{jacobianOutOut} \end{pmatrix} = \begin{pmatrix} \left.\frac{\partial}{\partial u_{\text{in}}} r^{\text{skel}}\right|_{\text{in}} & \left.\frac{\partial}{\partial u_{\text{out}}} r^{\text{skel}}\right|_{\text{in}} \\ \left.\frac{\partial}{\partial u_{\text{in}}} r^{\text{skel}}\right|_{\text{out}} & \left.\frac{\partial}{\partial u_{\text{out}}} r^{\text{skel}}\right|_{\text{out}} \end{pmatrix}.$$

For example, jacobianInOut contains the derivatives of the algebraic residual entries on the *inside* element with respect to the coefficients of the local configuration on the *outside* element. As usual, the local residual and configuration vectors contain the entries of all degrees of freedom on the element, not just those associated to the common intersection.

11.2.6. Example: Discontinuous Galerkin Methods

We conclude the chapter on local operators with a complete example that combines volume, skeleton, and boundary terms. For this we revisit the linear reaction–diffusion example of Chapter 11.1. The strong form of the boundary value problem there was

$$-\Delta u + cu = f \qquad \text{in } \Omega = [0,1]^2,$$

with constant reaction strength $c \equiv 10$, source function

$$f = \begin{cases} -10 & \text{on a circle of radius } 1/4 \text{ around the center of } \Omega \\ 10 & \text{elsewhere,} \end{cases}$$

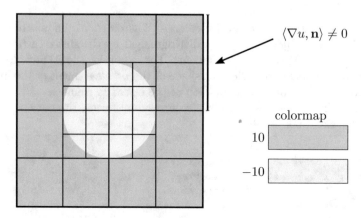

Figure 11.4.: Problem setting and grid of the DG example

and it was posed with zero Neumann boundary conditions. We keep the equation and the domain, but now we modify the Neumann boundary conditions to be

$$\langle \nabla u, \mathbf{n} \rangle = \mathrm{j} := \begin{cases} -2 & \text{if } x_1 = 1 \text{ and } x_2 \in \left[\frac{1}{2}, 1 \right], \\ 0 & \text{otherwise.} \end{cases}$$

The problem setting is illustrated in Figure 11.4.

The residual formulation of the boundary value problem is

$$\text{Find } u \in H^1(\Omega) : \quad r(u, v) = 0 \quad \text{for all } v \in H^1(\Omega),$$

with the residual form

$$r(u, v) := \int_\Omega \left[\langle \nabla u, \nabla v \rangle + cuv - fv \right] dx - \int_{\partial\Omega} \mathrm{j} v \, dx.$$

Unlike the previous examples, this form contains integrals over the domain boundary $\partial\Omega$ in addition to the volume integrals.

We will discretize the problem using the *Symmetric Interior Penalty Galerkin (SIPG)* method, a particular Discontinuous Galerkin method [4, 58]. In the literature, this method is also known simply as *Interior Penalty (IP)*. Let $\mathcal{T}$ be a grid of Ω, and define the nonconforming finite element space

$$\mathrm{DG}^k := \left\{ u \in L_2(\Omega) : u|_T \in \Pi_k \; \forall T \in \mathcal{T} \right\}.$$

The functions in this space are not necessarily continuous at the element boundaries, and we interpret them to be multi-valued there. Let γ be an intersection between two adjacent elements T_{in} and T_{out}. For a function $u_h \in DG^k$ we define u_h^{in} to be the restriction onto T_{in}, and u_h^{out} to be the restriction onto T_{out}. The jump across γ is

$$[\![u_h]\!]_\gamma(x) := u_h^{in}(x) - u_h^{out}(x) \qquad \forall x \in \gamma.$$

Likewise, we define the average as

$$\{u_h\}_\gamma(x) := \frac{1}{2}(u_h^{in}(x) + u_h^{out}(x)) \qquad \forall x \in \gamma.$$

Note that the sign of the jump $[\![u_h]\!]_\gamma$ depends on the orientation of γ, but the sign of the average does not. With these definitions, the discrete problem for the SIPG method reads

$$\text{Find } u_h \in DG^k : \quad r_h^{SIPG}(u_h, v_h) = 0 \qquad \forall v_h \in DG^k, \tag{11.15}$$

where

$$r_h^{SIPG}(u_h, v_h) := \int_\Omega \nabla u_h \nabla v_h \, dx + \int_\Omega c u_h v_h \, dx \tag{11.16}$$

$$- \int_\mathcal{F} \{\langle \nabla u_h, \mathbf{n}\rangle\}[\![v_h]\!] \, ds - \int_\mathcal{F} \{\langle \nabla v_h, \mathbf{n}\rangle\}[\![u_h]\!] \, ds$$

$$+ \kappa \sum_{\gamma \in \mathcal{F}} \frac{1}{h_\gamma} \int_\gamma [\![u_h]\!][\![v_h]\!] \, ds$$

$$- \int_\Omega f v_h \, dx - \int_{\partial\Omega} j v_h \, ds,$$

and $\mathcal{F}$ is the union of all skeleton intersections. Note the separation into volume, skeleton and boundary terms. The volume terms of this residual are identical to the ones of the conforming discretization of Chapter 11.1. The terms in the second row of the definition of $r_h^{SIPG}(\cdot, \cdot)$ penalize the jumps $[\![u_h]\!]$ at the inter-element boundaries. The third-row term makes sure that the quadratic part of $r_h^{SIPG}(\cdot, \cdot)$ remains elliptic, for a penalization parameter κ large enough. The final row contains the u_h-independent parts of the residual form. The volume integral over f is the source term, whereas the integral over the domain boundary $\partial\Omega$ is the weak form of the Neumann boundary condition $\langle \nabla u, \mathbf{n}\rangle = j$.

Provided that the penalty parameter κ is large enough, the residual equation (11.15) has a unique solution [4, 55]. This solution converges with optimal orders to the exact solution as the grid is refined [4, 55].

The `main` Method

As usual the example program is contained in a single file. It is printed in Appendix B.13, and it is available through the icon in the margin. The file contains the `main` method, which sets up the problem and calls the algebraic solver, and the implementation of a local operator that computes the SIPG residual form (11.16), and its Jacobian.

We discuss both parts in turn. The `main` method is very similar to the one of the reaction–diffusion system with Lagrange elements of Chapter 11.1. It begins by initializing MPI, and then setting up the grid:

```
680  int main(int argc, char *argv[])
681  {
682    // Initialize MPI, if present
683    MPIHelper::instance(argc, argv);
684
685    constexpr int dim = 2;
686    using Grid = UGGrid<dim>;
687    std::shared_ptr<Grid> grid
688      = StructuredGridFactory<Grid>::createCubeGrid({0.0,0.0},
689                                                    {1.0,1.0},
690                                                    {  4,  4});
691
692    // Nonconforming refinement near the domain center
693    grid->setClosureType(Grid::NONE);   // UGGrid-specific:
694                                        // turn off green closure
695
696    using Position = FieldVector<Grid::ctype,dim>;
697    for (const auto& element : elements(grid->leafGridView()))
698      if ( (element.geometry().center()-Position(0.5)).two_norm() < 0.25 )
699        grid->mark(1,element);
700
701    grid->preAdapt();
702    grid->adapt();
703    grid->postAdapt();
704
705    using GridView = Grid::LeafGridView;
706    const GridView gridView = grid->leafGridView();
```

Unlike in Chapter 11.1, the code does not use the `YaspGrid` grid implementation. Rather, to show how DG methods naturally handle non-conforming grids, the code uses the `UGGrid` implementation with some adaptive red refinement, as shown in Figure 11.4. The calls to `preAdapt` and `postAdapt` are mandated by the DUNE grid interface, even though no data is transferred here.

The next step is to set up the grid function space. The code uses the function space basis `LagrangeDGBasis` from the `dune-functions` module. It implements the standard discontinuous Lagrange basis for the space DG^k with a compile-time order k. The code sets this order to $k = 2$. The basis object is then used to set up a `GridFunctionSpace` object, which is the by now well-known `dune-pdelab` abstraction of a finite element space:

```
710    // Make dune-functions basis
711    using DGBasis = Functions::LagrangeDGBasis<GridView,2>;
712    auto dgBasis = std::make_shared<DGBasis>(gridView);
713
```

434

```
714    // Make dune-pdelab grid function space
715    using VectorBackend = PDELab::ISTL::VectorBackend<>;
716    using GridFunctionSpace
717        = PDELab::Experimental::GridFunctionSpace<DGBasis,
718                                                  VectorBackend,
719                                                  PDELab::NoConstraints>;
720
721    GridFunctionSpace gridFunctionSpace(dgBasis);
```

Then, the code sets up the assembler. First, the three parameter functions c, f, and j are defined. For simplicity, the code implements all three as small lambda objects that are later handed to the LocalOperator implementation:

```
725    // Reaction strength
726    auto c = [](const FieldVector<double,dim>& x)
727    {
728      return 10.0;
729    };
730
731    // Source term
732    auto f = [](const FieldVector<double,dim>& x)
733    {
734      const auto center = FieldVector<double,dim>(0.5);
735      auto distanceToCenter = (x - center).two_norm();
736      return (distanceToCenter <= 0.25) ? -10.0 : 10.0;
737    };
738
739    // Neumann boundary data
740    auto neumann = [](const FieldVector<double,dim>& x)
741    {
742      return (x[0] >= 0.999 && x[1] >= 0.5) ? -2 : 0;
743    };
```

Then the actual assembler is set up. The LocalOperator implementation is called ReactionDiffusionDG, and it is defined earlier in the example file. It is given the three parameter functions c, f, neumann, and the coercivity parameter kappa. Together with the function space bases and a matrix backend, the ReactionDiffusionDG local operator is then used to set up the GridOperator object:

```
747    // Make local operator
748    using LocalOperator = ReactionDiffusionDG<GridView>;
749    double kappa = 3;
750    LocalOperator localOperator(c,f,neumann,kappa);
751    using MatrixBackend = PDELab::ISTL::BCRSMatrixBackend<>;
752    MatrixBackend matrixBackend(45);      // 45: Expected approximate number
753                                          // of nonzeros per matrix row
754
755    using GridOperator
756        = PDELab::GridOperator<GridFunctionSpace,    // Trial function space
```

```
757                                GridFunctionSpace,   // Test function space
758                                LocalOperator,       // Element assembler
759                                MatrixBackend,       // Data structure
760                                                     // for the stiffness matrix
761                                double,              // Number type for
762                                                     // solution vector entries
763                                double,              // Number type for
764                                                     // residual vector entries
765                                double>;             // Number type for
766                                                     // stiffness matrix entries
767
768    GridOperator gridOperator(gridFunctionSpace,
769                              gridFunctionSpace,
770                              localOperator,
771                              matrixBackend);
```

The value $\kappa = 3$ has been chosen by experimentation. The GridOperator class is set up just like in previous examples.

At this point the assembler is ready, and the algebraic solver can be constructed and run:

```
775    // Make a vector of degrees of freedom, and initialize it with 0
776    using VectorContainer = typename GridOperator::Traits::Domain;
777    VectorContainer u(gridFunctionSpace,0.0);
778
779    // Make linear solver and solve problem
780    using LinearSolverBackend = PDELab::ISTLBackend_SEQ_CG_ILU0;
781    LinearSolverBackend linearSolverBackend(50,     // Maximum number
782                                                    // of iterations
783                                            2);     // Solver verbosity
784
785    using LinearProblemSolver
786        = PDELab::StationaryLinearProblemSolver<GridOperator,
787                                                LinearSolverBackend,
788                                                VectorContainer>;
789
790    LinearProblemSolver linearProblemSolver(gridOperator,
791                                            linearSolverBackend,
792                                            u,
793                                            1e-10);  // Desired relative
794                                                     // residual reduction
795    linearProblemSolver.apply();
```

As the final step, the result is written to a VTK file. There is no difference here to earlier examples:

```
799    // Make a discrete function from the FE basis and the coefficient vector
800    auto uFunction
801        = Functions::makeDiscreteGlobalBasisFunction<double>(
802                                           *dgBasis,
803                                           PDELab::Backend::native(u));
```

```
804
805    // We need to subsample, because VTKWriter
806    // cannot natively display second-order functions.
807    SubsamplingVTKWriter<GridView> vtkWriter(gridView,
808                                    refinementLevels(5));
809    vtkWriter.addVertexData(uFunction,
810                       VTK::FieldInfo("u",
811                                 VTK::FieldInfo::Type::scalar,
812                                 1));
813    vtkWriter.write("pdelab-dg-diffusion-result");
```

Remember that SubsamplingVTKWriter always uses the VTK::nonconforming data mode (Chapters 5.8.1 and 10.7), and it therefore captures the jumps of the DG functions automatically.

The Local Operator

The LocalOperator implementation ReactionDiffusionDG forms the central part of this example. It assembles the discrete SIPG residual (11.15) and its Jacobian matrix. As this DG discretization is more involved than a pure Lagrange discretization, the LocalOperator implementation is more complex, too. In total, the ReactionDiffusionDG class implements six methods: alpha_volume and jacobian_volume for the diffusion, reaction, and source terms, alpha_skeleton and jacobian_skeleton for the additional terms of the DG discretization, and alpha_boundary and jacobian_boundary for the Neumann boundary conditions.

The volume methods are straightforward to understand, once the p-Laplace example of Chapter 11.2.4 has been studied. They are therefore omitted from the presentation here (but they are shown in Appendix B.13). Only the latter four methods will be discussed.

The ReactionDiffusionDG class has the following signature:

```
21    template<class GridView>
22    class ReactionDiffusionDG
23    : public PDELab::LocalOperatorDefaultFlags,
24      public Dune::PDELab::FullVolumePattern,
25      public Dune::PDELab::FullSkeletonPattern,
26      public Dune::PDELab::FullBoundaryPattern
27    {
```

The base classes inject code that set up the matrix sparsity pattern for the three contributions. The class then contains a short helper method that computes the diameter of a geometry:

```
31    template<class Geometry>
32    static auto diameter (const Geometry& geometry)
33    {
34      typename Geometry::ctype h = 0.0;
35      for (int i=0; i<geometry.corners(); i++)
```

```
36        for (int j=i+1; j<geometry.corners(); j++)
37          h = std::max(h, (geometry.corner(j)-geometry.corner(i)).two_norm());
38
39      return h;
40    }
```

It is used below to compute the diameter h_γ of grid intersections.

The class then declares four data member variables. Besides the coercivity parameter κ, these are the three parameter functions c, f, and j. They are stored in objects of type std::function, to limit the number of template parameters of the class:

```
44    double kappa_;
45
46    static constexpr int dim = GridView::dimension;
47
48    std::function<double(const FieldVector<double,dim>&)> c_;
49    std::function<double(const FieldVector<double,dim>&)> f_;
50    std::function<double(const FieldVector<double,dim>&)> neumann_;
```

Afterwards, the LocalOperator implementation declares the set of interface methods it implements. By setting the flags doAlphaVolume, doAlphaSkeleton, and doAlphaBoundary to **true**, the class states that it will implement alpha and jacobian methods for volume, skeleton, and boundary terms. The doPattern flags will make the GridOperator object initialize the Jacobian matrix with the default pattern for volume, skeleton, and boundary terms:

```
55    // Residual assembly flags
56    enum { doAlphaVolume   = true };
57    enum { doAlphaSkeleton  = true };
58    enum { doAlphaBoundary  = true };
59
60    // Pattern assembly flags
61    enum { doPatternVolume = true };
62    enum { doPatternSkeleton = true };
63    enum { doPatternBoundary = true };
```

Next in the class is the constructor. It simply accepts the problem parameters and stores them in the class data members:

```
67    ReactionDiffusionDG(std::function<double(const FieldVector<double,dim>&)> c,
68                        std::function<double(const FieldVector<double,dim>&)> f,
69                        std::function<double(const FieldVector<double,dim>&)>
70                                                                      neumann,
71                        double kappa)
72    : kappa_(kappa),
73      c_(c),
74      f_(f),
75      neumann_(neumann)
76    {}
```

The code then contains the `alpha_volume` and `alpha_jacobian` methods, which we skip.

The first new method of the `ReactionDiffusionDG` class is `alpha_skeleton`, which assembles the algebraic formulation of the skeleton part of the SIPG residual given in (11.16), for a given discrete function u_h. More precisely, the method implements the contributions of a single intersection γ between two neighboring elements T_{in} and T_{out}:

$$\left(\mathcal{R}_\gamma^{\mathrm{SIPG,skel}}(u_h)\right)_i := -\int_\gamma \{\langle \nabla u_h, \mathbf{n}\rangle\}[\![\theta_i]\!]\,ds - \int_\gamma \{\langle \nabla \theta_i, \mathbf{n}\rangle\}[\![u_h]\!]\,ds \qquad (11.17)$$

$$+ \frac{\kappa}{h_\gamma}\int_\gamma [\![u_h]\!][\![\theta_i]\!]\,ds \qquad i = 0, \ldots, n-1.$$

These contributions can only be nonzero for test functions θ_i whose support intersects the elements T_{in} or T_{out}. We call these test functions θ_i^{in} and θ_i^{out}, respectively. Likewise, we write ϕ_i^{in} and ϕ_i^{out} for restrictions of trial basis functions to the two elements T_{in} and T_{out}, and ϕ_i for the entire multi-valued DG basis function.

The interface of the `alpha_skeleton` method is

```
223    template<class IntersectionGeometry,
224            class LocalFunctionSpaceU, class Vector,
225            class LocalFunctionSpaceV, class Residual>
226    void alpha_skeleton(const IntersectionGeometry& intersectionGeometry,
227                        const LocalFunctionSpaceU& localFunctionSpaceUIn,
228                        const Vector& xIn,
229                        const LocalFunctionSpaceV& localFunctionSpaceVIn,
230                        const LocalFunctionSpaceU& localFunctionSpaceUOut,
231                        const Vector& xOut,
232                        const LocalFunctionSpaceV& localFunctionSpaceVOut,
233                        Residual& residualIn, Residual& residualOut) const
234    {
```

As described in Chapter 11.2.5, `intersectionGeometry` is a wrapper around the current intersection, `localFunctionSpaceUIn` and `localFunctionSpaceUOut` are the local finite elements on the *inside* and *outside* elements, respectively, and `xIn` and `xOut` are the respective corresponding coefficient vectors of u_h. The parameters `localFunctionSpaceVIn` and `localFunctionSpaceVOut` hold the local finite elements of the test function spaces on the two elements T_{in} and T_{out}. The resulting residual, finally, is returned in the two containers `residualIn` and `residualOut`.

The implementation of the method is long, because all setup code is done twice—once for each of the two elements involved. The method first stores various types and geometric objects for later use:

```
236    using TrialFE = typename LocalFunctionSpaceU::Traits::FiniteElementType;
237    using TestFE  = typename LocalFunctionSpaceV::Traits::FiniteElementType;
238    using LocalBasisU = typename TrialFE::Traits::LocalBasisType;
239    using LocalBasisV = typename TestFE::Traits::LocalBasisType;
```

```
240    using RangeU = typename LocalBasisU::Traits::RangeType;
241    using RangeV = typename LocalBasisV::Traits::RangeType;
242    using GradientU = typename LocalBasisU::Traits::JacobianType;
243    using GradientV = typename LocalBasisV::Traits::JacobianType;
244    using size_type = typename LocalFunctionSpaceV::Traits::SizeType;
245
246    // References to inside and outside elements
247    const auto& elementInside = intersectionGeometry.inside();
248    const auto& elementOutside = intersectionGeometry.outside();
249
250    // Intersection and element geometries
251    auto geo = intersectionGeometry.geometry();
252    auto geoIn  = elementInside.geometry();
253    auto geoOut = elementOutside.geometry();
254
255    // Geometries of intersection in local coordinates
256    // of elementInside and elementOutside
257    auto geoInInside  = intersectionGeometry.geometryInInside();
258    auto geoInOutside = intersectionGeometry.geometryInOutside();
```

Then, the code precomputes the intersection diameter h_γ using the method `diameter` defined in Line 32:

```
263    auto h = diameter(geo);
```

After this, it creates containers to store shape function values and gradients on both the *inside* and *outside* element. Separate containers are used for gradients with respect to reference element coordinates and grid element coordinates:

```
267    // Shape function values
268    std::vector<RangeU> phiIn(localFunctionSpaceUIn.size());
269    std::vector<RangeV> thetaIn(localFunctionSpaceVIn.size());
270    std::vector<RangeU> phiOut(localFunctionSpaceUOut.size());
271    std::vector<RangeV> thetaOut(localFunctionSpaceVOut.size());
272
273    // Shape function gradients on the reference element
274    std::vector<GradientU> localGradPhiIn(localFunctionSpaceUIn.size());
275    std::vector<GradientV> localGradThetaIn(localFunctionSpaceVIn.size());
276    std::vector<GradientU> localGradPhiOut(localFunctionSpaceUOut.size());
277    std::vector<GradientV> localGradThetaOut(localFunctionSpaceVOut.size());
278
279    // Shape function gradients on the grid element
280    std::vector<GradientU> gradPhiIn(localFunctionSpaceUIn.size());
281    std::vector<GradientV> gradThetaIn(localFunctionSpaceVIn.size());
282    std::vector<GradientU> gradPhiOut(localFunctionSpaceUOut.size());
283    std::vector<GradientV> gradThetaOut(localFunctionSpaceVOut.size());
```

Next is the quadrature loop. For simplicity, the code uses twice the polynomial order of the basis on the *inside* element as an educated guess of a suitable quadrature order:

```
288        const auto quadOrder
289              = 2 * localFunctionSpaceUIn.finiteElement().localBasis().order();
290        constexpr auto intersectionDim = IntersectionGeometry::mydimension;
291        const auto& quadRule
292              = QuadratureRules<double,intersectionDim>::
293                                    rule(intersectionGeometry.geometry().type(),
294                                          quadOrder);
295
296        for (const auto& quadPoint : quadRule)
297        {
```

The quadrature loop extends all the way to the end of the method. At the beginning of the loop body, the code evaluates the shape functions and their Jacobians. The geoInInside and geoInOutside objects are used to transform the position of the quadrature point from coordinates on the intersection to coordinates on the adjacent elements (Chapter 5.4). Four calls are needed for the values and the gradients, because there are basis functions on two elements involved, and trial and test functions are treated separately:[9]

```
299        // Quadrature point position in local coordinates of adjacent elements
300        auto quadPointLocalIn = geoInInside.global(quadPoint.position());
301        auto quadPointLocalOut = geoInOutside.global(quadPoint.position());
302
303        // Evaluate shape functions
304        localFunctionSpaceUIn.finiteElement().localBasis()
305                                .evaluateFunction(quadPointLocalIn,phiIn);
306        localFunctionSpaceUOut.finiteElement().localBasis()
307                                .evaluateFunction(quadPointLocalOut,phiOut);
308        localFunctionSpaceVIn.finiteElement().localBasis()
309                                .evaluateFunction(quadPointLocalIn,thetaIn);
310        localFunctionSpaceVOut.finiteElement().localBasis()
311                                .evaluateFunction(quadPointLocalOut,thetaOut);
312
313        // Evaluate gradients of shape functions
314        localFunctionSpaceUIn.finiteElement().localBasis()
315                                .evaluateJacobian(quadPointLocalIn,
316                                                   localGradPhiIn);
317        localFunctionSpaceUOut.finiteElement().localBasis()
318                                .evaluateJacobian(quadPointLocalOut,
319                                                   localGradPhiOut);
320        localFunctionSpaceVIn.finiteElement().localBasis()
321                                .evaluateJacobian(quadPointLocalIn,
322                                                   localGradThetaIn);
323        localFunctionSpaceVOut.finiteElement().localBasis()
324                                .evaluateJacobian(quadPointLocalOut,
325                                                   localGradThetaOut);
```

---

[9] Again, real implementations will want to test whether trial and test bases are identical, to save evaluation time.

The shape function gradients are then transformed to coordinates of the two grid elements T_{in} and T_{out}. This happens in the usual way, but as there are two different elements involved, two different transformation matrices are used as well:

```
330     auto jacInvTranspIn
331        = geoIn.jacobianInverseTransposed(quadPointLocalIn);
332     for (size_type i=0; i<localFunctionSpaceUIn.size(); i++)
333        jacInvTranspIn.mv(localGradPhiIn[i][0], gradPhiIn[i][0]);
334     for (size_type i=0; i<localFunctionSpaceVIn.size(); i++)
335        jacInvTranspIn.mv(localGradThetaIn[i][0], gradThetaIn[i][0]);
336
337     auto jacInvTranspOut
338        = geoOut.jacobianInverseTransposed(quadPointLocalOut);
339     for (size_type i=0; i<localFunctionSpaceUOut.size(); i++)
340        jacInvTranspOut.mv(localGradPhiOut[i][0], gradPhiOut[i][0]);
341     for (size_type i=0; i<localFunctionSpaceVOut.size(); i++)
342        jacInvTranspOut.mv(localGradThetaOut[i][0], gradThetaOut[i][0]);
```

The following block then computes u_h and ∇u_h at the current integration point, as linear combinations of the basis function values and their gradients, respectively:

```
346     // Compute values of u_h
347     RangeU uIn(0.0);
348     for (size_type i=0; i<localFunctionSpaceUIn.size(); i++)
349        uIn += xIn(localFunctionSpaceUIn,i)*phiIn[i];
350     RangeU uOut(0.0);
351     for (size_type i=0; i<localFunctionSpaceUOut.size(); i++)
352        uOut += xOut(localFunctionSpaceUOut,i)*phiOut[i];
353
354     // Compute gradients of u_h
355     GradientU gradUIn(0.0);
356     for (size_type i=0; i<localFunctionSpaceUIn.size(); i++)
357        gradUIn.axpy(xIn(localFunctionSpaceUIn,i), gradPhiIn[i]);
358     GradientU gradUOut(0.0);
359     for (size_type i=0; i<localFunctionSpaceUOut.size(); i++)
360        gradUOut.axpy(xOut(localFunctionSpaceUOut,i), gradPhiOut[i]);
```

As u_h is not continuous at the quadrature point, there are two values of u_h to be computed, and two values of ∇u_h.

Finally, three loops compute the values of the three terms of $\mathcal{R}_\gamma^{\text{SIPG,skel}}$ as defined in (11.17):

```
364     // Unit normal from T_in to T_out
365     auto n_F = intersectionGeometry.unitOuterNormal(quadPoint.position());
366
367     // Integration factor
368     auto factor
369        = quadPoint.weight()*geo.integrationElement(quadPoint.position());
370
```

```
371        // Interior penalty term
372        auto interiorPenaltyTerm =  -0.5*(gradUIn[0]*n_F + gradUOut[0]*n_F);
373        for (size_type i=0; i<localFunctionSpaceVIn.size(); i++)
374          residualIn.accumulate(localFunctionSpaceVIn,
375                                i,
376                                interiorPenaltyTerm * thetaIn[i] * factor);
377        for (size_type i=0; i<localFunctionSpaceVOut.size(); i++)
378          residualOut.accumulate(localFunctionSpaceVOut,
379                                 i,
380                                 -interiorPenaltyTerm * thetaOut[i] * factor);
381
382        // Symmetric interior penalty term
383        for (size_type i=0; i<localFunctionSpaceVIn.size(); i++)
384          residualIn.accumulate(localFunctionSpaceVIn,
385                                i,
386                                -0.5 * (gradThetaIn[i][0]*n_F)
387                                     * (uIn-uOut) * factor);
388        for (size_type i=0; i<localFunctionSpaceVOut.size(); i++)
389          residualOut.accumulate(localFunctionSpaceVOut,
390                                 i,
391                                 -0.5 * (gradThetaOut[i][0]*n_F)
392                                      * (uIn-uOut) * factor);
393
394        // Coercivity term
395        auto coercivityTerm = (kappa_/h) * (uIn-uOut);
396        for (size_type i=0; i<localFunctionSpaceVIn.size(); i++)
397          residualIn.accumulate(localFunctionSpaceVIn,
398                                i,
399                                coercivityTerm * thetaIn[i] * factor);
400        for (size_type i=0; i<localFunctionSpaceVOut.size(); i++)
401          residualOut.accumulate(localFunctionSpaceVOut,
402                                 i,
403                                 -coercivityTerm * thetaOut[i] * factor);
404      }
405    }
```

Note how the output residual contribution is now split into two parts, one containing the results of testing with test functions θ_i^{in} of the *inside* element, and one for the test functions θ_i^{out} on the *outside* element:

- The first pair of loops adds $-\{\langle \nabla u_h, \mathbf{n}_\gamma \rangle\}\theta_i^{\text{in}}$ to the residual of the *inside* element, and $\{\langle \nabla u_h, \mathbf{n}_\gamma \rangle\}\theta_i^{\text{out}}$ to the residual of the *outside* element. This corresponds to the term $-\{\langle \nabla u_h, \mathbf{n}_\gamma \rangle\}[\![\theta_i]\!]$ of (11.17).

- The second pair of loops adds $-\frac{1}{2}\langle \nabla \theta_i^{\text{in}}, \mathbf{n}_\gamma \rangle[\![u_h]\!]$ to the residual of the *inside* element, and $-\frac{1}{2}\langle \nabla \theta_i^{\text{out}}, \mathbf{n}_\gamma \rangle[\![u_h]\!]$ to the residual of the *outside* element. This corresponds to the term $-\{\langle \nabla \theta, \mathbf{n}_\gamma \rangle\}[\![u_h]\!]$.

- The third pair of loops adds $\frac{\kappa}{h_\gamma} \llbracket u_h \rrbracket \theta_i^{\mathrm{in}}$ to the *inside* element, and $-\frac{\kappa}{h_\gamma} \llbracket u_h \rrbracket \theta_i^{\mathrm{out}}$ to the *outside* element. This corresponds to the term $\frac{\kappa}{h_\gamma} \llbracket u_h \rrbracket \llbracket \theta_i \rrbracket$.

This concludes the `alpha_skeleton` method.

The next method is `jacobian_skeleton`, which computes the derivative of the intersection residual $\mathcal{R}_\gamma^{\mathrm{SIPG,skel}}(u_h)$ with respect to the coefficients $\bar{u}^{\mathrm{in}} \in \mathbb{R}^{n_{T_{\mathrm{in}}}}$ and $\bar{u}^{\mathrm{out}} \in \mathbb{R}^{n_{T_{\mathrm{out}}}}$ of u_h on T_{in} and T_{out}, respectively. As the residual is linear in u_h, the derivative is a matrix that is independent of u_h. The derivatives are

$$\frac{\partial \mathcal{R}_\gamma^{\mathrm{SIPG,skel}}(\bar{u}))_i}{\partial \bar{u}_j^{\mathrm{in}}} = -\frac{1}{2} \int_\gamma \langle \nabla \phi_j^{\mathrm{in}}, \mathbf{n}_\gamma \rangle \llbracket \theta_i \rrbracket \, ds - \int_\gamma \{\langle \nabla \theta_i, \mathbf{n}_\gamma \rangle\} \phi_j^{\mathrm{in}} \, ds + \frac{\kappa}{h_\gamma} \int_\gamma \phi_j^{\mathrm{in}} \llbracket \theta_i \rrbracket \, ds \tag{11.18a}$$

$$\frac{\partial \mathcal{R}_\gamma^{\mathrm{SIPG,skel}}(\bar{u}))_i}{\partial \bar{u}_j^{\mathrm{out}}} = -\frac{1}{2} \int_\gamma \langle \nabla \phi_j^{\mathrm{out}}, \mathbf{n}_\gamma \rangle \llbracket \theta_i \rrbracket \, ds \tag{11.18b}$$

$$+ \int_\gamma \{\langle \nabla \theta_i, \mathbf{n}_\gamma \rangle\} \phi_j^{\mathrm{out}} \, ds - \frac{\kappa}{h_\gamma} \int_\gamma \phi_j^{\mathrm{out}} \llbracket \theta_i \rrbracket \, ds,$$

where (11.18a) shows the derivatives with respect to the degrees of freedom of the *inside* element, and (11.18b) shows the corresponding derivatives for the *outside* element. The matrix rows can be classified according to whether the corresponding residual vector entry is associated to the *inside* or *outside* element. For (11.18a), using that

$$\llbracket \theta_i \rrbracket = \theta_i^{\mathrm{in}} - \theta_i^{\mathrm{out}} \qquad \text{and} \qquad \{\nabla \theta_i\} = \frac{1}{2}(\nabla \theta_i^{\mathrm{in}} + \nabla \theta_i^{\mathrm{out}}),$$

we get

$$\frac{\partial \mathcal{R}_\gamma^{\mathrm{SIPG,skel}}(\bar{u}^{\mathrm{in}}))_i}{\partial \bar{u}_j^{\mathrm{in}}} = -\frac{1}{2}\left[\langle \nabla \phi_j^{\mathrm{in}}, \mathbf{n}_{\gamma^{\mathrm{i}}} \rangle \theta_i^{\mathrm{in}} + \langle \nabla \theta_i^{\mathrm{in}}, \mathbf{n}_{\gamma^{\mathrm{i}}} \rangle \phi_j^{\mathrm{in}} \right] + \frac{\kappa}{h_{\gamma^{\mathrm{i}}}} \phi_j^{\mathrm{in}} \theta_i^{\mathrm{in}} \tag{11.19a}$$

and

$$\frac{\partial \mathcal{R}_\gamma^{\mathrm{SIPG,skel}}(\bar{u}^{\mathrm{out}}))_i}{\partial \bar{u}_j^{\mathrm{in}}} = \frac{1}{2}\left[\langle \nabla \phi_j^{\mathrm{in}}, \mathbf{n}_{\gamma^{\mathrm{i}}} \rangle \theta_i^{\mathrm{out}} - \langle \nabla \theta_i^{\mathrm{out}}, \mathbf{n}_{\gamma^{\mathrm{i}}} \rangle \phi_j^{\mathrm{in}} \right] - \frac{\kappa}{h_{\gamma^{\mathrm{i}}}} \phi_j^{\mathrm{in}} \theta_i^{\mathrm{out}}. \tag{11.19b}$$

Similarly, for (11.18b) we obtain

$$\frac{\partial \mathcal{R}_\gamma^{\mathrm{SIPG,skel}}(\bar{u}^{\mathrm{in}}))_i}{\partial \bar{u}_j^{\mathrm{out}}} = \frac{1}{2}\left[-\langle \nabla \phi_j^{\mathrm{out}}, \mathbf{n}_{\gamma^{\mathrm{i}}} \rangle \theta_i^{\mathrm{in}} + \langle \nabla \theta_i^{\mathrm{in}}, \mathbf{n}_{\gamma^{\mathrm{i}}} \rangle \phi_j^{\mathrm{out}} \right] - \frac{\kappa}{h_{\gamma^{\mathrm{i}}}} \phi_j^{\mathrm{out}} \theta_i^{\mathrm{in}} \tag{11.19c}$$

and

$$\frac{\partial \mathcal{R}_\gamma^{\mathrm{SIPG,skel}}(\bar{u}^{\mathrm{out}}))_i}{\partial \bar{u}_j^{\mathrm{out}}} = \frac{1}{2}\left[\langle \nabla \phi_j^{\mathrm{out}}, \mathbf{n}_{\gamma^{\mathrm{i}}} \rangle \theta_i^{\mathrm{out}} + \langle \nabla \theta_i^{\mathrm{out}}, \mathbf{n}_{\gamma^{\mathrm{i}}} \rangle \phi_j^{\mathrm{out}} \right] + \frac{\kappa}{h_{\gamma^{\mathrm{i}}}} \phi_j^{\mathrm{out}} \theta_i^{\mathrm{out}}. \tag{11.19d}$$

This natural 2×2 matrix block structure is visible in the method interface:

```
409    template<class IntersectionGeometry,
410            class LocalFunctionSpaceU, class Vector,
411            class LocalFunctionSpaceV, class Jacobian>
412    void jacobian_skeleton(const IntersectionGeometry& intersectionGeometry,
413                           const LocalFunctionSpaceU& localFunctionSpaceUIn,
414                           const Vector& xIn,
415                           const LocalFunctionSpaceV& localFunctionSpaceVIn,
416                           const LocalFunctionSpaceU& localFunctionSpaceUOut,
417                           const Vector& xOut,
418                           const LocalFunctionSpaceV& localFunctionSpaceVOut,
419                           Jacobian& jacobianInIn, Jacobian& jacobianInOut,
420                           Jacobian& jacobianOutIn, Jacobian& jacobianOutOut)
421                                                                   const
```

where the four output matrices jacobianInIn, jacobianInOut, jacobianOutIn, and jacobianOutOut correspond to the four submatrices of $\nabla \mathcal{R}_\gamma^{\text{SIPG,skel}}$. All method parameters have been described in Chapter 11.2.5.

The implementation of the method starts similarly to the alpha_skeleton method, by precomputing various geometric objects, and allocating containers for shape function values and gradients:

```
425    // Define types
426    using TrialFE = typename LocalFunctionSpaceU::Traits::FiniteElementType;
427    using TestFE = typename LocalFunctionSpaceV::Traits::FiniteElementType;
428    using LocalBasisU = typename TrialFE::Traits::LocalBasisType;
429    using LocalBasisV = typename TestFE::Traits::LocalBasisType;
430    using RangeU = typename LocalBasisU::Traits::RangeType;
431    using RangeV = typename LocalBasisV::Traits::RangeType;
432    using GradientU = typename LocalBasisU::Traits::JacobianType;
433    using GradientV = typename LocalBasisV::Traits::JacobianType;
434    using size type = typename LocalFunctionSpaceV::Traits::SizeType;
435
436    // References to inside and outside elements
437    const auto& elementInside = intersectionGeometry.inside();
438    const auto& elementOutside = intersectionGeometry.outside();
439
440    // Get geometries
441    auto geo = intersectionGeometry.geometry();
442    auto geoIn = elementInside.geometry();
443    auto geoOut = elementOutside.geometry();
444
445    // Geometries of intersection in local coordinates
446    // of elementInside and elementOutside
447    auto geoInInside = intersectionGeometry.geometryInInside();
448    auto geoInOutside = intersectionGeometry.geometryInOutside();
449
450    // Intersection diameter
451    auto h = diameter(geo);
```

```
452
453     // Shape function values
454     std::vector<RangeU> phiIn(localFunctionSpaceUIn.size());
455     std::vector<RangeU> phiOut(localFunctionSpaceUOut.size());
456     std::vector<RangeV> thetaIn(localFunctionSpaceVIn.size());
457     std::vector<RangeV> thetaOut(localFunctionSpaceVOut.size());
458
459     // Shape function gradients on the reference element
460     std::vector<GradientU> localGradPhiIn(localFunctionSpaceUIn.size());
461     std::vector<GradientU> localGradPhiOut(localFunctionSpaceUOut.size());
462     std::vector<GradientV> localGradThetaIn(localFunctionSpaceVIn.size());
463     std::vector<GradientV> localGradThetaOut(localFunctionSpaceVOut.size());
464
465     // Shape function gradients on the grid element
466     std::vector<GradientU> gradPhiIn(localFunctionSpaceUIn.size());
467     std::vector<GradientU> gradPhiOut(localFunctionSpaceUOut.size());
468     std::vector<GradientV> gradThetaIn(localFunctionSpaceVIn.size());
469     std::vector<GradientV> gradThetaOut(localFunctionSpaceVOut.size());
```

Right after this begins the quadrature loop, which first computes the values and gradients of the trial basis functions $\{\phi_i^{\text{in}}\}_i$ and $\{\phi_i^{\text{out}}\}_i$ and the test basis functions $\{\theta_i^{\text{in}}\}_i$ and $\{\theta_i^{\text{out}}\}_i$ at the current quadrature point:

```
474     auto quadOrder
475          = 2*localFunctionSpaceUIn.finiteElement().localBasis().order();
476     constexpr auto intersectionDim = IntersectionGeometry::mydimension;
477     const auto& quadRule
478          = QuadratureRules<double,intersectionDim>::
479                               rule(intersectionGeometry.geometry().type(),
480                                    quadOrder);
481
482     for (const auto& quadPoint : quadRule)
483     {
484       // Position of quadrature point in local coordinates
485       // of inside and outside elements
486       auto quadPointLocalIn = geoInInside.global(quadPoint.position());
487       auto quadPointLocalOut = geoInOutside.global(quadPoint.position());
488
489       // Evaluate shape functions
490       localFunctionSpaceUIn.finiteElement().localBasis()
491                       .evaluateFunction(quadPointLocalIn,phiIn);
492       localFunctionSpaceUOut.finiteElement().localBasis()
493                       .evaluateFunction(quadPointLocalOut,phiOut);
494
495       localFunctionSpaceVIn.finiteElement().localBasis()
496                       .evaluateFunction(quadPointLocalIn,thetaIn);
497       localFunctionSpaceVOut.finiteElement().localBasis()
498                       .evaluateFunction(quadPointLocalOut,thetaOut);
```

```
499
500          // Evaluate gradients of shape functions
501          localFunctionSpaceUIn.finiteElement().localBasis()
502                              .evaluateJacobian(quadPointLocalIn,
503                                                localGradPhiIn);
504          localFunctionSpaceUOut.finiteElement().localBasis()
505                              .evaluateJacobian(quadPointLocalOut,
506                                                localGradPhiOut);
507
508          localFunctionSpaceVIn.finiteElement().localBasis()
509                              .evaluateJacobian(quadPointLocalIn,
510                                                localGradThetaIn);
511          localFunctionSpaceVOut.finiteElement().localBasis()
512                              .evaluateJacobian(quadPointLocalOut,
513                                                localGradThetaOut);
514
515          // Transform gradients of shape functions to element coordinates
516          auto jacInvTranspIn
517              = geoIn.jacobianInverseTransposed(quadPointLocalIn);
518          for (size_type i=0; i<localFunctionSpaceUIn.size(); i++)
519            jacInvTranspIn.mv(localGradPhiIn[i][0],gradPhiIn[i][0]);
520          for (size_type i=0; i<localFunctionSpaceVIn.size(); i++)
521            jacInvTranspIn.mv(localGradThetaIn[i][0],gradThetaIn[i][0]);
522
523          auto jacInvTranspOut
524              = geoOut.jacobianInverseTransposed(quadPointLocalOut);
525          for (size_type i=0; i<localFunctionSpaceUOut.size(); i++)
526            jacInvTranspOut.mv(localGradPhiOut[i][0],gradPhiOut[i][0]);
527          for (size_type i=0; i<localFunctionSpaceVOut.size(); i++)
528            jacInvTranspOut.mv(localGradThetaOut[i][0],gradThetaOut[i][0]);
```

This is all very similar to the alpha_skeleton method.

Then, four loops fill the four submatrices of the result with the derivatives of $\mathcal{R}_\gamma^{\text{SIPG,skel}}(\bar{u}^{\text{in}})$ and $\mathcal{R}_\gamma^{\text{SIPG,skel}}(\bar{u}^{\text{out}})$ with respect to the entries of $\bar{u}^{\text{in}}$ and $\bar{u}^{\text{out}}$, as given by the expressions (11.19a)–(11.19d):

```
532          // Unit normal from T_in to T_out
533          auto n_F = intersectionGeometry.unitOuterNormal(quadPoint.position());
534
535          // Integration factor
536          auto factor = quadPoint.weight()
537                      * geo.integrationElement(quadPoint.position());
538
539          // Fill jacobianInIn
540          for (size_type i=0; i<localFunctionSpaceVIn.size(); i++)
541          {
542            for (size_type j=0; j<localFunctionSpaceUIn.size(); j++)
543            {
```

```
544              jacobianInIn.accumulate(localFunctionSpaceVIn,
545                                      i,
546                                      localFunctionSpaceUIn,
547                                      j,
548                                      (-0.5 * (gradPhiIn[j][0]*n_F) * thetaIn[i]
549                                       - 0.5 * phiIn[j] * (gradThetaIn[i][0]*n_F)
550                                       + (kappa_ / h) * phiIn[j] * thetaIn[i] )
551                                                                    * factor);
552         }
553       }
554
555       // Fill jacobianInOut
556       for (size_type i=0; i<localFunctionSpaceVOut.size(); i++)
557       {
558         for (size_type j=0; j<localFunctionSpaceUIn.size(); j++)
559         {
560           jacobianInOut.accumulate(localFunctionSpaceVIn,
561                                    i,
562                                    localFunctionSpaceUOut,
563                                    j,
564                                    (-0.5 * (n_F*gradPhiOut[j][0]) * thetaIn[i]
565                                     + 0.5 * phiOut[j] * (n_F*gradThetaIn[i][0])
566                                     - (kappa_ / h) * phiOut[j] * thetaIn[i])
567                                                                  * factor);
568         }
569       }
570
571       // Fill jacobianOutIn
572       for (size_type i=0; i<localFunctionSpaceVIn.size(); i++)
573       {
574         for (size_type j=0; j<localFunctionSpaceUOut.size(); j++)
575         {
576           jacobianOutIn.accumulate(localFunctionSpaceVOut,
577                                    i,
578                                    localFunctionSpaceUIn,
579                                    j,
580                                    (0.5 * (n_F*gradPhiIn[j][0]) * thetaOut[i]
581                                     - 0.5 * phiIn[j] * (n_F*gradThetaOut[i][0])
582                                     - (kappa_ / h) * phiIn[j] * thetaOut[i])
583                                                                  * factor);
584         }
585       }
586
587       // Fill jacobianOutOut
588       for (size_type i=0; i<localFunctionSpaceVOut.size(); i++)
589       {
590         for (size_type j=0; j<localFunctionSpaceUOut.size(); j++)
```

```
591                {
592                    jacobianOutOut.accumulate(localFunctionSpaceVOut,
593                                              i,
594                                              localFunctionSpaceUOut,
595                                              j,
596                                              (0.5 * (n_F*gradPhiOut[j][0]) * thetaOut[i]
597                                              + 0.5 * phiOut[j] * (n_F*gradThetaOut[i][0])
598                                              + (kappa_ / h) * phiOut[j] * thetaOut[i])
599                                                                         * factor);
600                }
601            }
602        }
603    }
```

This concludes the `jacobian_skeleton` method.

The last methods of the `LocalOperator` implementation for the reaction–diffusion problem are `alpha_boundary` and `jacobian_boundary`, which compute the Neumann boundary term $-\int_{\partial\Omega} j\theta\, ds$ and its Jacobian, respectively. They are much simpler than the methods for the skeleton terms, because they only involve a single element. Additionally, as the Neumann term does not depend on u_h, the trial function space `localFunctionSpaceUIn` is not used at all:

```
607    // Boundary integral implementing the Neumann term
608    template<class IntersectionGeometry,
609             class LocalFunctionSpaceU, class Vector,
610             class LocalFunctionSpaceV, class Residual>
611    void alpha_boundary(const IntersectionGeometry& intersectionGeometry,
612                        const LocalFunctionSpaceU& localFunctionSpaceUIn,
613                        const Vector& xIn,
614                        const LocalFunctionSpaceV& localFunctionSpaceVIn,
615                        Residual& residualIn) const
616    {
617      using TestFE = typename LocalFunctionSpaceV::Traits::FiniteElementType;
618      using RangeV = typename TestFE::Traits::LocalBasisType::Traits::RangeType;
619      using size_type = typename LocalFunctionSpaceV::Traits::SizeType;
620
621      // Get geometry of intersection in local coordinates of element_inside
622      auto geoInInside = intersectionGeometry.geometryInInside();
623
624      std::vector<RangeV> thetaIn(localFunctionSpaceVIn.size());
625
626      // Loop over quadrature points
627      const int quadOrder
628          = 2*localFunctionSpaceVIn.finiteElement().localBasis().order();
629      constexpr auto intersectionDim = IntersectionGeometry::mydimension;
630      const auto& quadRule
631          = QuadratureRules<double,intersectionDim>::
```

```
632                                 rule(intersectionGeometry.geometry().type(),
633                                      quadOrder);
634
635     for (const auto& quadPoint : quadRule)
636     {
637       // Position of quadrature point in local coordinates of elements
638       auto quadPointLocalIn = geoInInside.global(quadPoint.position());
639
640       // Evaluate basis functions
641       localFunctionSpaceVIn.finiteElement().localBasis()
642                       .evaluateFunction(quadPointLocalIn,thetaIn);
643
644       // Integration factor
645       auto factor
646           = quadPoint.weight()
647           * intersectionGeometry.geometry()
648                           .integrationElement(quadPoint.position());
649
650       // Evaluate Neumann boundary condition
651       auto neumannValue
652           = neumann_(intersectionGeometry.geometry()
653                               .global(quadPoint.position()));
654
655       // Integrate
656       for (size_type i=0; i<localFunctionSpaceVIn.size(); i++)
657         residualIn.accumulate(localFunctionSpaceVIn,
658                               i,
659                               -1 * neumannValue * thetaIn[i] * factor);
660     }
661   }
```

What is more, the derivative of the Neumann term with respect to u_h vanishes, and the jacobian_boundary method can simply be empty:

```
665   template<class IntersectionGeometry,
666           class LocalFunctionSpaceU, class Vector,
667           class LocalFunctionSpaceV, typename Jacobian>
668   void jacobian_boundary(const IntersectionGeometry& intersectionGeometry,
669                       const LocalFunctionSpaceU& localFunctionSpaceUIn,
670                       const Vector& xIn,
671                       const LocalFunctionSpaceV& localFunctionSpaceVIn,
672                       Jacobian& jacobianInIn) const
673   {
674     // Does not do anything---the Jacobian of the boundary term is zero.
675   }
```

Note, however, that one cannot simply leave out the method. As the doAlphaBoundary flag is set, the GridOperator assumes that method to be present. Completely omitting it from the code leads to compile-time errors.

Figure 11.5.: Result of the DG example program, at two different resolutions

Running the Program

When the program is run, it prints the following screen output:

```
=== matrix setup (max) 0.0141681 s
=== matrix assembly (max) 0.040603 s
=== residual assembly (max) 0.0111276 s
=== solving (reduction: 1e-10) === Dune::IterativeSolver
 Iter         Defect          Rate
   0          1.18145
   1          0.558789        0.472969
   2          0.353185        0.632054
   3          0.292861        0.829199
   4          0.0752357       0.256899
[...]
  18          1.88705e-09     0.279193
  19          4.09746e-10     0.217135
  20          9.82223e-11     0.239715
=== rate=0.313321, T=0.0352053, TIT=0.00176027, IT=20
0.0531138 s
```

It also writes a file called pdelab-dg-diffusion-result.vtu to disk, which contains the subsampled result function u_h. A visualization of the result as a height field is shown on the left of Figure 11.5. The discontinuities of the finite element solution are clearly seen. The right side of the same figure shows the result of the program when run using a finer grid. As predicted by the DG theory, the discontinuities diminish with increasing grid resolution.

11.3. Dirichlet Boundary Conditions

In all previous example problems of this chapter we have deliberately omitted Dirichlet boundary conditions, i.e., constraints of the form

$$u = g \qquad \text{on } \Gamma_D$$

for a given function g on a subset Γ_D of the domain boundary $\partial\Omega$. Now we make up for this.

In dune-pdelab, Dirichlet boundary conditions are implemented as a special case of a more general mechanism that allows to apply arbitrary linear constraints to the coefficients of finite element functions. This has interesting applications, like constructing conforming finite element spaces on nonconforming grids, conforming p-adaptive methods, and periodic boundary conditions. However, to limit the overall size of the book, only Dirichlet boundary conditions will be discussed. Readers interested in more general constraints should consult the online documentation.

11.3.1. Dirichlet Boundary Conditions and the Residual Form

Nonzero Dirichlet boundary conditions do not fit into the framework of Chapter 11.2.1, which we therefore generalize slightly. Let Ω be an open bounded subset of $\mathbb{R}^d$, and $f : \Omega \to \mathbb{R}$ a given function of sufficient smoothness. As in Chapter 11.2.1, we look for a function $u : \Omega \to \mathbb{R}$ that solves the equation

$$- \operatorname{div} \mathcal{A}(x, u, \nabla u) + h(u) = f \qquad \text{in } \Omega \subset \mathbb{R}^d,$$

where $\mathcal{A} : \Omega \times \mathbb{R} \times \mathbb{R}^d \to \mathbb{R}^d$ is an algebraic representation of a first-order elliptic differential operator, and $h : \mathbb{R} \to \mathbb{R}$ is a function describing the reaction term. In contrast to Chapter 11.2.1 we partition the domain boundary into two parts Γ_D and Γ_N, and we replace the pure Neumann boundary condition (11.2b) by the mixed condition

$$u = \mathrm{g} \qquad \text{on } \Gamma_D, \tag{11.20a}$$

$$\langle \mathcal{A}(x, u, \nabla u), \mathbf{n} \rangle = \mathrm{j} \qquad \text{on } \Gamma_N, \tag{11.20b}$$

where $\mathrm{g} : \Gamma_D \to \mathbb{R}$ is a given Dirichlet boundary value function of sufficient smoothness. To state the weak form of this problem we introduce the new Sobolev spaces

$$H_{D,\mathrm{g}}^1 := \left\{ v \in H^1(\Omega) \; : \; v = \mathrm{g} \text{ on } \Gamma_D \right\}$$

and

$$H_{D,0}^1 := \left\{ v \in H^1(\Omega) \; : \; v = 0 \text{ on } \Gamma_D \right\},$$

with the boundary conditions to be understood in the sense of traces [37, 42]. Using the standard procedure we find the weak form of the model problem to be

$$\text{Find } u \in H_{D,\mathrm{g}}^1 \; : \quad \int_{\Omega} \langle \mathcal{A}(x, u, \nabla u), \nabla v \rangle \, dx + \int_{\Omega} h(u) v \, dx$$
$$= \int_{\Omega} f v \, dx + \int_{\Gamma_N} \mathrm{j} v \, ds \qquad \forall v \in H_{D,0}^1.$$

The residual form of this equation is

$$\text{Find } u \in H_{D,\mathrm{g}}^1 \; : \quad r(u, v) = 0 \qquad \forall v \in H_{D,0}^1 \tag{11.21}$$

with

$$r(u,v) := \int_\Omega \langle \mathcal{A}(x,u,\nabla u), \nabla v \rangle \, dx + \int_\Omega h(u)v \, dx - \int_\Omega fv \, dx - \int_{\Gamma_N} jv \, ds.$$

This problem is not of the type (11.5), because the solution u is not taken from a linear space. However, the space $H^1_{D,\mathbf{g}}$ is affine, and can therefore be written as $H^1_{D,\mathbf{g}} = w_\mathbf{g} + H^1_{D,0}$ for any $w_\mathbf{g} \in H^1_{D,\mathbf{g}}$. Using this, the residual formulation (11.21) can be reformulated with only linear spaces. We obtain

$$\text{Find } \tilde{u} \in H^1_{D,0} \; : \quad r(w_\mathbf{g} + \tilde{u}, v) = 0 \qquad \forall v \in H^1_{D,0}.$$

To discretize the problem we pick two finite element spaces U_h and V_h for the trial and test functions, respectively. The function $w_\mathbf{g}$ is approximated by a discrete function $w_{h,\mathbf{g}}$. Usually, $w_{h,\mathbf{g}}$ approximates $\mathbf{g}$ on the Dirichlet boundary, and is zero away from that boundary.

To allow for non-conforming finite elements, that is, finite element spaces U_h, V_h that are not subspaces of $H^1_{D,0}$, we formally replace the residual with a discrete approximation r_h. As previously, the residual form $r_h : U_h \times V_h \to \mathbb{R}$ may be nonlinear in its first argument, but is always linear in its second argument. Altogether, we obtain the affine residual formulation, which generalizes the linear one given by Assumption 11.1.

Assumption 11.3 *All relevant boundary value problems can be written in the form*

$$\text{Find } u_h \in w_{h,\mathbf{g}} + U_h \; : \qquad r_h(u_h, v_h) = 0 \qquad \forall v_h \in V_h. \tag{11.22}$$

for suitable vector spaces U_h, V_h, functions $w_{h,\mathbf{g}}$, and forms $r_h(\cdot,\cdot)$.

We assume again U_h, V_h, $w_{h,\mathbf{g}}$, and $r_h(\cdot,\cdot)$ to be such that (11.22) has at least one solution.

In `dune-pdelab`, discretizations of a space with homogeneous Dirichlet boundary conditions such as $H^1_{D,0}$ are called *constrained spaces*, because they are seen as the combination of a "full" space like H^1 with homogeneous linear algebraic constraints $v = 0$ on Γ_D. An iterative method (like the Newton method which is the standard `dune-pdelab` way to solve the nonlinear systems (11.22)), when started from the initial iterate $w_{h,\mathbf{g}}$, will produce iterates with the correct boundary conditions if the corrections are computed in a space that has homogeneous Dirichlet boundary conditions on Γ_D.

The `dune-pdelab` implementation of an affine residual formulation proceeds in two steps. It first requires a class that represents the type of constraint applied to the finite element space. The `dune-pdelab` module offers several choices; the previous examples have used

```
PDELab::NoConstraints
```

which specifies the absence of constraints. The appropriate type for Dirichlet boundary conditions is

```
PDELab::ConformingDirichletConstraints
```

Both types are used as tag classes: The user is not expected to construct objects of this type; rather, the type is used to parametrize the GridFunctionSpace class. For example, the setup code for a GridFunctionSpace object for a first-order Lagrange space with Dirichlet constraints is:

```
63    using Basis = Functions::LagrangeBasis<GridView,1>;
64    auto basis = std::make_shared<Basis>(gridView);
65
66    using VectorBackend = PDELab::ISTL::VectorBackend<>;
67    using Constraints = PDELab::ConformingDirichletConstraints;
68
69    using GridFunctionSpace
70        = PDELab::Experimental::GridFunctionSpace<Basis,
71                                                  VectorBackend,
72                                                  Constraints>;
73
74    GridFunctionSpace gridFunctionSpace(basis);
```

Secondly, implementations need to have a so-called ConstraintsContainer object, which stores the actual constraints. As constraints in dune-pdelab are linear, a ConstraintsContainer can be conceptualized as a sparse matrix. The kernel of this matrix consists precisely of those coefficient vectors that fulfill the constraints. Unlike for the Constraints tag classes above, user programs are expected to create a ConstraintsContainer object themselves. The precise type is exported by the GridFunctionSpace class as

```
template<class T>
class GridFunctionSpace::ConstraintsContainer<T>::Type
```

where T is the number type used to store the constraints matrix entries. An example of how to access this type is given in Line 86 of the example code in the next section.

Assembly of the constraints is triggered by the free method

```
template<class Predicate, class GridFunctionSpace, class ConstraintsContainer>
void PDELab::constraints(const Predicate& predicate,
                         const GridFunctionSpace& gridFunctionSpace,
                         ConstraintsContainer& constraintsContainer,
                         const bool verbose = false)
```

The type Predicate must be a predicate class that determines the set of constrained degrees of freedom. For constraints of type ConformingDirichletConstraints, it must implement a method

```
template<class IntersectionGeometry>
bool isDirichlet(const IntersectionGeometry& intersection,
                 const IntersectionGeometry::Geometry::LocalCoordinate& xi)
                                                                         const
```

that returns **true** if the point with coordinates `xi` on the boundary intersection intersection is part of the Dirichlet boundary Γ_D.

The `ConstraintsContainer` type and object must then be handed over when the `GridOperator` is constructed. For this, the class has an extended template signature

```
template<class GridFunctionSpaceU, class GridFunctionSpaceV,
         class LocalOperator,
         class MatrixBackend,
         class DomainNumberType, class RangeNumberType,
         class JacobianNumberType,
         class ConstraintsU=PDELab::NoConstraints,
         class ConstraintsV=PDELab::NoConstraints>
class GridOperator
```

`ConstraintsContainer` objects for the trial and test spaces can then be handed to a dedicated constructor:

```
GridOperator(const GridFunctionSpaceU& gridFunctionSpaceU,
             const ConstraintsContainerU& constraintsContainerU,
             const GridFunctionSpaceV& gridFunctionSpaceV,
             const ConstraintsContainerV& constraintsContainerV,
             LocalOperator& localOperator,
             const MatrixBackend& matrixBackend = MatrixBackend())
```

`GridOperator` objects set up with this constructor will assemble the algebraic finite element problems in the constrained spaces.

The final step of setting up an algebraic finite element problem with Dirichlet constraints is to set or modify the initial iterate such that it fulfills the Dirichlet boundary conditions. This is usually done by projecting the Dirichlet values function w_g onto the space U_h. While dune-pdelab provides ways to do that, all examples in this book use the corresponding functionality of dune-functions directly (Chapter 10.6.3). See the example in Chapter 11.3.2 for how it is done.

Besides the constraint-aware `GridFunctionSpace` and `GridOperator` objects, the dune-pdelab module offers a range of further useful functionality for the handling of Dirichlet constraints. For example, the method

```
template<class ConstraintsContainer, class VectorContainer>
void set_constrained_dofs(const ConstraintsContainer& constraintsContainer,
                          VectorContainer::ElementType v,
                          VectorContainer& vector)
```

sets all constrained entries of the coefficient vector vector to the scalar value v. The object constraintsContainer is the constraints container that describes the Dirichlet constraints. The type VectorContainer is an array type determined by the current linear algebra backend (Chapter 11.4), and VectorContainer::ElementType is the corresponding number type. Similarly, the method

```
template<class ConstraintsContainer, class VectorContainer>
void copy_constrained_dofs(const ConstraintsContainer& constraintsContainer,
                           const VectorContainer& xIn, VectorContainer& xOut)
```

copies only the constrained degrees of freedom from the vector xIn to the vector xOut. Conversely, the methods

```
template<class ConstraintsContainer, class VectorContainer>
void set_nonconstrained_dofs(const ConstraintsContainer& constraintsContainer,
                             VectorContainer::ElementType v,
                             VectorContainer& vector)
```

and

```
template<class ConstraintsContainer, class VectorContainer>
void copy_nonconstrained_dofs(const ConstraintsContainer& constraintsContainer,
                              const VectorContainer& xIn,
                              VectorContainer& xOut)
```

do the same for the unconstrained degrees of freedom. Further convenience methods are defined in the file dune/pdelab/constraints/common/constraints.hh. See its online documentation for more details.

11.3.2. Example: The Poisson Equation with Dirichlet Boundary Conditions

To demonstrate how to implement Dirichlet boundary conditions in dune-pdelab, we revisit the Poisson example of Chapter 3.3. While the implementation in that chapter used only dune-functions and a hand-written assembler, we now use the full infrastructure of dune-pdelab. This results in markedly shorter code.

The problem to be solved is the Poisson equation

$$-\Delta u = -5 \tag{11.23}$$

on the L-shaped domain $\Omega = (0,1)^2 \setminus [0.5, 1)^2$, with Dirichlet boundary conditions

$$u = \mathbf{g} := \begin{cases} 0 & \text{on } \{0\} \times [0,1] \cup [0,1] \times \{0\}, \\ 0.5 & \text{on } \{0.5\} \times [0.5, 1] \cup [0.5, 1] \times \{0.5\}, \end{cases} \tag{11.24}$$

and homogeneous Neumann conditions on the remainder of the boundary. Domain and boundary conditions have been shown in Figure 3.1, and the grid can be seen in Figure 3.2.

The complete source code for the new example is contained in a single file, which is printed in Appendix B.14. Readers of an electronic version of this text can also access the file by clicking on the icon in the margin. The file only contains the `main` method and a small parameter class called `PoissonProblem`. The local operator that assembles the weak form of the Poisson problem (11.23) is provided by dune-pdelab, in form of a `LocalOperator` implementation called `ConvectionDiffusionFEM` (in the header file dune/pdelab/localoperator/convectiondiffusionfem.hh).[10] The parameter class is used to control this `LocalOperator` implementation. It derives from `PDELab::ConvectionDiffusionModelProblem`, and implements methods for the source term and for the type of boundary condition:

```
19  template<class GridView, class Range>
20  class PoissonProblem
21  : public PDELab::ConvectionDiffusionModelProblem<GridView,Range>
22  {
23  public:
24    using Traits
25        = typename PDELab::ConvectionDiffusionModelProblem<GridView,Range>::
26                                                              Traits;
27    // Source term
28    auto f(const typename Traits::ElementType& element,
29           const typename Traits::DomainType& xi) const
30    {
31      return -5.0;
32    }
33
34    //! Boundary condition type function
35    auto bctype (const typename Traits::IntersectionType& intersection,
36                 const typename Traits::IntersectionDomainType& xi) const
37    {
38      auto x = intersection.geometry().global(xi);
39      return ( x[0]< 1e-8 || x[1] < 1e-8 || (x[0] > 0.4999 && x[1] > 0.4999) )
40        ? PDELab::ConvectionDiffusionBoundaryConditions::Dirichlet
41        : PDELab::ConvectionDiffusionBoundaryConditions::Neumann;
42    }
43  };
```

The class receives template arguments that represent the grid view and the number type used for function values. It exports a type called `Traits`, which in turn exports compile-time information required by the local operator.

The method `f` expects a grid element and a local position in that element. In this particular implementation it always returns the number -5, which is the right-hand side of the equation (11.23). The second method `bctype` implements the partition of the domain boundary $\partial\Omega$ into the two parts Γ_D and Γ_N. It receives a boundary intersection and a local position `xi` on that intersection, and returns

---

[10]The local operator `ConvectionDiffusionFEM` can assemble much more general problems, which require more elaborate parameter classes. See its online documentation for details.

one of the enumeration values `Dirichlet` or `Neumann`. (These enumeration values live in the `PDELab::ConvectionDiffusionBoundaryConditions` namespace.) The `alpha_boundary` method of `ConvectionDiffusionFEM` will only assemble Neumann contributions on intersections for which `bctype` returns `Neumann`, and even though our specific example has no nonzero Neumann contributions, it is cleaner to implement the splitting correctly. The particular code here implements the partition given in (11.24).

The `main` method begins by setting up MPI, creating the unstructured grid from a GMSH file, and globally refining the grid twice:

```
47  int main(int argc, char *argv[])
48  {
49      // Initialize MPI, if available
50      MPIHelper::instance(argc, argv);
51
52      constexpr int dim = 2;
53      using Grid = UGGrid<dim>;
54      std::shared_ptr<Grid> grid = GmshReader<Grid>::read("l-shape.msh");
55
56      grid->globalRefine(2);
57
58      using GridView = Grid::LeafGridView;
59      GridView gridView = grid->leafGridView();
```

There is no difference so far to examples without Dirichlet constraints. The next code block then sets up the constrained grid function space. It will be used for trial and test functions, and therefore corresponds to the constrained linear spaces $U_h = V_h \subset H^1_{D,0}$ of Chapter 11.3.1:

```
63      using Basis = Functions::LagrangeBasis<GridView,1>;
64      auto basis = std::make_shared<Basis>(gridView);
65
66      using VectorBackend = PDELab::ISTL::VectorBackend<>;
67      using Constraints = PDELab::ConformingDirichletConstraints;
68
69      using GridFunctionSpace
70          = PDELab::Experimental::GridFunctionSpace<Basis,
71                                                    VectorBackend,
72                                                    Constraints>;
73
74      GridFunctionSpace gridFunctionSpace(basis);
```

In contrast to earlier examples, the third template argument of the `GridFunctionSpace` class is now `PDELab::ConformingDirichletConstraints`, a nontrivial constraints assembler object.

The actual linear constraints on the values of the coefficient vector are assembled next, using a `ConstraintsContainer` object. As we have set the constraints type to `ConformingDirichletConstraints`, the constraints container can only hold constraints of the type

$$u_i = 0, \qquad i \in \mathcal{I},$$

where u_i is a coefficient of the configuration vector $\bar{u} \in \mathbb{R}^n$, and $\mathcal{I} \subset \{0, \ldots, n-1\}$ is an index set. Therefore, the only information required is this index set, which is computed from the set of boundary intersections that form the Dirichlet boundary Γ_D. For consistency, it is taken from the PoissonProblem class. As the predicate method bctype is implemented as a member method of the parameter class PoissonProblem, it has to be extracted from there by the wrapper method called ConvectionDiffusionBoundaryConditionAdapter:

```
79   using Problem = PoissonProblem<GridView,double>;
80   Problem problem;
81   PDELab::ConvectionDiffusionBoundaryConditionAdapter<Problem> bctype(problem);
```

The grid function space exports the type of the ConstraintsContainer attached to it as GridFunctionSpace::ConstraintsContainer<double>::Type. The double argument is the number type used for the coefficients of the linear constraints. The call to the constraints method does the actual assembly of the constraints matrix:

```
85   using ConstraintsContainer
86       = GridFunctionSpace::ConstraintsContainer<double>::Type;
87   ConstraintsContainer constraintsContainer;
88   PDELab::constraints(bctype, gridFunctionSpace, constraintsContainer);
```

The next step is to set up a GridOperator object for the constrained trial and test function spaces. This is easy: The LocalOperator object is obtained by calling the ConvectionDiffusionFEM constructor with the PoissonProblem object.

In addition to the unconstrained code, the GridOperator receives the corresponding ConstraintsContainer, both as a template argument and as a constructor argument:

```
93   using LocalOperator
94       = PDELab::ConvectionDiffusionFEM<Problem,
95                                        GridFunctionSpace::Traits::
96                                                          FiniteElementMap>;
97   LocalOperator localOperator(problem);
98
99   using MatrixBackend = PDELab::ISTL::BCRSMatrixBackend<>;
100  MatrixBackend matrixBackend(7);
101
102  using GridOperator
103      = PDELab::GridOperator<GridFunctionSpace,
104                             GridFunctionSpace,
105                             LocalOperator,
106                             MatrixBackend,
107                             double,double,double,
108                             ConstraintsContainer,    // For trial space
109                             ConstraintsContainer>;   // For test space
110
```

```
111    GridOperator gridOperator(gridFunctionSpace,
112                               constraintsContainer,   // Trial space
113                               gridFunctionSpace,
114                               constraintsContainer,   // Test space
115                               localOperator,
116                               matrixBackend);
```

The `ConstraintsContainer` appears twice: once for the trial space and once for the test space.

After the constrained finite element space has been set up, the code constructs a coefficient vector u to hold the iterates of the solver, and initializes it with the Dirichlet boundary values:

```
121    using U = PDELab::Backend::Vector<GridFunctionSpace,double>;
122    U u(gridFunctionSpace,0.0);
123    auto g = [](auto x){return  (x[0]< 1e-8 || x[1] < 1e-8) ? 0 : 0.5;};
124    Functions::interpolate(*basis, PDELab::Backend::native(u), g);
```

The lambda g implements the function $w_g : \Omega \to \mathbb{R}$, whose restriction to the Dirichlet boundary Γ_D is the Dirichlet boundary value function g of (11.24). The linear solver computes corrections in U_h, which is a subspace of $H^1_{D,0}$, and hence contains only functions with zero Dirichlet values on Γ_D. Therefore, all iterates have the correct Dirichlet values. The call to `PDELab::native` is necessary to extract the actual dune-istl container from u. See Chapter 11.4 for details.

After this, setting up and invoking the algebraic solver is straightforward:

```
128    // Select a linear solver backend
129    using LinearSolverBackend = PDELab::ISTLBackend_SEQ_CG_SSOR;
130    LinearSolverBackend linearSolverBackend(50,2);
131
132    // Select linear problem solver
133    using LinearProblemSolver
134        = PDELab::StationaryLinearProblemSolver<GridOperator,
135                                                LinearSolverBackend,
136                                                U>;
137    LinearProblemSolver linearProblemSolver(gridOperator,
138                                            linearSolverBackend,
139                                            u,
140                                            1e-5);
141    // Solve linear problem.
142    linearProblemSolver.apply();
```

Similarly, writing the result to a VTK file remains unchanged from previous examples:

```
147    VTKWriter<GridView> vtkwriter(gridView);
148    auto uFunction
149        = Functions::makeDiscreteGlobalBasisFunction<double>(
150                                        *basis,
151                                        PDELab::Backend::native(u));
```

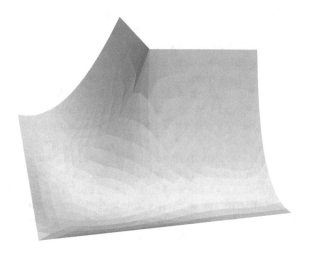

Figure 11.6.: Result of the Poisson problem (11.23) with Dirichlet boundary conditions (11.24). This is the same function as in Figure 3.3.

```
152   vtkwriter.addVertexData(uFunction,
153                   VTK::FieldInfo("u",
154                                   VTK::FieldInfo::Type::scalar,
155                                   1));
156   vtkwriter.write("pdelab-poisson-dirichlet-result");
```

This concludes the discussion of the example. Figure 11.6 shows visualizations of the result computed with this example.

11.4. Linear Algebra Backends

The result of a finite element assembler like the one implemented as the `GridOperator` class is a system of linear or nonlinear algebraic equations. As such, it needs data structures for the objects of numerical linear algebra: sparse matrices and vectors. Any finite element or finite volume code has to have them in one form or the other, and their efficiency can crucially influence the overall speed of the simulation.

While the design space for data structures for matrices and vectors is a bit smaller than the one for finite element grids, there nevertheless exists quite a number of different linear algebra implementations. Each has its strengths and weaknesses, and different people choose different libraries for their purposes. The situation is comparable to the choice of a grid data structure, where no single implementation

461

can fit all needs. To be truly flexible, a finite element software must therefore not tie the user to a particular linear algebra implementation.

DUNE attempts to achieve this flexibility at several levels. The grid interface from the dune-grid module meticulously avoids requiring a particular linear algebra software (Chapter 5.6). On a higher level of abstraction, dune-pdelab decouples the assembly process from linear algebra data structures. It defines an abstract interface for matrices and vectors, and provides several implementations for this interface. In the language of dune-pdelab, these different implementations are called *linear algebra backends.*

It would be more in the DUNE spirit to have dune-istl, the dedicated DUNE linear algebra module, declare the abstract interface. However, constructing an interface that is as rich and general as the one from dune-istl, and that is a the same time efficiently implementable by more than a single linear algebra library proved to be too difficult. The dune-pdelab module escapes from this dilemma by declaring an interface that is geared towards only a single user: dune-pdelab. It is more limited than a general-purpose linear algebra interface, and can therefore be implemented more easily. The dune-pdelab assemblers and iterative solvers only use methods from this interface. That way, the linear algebra is completely decoupled from the discretization, and different linear algebra libraries can be used, even without a deeper understanding of them. Currently, dune-pdelab provides backends for dune-istl, the EIGEN library[11][87], and for using data structures from the C++ standard library. These backends are all described in this chapter.

11.4.1. The ISTL Backend

In previous chapters of this book we have always tried to describe DUNE interfaces thoroughly, and systematically. For the linear algebra backends we deviate a little from this approach: As the interface exists only to be used by dune-pdelab, users of dune-pdelab will typically manipulate only a small part of it directly. Also, not all aspects of the interface are set in stone, and some may change in later revisions of dune-pdelab. As only dune-pdelab uses the interfaces, this instability is usually not a problem.

Rather than discussing the interface methods one by one, we therefore explain the linear algebra backends by discussing a typical use case. For this we revisit the code for the linear reaction–diffusion equation presented in Chapter 11.1. This code will repeatedly appear in this chapter, each time using a different linear algebra backend. All incarnations are taken from a single source file, which has been printed in Appendix B.15. Additionally, electronic versions of this document contain an embedded copy of the file.

Linear algebra backends provide three things for dune-pdelab: vectors, matrices, and linear solvers. The first example uses the ISTL backend, as it is both the most elaborate and the most widely used. The example consists of a single method that

---

[11]eigen.tuxfamily.org

receives a grid view and a file name for the result file. The first part of the code constructs the basis of a second-order Lagrange space on the given grid view, and uses it to construct a GridFunctionSpace, i.e., a function space in the dune-pdelab sense:

```
39  template<class GridView>
40  void solveReactionDiffusionProblemISTL(const GridView& gridView,
41                                          std::string filename)
42  {
43    // Make grid function space
44    using Basis = Functions::LagrangeBasis<GridView,2>;
45    auto basis = std::make_shared<Basis>(gridView);
46
47    using VectorBackend = PDELab::ISTL::VectorBackend<>;
48
49    using GridFunctionSpace
50       = PDELab::Experimental::GridFunctionSpace<Basis,
51                                                 VectorBackend,
52                                                 PDELab::NoConstraints>;
53
54    GridFunctionSpace gridFunctionSpace(basis);
```

The crucial part here is Line 47, which selects the vector backend, i.e., the data structure to use for coefficient vectors. The fact that it is taken from the PDELab::ISTL namespace selects a vector data structure from the dune-istl module. The currently available backend for such vectors is

```
template<Blocking blocking = Blocking::none, std::size_t blocksize = 1>
class VectorBackend
```

It selects the data type BlockVector<FieldVector<NumberType,blocksize> > from the dune-istl module to be used for coefficient vectors, where blocksize is the number given to the VectorBackend template, and the NumberType is later determined by the GridOperator. For the first template parameter of ISTL::VectorBackend, there is the choice between Blocking::none, Blocking::bcrs, and Blocking::fixed, but everything besides the default value none is beyond the scope of this book. The VectorBackend is given to the GridFunctionSpace in Line 51, and from there on everything works automagically.[12]

The next part of the code constructs the finite element assembler, i.e., the device that turns the finite element problem into a linear algebra one:

```
58    // Make grid operator
59    LinearReactionDiffusionProblem<GridView,double> problem;
60    using LocalOperator
61       = PDELab::ConvectionDiffusionFEM<decltype(problem),
```

---

[12]Discrete function spaces in dune-pdelab know what kind of container is used for coefficient vectors. This distinguishes GridFunctionSpace objects from the basis objects in dune-functions, which exist separately from any implementation of coefficient containers.

```
62                                       typename GridFunctionSpace::Traits::
63                                                       FiniteElementMap>;
64      LocalOperator localOperator(problem);
65
66      using MatrixBackend = PDELab::ISTL::BCRSMatrixBackend<>;
67      using GridOperator
68          = PDELab::GridOperator<GridFunctionSpace,    // Trial function space
69                                 GridFunctionSpace,    // Test function space
70                                 LocalOperator,        // Element assembler
71                                 MatrixBackend,        // Data structure
72                                                       // for the stiffness matrix
73                                 double,               // Number type for
74                                                       // solution vector entries
75                                 double,               // Number type for
76                                                       // residual vector entries
77                                 double>;              // Number type for
78                                                       // stiffness matrix entries
79
80      MatrixBackend matrixBackend(25);
81      GridOperator gridOperator(gridFunctionSpace,
82                                gridFunctionSpace,
83                                localOperator,
84                                matrixBackend);
```

The GridOperator object needs to know the data structure to assemble to stiffness matrix into. Line 66 selects an appropriate matrix backend. For the ISTL backend, the standard choice is

```
template<class EntriesPerRow = std::size_t>
class ISTL::BCRSMatrixBackend
```

which uses a BCRSMatrix<FieldMatrix> as the underlying container. The number type is set by the grid operator; it is the seventh entry from the GridOperator template list.

The type of the matrix backend is used in the construction of the global assembler, implemented by the GridOperator class. This is not surprising: the global assembler has to know the matrix type it assembles into. It also needs to know the vector types, but it does not receive them directly: rather the two GridFunctionSpace objects hand over the vector types.

The vector backend was selected by choosing a particular C++ type. In contrast, one may have to construct a MatrixBackend *object*, and hand it to the GridOperator. This allows to give the MatrixBackend additional information. For example, the dune-istl matrix backend needs an estimate of the average number of matrix entries per row, because it uses the implicit build mode of Chapter 7.3.2 internally. This is implementation-dependent; other backends may require different information here.

Finally, the code selects a linear solver, and solves the resulting linear system:

```
88   // Select vector data type to hold the iterates
89   using VectorContainer = PDELab::Backend::Vector<GridFunctionSpace,double>;
90   VectorContainer u(gridFunctionSpace,0.0);      // Initial iterate
91
92   // Select a linear solver backend
93   using LinearSolverBackend = PDELab::ISTLBackend_SEQ_CG_SSOR;
94   LinearSolverBackend linearSolverBackend(5000,    // Maximal number
95                                                    // of iterations
96                                            2);     // Verbosity level
97
98   // Solve linear problem
99   using LinearProblemSolver
100      = PDELab::StationaryLinearProblemSolver<GridOperator,
101                                              LinearSolverBackend,
102                                              VectorContainer>;
103  LinearProblemSolver linearProblemSolver(gridOperator,
104                                          linearSolverBackend,
105                                          u,
106                                          1e-10);
107  linearProblemSolver.apply();
```

Line 89 sets up the container that holds the current iterate of the iterative solver. The precise type of this is determined by the vector backend. The ISTL vector backend used in this example uses a `BlockVector<FieldVector>` with a correct `FieldVector` size internally. Lines 93–96 select the linear solver backend, in this case a (sequential) conjugate gradient algorithm preconditioned by an SSOR method [138]. Alternative choices are `ISTLBackend_SEQ_BCGS_SSOR`, a stabilized bi-conjugate gradient method with the same preconditioner, and `ISTLBackend_SEQ_SuperLU`, which uses the direct sparse solver SUPERLU[13] [116]. The available solver backends are listed in Table 11.2. Additionally, there is a choice of parallel solvers, which are however beyond the scope of this book.

A linear solver backend object needs to be created by the user, and usually it will need some implementation-dependent settings. In the present example, the numbers are the maximum number of iterations, and the verbosity level. Both linear solver backend type and object are then handed to the `StationaryLinearProblemSolver` class, which assembles the problem and calls the linear solver. The precise interface of the linear solver backend implementations is typically never seen by the user, because it is only used by `StationaryLinearProblemSolver` and the Newton solver.

For completeness, we show the final part of the example, which writes the result to a VTK file. Code like this has already appeared a few times in earlier examples of this chapter. Observe that `makeDiscreteGlobalBasisFunction` from dune-functions requires an actual array rather than an object of the dune-pdelab vector backend. The free method

---

[13]https://portal.nersc.gov/project/sparse/superlu

solver backend	solver	preconditioner
`ISTLBackend_SEQ_LOOP_Jac`	Loop	Jacobi
`ISTLBackend_SEQ_BCGS_Jac`	BiCGStab	Jacobi
`ISTLBackend_SEQ_BCGS_SSOR`	BiCGStab	SSOR
`ISTLBackend_SEQ_CG_SSOR`	CG	SSOR
`ISTLBackend_SEQ_BCGS_ILU0`	BiCGStab	ILU0
`ISTLBackend_SEQ_CG_ILU0`	CG	ILU0
`ISTLBackend_SEQ_BCGS_ILUn`	BiCGStab	ILU(n)
`ISTLBackend_SEQ_CG_ILUn`	CG	ILU(n)
`ISTLBackend_SEQ_MINRES_SSOR`	MinRes	SSOR
`ISTLBackend_SEQ_CG_AMG_SSOR`	CG	AMG
`ISTLBackend_SEQ_BCGS_AMG_SSOR`	BiCGStab	AMG
`ISTLBackend_SEQ_SuperLU`	SuperLU	—
`ISTLBackend_SEQ_UMFPack`	UMFPack	—

Table 11.2.: List of available solvers from the ISTL backend. Most combine a preconditioner with an iterative method. Others wrap third-party direct solver implementations.

```
NativeVector& PDELab::Backend::native(const VectorContainer& u)
```

produces a `dune-istl` vector from the vector backend. More on this kind of direct access is given in Chapter 11.4.3:

```
111    // Output as VTK file
112    SubsamplingVTKWriter<GridView> vtkwriter(gridView,refinementLevels(2));
113    auto uFunction
114        = Functions::makeDiscreteGlobalBasisFunction<double>(
115                                    *basis,
116                                    PDELab::Backend::native(u));
117    vtkwriter.addVertexData(uFunction,
118                    VTK::FieldInfo("u",
119                                VTK::FieldInfo::Type::scalar,
120                                1));
121    vtkwriter.write(filename);
122  }
```

This marks the end of the `solveReactionDiffusionProblemISTL` method. The following sections discuss variations of it that use other linear algebra backends.

solver backend	solver	preconditioner
`EigenBackend_BiCGSTAB_IILU`	BiCGSTAB	ILU
`EigenBackend_BiCGSTAB_Diagonal`	BiCGSTAB	Jacobi
`EigenBackend_CG_IILU_Up`	CG	ILU
`EigenBackend_CG_Diagonal_Up`	CG	Jacobi
`EigenBackend_CG_IILU_Lo`	CG	ILU
`EigenBackend_CG_Diagonal_Lo`	CG	Jacobi
`EigenBackend_SimplicialCholesky_Up`	Cholesky	—
`EigenBackend_SimplicialCholesky_Lo`	Cholesky	—

Table 11.3.: List of available solvers from the EIGEN backend

11.4.2. The Eigen Backend

The Eigen backend uses the MPL2-licensed linear algebra library EIGEN[14][87]. EIGEN puts a lot of attention on execution speed, and makes heavy use of expression templates [1, 156].

The Eigen backend offers one vector backend and one matrix backend. The vector backend is

```
class PDELab::EigenVectorBackend
```

and offers no parameters. Internally, it uses the

```
Eigen::Matrix<double, Eigen::Dynamic, 1>
```

data structure from the EIGEN library (a one-column matrix with dynamic number of rows). This is the standard EIGEN type for dense vectors.

The matrix backend is

```
template<int Options = Eigen::RowMajor>
class PDELab::EigenMatrixBackend
```

It uses the `SparseMatrix` matrix class from the EIGEN library, which is a sparse matrix stored in a variant of the compressed-row-storage format. The `Options` argument is directly handed over to EIGEN. It selects whether the matrix uses row-major or column-major storage internally.

The Eigen backend also wraps several sparse linear solvers from the EIGEN library. A selection is listed in Table 11.3. The `_Up` and `_Lo` suffixes indicate whether the `Eigen::upper` or `Eigen::lower` flags are passed to the underlying EIGEN solvers. These indicate whether the triangular decompositions take place in the upper or lower triangular part of the matrix. Each solver backend has a constructor that takes the maximum number of iterations as its single argument.

---

[14] eigen.tuxfamily.org

Switching the reaction–diffusion example from the ISTL backend to its EIGEN counterpart requires only a few small changes, which demonstrates the strengths of the backend concept. We print the entire modified method here for completeness. The only differences are in Lines 136, 153, and 179, where the vector, matrix, and solver backends are selected. As for the ISTL backend, the MatrixBackend object needs to be given the expected average number of nonzero entries per row:

```
128   template<class GridView>
129   void solveReactionDiffusionProblemEigen(const GridView& gridView,
130                                           std::string filename)
131   {
132     // Make grid function space
133     using Basis = Functions::LagrangeBasis<GridView,2>;
134     auto basis = std::make_shared<Basis>(gridView);
135
136     using VectorBackend = PDELab::Eigen::VectorBackend;
137
138     using GridFunctionSpace
139         = PDELab::Experimental::GridFunctionSpace<Basis,
140                                                   VectorBackend,
141                                                   PDELab::NoConstraints>;
142
143     GridFunctionSpace gridFunctionSpace(basis);
144
145     // Make grid operator
146     LinearReactionDiffusionProblem<GridView,double> problem;
147     using LocalOperator
148         = PDELab::ConvectionDiffusionFEM<decltype(problem),
149                                 typename GridFunctionSpace::Traits::
150                                               FiniteElementMap>;
151     LocalOperator localOperator(problem);
152
153     using MatrixBackend = PDELab::Eigen::MatrixBackend<>;
154     using GridOperator
155         = PDELab::GridOperator<GridFunctionSpace,    // Trial function space
156                                GridFunctionSpace,    // Test function space
157                                LocalOperator,        // Element assembler
158                                MatrixBackend,        // Data structure
159                                                      // for the stiffness matrix
160                                double,               // Number type for
161                                                      // solution vector entries
162                                double,               // Number type for
163                                                      // residual vector entries
164                                double>;              // Number type for
165                                                      // stiffness matrix entries
166
167
168     MatrixBackend matrixBackend(25);
```

```
169    GridOperator gridOperator(gridFunctionSpace,
170                              gridFunctionSpace,
171                              localOperator,
172                              matrixBackend);
173
174    // Select vector data type to hold the iterate
175    using VectorContainer = PDELab::Backend::Vector<GridFunctionSpace,double>;
176    VectorContainer u(gridFunctionSpace,0.0);    // Initial iterate
177
178    // Select a linear solver backend
179    using LinearSolverBackend = PDELab::EigenBackend_CG_Diagonal_Up;
180    LinearSolverBackend linearSolverBackend(5000);
181
182    // Solve linear problem
183    using LinearProblemSolver
184        = PDELab::StationaryLinearProblemSolver<GridOperator,
185                                                LinearSolverBackend,
186                                                VectorContainer>;
187    LinearProblemSolver linearProblemSolver(gridOperator,
188                                            linearSolverBackend,
189                                            u,
190                                            1e-10);
191    linearProblemSolver.apply();
192
193    // Output as VTK file
194    SubsamplingVTKWriter<GridView> vtkwriter(gridView,refinementLevels(2));
195    auto uFunction
196        = Functions::makeDiscreteGlobalBasisFunction<double>(
197                                                *basis,
198                                                PDELab::Backend::native(u));
199    vtkwriter.addVertexData(uFunction,
200                            VTK::FieldInfo("u",
201                                           VTK::FieldInfo::Type::scalar,
202                                           1));
203    vtkwriter.write(filename);
204 }
```

Notice the similarity to the corresponding ISTL example.

11.4.3. Working with the Actual Data Structures

In the standard use case presented so far, a lot of the details happen behind the scenes. The linear algebra library is controlled by dune-pdelab, and the user never gets to see the underlying data structures. For example, no actual dune-istl vector and matrix objects ever appear in the code of the ISTL backend example in Chapter 11.4.1, except in Line 116 where the coefficient vector of the solution is handed to dune-functions. This is of course very convenient: The linear algebra implementation can be switched by changing only a few type definitions. Only minimal knowledge of the underlying libraries is needed, and the user code is short.

On the other hand, sometimes it is necessary to have more control. For example, one may want to use the nested dune-istl data structures together with a solver that is not available in dune-istl. For such a purpose the native data structures can be accessed directly. Doing this, the code becomes tied to a particular linear algebra implementation, but the user gains back full control in exchange.

The dune-pdelab abstraction of a vector is called a *vector container*. Such a vector container is always associated to a GridFunctionSpace. For a given GridFunctionSpace, the corresponding vector container type is available as

```
template<class GridFunctionSpace, class NumberType>
class PDELab::Backend::Vector
```

We do not discuss the full interface of this class, but let us mention that for any vector container type VectorContainer corresponding to some GridFunctionSpace there are constructors

```
VectorContainer(const GridFunctionSpace& gridFunctionSpace)
```

and

```
VectorContainer(const GridFunctionSpace& gridFunctionSpace,
                const NumberType& v)                 // Initialize with scalar v
```

that allow to construct a coefficient vector with or without initializing the content. There is also a constructor that takes a native vector data structure

```
VectorContainer(const GridFunctionSpace& gridFunctionSpace,
                Container& container)
```

With this constructor, the VectorContainer will use the given array container as its storage. The type Container is the actual vector container type used by the backend, without further encapsulation. The VectorContainer abstraction also has a limited interface for linear algebra operations, which will make a brief appearance in Chapter 11.4.4.

To get access to the actual array data structure of a VectorContainer there is the free method

```
template<class VectorContainer>
Vector& PDELab::Backend::native(VectorContainer& vectorContainer)
```

For example, suppose that x is an object of type VectorContainer. Then write

```
auto& xNative = PDELab::Backend::native(x);
```

for a reference to its content. When using the ISTL backend, then xNative will be a reference to a BlockVector<FieldVector<NumberType,blocksize> > object, where the number blocksize is determined by the template argument given to the

VectorBackend class (Chapter 11.4.1), and NumberType is specified as part of the PDELab::Backend::Vector type, i.e.,

```
using VectorContainer = PDELab::Backend::Vector<GridFunctionSpace,NumberType>;
```

Accessing the native matrix data structure works in a similar way. The dune-pdelab abstraction of a matrix is called MatrixContainer, and it is always tied to the types for vectors for the domain and range spaces of the matrix. A matrix container for a given backend is available as

```
template<class Backend,
         class VectorContainerU,
         class VectorContainerV,
         class NumberType>
class PDELab::Backend::Matrix
```

The first of the four template parameters is the MatrixBackend type. The following two are VectorContainer types for domain and range space vectors respectively, i.e., for vectors that can be multiplied from the right and from the left to the matrix, respectively. For backends like the ISTL one, this means in particular that the vector nesting and blocking patterns are appropriate. Finally, the fourth parameter is the number type used for scalar matrix entries.

Constructing MatrixContainer objects requires a GridOperator, i.e., a global assembler. The following constructors are available:

```
template<class GridOperator>
MatrixContainer(const GridOperator& gridOperator)

template<class GridOperator>
MatrixContainer(const GridOperator& gridOperator,
                const NumberType& v)     // Initial value for all entries

template<class GridOperator>
MatrixContainer(const GridOperator& gridOperator, Container& container)
```

All three constructors initialize the matrix pattern using the gridOperator object. The first one leaves the values uninitialized, whereas the second one fills all matrix entries with the scalar v. The third constructor takes a native matrix, and uses that for storage, rather than allocating its own. Access to the underlying matrix data structure is possible using the free method

```
template<class MatrixContainer>
Matrix& PDELab::Backend::native(MatrixContainer& matrixContainer)
```

again.

When constructing a grid operator, it has to be given the matrix backend, and two vector backends for trial and test vectors, respectively (Chapter 11.2). Conversely, given a GridOperator, the vector and matrix types are also available as:

```
GridOperator::Traits::Domain    // For primal vectors, e.g., iterates
GridOperator::Traits::Range     // For dual vectors, e.g., residuals
GridOperator::Jacobian          // Tangent stiffness matrix type
```

The GridOperator object has member methods for assembling the residual and its Jacobian directly. First, the method

```
void residual(const Domain& x, Range& r) const
```

assembles the residual vector $\mathcal{R}(\bar{u}) \in \mathbb{R}^n$ at a given configuration $\bar{u} \in \mathbb{R}^n$, and adds it to the coefficient vector given in the parameter r. Both arguments Domain and Range are implementations of the VectorContainer interface. Note that, by (11.8), the residual of the linear system $Ax = b$ is $r(x) = Ax - b$. Hence, to obtain the vector b, the residual $\mathcal{R}(0)$ must be multiplied by -1. Likewise, the method

```
void jacobian(const Domain& u, Jacobian& nablaR) const
```

assembles the Jacobian $\nabla \mathcal{R}(\bar{u}) \in \mathbb{R}^{n \times n}$ of the residual $\mathcal{R}$ at a given configuration $\bar{u} \in \mathbb{R}^n$. If the problem is linear, then this method yields the stiffness matrix. The result is added to the nablaR variable, which is an implementation of the MatrixContainer interface.

We show how to use these features in another incarnation of the reaction–diffusion example. The code uses the ISTL backend again to assemble the linear algebraic problem, but for solving this problem it acquires the vector and matrix data structures directly, and calls a dune-istl solver with them without going through the corresponding dune-pdelab linear solver backend.

As the example is very similar to the dune-istl example of Chapter 11.4.1, only the differences will be discussed. In particular, we skip the construction of the grid function space and the grid operator. The first difference can be found at Line 255 (which corresponds to Line 89 in Chapter 11.4.1), where the code creates a vector for the iterates of the linear solver, and initializes it with zero:

```
255    using VectorContainer = typename GridOperator::Traits::Domain;
256    VectorContainer xContainer(gridFunctionSpace, 0.0);
```

Unlike in Line 89, the vector container type is extracted from the GridOperator for a change. Then, the GridOperator is called to assemble the current residual and its Jacobian:

```
260    // Evaluate residual at the zero configuration
261    typename GridOperator::Traits::Range rContainer(gridFunctionSpace, 0.0);
262
263    gridOperator.residual(xContainer,rContainer);
264
265    // Compute stiffness matrix
266    using MatrixContainer = typename GridOperator::Jacobian;
```

```
267    MatrixContainer matrixContainer(gridOperator, 0.0);
268
269    gridOperator.jacobian(xContainer,matrixContainer);
```

As the problem is linear, the matrix does not actually depend on xContainer, but the interface requires that vector container to be handed over anyway.

For the actual solving, the code retrieves the native dune-istl data structures that contain the current iterate, the negative residual at zero, and the stiffness matrix:

```
273    auto& x = PDELab::Backend::native(xContainer);
274    auto& b = PDELab::Backend::native(rContainer);
275    b *= -1.0;
276    const auto& stiffnessMatrix = PDELab::Backend::native(matrixContainer);
```

Note that, by (11.8), the residual of the linear system $Ax = b$ is $\mathcal{R}(x) = Ax - b$. Hence, to obtain the vector b, the residual $\mathcal{R}(0)$ must be multiplied by -1. This multiplication is performed using a method of the BlockVector interface.

For subsequent use the code also defines names for the native types:

```
280    using DomainVector = std::decay_t<decltype(x)>;   // Decay from reference
281                                                      // to value type
282    using RangeVector  = std::decay_t<decltype(b)>;
283    using Matrix       = std::decay_t<decltype(stiffnessMatrix)>;
```

Finally, the code calls a native dune-istl solver. This part of the code is similar to earlier examples like the one in Chapter 3.3:

```
287    // Turn the matrix into a linear operator
288    MatrixAdapter<Matrix,DomainVector,RangeVector>
289                                        linearOperator(stiffnessMatrix);
290
291    // SSOR as the preconditioner
292    SeqSSOR<Matrix,DomainVector,RangeVector> preconditioner(stiffnessMatrix,
293                                                    1,      // Number of
294                                                            // iterations
295                                                    1.0);   // Damping
296
297    // Preconditioned conjugate-gradient solver
298    CGSolver<DomainVector> cg(linearOperator,
299                              preconditioner,
300                              1e-4,   // Desired residual reduction factor
301                              50,     // Maximum number of iterations
302                              2);     // Verbosity level
303
304    // Object storing some statistics about the solving process
305    InverseOperatorResult statistics;
306
307    // Solve!
308    cg.apply(x, b, statistics);
```

While this example as it is does not provide any advantages over the one from Chapter 11.4.1, it demonstrates how the assembled algebraic problem can be extracted from dune-pdelab, to feed it into alternative unrelated solver implementations.

11.4.4. The Simple Backend

The Simple backend is, as the name says, intended to be simple. It uses no external linear algebra software at all, relying only on features of the C++ standard library. It provides no solvers, and therefore linear algebraic problems have to be fed to third-party solvers using the techniques of the previous section.

The Simple backend provides a single VectorContainer, which is available as

```
template<class Container = Simple::default_vector>
class PDELab::Simple::VectorBackend
```

It uses std::vector internally, but this choice can be overridden by setting the Container parameter.

There is one matrix backend for dense matrices and one for sparse ones. Since the C++ standard library does not provide matrices, the Simple backend actually contains simple matrix implementations itself. The first,

```
template<class Container = Simple::default_vector>
class PDELab::Simple::MatrixBackend
```

implements a dense matrix that stores its entries in a std::vector in row-wise ordering. The second one,

```
template<class Container = Simple::default_vector,
         class IndexType = std::size_t>
class PDELab::Simple::SparseMatrixBackend
```

is a simple compressed-row-storage sparse matrix [133]. The first template parameter is the array type used for all internal arrays (again defaulting to std::vector). The second parameter is the type used for row and column indices.

The reaction–diffusion example using the Simple backend differs from the others mainly as far as the solver is concerned. The first part sets up the GridFunctionSpace and selects Simple::VectorBackend<> as the vector data type:

```
320  template<class GridView>
321  void solveReactionDiffusionProblemSimple(const GridView& gridView,
322                                           std::string filename)
323  {
324    // Make grid function space
325    using Basis = Functions::LagrangeBasis<GridView,2>;
326    auto basis = std::make_shared<Basis>(gridView);
327
328    using VectorBackend = PDELab::Simple::VectorBackend<>;
329
```

```
330    using GridFunctionSpace
331        = PDELab::Experimental::GridFunctionSpace<Basis,
332                                                   VectorBackend,
333                                                   PDELab::NoConstraints>;
334
335    GridFunctionSpace gridFunctionSpace(basis);
```

Then, the code sets up the GridOperator, using Simple::SparseMatrixBackend as the matrix backend:

```
340    LinearReactionDiffusionProblem<typename Basis::GridView,double> problem;
341    using LocalOperator
342        = PDELab::ConvectionDiffusionFEM<decltype(problem),
343                                          typename GridFunctionSpace::Traits::
344                                                       FiniteElementMap>;
345    LocalOperator localOperator(problem);
346
347    using MatrixBackend = PDELab::Simple::SparseMatrixBackend<>;
348    using GridOperator
349        = PDELab::GridOperator<GridFunctionSpace,      // Trial function space
350                               GridFunctionSpace,      // Test function space
351                               LocalOperator,          // Element assembler
352                               MatrixBackend,          // Data structure
353                                                       // for the stiffness matrix
354                               double,                 // Number type for
355                                                       // solution vector entries
356                               double,                 // Number type for
357                                                       // residual vector entries
358                               double>;                // Number type for
359                                                       // stiffness matrix entries
360
361
362    GridOperator gridOperator(gridFunctionSpace,
363                              gridFunctionSpace,
364                              localOperator);
```

No parameters need to be handed to the matrix backend, and therefore no matrix backend object needs to be constructed in the user code at all.

The code then proceeds similarly to Chapter 11.4.3: It constructs a coefficient vector filled with zeros, and uses that to explicitly compute the stiffness matrix and the right hand side of the linear system of equations:

```
368    // Vector for the iterates
369    typename GridOperator::Traits::Domain xContainer(gridFunctionSpace,0.0);
370
371    // Evaluate residual at the zero vector
372    typename GridOperator::Traits::Range rContainer(gridFunctionSpace, 0.0);
373
374    gridOperator.residual(xContainer,rContainer);
```

```
375    rContainer *= -1.0;
376
377    // Compute stiffness matrix
378    using MatrixContainer = typename GridOperator::Traits::Jacobian;
379    MatrixContainer matrixContainer(gridOperator, 0.0);
380
381    gridOperator.jacobian(xContainer,matrixContainer);
```

Again, as the residual at $x \in \mathbb{R}^n$ is $\mathcal{R}(x) = Ax - b$ (see (11.8)), the Jacobian is the stiffness matrix, and the residual at $x = 0$ is the *negative* load vector b. This time, however, the code uses a method of the dune-pdelab VectorContainer interface to compute the sign change (in Line 375).

At this point one could extract the native data structures and hand them to the desired solver, as shown in the previous section. To keep this exposition short we instead implement a simple damped Richardson iteration [138]

$$x_{k+1} = x_k + \omega(b - Ax_k) \qquad \omega > 0,$$

using the linear algebra methods of the vector and matrix backends:

```
385    double omega = 0.2;
386    for (int i=0; i<200; i++)
387    {
388      // Damped Richardson iteration
389      auto correction = rContainer;
390      matrixContainer.usmv(-1.0,xContainer,correction);
391      xContainer.axpy(omega,correction);
392    }
```

The method usmv is a method from the dune-pdelab matrix container interface. It computes $-1 \cdot Ax$, and adds the result to the vector correction. It is not a method from the dune-istl matrix interface, even though the same method exists there.

As a final remark, note how easy it would be to write a matrix-free variant of this code: Rather than computing the residual $b - Ax_k$ by matrix–vector multiplication, it could also be computed directly at each iteration by calling the GridOperator::residual method. Then, the stiffness matrix would not be needed at all, which would possibly save a considerable amount of memory.

11.5. Local Grid Adaptivity

When the solution of a partial differential equation shows very localized features, local grid adaptivity can help to reduce the computation time without sacrificing accuracy. A prime example is the Poisson problem of Chapter 3.3, where the inward corner of the L-shaped domain leads to a singularity in the solution. Standard Lagrange discretizations will not converge with the optimal order, unless the degrees of freedom are concentrated around the singularity. Example grids with such locally varying resolutions have been shown in Figures 2.15, 2.19, and 2.20.

Grid adaptivity has already been discussed in Chapter 5.9 in the context of the DUNE grid interface. The DUNE grid interface focuses on *h*-refinement, i.e., splitting and merging of elements to control the local resolution (Chapter 2.3.2). There are interface methods for local refinement and coarsening of elements, but grid implementations are not required to implement them, as they do not make sense for all kinds of grid data structures. Most unstructured grids available for DUNE do support grid adaptation, though—more details are given in Chapter 5.10.

Besides the actual modifications of the grid data structure, the challenge when implementing local grid adaptivity is to retain simulation data attached to the grid across the grid modification. Any discrete function attached to the grid should still be available after a grid refinement or coarsening if desired, with as little loss of approximation quality as the new grid permits. Chapter 5.9 showed how this can be done on the level of the grid interface. This approach gives the programmer full control over data handling, interpolation methods, and other details of the algorithm. However, it is tedious, because it operates on a very low abstraction level, and quite a bit of code is needed even for simple finite element spaces. For easier coding of standard situations, dune-pdelab provides a higher-level interface for grid adaptivity and persistent functions. It offers implementations of standard marking strategies, and provides persistent finite element functions even for complicated finite element spaces. This chapter briefly presents some of this functionality, and then discusses a complete example.

11.5.1. Local Adaptivity in `dune-pdelab`

On a very abstract level, solving stationary PDEs in dune-pdelab with adaptive *h*-refinement follows the classical refinement loop, which has already been explained in Chapter 2.3. Figure 11.7 shows shows the loop as it is seen by dune-pdelab. The first two steps, **assemble** and **solve**, set up the algebraic formulation of the PDE problem, and compute its solution. Earlier parts of this chapter have already shown how to do that with dune-pdelab. The **estimate** step does two things: first of all it tries to estimate the overall error. If this global error is sufficiently low then the loop can be terminated. Secondly, the **estimate** step computes local indicators of the error, which are associated to grid elements. Error estimation is highly problem-specific, and needs to be provided by the user. The example in Chapter 11.5.2 will show how to use the `LocalOperator` machinery for this purpose. The final three steps modify the grid as suggested by the error indicators. In contrast to the classical refinement loop in Figure 2.12, the dune-pdelab loop has a separate **judge** step, which is seen as part of **mark** elsewhere. The **judge**, **mark**, and **adapt** steps are provided generically by dune-pdelab in form of several methods in the file dune/pdelab/adaptivity/adaptivity.hh. We suppose that the **estimate** step has produced a vector $\eta^2 = (\eta_T^2)_{T \in \mathcal{T}}$ of squared error indicators associated to the grid elements, and discuss the following three steps in turn.

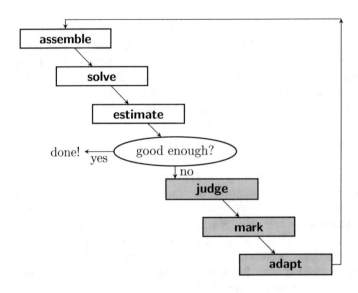

Figure 11.7.: Local grid adaptivity workflow in dune-pdelab

judge: Determining Which Elements to Modify

The **judge** step determines the set of elements to refine and coarsen, based on the error indicator produced by the **estimate** step. The dune-pdelab interface assumes that local error indicators are always associated to grid elements. Let therefore $n_\mathcal{T}$ be the number of grid elements, and $\eta^2 \in \mathbb{R}^{n_\mathcal{T}}$ the vector of element error indicators. Several strategies exist to turn such indicators into refinement marks. For example, one may want to mark a fixed fraction of the elements for refinement, and a fixed fraction for coarsening. This strategy is implemented by the method

```
template<class VectorContainer>
void element_fraction(const VectorContainer& etaSquared,
                      VectorContainer::ElementType alpha,
                      VectorContainer::ElementType beta,
                      VectorContainer::ElementType& refineThreshold,
                      VectorContainer::ElementType& coarsenThreshold,
                      int verbose=0)
```

The first parameter, etaSquared, is the array of element error indicators η^2, given as an object of type VectorContainer from the linear algebra backend (Chapter 11.4). This VectorContainer must be associated to a GridFunctionSpace of piecewise constant functions with respect to the current leaf grid view, which determines the ordering of the vector entries.

The parameters alpha and beta are the fractions of elements to refine and coarsen, respectively. For given values $0 \le$ alpha, beta ≤ 1, the element_fraction method marks the (approximately) alpha $\cdot n_{\mathcal{T}}$ elements with the largest error indicator values for refinement, and the (approximately) beta $\cdot n_{\mathcal{T}}$ elements with lowest values for coarsening. Note, however, that the method does not actually mark any elements in the sense of the DUNE grid interface (Chapter 5.9). Rather, it returns two threshold values in the output parameters refineThreshold and coarsenThreshold. After completion of the element_fraction method, they contain the error indicator values above and below which elements should be marked for refinement and coarsening, respectively, to achieve the desired fractions. The two numbers serve as input for the **mark** stage of the refinement loop, which does the actual marking. Finally, the verbose parameter allows to control the amount of screen output produced by the method.

Instead of selecting the top alpha $\cdot n_{\mathcal{T}}$ elements for refinement, one may also have the fraction alpha refer to the accumulated squared error values. In other words, a set $\mathcal{M}$ of elements is marked for refinement such that

$$\sum_{T \in \mathcal{M}} \eta_T^2 \geq \text{alpha} \sum_{T \in \mathcal{T}} \eta_T^2.$$

This approach is known in the literature as *Dörfler marking* [57, 157]. It is provided by dune-pdelab in form of the method

```
template<class VectorContainer>
void error_fraction(const VectorContainer& etaSquared,
                VectorContainer::ElementType alpha,
                VectorContainer::ElementType beta,
                VectorContainer::ElementType& refineThreshold,
                VectorContainer::ElementType& coarsenThreshold,
                int verbose=0)
```

The parameters of this method have precisely the same meaning as the corresponding ones of element_fraction.

mark: Marking the Grid for Local Modification

In the **mark** step, the grid elements on the leaf grid view are marked for refinement or coarsening, without actually modifying the grid yet. The step is implemented in form of the free method

```
template<class Grid, class VectorContainer>
void mark_grid(Grid &grid, const VectorContainer& etaSquared,
            VectorContainer::ElementType refineThreshold,
            VectorContainer::ElementType coarsenThreshold,
            int minLevel = 0,
            int maxLevel = std::numeric_limits<int>::max(),
            int verbose=0)
```

The first parameter of this is the (hierarchical) grid to modify. The second parameter is a vector with the squared element error indicators, just as for the element_fraction and `error_fraction` methods. The third and fourth parameters are the refinement and coarsening thresholds computed during the **judge** step. An element is marked for refinement if its error indicator given in `etaSquared` is larger than the refinement threshold given in `refineThreshold`. Likewise, the element is marked for coarsening if the indicator is below the value given in `coarsenThreshold`. Usually, the values computed by `element_fraction` or `error_fraction` are used here. Additionally, two optional parameters `minLevel` and `maxLevel` allow to set upper and lower bounds on the grid resolution. An element will only be marked for coarsening if its level in the grid refinement hierarchy is strictly larger than `minLevel`. Likewise, refinement marks are only applied to elements strictly below `maxLevel`. The default values do not impose any restrictions at all.

Even after `mark_grid` has been called, the grid is still unchanged. However, its leaf elements now carry the refinement and coarsening marks, so that calls to `preAdapt`, `adapt`, and `postAdapt` as explained in Chapter 5.9 will correctly modify the grid.

adapt: Modifying the Grid and Adjusting Grid Functions

When a discrete function is attached to the grid and needs to be retained across the grid modification step, then simply calling `adapt` on the grid is not enough. Chapter 5.9 has shown how to keep data across grid modifications on the level of algebraic data attached to grid entities. The `dune-pdelab` module hides a lot of that complexity behind a simple interface. The relevant method is

```
template<class Grid, class GridFunctionSpace, class VectorContainer>
void adapt_grid(Grid& grid,
                GridFunctionSpace& gridFunctionSpace,
                VectorContainer& x,
                int intOrder)
```

Here, `grid` is the (hierarchical) grid, `gridFunctionSpace` is the object that describes the finite element space, and `x` is a vector of coefficients with respect to that space that defines the function that needs to be modified along with the grid. Note that only one function can be handled by the method—to handle several functions the `GridFunctionSpace` object needs to combine them in a tree structure as explained in Chapter 10.1.

The `intOrder` parameter allows to control a particular aspect of the transfer of the discrete function from the old grid to the new one. When grid elements are refined, then existing functions on the unrefined elements must be represented in the larger space defined with respect to the refined elements. For that, the `LocalInterpolation` infrastructure of the `dune-localfunctions` module is used (Chapter 8.2.2). Conversely, when elements are coarsened, existing functions must be represented in smaller spaces. For this, the `adapt_grid` method uses an L^2-projection, and `intOrder` sets the integration order for the assembly of the mass matrix that implements the projection. For a perfect projection, twice the polynomial order has to be chosen, but smaller numbers are frequently sufficient.

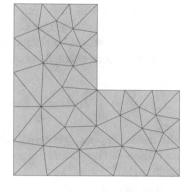

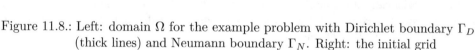

Figure 11.8.: Left: domain Ω for the example problem with Dirichlet boundary Γ_D (thick lines) and Neumann boundary Γ_N. Right: the initial grid

11.5.2. Example: Adaptive Finite Elements for the Poisson Equation

We demonstrate the adaptivity features of dune-pdelab by giving a complete example. For this, we extend the Poisson example from the introductory Chapter 3.3 to include local grid adaptivity based on a residual-type error estimator [157]. The complete code is printed in Appendix B.16, and it is available in electronic form through the icon in the margin. This is the program that was used to produce Figure 2.20 in the chapter on grid adaptivity.

Recall first that the example in Chapter 3.3 computed the weak solution of the equation

$$- \Delta u = f \tag{11.25}$$

with $f \equiv 5$ on the L-shaped domain $\Omega = (0,1)^2 \setminus [0.5,1)^2$, with Dirichlet boundary conditions

$$u = \begin{cases} 0 & \text{on } \{0\} \times [0,1] \cup [0,1] \times \{0\}, \\ 0.5 & \text{on } \{0.5\} \times [0.5,1] \cup [0.5,1] \times \{0.5\}, \end{cases}$$

and Neumann conditions $j \equiv 0$ on the rest of the boundary. The domain and boundary conditions were shown in Figure 3.1, but we repeat them in Figure 11.8 for convenience. As the domain has an inward corner, the problem is not H^2-regular, and unbounded derivatives of the solution are observed near that corner (Figure 2.11). For the discretization, we use the same conforming triangle grid as in Chapter 3.3 (shown again in Figure 11.8), and we discretize the problem using first-order Lagrange finite elements. The grid will be stored using the UGGrid grid manager, which implements red–green h-refinement, as explained in Chapter 2.3.2.[15] It has been shown in previous

---

[15]Replacing UGGrid by AlbertaGrid produces the bisection refinement grid shown in Figure 2.15.

chapters how to solve the Poisson problem with Dirichlet boundary conditions using dune-pdelab, and we will make full use of this knowledge here.

To control the adaptivity the code implements a standard residual-type error estimator. Let u be the weak solution of the boundary value problem (11.25), and u_h the corresponding finite element solution. The idea is that the error $u - u_h$ is related to the residual

$$r(u_h, v) = \int_\Omega fv\, dx - \int_\Omega \langle \nabla u_h, \nabla v \rangle\, dx + \int_{\Gamma_N} jv\, ds, \qquad v \in H^1_{D,0}.$$

Indeed, the residual $r_h(u_h, \cdot)$ is zero if and only if u_h weakly solves the equation (11.25) and it can be shown that the H^1-norm of the error is equivalent to the operator norm of the residual [157].

This insight can be turned into an error estimator by approximating the operator norm of $r(u_h, \cdot)$ by computable quantities. For this, we localize the residual by splitting the integrals along the elements

$$r_h(u_h, v) = \sum_{T \in \mathcal{T}} \left[\int_T fv\, dx - \int_T \langle \nabla u_h, \nabla v \rangle\, dx + \int_{T \cap \Gamma_N} jv\, ds \right],$$

and use Green's identity on each element to obtain

$$r_h(u_h, v) = \sum_{T \in \mathcal{T}} \left[\int_T (f + \Delta u_h)v\, dx - \int_{\partial T} \langle \nabla u_h, \mathbf{n} \rangle v\, ds + \int_{T \cap \Gamma_N} jv\, ds \right].$$

The quantity $\int_T (f + \Delta u_h)v\, dx$ corresponds to the weak form of the Poisson equation on the element T. The sum over the integrals over the element boundaries can be rewritten as a sum over the element intersections of the grid. We get

$$r_h(u_h, v) = \sum_{T \in \mathcal{T}} \int_T (f + \Delta u_h)v\, dx$$
$$+ \sum_{\gamma^b \in \mathcal{B} \cap \Gamma_N} \int_{\gamma^b} (j - \langle \nabla u_h, \mathbf{n} \rangle)v\, ds + \sum_{\gamma^i \in \mathcal{F}} \int_{\gamma^i} [\![\langle \mathbf{n}, \nabla u_h \rangle]\!]v\, ds,$$

where $\mathcal{B} \cap \Gamma_N$ is the set of element facets on the Neumann boundary, and $\mathcal{F}$ is the grid skeleton. Integrals over facets on the Dirichlet boundary drop out, because the test functions v are zero there.

Residual-based estimators combine these terms to a local error indicator

$$\eta_T^2 := h_T^2 \|f + \Delta u_h\|_{0,T}^2$$
$$+ \frac{1}{2} \sum_{\gamma^i \in \mathcal{F}} h_{\gamma^i} \|[\![\langle \nabla u_h, \mathbf{n} \rangle]\!]\|_{0,\gamma^i}^2 + \sum_{\gamma^b \in \mathcal{B} \cap \Gamma_N} h_{\gamma^b} \|j - \langle \nabla u_h, \mathbf{n} \rangle\|_{0,\gamma^b}^2, \quad (11.26)$$

where h_T is the diameter of element T, h_γ is the diameter of the facet γ, and $\|\cdot\|_0$ denotes the L^2-norm. It can be shown that the quantity $\eta := \sqrt{\sum_T \eta_T^2}$ combined with additional data oscillation terms bounds the H^1-error of the discrete solution u_h from above and below [157]. At the same time, adaptive refinement strategies using the η_T^2 as refinement indicators can be shown to converge [57, 128], and work well in practice. As the example uses piecewise constant data f and j only, we can disregard the data oscillation terms in the following.

The `main` Method

The code of the example program consists of three parts: A class `PoissonProblem` that serves as problem description for the `ConvectionDiffusionFEM` local operator, a local operator `ResidualErrorEstimator` that computes the error indicators η_T^2, $T \in \mathcal{T}$, and a main method. The `PoissonProblem` class is identical to the one used in Chapter 11.3, and we do not repeat it here. Rather, we discuss the `main` method, and show the implementation of the local error estimator.

The `main` method starts by initializing `MPIHelper`, and loading the unstructured grid shown in Figure 11.8 from a GMSH file into a `UGGrid` object. Electronic versions of this document contain the grid as a file attachment, accessible through the icon in the text margin:

```
288  int main(int argc, char *argv[])
289  {
290    // Initialize MPI, if available
291    MPIHelper::instance(argc, argv);
292
293    constexpr int dim = 2;
294    using Grid = UGGrid<dim>;
295    std::shared_ptr<Grid> grid = GmshReader<Grid>::read("l-shape.msh");
296    using GridView = Grid::LeafGridView;
297    auto gridView = grid->leafGridView();
```

Directly afterwards, the method defines a variable storing an error threshold:

```
301    const double estimatedErrorThreshold = 0.1;
```

The adaptive refinement loop of Figure 11.7 will iterate until the estimated total error drops below this threshold.

Next, the code sets up the discrete function space; a first-order Lagrange space with Dirichlet constraints and the ISTL vector backend:

```
305    // Make grid function space
306    using Basis = Functions::LagrangeBasis<GridView,1>;
307    auto basis = std::make_shared<Basis>(gridView);
308
309    using Constraints = PDELab::ConformingDirichletConstraints;
310    using VectorBackend = PDELab::ISTL::VectorBackend<>;
311
312    using GridFunctionSpace
313        = PDELab::Experimental::GridFunctionSpace<Basis,
```

```
314                                              VectorBackend,
315                                              Constraints>;
316
317    GridFunctionSpace gridFunctionSpace(basis);
```

The Dirichlet boundary conditions are treated just like in Chapter 11.3:

```
322    PoissonProblem<GridView,double> poissonProblem;
323    PDELab::
324      ConvectionDiffusionBoundaryConditionAdapter<decltype(poissonProblem)>
325                                              bctype(poissonProblem);
326
327    using ConstraintsContainer
328        = GridFunctionSpace::ConstraintsContainer<double>::Type;
329    ConstraintsContainer constraintsContainer;
330    PDELab::constraints(bctype, gridFunctionSpace, constraintsContainer);
```

The method bctype, the predicate that specifies the degrees of freedom on the Dirichlet boundary, is imported from the problem description in the PoissonProblem class given earlier in the file.

Next, the main method creates a coefficient vector u for a discrete function from the GridFunctionSpace. This vector will hold the coefficients of the solution u_h of the discrete problem on the individual grids, and it therefore has to be defined before the beginning of the refinement loop. The Dirichlet values are set by sampling the function

$$g(x) = \begin{cases} 0 & \text{if } x_0 \le 0 \text{ or } x_1 \le 0 \\ 0.5 & \text{otherwise} \end{cases}$$

at the Lagrange nodes:

```
335    using U = PDELab::Backend::Vector<GridFunctionSpace,double>;
336    U u(gridFunctionSpace,0.0);
337    auto g = [](auto x){return  (x[0]< 1e-8 || x[1] < 1e-8) ? 0 : 0.5;};
338    Functions::interpolate(*basis, PDELab::Backend::native(u), g);
```

Observe how the actual sampling in Line 338 is done by a method from dune-functions, operating directly on the dune-istl container accessible through the method PDELab::Backend::native.

After these preparations starts the actual refinement loop:

```
342    std::size_t i=0;        // Loop variable, for data output
343    while (true)            // Loop termination condition
344                           // is near the middle of the loop body
345    {
346      std::cout << "Iteration: " << i
347              << ", highest level in grid: " << grid->maxLevel()
348              << ", degrees of freedom: " << gridFunctionSpace.globalSize()
349              << std::endl;
```

It is implemented as a **while** loop with a loop condition that never fails, hence the loop can only be terminated by a **break** statement. This construction is chosen because the test for whether the iteration should terminate happens in the middle of the refinement loop, see Figure 11.7.

Following the sequence of steps in that figure, we first need to assemble the problem on the given grid, and solve the resulting algebraic problem. We create an assembler and a linear solver from the linear algebra backend:

```
354    using LocalOperator
355        = PDELab::ConvectionDiffusionFEM<decltype(poissonProblem),
356                                         GridFunctionSpace::Traits::
357                                                        FiniteElementMap>;
358    LocalOperator localOperator(poissonProblem);
359
360    using MatrixBackend = PDELab::ISTL::BCRSMatrixBackend<>;
361    MatrixBackend matrixBackend(7);      // Expected average number
362                                         // of matrix entries per row
363    using GridOperator = PDELab::GridOperator<GridFunctionSpace,
364                                              GridFunctionSpace,
365                                              LocalOperator,
366                                              MatrixBackend,
367                                              double,double,double,
368                                              ConstraintsContainer,
369                                              ConstraintsContainer>;
370    GridOperator gridOperator(gridFunctionSpace, constraintsContainer,
371                              gridFunctionSpace, constraintsContainer,
372                              localOperator,
373                              matrixBackend);
374
375    // Select a linear solver backend
376    using LinearSolverBackend = PDELab::ISTLBackend_SEQ_CG_SSOR;
377    LinearSolverBackend linearSolverBackend(5000,    // Maximum number
378                                                     // of iterations
379                                            1);      // Verbosity level
380
381    // Select linear problem solver
382    using LinearProblemSolver
383        = PDELab::StationaryLinearProblemSolver<GridOperator,
384                                                LinearSolverBackend,
385                                                U>;
386    LinearProblemSolver linearProblemSolver(gridOperator,
387                                            linearSolverBackend,
388                                            u,
389                                            1e-10);
390
391    // Solve linear problem
392    linearProblemSolver.apply();
```

At this point, the vector container u contains the finite element solution u_h on the current grid. This current solution is written to a file, with a file name that gets the refinement iteration number as a suffix. That way, all intermediate grids can be easily inspected later:

```
397    VTKWriter<GridView> vtkwriter(gridView);
398    auto uFunction
399        = Functions::makeDiscreteGlobalBasisFunction<double>(
400                                    *basis,
401                                    PDELab::Backend::native(u));
402    vtkwriter.addVertexData(uFunction,
403                    VTK::FieldInfo("u",
404                            VTK::FieldInfo::Type::scalar,
405                            1));
406    vtkwriter.write("pdelab-adaptivity-result-" + std::to_string(i));
```

We now have computed the solution of the model problem with respect to the given grid. From there, we proceed to the actual error estimation step. Somewhat surprisingly, the code uses a particular LocalOperator implementation to compute the squared error indicators η_T^2. The trick is to realize that the vector of squared indicators can be written as the algebraic representation of a linear form on the space of piecewise constant functions. Let θ_T be the function that has the value 1 on element T, and is zero elsewhere. Then, the error indicator (11.26) can be rewritten as

$$\eta_T^2 = r^{\mathrm{err}}(u_h, \theta_T)$$

with

$$
\begin{aligned}
r^{\mathrm{err}}(u_h, \theta) := & \sum_{T \in \mathcal{T}} h_T^2 \int_T |(f + \Delta u_h)|^2 \theta \, dx \\
& + \sum_{\gamma^{\mathrm{i}} \in \mathcal{F}} h_{\gamma^{\mathrm{i}}} \int_{\gamma^{\mathrm{i}}} [\![\langle \nabla u_h, \mathbf{n} \rangle]\!]^2 \{\theta\} \, ds \\
& + \sum_{\gamma^{\mathrm{b}} \in \mathcal{B} \cap \Gamma_N} h_{\gamma^{\mathrm{b}}} \int_{\gamma^{\mathrm{b}}} |\mathbf{j} - \langle \nabla u_h, \mathbf{n} \rangle|^2 \theta \, ds.
\end{aligned}
\tag{11.27}
$$

This is indeed linear in θ, and it has the typical three-way splitting of Assumption 11.2 that is at the heart of the dune-pdelab assembler interface. We therefore define a function space of piecewise constant test functions, and a vector of coefficients etaSquared defined with respect to its standard basis, which will hold the local error indicators η_T^2:

```
411    using P0Basis = Functions::LagrangeBasis<GridView,0>;
412    auto p0Basis = std::make_shared<P0Basis>(gridView);
413
414    using P0GridFunctionSpace
415        = PDELab::Experimental::GridFunctionSpace<P0Basis,
```

```
416                                           VectorBackend,
417                                           PDELab::NoConstraints>;
418
419     P0GridFunctionSpace p0GridFunctionSpace(p0Basis);
420     using U0 = PDELab::Backend::Vector<P0GridFunctionSpace,double>;
421     U0 etaSquared(p0GridFunctionSpace,0.0);
```

Then, we set up a new `GridOperator` object with a custom `LocalOperator` called `ResidualErrorEstimator` that implements the residual formulation (11.27) of the error estimator. This `LocalOperator` implementation will be described in detail in Chapter 11.5.2. It is given the problem description, because it has to know the source term and the boundary conditions:

```
426     using EstimatorLocalOperator
427         = ResidualErrorEstimator<decltype(poissonProblem)>;
428     EstimatorLocalOperator estimatorLocalOperator(poissonProblem);
429     using EstimatorGridOperator = PDELab::GridOperator<GridFunctionSpace,
430                                               P0GridFunctionSpace,
431                                               EstimatorLocalOperator,
432                                               MatrixBackend,
433                                               double,double,double>;
434     EstimatorGridOperator estimatorGridOperator(gridFunctionSpace,
435                                               p0GridFunctionSpace,
436                                               estimatorLocalOperator,
437                                               matrixBackend);
```

Note how this time the `GridOperator` receives two different function spaces, namely the first-order Lagrange space for the finite element solution u_h, and the space of piecewise constant test functions θ. The implementation of `estimatorLocalOperator` is discussed in detail below.

We then evaluate the indicator by calling the `residual` method of the grid operator for the current finite element solution in `u`:

```
441     estimatorGridOperator.residual(u,etaSquared);
```

We have now completed the **estimate** step of the refinement loop. The vector container `etaSquared` now contains η_T^2 for each element T of the leaf grid view. An estimation of the squared total error is obtained by summing over the η_T^2 [157], and this estimate is used to decide whether the refinement loop should terminate:

```
445     // Terminate if desired accuracy has been reached
446     double totalEstimatedError = std::sqrt(std::accumulate(etaSquared.begin(),
447                                               etaSquared.end(),
448                                               0.0));
449     std::cout << "Total estimated error: " << totalEstimatedError << std::endl;
450
451     if (totalEstimatedError < estimatedErrorThreshold)
452       break;
```

If the estimated error is not sufficiently small, we proceed by using the element error indicators in the `etaSquared` array to determine which elements to mark for refinement. The code uses Dörfler marking, as implemented by the `error_fraction` method. We set the parameter `alpha` to a positive value and `beta` to zero, to allow only refinement but no coarsening:

```
457    double alpha = 0.4;        // Refinement fraction
458    double refineThreshold;    // Refinement threshold
459    double beta = 0.0;         // Coarsening fraction
460    double coarsenThreshold;   // Coarsening threshold
461    int verbose = 0;           // No screen output
462
463    PDELab::error_fraction(etaSquared,
464                           alpha, beta,
465                           refineThreshold, coarsenThreshold,
466                           verbose);
```

After this call, the variable `refineThreshold` contains the error value that separates the elements that make up the upper 40 % of the error from the rest; `coarsenThreshold` contains 0 because by setting `beta` to zero no elements will be marked for coarsening.

The actual marking is done by the method `mark_grid`:

```
470    PDELab::mark_grid(*grid, etaSquared, refineThreshold, coarsenThreshold);
```

Finally, a call to `adapt_grid` does the actual grid modification, and transfers the current discrete solution in u from the old grid to the new one:

```
473    PDELab::adapt_grid(*grid, gridFunctionSpace, u, 2 );
```

The argument 2 here is the quadrature order used for the L^2-projection that is used for transferring data to coarser grids (Chapter 11.5.1). As no coarsening happens in this example, the number has no effect.

After the grid function space and the solution have been adapted to the new grid, the Dirichlet boundary constraints need to be re-applied on the adapted solution vector u. This happens in a two-part process. First, the Dirichlet values function g is sampled *on the entire grid* into a temporary vector `dirichletValues`, using a method of dune-functions. Then, the degrees of freedom constrained by the Dirichlet boundary conditions are copied from the temporary vector into u. That way, only Dirichlet boundary values of u are modified:

```
477    // Reassemble the Dirichlet constraints
478    PDELab::constraints(bctype,gridFunctionSpace,constraintsContainer);
479
480    // Interpolate the Dirichlet values function on the entire domain!
481    PDELab::Backend::Vector<GridFunctionSpace,double>
482                               dirichletValues(gridFunctionSpace);
483    Functions::interpolate(*basis,
484                           PDELab::Backend::native(dirichletValues),
485                           g);
```

```
486
487      // Copy all Dirichlet degrees of freedom into the actual solution vector
488      PDELab::copy_constrained_dofs(constraintsContainer,dirichletValues,u);
489
490      // Increment loop counter
491      i++;
492    }
```

This ends the refinement loop, and the main method with it.

Local Error Indicator

What is left to discuss is the LocalOperator implementation that computes the vector of error indicators η_T^2. Recall that we reformulated the textbook residual estimator (11.26) as a linear functional on the space of piecewise constant functions

$$\eta_T^2 = r^{\mathrm{err}}(u_h, \theta_T)$$

with

$$
r^{\mathrm{err}}(u_h, \theta) := \sum_{T \in \mathcal{T}} h_T^2 \int_T |(f + \Delta u_h)|^2 \theta \, dx
$$
$$
+ \sum_{\gamma^{\mathrm{i}} \in \mathcal{F}} h_{\gamma^{\mathrm{i}}} \int_{\gamma^{\mathrm{i}}} [\![\langle \nabla u_h, \mathbf{n} \rangle]\!]^2 \{\theta\} \, ds
$$
$$
+ \sum_{\gamma^{\mathrm{b}} \in \mathcal{B} \cap \Gamma_N} h_{\gamma^{\mathrm{b}}} \int_{\gamma^{\mathrm{b}}} |\mathrm{j} - \langle \nabla u_h, \mathbf{n} \rangle|^2 \theta \, ds.
$$

Written in this form, the set of indicators is composed of element contributions, of contributions from element intersections from the skeleton, and from Neumann boundary intersections. Consequently, the LocalOperator implementation implements the methods alpha_volume, alpha_skeleton, and alpha_boundary. As the derivatives of the η_T^2 with respect to u_h are not required for adaptive finite elements computations, the corresponding jacobian methods are omitted.

The class signature is simple: Its only template parameter is the problem description class, which supplies the source term function and the boundary conditions. The class only inherits from PDELab::LocalOperatorDefaultFlags, because no Jacobians are ever computed, and hence matrix occupation patterns are not required:

```
45   template<class Problem>
46   class ResidualErrorEstimator
47     : public PDELab::LocalOperatorDefaultFlags
48   {
49     Problem problem_;
50
51   public:
52     ResidualErrorEstimator(const Problem& problem)
```

```
53      : problem_(problem)
54      {}
```

The class then sets the three standard flags to announce to the `GridOperator` that the residual contains volume, skeleton, and boundary terms:

```
59      static const bool doAlphaVolume   = true;
60      static const bool doAlphaSkeleton = true;
61      static const bool doAlphaBoundary = true;
```

For the reason given above the doPattern flags are not required.

Then comes the method `alpha_volume`, which computes the element part

$$
\eta_{T,\mathrm{vol}}^2 := h_T^2 \int_T |(f + \Delta u_h)|^2 \theta_T \, dx
$$

of the error indicator. The code does not actually evaluate the test function θ_T, because it is known to have the value 1 on T anyway. For simplicity we also exploit the fact that u_h is an affine function on T, and hence $\Delta u_h = 0$:

```
65      template<class EntityGeometry,
66              class LocalFunctionSpaceU, class Vector,
67              class LocalFunctionSpaceV, class Residual>
68      void alpha_volume(const EntityGeometry& entityGeometry,
69                      const LocalFunctionSpaceU& localFunctionSpaceU,
70                      const Vector& x,
71                      const LocalFunctionSpaceV& localFunctionSpaceV,
72                      Residual& residual) const
73      {
74        using RangeField
75          = typename LocalFunctionSpaceU::Traits::
76                      FiniteElement::Traits::LocalBasisType::Traits::RangeType;
77
78        // Element diameter
79        auto h_T = diameter(entityGeometry.geometry());
80
81        // Loop over quadrature points
82        int intOrder = localFunctionSpaceU.finiteElement().localBasis().order();
83        constexpr auto entityDim = EntityGeometry::Entity::mydimension;
84        const auto& quadRule
85          = QuadratureRules<double,entityDim>::rule(entityGeometry.geometry()
86                                                              .type(),
87                                                    intOrder);
88
89        for (const auto& quadPoint : quadRule)
90        {
91          // Laplace of u_h is always zero,
92          // because we are using first-order elements on simplices
93          RangeField laplaceU = 0.0;
94
95          // Evaluate source term function f
```

```
96          auto f = problem_.f(entityGeometry.entity(),quadPoint.position());
97
98          // Integrate h_T^2(f + Δu_h)^2
99          auto factor = quadPoint.weight()
100                      * entityGeometry.geometry()
101                            .integrationElement(quadPoint.position());
102          auto value = std::pow(h_T * (f + laplaceU), 2);
103          residual.accumulate(localFunctionSpaceV, 0, value * factor);
104       }
105     }
```

Next is the method `alpha_skeleton`, which computes the skeleton contributions

$$\eta^2_{\gamma^i,\text{skel}} := h_{\gamma^i} \int_{\gamma^i} [\![\langle \nabla u_h, \mathbf{n} \rangle]\!]^2 \{\theta\}\, ds$$

for each intersection γ^i that is not on the domain boundary. Note that the test function θ is either 1 on the *inside* element and 0 on the *outside* one or vice versa. Hence the averaging expression $\{\theta\}$ can be replaced by the value $\frac{1}{2}$:

```
109   template<class IntersectionGeometry,
110           class LocalFunctionSpaceU, class Vector,
111           class LocalFunctionSpaceV, class Residual>
112   void alpha_skeleton(const IntersectionGeometry& intersectionGeometry,
113                       const LocalFunctionSpaceU& localFunctionSpaceUIn,
114                       const Vector& xIn,
115                       const LocalFunctionSpaceV& localFunctionSpaceVIn,
116                       const LocalFunctionSpaceU& localFunctionSpaceUOut,
117                       const Vector& xOut,
118                       const LocalFunctionSpaceV& localFunctionSpaceVOut,
119                       Residual& residualIn, Residual& residualOut) const
120   {
121     // Extract type of shape function gradients
122     using TrialFE = typename LocalFunctionSpaceU::Traits::FiniteElementType;
123     using LocalBasisU = typename TrialFE::Traits::LocalBasisType;
124     using GradientU = typename LocalBasisU::Traits::JacobianType;
125     using size_type = typename LocalFunctionSpaceU::Traits::SizeType;
126
127     auto insideGeometry = intersectionGeometry.inside().geometry();
128     auto outsideGeometry = intersectionGeometry.outside().geometry();
129
130     auto h_F = diameter(intersectionGeometry.geometry());
131
132     auto geometryInInside = intersectionGeometry.geometryInInside();
133     auto geometryInOutside = intersectionGeometry.geometryInOutside();
134
135     std::vector<GradientU> localGradPhiIn(localFunctionSpaceUIn.size());
136     std::vector<GradientU> localGradPhiOut(localFunctionSpaceUOut.size());
137
```

```
138      // Loop over quadrature points and integrate the jump term
139      const int intOrder
140          = 2*localFunctionSpaceUIn.finiteElement().localBasis().order();
141      constexpr auto intersectionDim = IntersectionGeometry::mydimension;
142      const auto& quadRule
143          = QuadratureRules<double,intersectionDim>::rule(
144                                    intersectionGeometry.geometry().type(),
145                                    intOrder);
146
147      for (const auto& quadPoint : quadRule)
148      {
149        // Position of quadrature point in local coordinates of elements
150        auto quadPointLocalIn = geometryInInside.global(quadPoint.position());
151        auto quadPointLocalOut = geometryInOutside.global(quadPoint.position());
152
153        // Evaluate gradient of basis functions
154        localFunctionSpaceUIn.finiteElement().localBasis()
155                        .evaluateJacobian(quadPointLocalIn,localGradPhiIn);
156        localFunctionSpaceUOut.finiteElement().localBasis()
157                        .evaluateJacobian(quadPointLocalOut,localGradPhiOut);
158
159        // Compute gradient of u
160        GradientU localGradUIn(0.0);
161        for (size_type i=0; i<localFunctionSpaceUIn.size(); i++)
162          localGradUIn.axpy(xIn(localFunctionSpaceUIn,i),localGradPhiIn[i]);
163        GradientU localGradUOut(0.0);
164        for (size_type i=0; i<localFunctionSpaceUOut.size(); i++)
165          localGradUOut.axpy(xOut(localFunctionSpaceUOut,i),localGradPhiOut[i]);
166
167        GradientU gradUIn;
168        auto jacIn = insideGeometry.jacobianInverseTransposed(quadPointLocalIn);
169        jacIn.mv(localGradUIn[0],gradUIn[0]);
170
171        GradientU gradUOut;
172        auto jacOut
173            = outsideGeometry.jacobianInverseTransposed(quadPointLocalOut);
174        jacOut.mv(localGradUOut[0],gradUOut[0]);
175
176        // Integrate
177        const auto n_F
178            = intersectionGeometry.unitOuterNormal(quadPoint.position());
179        auto jumpSquared = std::pow((n_F*gradUIn[0])-(n_F*gradUOut[0]), 2);
180        auto factor = quadPoint.weight()
181                    * intersectionGeometry.geometry()
182                                .integrationElement(quadPoint.position());
183
184        // Accumulate indicator
```

```
185        residualIn.accumulate(localFunctionSpaceVIn,
186                               0,
187                               0.5 * h_F * jumpSquared * factor);
188        residualOut.accumulate(localFunctionSpaceVOut,
189                               0,
190                               0.5 * h_F * jumpSquared * factor);
191     }
192   }
```

This should all be straightforward for readers with a working knowledge of the `alpha_skeleton` interface method (Chapter 11.2.5). As the only slight deviation from standard procedure, the method does not transform the trial shape function gradients $\{\nabla\phi_i^{\mathrm{in}}\}$ and $\{\nabla\phi_i^{\mathrm{out}}\}$ from reference element coordinates to grid element coordinates. As they are only used to compute the function gradients $\nabla u_h|_{T_{\mathrm{in}}}$ and $\nabla u_h|_{T_{\mathrm{out}}}$, these latter quantities are computed on the reference element first, and transformed to the grid element.

Finally, there is the method `alpha_boundary`, which is called for all boundary intersections, and which has to assemble the Neumann boundary contributions

$$\eta^2_{\gamma^{\mathrm{b}},\mathrm{Neu}} := h_{\gamma^{\mathrm{b}}} \int_{\gamma^{\mathrm{b}}} |\mathbf{j} - \langle\nabla u_h, \mathbf{n}\rangle|^2 \theta\, ds.$$

As the test function θ takes the value 1 whenever that contribution is nonzero, we can again omit θ from the implementation:

```
196   template<class IntersectionGeometry,
197           class LocalFunctionSpaceU, class Vector,
198           class LocalFunctionSpaceV, class Residual>
199   void alpha_boundary(const IntersectionGeometry& intersectionGeometry,
200                       const LocalFunctionSpaceU& localFunctionSpaceUIn,
201                       const Vector& xIn,
202                       const LocalFunctionSpaceV& localFunctionSpaceVIn,
203                       Residual& residualIn) const
204   {
205     using TrialFE = typename LocalFunctionSpaceU::Traits::FiniteElementType;
206     using LocalBasisU = typename TrialFE::Traits::LocalBasisType;
207     using GradientU = typename LocalBasisU::Traits::JacobianType;
208     using size_type = typename LocalFunctionSpaceU::Traits::SizeType;
209
210     auto insideGeometry = intersectionGeometry.inside().geometry();
211
212     auto geometryInInside = intersectionGeometry.geometryInInside();
213
214     // Intersection diameter
215     auto h_F = diameter(intersectionGeometry.geometry());
216
217     std::vector<GradientU> localGradPhi(localFunctionSpaceUIn.size());
218
```

```
219    // Loop over quadrature points and integrate the jump term
220    const int intOrder
221        = localFunctionSpaceUIn.finiteElement().localBasis().order();
222    constexpr auto intersectionDim = IntersectionGeometry::mydimension;
223    const auto& quadRule
224        = QuadratureRules<double,intersectionDim>::rule(
225                                    intersectionGeometry.geometry().type(),
226                                    intOrder);
227
228    for (const auto& quadPoint : quadRule)
229    {
230        // Skip if quadrature point is not on Neumann boundary
231        auto bcType = problem_.bctype(intersectionGeometry.intersection(),
232                                    quadPoint.position());
233        if (bcType != PDELab::ConvectionDiffusionBoundaryConditions::Neumann)
234            return;
235
236        // Position of quadrature point in local coordinates of element
237        auto quadPointLocalIn = geometryInInside.global(quadPoint.position());
238
239        // Evaluate gradient of trial shape functions
240        localFunctionSpaceUIn.finiteElement().localBasis()
241                                        .evaluateJacobian(quadPointLocalIn,
242                                                    localGradPhi);
243
244        // Evaluate gradient of u_h
245        GradientU localGradU(0.0);
246        for (size_type i=0; i<localFunctionSpaceUIn.size(); i++)
247            localGradU.axpy(xIn(localFunctionSpaceUIn,i), localGradPhi[i]);
248
249        GradientU gradU;
250        auto jac = insideGeometry.jacobianInverseTransposed(quadPointLocalIn);
251        jac.mv(localGradU[0], gradU[0]);
252
253        // Evaluate Neumann boundary value function
254        auto neumann = problem_.j(intersectionGeometry.intersection(),
255                                quadPoint.position());
256
257        // Integrate
258        auto factor = quadPoint.weight()
259                    * intersectionGeometry.geometry()
260                                        .integrationElement(quadPoint.position());
261
262        const auto n
263            = intersectionGeometry.unitOuterNormal(quadPoint.position());
264
265        residualIn.accumulate(localFunctionSpaceVIn,
```

```
266                          0,
267                          h_F * std::pow((neumann - n*gradU[0]), 2)
268                                                    * factor);
269      }
270    }
```

Finally, the class contains the small helper method `diameter`, that computes the diameters h_T and h_γ of elements and intersections. This method is a verbatim copy of the method of the same name that has been used in the Discontinuous Galerkin example in Chapter 11.2.6, and it is therefore not shown here again. Since both element and intersection geometries are represented in DUNE as `Geometry` objects, only a single method is needed.

Running the Program

When run, the program produces the following screen output:

```
Reading 2d Gmsh grid...
version 2.2 Gmsh file detected
file contains 43 nodes
file contains 90 elements
number of real vertices = 43
number of boundary elements = 22
number of elements = 62
Iteration: 0, highest level in grid: 0, degrees of freedom: 43
=== matrix setup (max) 0.000568451 s
=== matrix assembly (max) 0.00179137 s
=== residual assembly (max) 0.00117358 s
=== solving (reduction: 1e-10) === Dune::IterativeSolver
=== rate=0.017215, T=0.00123597, TIT=0.000205994, IT=6
0.00126493 s
Total estimated error: 1.26612
Iteration: 1, highest level in grid: 1, degrees of freedom: 64
=== matrix setup (max) 0.000901343 s
=== matrix assembly (max) 0.00300309 s
=== residual assembly (max) 0.00206459 s
=== solving (reduction: 1e-10) === Dune::IterativeSolver
=== rate=0.0551536, T=0.00272761, TIT=0.000303067, IT=9
0.00275909 s
Total estimated error: 1.072

[...]

Iteration: 12, highest level in grid: 11, degrees of freedom: 7227
=== matrix setup (max) 0.126007 s
=== matrix assembly (max) 0.411516 s
=== residual assembly (max) 0.26253 s
=== solving (reduction: 1e-10) === Dune::IterativeSolver
```

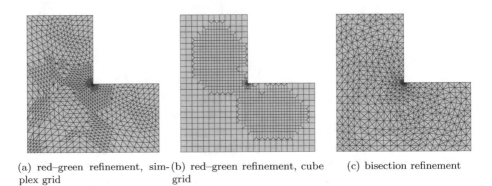

(a) red–green refinement, simplex grid

(b) red–green refinement, cube
grid

(c) bisection refinement

Figure 11.9.: Intermediate iterates of the adaptivity example for different refinement
strategies

```
=== rate=0.746897, T=2.91045, TIT=0.0368411, IT=79
2.91064 s
Total estimated error: 0.0973319
```

It can be seen how the refinement loop is performed 13 times, before stopping because
the estimated error has dropped below the requested threshold of 0.1.

Figure 11.9, left, shows the output of the program. It can be seen that the grid is
refined near the inward corner, where the solution becomes singular. For comparison we
also show the grid when starting from a grid with quadrilateral elements (Figure 11.9,
center), and when using bisection refinement (Figure 11.9, right). Bisection refinement
was implemented by simply switching the grid manager definition in Line 294 from
UGGrid to AlbertaGrid. As bisection refinement introduces less new elements than
red–green refinement, the refinement loop iterated 23 rather than 13 times to arrive
at the same level of estimated error.

Part IV.

Appendix

A. The Dune Build System

Any nontrivial piece of software in a compiled language requires a build system. Such a system not only compiles, links, and installs the code, but it also tracks external dependencies, provides ways to steer the build process via parameters, and may help to organize automated code testing.

Several standard build systems are currently on the market, and they fulfill the requirements of most software projects. However, as DUNE is a set of inter-dependent modules rather than a monolithic code, building and installing DUNE poses a few special challenges. For example, dependency tracking differs a bit from what is done for other software libraries, and it is frequently required to build an entire set of modules together with the same set of parameters. Therefore, DUNE has always had its own build system, which nowadays is written on top of the standard tool cmake.[1]

An introductory example of how to use the DUNE build system has already appeared in Chapter 3. Readers that are new to DUNE and want to build their first example should start there. The present chapter explains the DUNE build system and the internal structure of DUNE modules a bit more thoroughly.

A.1. Building and Installing Dune Modules

DUNE modules are the building blocks of the DUNE software system. Each module is a separate C++ source tree from which libraries and/or executables are built. Additionally, as DUNE uses a great deal of C++ template programming, a lot of code resides in C++ header files, of which there are a lot in most DUNE modules.[2] The directory hierarchies of the source trees must conform to a predefined structure, and each module must contain a bit of meta data and a few cmake configuration files. Figure A.1 shows a schematic drawing of the DUNE modules discussed in this book, with build dependencies between modules shown as arrows. The advantages of this modular design have been discussed in Chapter 4.2.

To build a set of modules from source, each module has to be downloaded and built separately. Each module can only be built if its dependencies have already been built before. This implies that the modules have to be built in a suitable order. For itself, each DUNE module uses a standard cmake build system, only augmented by a few DUNE-specific macros provided by the dune-common module.

DUNE modules can be built and installed in two ways. Many important modules are now available in major Linux distributions, and can be installed easily using the

---

[1] https://cmake.org; Readers of this appendix should have a least some familiarity with it.
[2] In fact, in some modules all C++ files are header files, and no libraries are built at all.

© Springer Nature Switzerland AG 2020
O. Sander, *DUNE — The Distributed and Unified Numerics Environment*, Lecture Notes in Computational Science and Engineering 140, https://doi.org/10.1007/978-3-030-59702-3

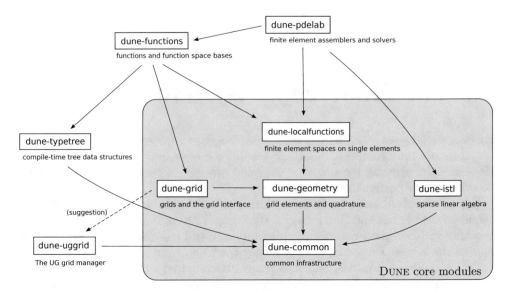

Figure A.1.: The DUNE modules discussed in this book, and their dependency graph. The modules contained in the grey box are what is commonly referred to as the "DUNE core".

standard package management systems.[3] If packages are not available for a given platform, or if bleeding-edge features are needed that are not in a release yet, the modules will need to be downloaded, built, and installed manually.

A.1.1. The `dunecontrol` Program

While the process of building and installing the modules can, in principle, be done by hand, the `dune-common` module provides the `dunecontrol` program, which makes building and installing DUNE modules much easier. The `dunecontrol` program works much like a Linux package manager in that it tracks the dependencies in a network of DUNE modules, computes an appropriate build order, and then builds the modules one by one using their `cmake` build systems. The main difference is that `dunecontrol` does not download the module source codes. This step has to be done separately. The rest of this section describes the major features of `dunecontrol`.

`dunecontrol` Build Commands

The `dunecontrol` program is a command-line tool. A call to `dunecontrol` has the following form:

---

[3]While this chapter focuses on unixoid operating systems, DUNE can also be built on Windows.

(a) Dependency tracking

print	Print the list of modules in an admissible build order
info	Same as print, but including whether it is a dependency or a suggestion
printdeps	Print recursive dependencies of a module
export	Prints colon-separated list of all DUNE source directories, preceded by export DUNE_CONTROL_PATH=; cf. Appendix A.1.2

(b) Building

cmake	Run cmake for each module, to configure each module's build system
make	Run make for each module, to build each module
all	Run the cmake and make steps for each module

(c) Version control tools

status	Show version control status for all modules (git or svn)
update	Update all modules from their repositories (git or svn)
git	Invoke a git command for each git-managed module
svn	Invoke a svn command for each svn-managed module

(d) Miscellaneous

exec	Execute an arbitrary command in each module source directory
bexec	Execute an arbitrary command in each module build directory
help	Show a help message

Table A.1.: Commands of the dunecontrol program

```
dunecontrol [OPTIONS] COMMAND(S) [COMMAND-OPTIONS]
```

The program is given a command (or a colon-separated list of commands), possibly followed by command-specific options. In addition, preceding the commands, dunecontrol also accepts general options. We only discuss the more important commands here; see dunecontrol -help or the program manpage for a complete list.

The most frequently used commands are cmake and make, which build a collection of DUNE modules in a suitable order.[4] For the following, assume that a set of DUNE module sources has been downloaded to a common directory. Assume further that all DUNE modules depended upon are either part of the set, or are available as system-wide installation. To build all modules in the common directory, enter that directory and run

```
dunecontrol cmake
dunecontrol make
```

---

[4]These commands have the same names as the programs they invoke. It should be clear from the context which meaning is intended.

This will build (but not install) all modules in the current directory. In more detail, the first of the two calls performs the following three steps:

1. For each module that `dunecontrol` finds, it checks (recursively) whether the modules that are listed as dependencies can be found. The build process aborts if not all required dependencies are found. Dependencies can be versioned (see Appendix A.2.2). Missing dependencies marked as suggestions merely lead to a warning. The dependency graph of modules is then sorted, which produces a valid module build order. The list of modules in build order can be seen by running `dunecontrol print`.

2. Unless specified otherwise, a build directory called `build-cmake` is created in each module source tree. Names and locations of these build directories can be controlled with the `-builddir` option (see below).

3. The program `cmake` is called for each module. This checks for the properties of the build toolchain, looks for the dependencies that are not DUNE modules, and creates the makefiles. The `cmake` program not finding certain external dependencies may result in hard errors, but it may also simply disable certain features.

The subsequent call to

`dunecontrol make`

then performs the following fourth step:

4. The `make` program is run for each module. This compiles all code of the module. All executables, libraries, and other generated files are placed in the module's build directory. If desired, the DUNE build system can be configured to use an alternative to the `make` program here, such as for example `ninja`,[5] which is reported to be faster than `make` (see Appendix A.1.3 for an example of how this is done). Also, the build tool may have to be replaced by something platform-specific on non-unix platforms.

As all four steps are usually performed together, there are shortcuts to trigger them. First, as the `dunecontrol` program can be called with more than one command (the commands then separated by a colon), all four steps can be triggered by

`dunecontrol cmake : make`

If this is still too much typing, another alternative is

`dunecontrol all`

Be aware, however, that the word `all` here refers to "all build steps", not "all modules".

---

[5]`https://ninja-build.org`

So far, the modules have been configured and built, but they have not been installed yet. Installing DUNE is a contentious subject. Indeed, installing DUNE modules is not required as long as all downstream code uses the `dunecontrol` program. If desired, DUNE modules can be installed by calling

```
dunecontrol make install
```

which installs the modules into the default system location. This may require administrator privileges—to install the modules locally with user rights, pass the installation path as a cmake option, as explained in Appendix A.1.3. At the time of writing, working with installed modules is unfortunately not well tested, and may need further improvements.

`dunecontrol` Build Options

The configuration and build process can be influenced by various options for the `dunecontrol` program. For example, to consider only a subset of the modules in the current directory, the `-only` and `-module` options restrict the set of modules to be considered by `dunecontrol`. In particular

```
dunecontrol --only=<module-name> all
```

will perform the four steps listed above for the single module called `<module-name>` only.[6] Similarly,

```
dunecontrol --module=<module-name> all
```

will do the four steps for the module `<module-name>` and for all of its direct and indirect dependencies.

By default, an individual build directory called `build-cmake` is created inside each module source tree. The name and relative location can be changed by passing a relative path as the `-builddir` option. For example,

```
dunecontrol --builddir=build-debug --module=dune-foo all
```

will build the module `dune-foo` and all of the DUNE modules it depends on in build directories called `build-debug` that are created in the root directories of the respective source trees. By choosing an absolute path instead, it is possible to specify a custom build directory at an arbitrary location in the file system:

```
dunecontrol --builddir=$HOME/dune-build-dir all
```

This will create a new directory `dune-build-dir` in the user's home directory, and the build directories of the individual modules will appear as subdirectories of `dune-build-dir`. Their names will be identical to the module names as specified in the `dune.module` files (Appendix A.2.2).

---

[6]The relevant name here is the name given in the `dune.module` file (Appendix A.2.2), not the name of the source tree directory.

Further `dunecontrol` Commands

There are further useful commands of `dunecontrol`. For example,

`dunecontrol exec <command>`

will loop over all source directories and call `<command>` in each of them. Likewise,

`dunecontrol bexec <command>`

will do the same for the build directories. So, for example,

`dunecontrol --module=dune-foo bexec "rm -rf *"`

will empty the build directory of the `dune-foo` module and of all of its direct and indirect dependencies. This particular example is useful for forcing `cmake` to do a full rebuild. Such a rebuild can become necessary when a dependency has changed, for example if a suggested but absent DUNE module has suddenly become available.

Finally, several commands exist to perform version control operations on sets of modules. Running

`dunecontrol status`

shows the version control status of each module, and

`dunecontrol update`

tries to update the sources from their upstream repositories. Both are currently implemented for the GIT and SUBVERSION version control systems. For more fine-grained control, the commands

`dunecontrol git <git command>`
`dunecontrol svn <svn command>`

allow to execute individual `git` or `svn` commands in each module source directory. See the online help of `dunecontrol` for a full list of commands and options.

A.1.2. Specifying the Module Search Path with the `DUNE_CONTROL_PATH` Variable

Above, it has been assumed for simplicity that all modules to be built are contained in a single directory. This is sometimes too restrictive, and therefore more general settings are supported. To understand how `dunecontrol` searches for modules, it is important to understand that `dunecontrol` must find DUNE modules for two slightly different reasons: It needs to find the modules it is supposed to build, and it needs to find modules that serve as dependencies for other modules. The latter ones may have been built by `dunecontrol` previously, but they may also have come preinstalled in binary form by the system package manager.

By default, `dunecontrol` looks for modules to be built in the current directory and its subdirectories. When looking for dependencies, the globally installed modules are considered in addition. If `dunecontrol` finds a local module and a global module with the same name, then the local module is preferred. If it finds two local modules with the same name, it will abort with an error message. Note that the name of the module is the corresponding string in the `dune.module` file (Appendix A.2.2), *not* the name of the module source directory (even though the two should usually coincide).

Alternatively, `dunecontrol` can be given a set of directories where to look for Dune modules in the `DUNE_CONTROL_PATH` environment variable. This allows to keep the set of Dune modules distributed across several directories, or to call `dunecontrol` from other directories than the one where the module source trees reside. Lists of directories must be separated by colons. When looking for build dependencies, globally installed modules will still be considered, even if the global installation path is not part of `DUNE_CONTROL_PATH`. On the other hand, similar to how other environment variables are handled on Unix systems, when `DUNE_CONTROL_PATH` is set then the current directory is not automatically considered. It must be explicitly added to the `DUNE_CONTROL_PATH` variable if desired.

The following small oddity is worth mentioning again: The traditional Unix way is that a piece of software gets built, installed, and can then serve as dependency for others. It is a peculiarity of Dune that the installation step can be skipped for modules: up-to-date build directories can serve as dependencies for other modules, even if they have not been installed. In other words, when looking for dependencies, `dunecontrol` makes no difference between installed modules and build directories. If all Dune modules are kept locally (i.e., in a directory tree that belongs to the user account), there is no need to actually install the modules. This can simplify developing Dune modules that many other modules depend upon, but it is not explicitly recommended.

A.1.3. Setting Build Options for Sets of Modules

In many cases there is the need to pass options to the build process for each module. The most common cases are nonstandard installation directories, external libraries located in nonstandard locations, or optimization options for the compiler. When building a single module directly, such options can be passed directly to the `cmake` build system as described in the `cmake` documentation. For example, the following commands build `dune-grid` with `-O3` optimization and install it in the folder `dune-inst` of the user's home directory:

```
cd dune-grid
mkdir build-cmake
cd build-cmake
cmake -DCMAKE_CXX_FLAGS='-O3' -DCMAKE_INSTALL_PREFIX=$HOME/dune-inst ..
make
make install
```

Passing options this way is convenient when working with a single module. However, it gets tedious when entire groups of modules have to be built, and it is usually important to ensure that all modules of such a group are built with the same set of options. When using `dunecontrol` to build modules, these options can be given to `dunecontrol`, which then passes them on to `cmake`. For this, `dunecontrol` allows to write

```
dunecontrol --[COMMAND]-opts=<options> COMMAND(S)
```

This passes options to the `cmake` or `make` commands emitted by `dunecontrol`. For example, the following commands will build and install all modules in the current directory with the same options as in the previous example:

```
dunecontrol --cmake-opts="-DCMAKE_CXX_FLAGS='-O3' \
                    -DCMAKE_INSTALL_PREFIX=$HOME/dune-inst" cmake
dunecontrol make
dunecontrol make install
```

Alternatively, it is possible to write the `cmake` options directly after the command:

```
dunecontrol cmake "-DCMAKE_CXX_FLAGS='-O3' \
                    -DCMAKE_INSTALL_PREFIX=$HOME/dune-inst"
dunecontrol make
dunecontrol make install
```

Similarly, using `-make-opts` allows to pass options to the `make` program. The most frequent use is to pass `-j <n>`, to build with n processes in parallel.

The set of options can be long, and it can get tedious to pass them all on the command line. A better way is then to store the options in a dedicated options file. Such a file can be given to `dunecontrol` using the `-opts` option:

```
dunecontrol --opts=<options file> COMMAND(S)
```

The options file is a text file in shell syntax that is run to set the values of certain variables. In most cases it will simply contain lines like

```
CMAKE_FLAGS="<...>"
```

or

```
MAKE_FLAGS="<...>"
```

together with comment lines starting with #. The values are then passed to the build environment. The following variables can currently be set:

`CMAKE:`	The precise `cmake` executable to use
`CMAKE_FLAGS:`	Flags to be passed to the `cmake` program
`CMAKE_MODULE_PATH:`	Semicolon-separated list of folders specifying a search path for `cmake` modules
`MAKE_FLAGS:`	Flags to be passed to the `make` program

Here is a small example:

```
# Use a particular compiler (clang++),
# install to a custom directory
# disable the external library SuperLU,
# use ninja instead of make as the build-tool
CMAKE_FLAGS="-DCMAKE_CXX_COMPILER=clang++ \
            -DCMAKE_INSTALL_PREFIX='$HOME/user/dune-inst' \
            -DCMAKE_DISABLE_FIND_PACKAGE_SuperLU=true \
            -GNinja"

# Build with four processes in parallel
MAKE_FLAGS="-j 4"
```

This example uses cmake's -D option to select a particular compiler and a local installation directory, and the -G option for a build system generator, in this case ninja. See the cmake documentation for details.

As can also be seen from the example, the -D option of cmake is also used to switch off features at configuration time. To disable an external dependency Foo, add

```
-DCMAKE_DISABLE_FIND_PACKAGE_Foo=true
```

to the CMAKE_FLAGS variable. The name of the dependency is case sensitive, but there is no canonical naming scheme. The correct package name can be found in the cmake output. For example, to have a sequential build despite an installed MPI library, the MPI dependency can be disabled explicitly by setting

```
-DCMAKE_DISABLE_FIND_PACKAGE_MPI=true
```

in the CMAKE_FLAGS.

Errors occurring during the build process get noted in a file called CMakeError.log in the directory CMakeFiles in the build directory of the corresponding DUNE module. This is not particular to DUNE, but rather a standard cmake feature. The content of CMakeError.log usually provides a hint at what the problem is. Note, though, that cmake caches aggressively, and running it after a modification to the build system or the build environment may not actually pick up the change unless cmake is forced to do a complete rerun. For this, delete the file CMakeCache.txt from the module's build directory. To be on the safe side, delete the entire build directory.

Many build errors actually result from faulty interplay between the build systems of several modules. When in doubt, one may want to trigger a rebuild of all DUNE modules currently in use. This can be done conveniently by calling

```
dunecontrol bexec "rm CMakeCache.txt"
```

before the build. It deletes the CMakeCache.txt files in all modules.

A.2. Dune Modules

This section describes the structure of a typical DUNE module. A DUNE module is a source tree that adheres to certain conventions regarding naming and the directory hierarchy. It must contain certain configuration files (mainly for cmake), and a file with meta information called dune.module.

A.2.1. The Structure of Dune Modules

A freshly created DUNE module dune-foo has the following directory structure:[7]

```
dune-foo
|-- cmake
|   '-- modules
|       |-- CMakeLists.txt
|       '-- DuneFooMacros.cmake
|-- CMakeLists.txt
|-- config.h.cmake
|-- doc
|   |-- CMakeLists.txt
|   '-- doxygen
|       |-- CMakeLists.txt
|       '-- Doxylocal
|-- dune
|   |-- CMakeLists.txt
|   '-- foo
|       |-- CMakeLists.txt
|       '-- foo.hh
|-- dune-foo.pc.in
|-- dune.module
|-- README
'-- src
    |-- CMakeLists.txt
    '-- dune-foo.cc
```

The source tree contains 7 directories and 15 files. These contain the build system configuration, module meta information, and a minimal C++ example. The main directory name typically matches the module name, but this is not strictly required. By convention, the directory name should start with a dune- prefix, but this is not a technical requirement.

The top level directory has four subdirectories:

1. The directory cmake contains build system macros particular to the module. These have to be located in the directory cmake/modules. Any cmake code in

---

[7]Such empty DUNE modules can be conveniently created with the duneproject program; see Chapter 3.

this directory can be used by the dune-foo module itself and by all downstream Dune modules that depend on dune-foo.

The file cmake/modules/DuneFooMacros.cmake is special: Its contents are always executed when configuring the module dune-foo, or any other Dune module that has dune-foo as a required or optional dependency. If the dune-foo module has dependencies that are not Dune modules themselves, then they should be checked for in the file DuneFooMacros.cmake (see also Appendix A.2.5). Typically, cmake's find_package can be used for this. In a Dune module freshly created by duneproject, the file is present but empty.

2. The doc directory is meant to hold module documentation. By default, it contains the infrastructure to build an online code documentation using the Doxygen tool[8] (Chapter A.2.6). Several core modules also use this directory to store manuals and tutorials in LaTeX format.

3. The subtree rooted at the dune directory keeps the code of the module. As Dune uses C++ templates extensively, most code is contained in header files, but code to be compiled into a library should be stored in the dune subtree, too.

   The dune directory itself is expected to be empty, except for a CMakeLists.txt file and another directory. This other directory contains the actual code, possibly in further subdirectories. It is to be named like the module name without the dune- prefix. So, for example, in the module dune-grid, this directory is called grid. The effect of this convention is that header **#include** directives have a standardized form: Code that includes headers from dune-grid does so by writing

   #include <dune/grid/...>

   and headers from dune-istl are included by

   #include <dune/istl/...>

   It is hence obvious, from reading the **#include** directive, what module a particular header comes from. C++ header files in Dune modules traditionally have a .hh suffix.

4. Finally, the src directory holds the code for executables contained in the module. These are typically programs that perform some sort of numerical simulation using the Dune modules being depended upon. Compilable C++ sources files should always end with .cc. Newly created Dune modules contain a short example program called dune-<modulename>.cc here. If a module only provides a library, then the src directory can be empty, too.

   Sometimes the executables are example programs for the features of the module, rather than serious simulation codes. In that case the src directory may be renamed to examples. This is the case, e.g., in the dune-functions module.

---

[8]http://doxygen.org

As for files in the main source tree directory, the two central ones dune.module and config.h.cmake are discussed in dedicated sections below. The file dune-foo.pc.in contains information used by the pkg-config tool[9] to find the dune-foo module.

Finally, every directory in the DUNE module contains a file called CMakeLists.txt, which is written in the cmake language. Upon calling cmake, the CMakeLists.txt file in the main module directory is executed. Whenever an add_subdirectory command is encountered, the CMakeLists.txt file of that sub-directory is executed. The top-level CMakeLists.txt file is special, because it sets up the entire DUNE module correctly. Users are free to add more code to it, but the parts generated by duneproject must not be deleted.

A.2.2. The dune.module File

Each DUNE module contains a small text file called dune.module, which contains meta data of the module. In particular, it lists the official module name (which could, in principle, differ from the name of the source directory), and the dependencies of the module.

A typical dune.module file looks like this:

```
#################################
# Dune module information file #
#################################

# Name of the module
Module: dune-foo
Version: 3.1.4
Maintainer: paul.atreides@dune-project.org
# Required build dependencies
Depends: dune-grid dune-localfunctions
# Optional build dependencies
#Suggests:
```

It consists of lines of the form

```
<keyword>: <value(s)>
```

Lines starting with a # are treated as comments. The possible keywords are:

- Module: The official module name for the dunecontrol build system. Dependencies of other modules on this one must use this name. Only changing this string is not enough to completely change the module name, however. The name is also hard-coded in the main CMakeLists.txt file of the module, and it is part of the name of the dune-foo.pc.in file.

- Version: A string that encodes the module version. Besides documentation value, it allows to depend on particular versions of a module. The version

---

[9]https://www.freedesktop.org/wiki/Software/pkg-config

string should follow the X.Y.Z scheme, but additional arbitrary suffixes may be appended. By convention, the version on the current development branch is X.Y-git, where X.Y is the planned number of the next release. The version number is also available in the C++ code via preprocessor variables defined in the file config.h (see Appendix A.2.3).

- Maintainer: Contact email address(es) of the person or group of persons responsible for module maintenance and bugfixing. Appears in the PACKAGE_BUGREPORT preprocessor variable defined in the file config.h.

- Depends: Whitespace-separated list of Dune modules that the present module needs to build and run properly. The names used here are the ones given in the dune.module files of the respective modules. Note that no distinction is made between build and run-time dependencies. Only the direct dependencies need to be listed here—indirect dependencies are resolved automatically by dunecontrol.

The dependencies can be versioned, meaning that only modules in a particular version or range of versions satisfy a dependency. Version requirements appear in parentheses after the module name, and have the form (<operator><version>). Valid operators are <=, =, >=, <, and >. For example, (>=2.7.0) states that the module must have version 2.7.0 or later. Two conditions can be combined by && for "and" and || for "or". For example, a version between 2.7.0 and 3.0.0 is requested by writing (>=2.7 && <3.0).

On the other hand, the dunecontrol option -skipversioncheck allows to disregard these checks.

- Suggests: Whitespace-separated list of Dune modules that the module can use, but that are not strictly necessary to build the module. Typically, such modules allow to enable additional features when present at build time. When a suggested dependency is not met, dunecontrol issues a warning, but does not abort the build process. For example, the dune-grid module lists dune-uggrid as a suggestion.

If a suggested dependency is found, this is signaled to cmake by defining the variable <module>_FOUND, where <module> is the name of the Dune module. Hence, for example, in a CMakeLists.txt file, compilation can be controlled by constructions like

```
if (dune-uggrid_FOUND)
  add_executable("unstructured-grid-example" unstructured-grid-example.cc)
endif ()
```

Presence of the suggested dependency is signaled to C++ code by defining the preprocessor variable HAVE_<MODULE_NAME> to 1. Here <MODULE_NAME> is the module name, but written in capitals and with hyphens replaced by underscores.

A.2.3. The Files `config.h` and `config.h.cmake`

As is common in C++ software on Unix systems, the build system of each DUNE module writes a file called `config.h` into the module's build directory. This file contains some of the build system knowledge about the build environment, transferred to the C++ code by a list of preprocessor macros. All non-header C++ files in a DUNE module are expected to include `config.h` as their first header.

The first block of information in a `config.h` file contains some of the meta data specified in the `dune.module` file. For example, the `config.h` file for the module with the `dune.module` file of Appendix A.2.2 will contain the definitions

```
#define PACKAGE_NAME "dune-foo"
#define PACKAGE_BUGREPORT "paul.atreides@dune-project.org"
#define PACKAGE_VERSION "3.1.4"
```

In addition, the file contains similar definitions for all direct and indirect dependencies. If the module depends on `dune-grid`, for example, then its `config.h` file will contain

```
#define HAVE_DUNE_GRID 1
```

as well as the `dune-grid` version number in form of

```
#define DUNE_GRID_VERSION "2.7.0"
#define DUNE_GRID_VERSION_MAJOR 2
#define DUNE_GRID_VERSION_MINOR 7
#define DUNE_GRID_VERSION_REVISION 0
```

The header `dune/common/version.hh` from the `dune-common` module contains tools to conveniently deal with this version information in C++ code.

In addition, any module can supply any kind of additional module-specific information in its `config.h` file. Furthermore, such information will appear in the `config.h` files of all downstream modules. Information intended to appear in `config.h` needs to be written into the template file `config.h.cmake`, which is part of each DUNE module. The process strictly follows `cmake`'s procedure, and will not be explained here. Please see the `cmake` documentation for details.

A.2.4. Adding New Code to a Dune Module

New code is added to a DUNE module roughly like to any other C++ software with a `cmake` build system, but there are a few DUNE-specific features.

As DUNE uses C++ templates almost everywhere, a lot of typical DUNE code is kept in header files. New headers should be placed in the folder `dune/<modulename>`, or in subfolders thereof. All headers in a template-heavy code are typically required by downstream modules, and therefore they have to be installed. The standard `cmake` machinery is used for this: In the `CMakeLists.txt` file of the directory of the header to be installed, add the header to the list of files given to the `install` directive:

```
install(FILES
  header1.hh
  header2.hh
  [...]
  DESTINATION ${CMAKE_INSTALL_INCLUDEDIR}/<header path in source tree>)
```

The header path given is relative to the main module installation directory, i.e., it should start with dune/.

Actually compiling code requires DUNE-specific cmake macros. To create a library, use the cmake command

```
dune_add_library(<basename> SOURCES <sourcefile>.cc)
```

This creates both static and shared libraries. On a Unix system they will be called lib<basename>.a and lib<basename>.so, respectively. The option SOURCES specifies a whitespace-separated list of source files that are compiled and linked into the library. Other libraries can be incorporated with the ADD_LIBS option. Consult the online documentation for further possibilities.

Files that are to be compiled into executables (such as simulation codes or examples) have to be announced to cmake by the standard cmake command

```
add_executable(<executablename> <sourcefile>.cc)
```

which compiles the source file called <sourcefile>.cc into an executable called <executablename>. Various options allow to influence the compilation and linking process. Source code for such executables is normally placed in the src directory in the main module source directory, not under dune.

Executables need to be given information about their dependencies, in form of include paths, libraries to link to, and compiler and linker options. All of this is handled automatically by a cmake macro called dune_enable_all_packages, which gets invoked in the top-level CMakeLists.txt file (before any calls to add_subdirectory). This macro adds paths, flags, etc. of all dependencies found by the build system to all executables. This means that all of them typically build and link out of the box as long as the corresponding add_executable call is there. The price is a long list of possibly unneeded paths and flags being handed to the compiler and linker when building executables, but that is not usually a problem.

For users that do want more control, the DUNE build system offers dedicated macros to add individual dependencies to an executable. For example, in any DUNE module that depends on dune-istl, the line

```
add_dune_umfpack_flags(<executablename>)
```

will add all flags and paths that are necessary to build <executablename> if it uses the direct sparse solver UMFPACK[10] [46]. Likewise, there are add_dune_mpi_flags, add_dune_parmetis_flags, and many others. No such macros are needed if the

---

[10]http://faculty.cse.tamu.edu/davis/suitesparse.html

dependency is a DUNE module itself! Manually linking libraries can be done through the `target_link_libraries` command. Even more control is possible using cmake features like `set_property` and `get_property`.

A.2.5. Dependencies that are not Dune Modules

Not all dependencies of a DUNE module are other DUNE modules. Such external dependencies are not handled automatically by the `dunecontrol` program, and need to be treated by cmake directly. The cmake program looks for such dependencies with its `find_package` command. If the external code is sufficiently cmake-aware, then it may have provided a *config-file package* in one of the standard locations. The `find_package` command will then find the external dependency's library, headers, and all further relevant information along with it.

If there is no config-file package, the DUNE module itself has to provide a so-called *find-module package*, which looks for the dependency, and determines the various compiler and linker flags it requires. Each custom find-module is a dedicated file, and must be kept in the `cmake/modules` folder of the first DUNE module that uses it. For an external package called SomePackage, the cmake find-module must be called `FindSomePackage.cmake`.[11] If cmake encounters a `find_package(SomePackage)` command and cannot find a corresponding config-file package, it searches for the file `FindSomePackage.cmake`, and executes the code it contains. This code should determine all relevant information, and write it into the standard cmake variables. Many such find-module packages are available on the web. Depending on how common the external dependency is, there may be no need to write the find-module from scratch.

The compiler and linker flags of the new dependency are applied to the building of all executables by virtue of the `dune_enable_all_packages` mechanism (see above). To make `dune_enable_all_packages` pick up the new dependency, the new flags need to be registered. This is done by means of the macro

```
dune_register_package_flags(INCLUDE_DIRS <list of directories>
                            LIBRARIES <list of libraries>)
```

which should be called near the end of the find-module. The options allow to add header paths and libraries to the global list of flags kept by `dune_enable_all_packages`. Furthermore, `COMPILE_OPTIONS` allows to set particular compiler options, and the option `COMPILE_DEFINITIONS` can be used to set preprocessor flags.

For each custom find-module that looks for a software called SomePackage, it is also good style to provide a macro `add_dune_somepackage_flags`, which should be defined in a file called `AddSomePackageFlags.cmake` in the folder `cmake/modules`. This macro should provide all necessary paths and options to an individual executable (see above), and is the tool to use when `dune_enable_all_packages` is not appropriate.

---

[11] Note that cmake package names are case sensitive.

A.2.6. Documentation

The Dune build system offers infrastructure to build documentation using the Doxygen [12] tool for code, and LaTeX for general documents. All documentation of a Dune module is created by calling

```
make doc
```

in the module's build directory.

The Doxygen program parses source code, extracts comment blocks from it, and builds documentation in various output formats from these comments. By default, the Dune build system will call Doxygen such as to produce HTML output. The HTML documentation will appear in the folder doc/doxygen/html in the build directory, and the main page is doc/doxygen/html/index.html.

Doxygen is controlled by a large set of options. These are stored in a text file which is usually called Doxyfile. In principle, each Dune module could have its own Doxyfile; however, typical Doxyfiles differ little from module to module. The Dune build system therefore contains the common part factored out in a separate file called Doxystyle, which is located in dune-common/doc/doxygen. Each downstream module only keeps a text file called Doxylocal with the module-specific bits. When make doc is called, the two files are joined, and placeholders for module names and directories are replaced with actual values. Here is the default Doxylocal file as written by duneproject, in slightly shortened form:

```
# The INPUT tag can be used to specify the files and/or directories that contain
# documented source files. You may enter file names or directories, separated
# by spaces.

INPUT               += @top_srcdir@/dune/

# The EXCLUDE tag can be used to specify files and/or directories that should
# be excluded from the INPUT source files.

# EXCLUDE             += @top_srcdir@/dune/foo/test

# The EXAMPLE_PATH tag can be used to specify one or more files or
# directories that contain example code fragments that are included
# with the \include command.

# EXAMPLE_PATH        += @top_srcdir@/src

# The IMAGE_PATH tag can be used to specify one or more files or
# directories that contain image that are included with the \image command.

# IMAGE_PATH          += @top_srcdir@/dune/foo/pics
```

---

[12] http://doxygen.org

The file mainly determines where to look for source files, images, and other input. The variable @top_srcdir is replaced by the absolute path of the module source directory. Further DOXYGEN options can be added to Doxylocal, and will then override the defaults in the Doxystyle file in the dune-common module.

By default, the doc directory of a module contains only the doxygen subdirectory and a trivial CMakeLists.txt file. However, it may also contain further subfolders with module information in other formats. For LaTeX documents, the DUNE build system ships the UseLatexMk.cmake tool,[13] which uses latexmk[14] to build pdf files from LaTeX sources. It offers a single command

```
add_latex_document(SOURCE <texsource>.tex)
```

where <texsource>.tex should be a single LaTeX file. As the latexmk program is used for invoking the TeX engine, all dependency tracking happens internally, and the required number of calls to pdflatex, bibtex, etc. is performed automatically without further user intervention. The add_latex_document macro has further options, which are described in the online documentation.

The UseLatexMk.cmake macro also introduces a make target called clean_latex, which removes all files created by the LaTeX toolchain. This allows to build the LaTeX documents of a DUNE module from scratch without touching the build of the C++ code.

A.2.7. Automated Tests

Many DUNE modules contain test programs that check individual parts of the module's functionality. These can range from small unit tests that verify the functioning of specific details of individual classes all the way to full-blown integration tests that run a complete numerical simulation and check whether the result is correct. Having such tests is an important part of professional software development, in particular for template-rich code like DUNE, where a lot of code is not really exercised when building the module. Without tests, a lot of bugs become visible only when using a module's code downstream. The test sources are usually collected in directories named test. There may be one or more of those, at different places in the directory hierarchy of a module.

The tests are not built automatically when building a module. Rather, compiling and linking tests needs to be triggered manually, using the command

```
make build_tests
```

in the build directory. This produces a set of test executables that can be run individually just like other executables. The entire set of tests can be run using the ctest command that is part of a normal cmake installation. It calls all available tests of a module, and reports the results in overview form. Check ctest -help for a lot

[13]https://github.com/dokempf/UseLatexMk
[14]https://www.ctan.org/pkg/latexmk

of useful options, such as choosing the set of tests to be run by matching regular expressions or showing the output of failed tests.

A newly written test is announced to the DUNE build system by the cmake macro

```
dune_add_test(SOURCES <testsource>.cc)
```

It adds the test to the build_tests target, supplies the necessary compiler and linker flags, and adds the test to the list of programs to be run by ctest. Test behavior can be controlled by various options of the dune_add_test command:

- COMPILE_ONLY: If set, the given test will be compiled during make build_tests, but it will not be run by ctest. This can be useful to reduce testing time.

- CMAKE_GUARD: This allows conditional building of unit tests based on cmake variables. Its argument is a boolean condition. If this condition evaluates to FALSE, the test sources will be replaced by a dummy source that only returns the value 77, and ctest will ignore the test result. This approach is preferable to guarding the call to dune_add_test with a cmake if-clause: The test at least appears in the list of test results, even if it could not be built or run.

- MPI_RANKS: To test features regarding distributed computing, a test may need to be run on more than one process. A list of numbers may be given to the MPI_RANKS option, and the test is run once for each number in this list. The number determines how many processes are used for that run. A global limit can be set in form of the environment variable DUNE_MAX_TEST_CORES.

- TIMEOUT: Maximum time permitted for running the test (in seconds). This prevents dead-locked distributed tests from completely locking up the system. It defaults to 5 minutes, but an explicit value has to be set when the MPI_RANKS option is used.

These are just the more prominent options; consult the online documentation for a full list.

B. Complete Source Codes of the Example Programs

In this appendix we print the example programs appearing in this book in their entirety. C++ programs traditionally come with a lot of boiler-plate code. When discussing example programs, showing all that boiler-plate makes the text too verbose. Still it is useful to have entire programs at hand. In electronic versions of this document, the source code files are also embedded in the document file, and can be accessed via the colored icons in the text margin. In addition, the icon right here contains an entire DUNE module called `dune-book-examples`, which has all the source files of this book, the corresponding build system, and all the grid files. It is to be used with DUNE version 2.7, and may require minor modifications when used with later versions. The dune-book-examples module can also be downloaded from the publisher website of the book.

B.1. The Geometry of the Example Domain

Various examples in this book use a grid on an L-shaped domain, and load it from files called `l-shape.msh` or `l-shape-refined.msh` in the format of the GMSH grid generator.[1] While the grid files themselves are too large to print them here, we show the GMSH input geometry file `l-shape.geo` that was used to create them:

```
// Scale global grid element size
Mesh.CharacteristicLengthFactor = 1.5;

// Corners of the domain
Point(0) = {  0,   0, 0};
Point(1) = {  1,   0, 0};
Point(2) = {  1, 0.5, 0};
Point(3) = {0.5, 0.5, 0};
Point(4) = {0.5,   1, 0};
Point(5) = {  0,   1, 0};

// Boundary edges of the domain
Line(1) = {0,1};
Line(2) = {1,2};
Line(3) = {2,3};
Line(4) = {3,4};
Line(5) = {4,5};
```

---

[1] http://gmsh.info

© Springer Nature Switzerland AG 2020
O. Sander, *DUNE — The Distributed and Unified Numerics Environment*, Lecture Notes in Computational Science and Engineering 140, https://doi.org/10.1007/978-3-030-59702-3

```
Line(6) = {5,0};

// Boundary of the domain
Line Loop(1) = {1,2,3,4,5,6};

// The domain itself
Plane Surface(1) = {1};
```

This file and all grid files are also contained in the example DUNE module dune-book-examples. Let us stress again that as of DUNE Version 2.7, the GmshReader class can only read GMSH files in format version 2, whereas the current default of GMSH is to write files in format version 4. When constructing grids in GMSH for use with DUNE, format 2 has to be explicitly selected when writing the grid file.

B.2. Finite Element Method for the Poisson Equation

In this and all following printed source codes, lines of the form

// { <label> }

are tags that are used by the typesetting machinery to import code sections directly into the text.

File:	getting-started-poisson-fem.cc
Equation:	Poisson equation
Discretization:	First-order Lagrange finite elements
Grid:	Unstructured triangle grid, implemented with UGGrid
Solver:	CG solver with ILU preconditioner
Distributed:	no
Discussed in:	Chapter 3.3

```
1   #include <config.h>
2
3   #include <vector>
4
5   #include <dune/geometry/quadraturerules.hh>
6
7   // { include_uggrid_begin }
8   #include <dune/grid/uggrid.hh>
9   #include <dune/grid/io/file/gmshreader.hh>
10  // { include_uggrid_end }
11  #include <dune/grid/io/file/vtk/vtkwriter.hh>
12
13  #include <dune/istl/matrix.hh>
14  // { include_matrix_vector_begin }
15  #include <dune/istl/bcrsmatrix.hh>
16  #include <dune/istl/bvector.hh>
17  // { include_matrix_vector_end }
18  #include <dune/istl/matrixindexset.hh>
19  #include <dune/istl/preconditioners.hh>
20  #include <dune/istl/solvers.hh>
21  #include <dune/istl/matrixmarket.hh>
22
23  #include <dune/functions/functionspacebases/lagrangebasis.hh>
24  #include <dune/functions/functionspacebases/interpolate.hh>
25
26
27  // { using_namespace_dune_begin }
```

```
28    using namespace Dune;
29    // { using_namespace_dune_end }
30
31    // Compute the stiffness matrix for a single element
32    // { local_assembler_signature_begin }
33    template<class LocalView, class Matrix>
34    void assembleElementStiffnessMatrix(const LocalView& localView,
35                                        Matrix& elementMatrix)
36    // { local_assembler_signature_end }
37    {
38    // { local_assembler_get_geometry_begin }
39      using Element = typename LocalView::Element;
40      constexpr int dim = Element::dimension;
41      auto element = localView.element();
42      auto geometry = element.geometry();
43    // { local_assembler_get_geometry_end }
44
45      // Get set of shape functions for this element
46    // { get_shapefunctions_begin }
47      const auto& localFiniteElement = localView.tree().finiteElement();
48    // { get_shapefunctions_end }
49
50      // Set all matrix entries to zero
51    // { init_element_matrix_begin }
52      elementMatrix.setSize(localView.size(),localView.size());
53      elementMatrix = 0;         // Fill the entire matrix with zeros
54    // { init_element_matrix_end }
55
56      // Get a quadrature rule
57    // { get_quadrature_rule_begin }
58      int order = 2 * (localFiniteElement.localBasis().order()-1);
59      const auto& quadRule = QuadratureRules<double, dim>::rule(element.type(),
60                                                               order);
61    // { get_quadrature_rule_end }
62
63      // Loop over all quadrature points        .
64    // { loop_over_quad_points_begin }
65      for (const auto& quadPoint : quadRule)
66      {
67    // { loop_over_quad_points_end }
68
69    // { get_quad_point_info_begin }
70        // Position of the current quadrature point in the reference element
71        const auto quadPos = quadPoint.position();
72
73        // The transposed inverse Jacobian of the map from the reference element
74        // to the grid element
75        const auto jacobian = geometry.jacobianInverseTransposed(quadPos);
76
77        // The determinant term in the integral transformation formula
78        const auto integrationElement = geometry.integrationElement(quadPos);
79    // { get_quad_point_info_end }
80
81    // { compute_gradients_begin }
82        // The gradients of the shape functions on the reference element
83        std::vector<FieldMatrix<double,1,dim> > referenceGradients;
84        localFiniteElement.localBasis().evaluateJacobian(quadPos,
85                                                         referenceGradients);
86
87        // Compute the shape function gradients on the grid element
88        std::vector<FieldVector<double,dim> > gradients(referenceGradients.size());
89        for (size_t i=0; i<gradients.size(); i++)
90          jacobian.mv(referenceGradients[i][0], gradients[i]);
91    // { compute_gradients_end }
92
93        // Compute the actual matrix entries
94    // { compute_matrix_entries_begin }
95        for (size_t p=0; p<elementMatrix.N(); p++)
96        {
97          auto localRow = localView.tree().localIndex(p);
98          for (size_t q=0; q<elementMatrix.M(); q++)
99          {
100            auto localCol = localView.tree().localIndex(q);
101            elementMatrix[localRow][localCol] += (gradients[p] * gradients[q])
102                                    * quadPoint.weight() * integrationElement;
103          }
104        }
105    // { compute_matrix_entries_end }
106      }
107    }
108
109
110    // Compute the source term for a single element
111    template<class LocalView>
112    void assembleElementVolumeTerm(
113            const LocalView& localView,
114            BlockVector<double>& localB,
```

```
115            const std::function<double(FieldVector<double,
116                                     LocalView::Element::dimension>)> volumeTerm)
117  {
118    using Element = typename LocalView::Element;
119    auto element = localView.element();
120    constexpr int dim = Element::dimension;
121
122    // Set of shape functions for a single element
123    const auto& localFiniteElement = localView.tree().finiteElement();
124
125    // Set all entries to zero
126    localB.resize(localFiniteElement.size());
127    localB = 0;
128
129    // A quadrature rule
130    int order = dim;
131    const auto& quadRule = QuadratureRules<double, dim>::rule(element.type(), order);
132
133    // Loop over all quadrature points
134    for (const auto& quadPoint : quadRule)
135    {
136      // Position of the current quadrature point in the reference element
137      const FieldVector<double,dim>& quadPos = quadPoint.position();
138
139      // The multiplicative factor in the integral transformation formula
140      const double integrationElement = element.geometry().integrationElement(quadPos);
141
142      double functionValue = volumeTerm(element.geometry().global(quadPos));
143
144      // Evaluate all shape function values at this point
145      std::vector<FieldVector<double,1> > shapeFunctionValues;
146      localFiniteElement.localBasis().evaluateFunction(quadPos, shapeFunctionValues);
147
148      // Actually compute the vector entries
149      for (size_t p=0; p<localB.size(); p++)
150      {
151        auto localIndex = localView.tree().localIndex(p);
152        localB[localIndex] += shapeFunctionValues[p] * functionValue
153                            * quadPoint.weight() * integrationElement;
154      }
155    }
156  }
157
158  // Get the occupation pattern of the stiffness matrix
159  template<class Basis>
160  void getOccupationPattern(const Basis& basis, MatrixIndexSet& nb)
161  {
162    nb.resize(basis.size(), basis.size());
163
164    auto gridView = basis.gridView();
165
166    // A loop over all elements of the grid
167    auto localView = basis.localView();
168
169    for (const auto& element : elements(gridView))
170    {
171      localView.bind(element);
172
173      for (size_t i=0; i<localView.size(); i++)
174      {
175        // The global index of the i-th vertex of the element
176        auto row = localView.index(i);
177
178        for (size_t j=0; j<localView.size(); j++ )
179        {
180          // The global index of the j-th vertex of the element
181          auto col = localView.index(j);
182          nb.add(row,col);
183        }
184      }
185    }
186  }
187
188
189  /** \brief Assemble the Laplace stiffness matrix on the given grid view */
190  // { global_assembler_signature_begin }
191  template<class Basis>
192  void assemblePoissonProblem(const Basis& basis,
193                              BCRSMatrix<double>& matrix,
194                              BlockVector<double>& b,
195                              const std::function<
196                                  double(FieldVector<double,
197                                         Basis::GridView::dimension>)
198                                      > volumeTerm
199  // { global_assembler_signature_end }
200  {
201    // { assembler_get_grid_info_begin }
```

```
202      auto gridView = basis.gridView();
203      // { assembler_get_grid_info_end }
204
205      // MatrixIndexSets store the occupation pattern of a sparse matrix.
206      // They are not particularly efficient, but simple to use.
207      // { assembler_matrix_pattern_begin }
208      MatrixIndexSet occupationPattern;
209      getOccupationPattern(basis, occupationPattern);
210      occupationPattern.exportIdx(matrix);
211      // { assembler_matrix_pattern_end }
212
213      // Set all entries to zero
214      // { assembler_zero_matrix_begin }
215      matrix = 0;
216      // { assembler_zero_matrix_end }
217
218      // { assembler_zero_vector_begin }
219      // Set b to correct length
220      b.resize(basis.dimension());
221
222      // Set all entries to zero
223      b = 0;
224      // { assembler_zero_vector_end }
225
226      // A loop over all elements of the grid
227      // { assembler_element_loop_begin }
228      auto localView = basis.localView();
229
230      for (const auto& element : elements(gridView))
231      {
232      // { assembler_element_loop_end }
233
234          // Now let's get the element stiffness matrix
235          // A dense matrix is used for the element stiffness matrix
236      // { assembler_assemble_element_matrix_begin }
237          localView.bind(element);
238
239          Matrix<double> elementMatrix;
240          assembleElementStiffnessMatrix(localView, elementMatrix);
241      // { assembler_assemble_element_matrix_end }
242
243      // { assembler_add_element_matrix_begin }
244          for(size_t p=0; p<elementMatrix.N(); p++)
245          {
246            // The global index of the p-th degree of freedom of the element
247            auto row = localView.index(p);
248
249            for (size_t q=0; q<elementMatrix.M(); q++ )
250            {
251              // The global index of the q-th degree of freedom of the element
252              auto col = localView.index(q);
253              matrix[row][col] += elementMatrix[p][q];
254            }
255          }
256      // { assembler_add_element_matrix_end }
257
258          // Now get the local contribution to the right-hand side vector
259          BlockVector<double> localB;
260          assembleElementVolumeTerm(localView, localB, volumeTerm);
261
262          for (size_t p=0; p<localB.size(); p++)
263          {
264            // The global index of the p-th vertex of the element
265            auto row = localView.index(p);
266            b[row] += localB[p];
267          }
268      }
269   }
270
271
272   int main(int argc, char *argv[])
273   {
274   // { mpi_setup_begin }
275      // Set up MPI, if available
276      MPIHelper::instance(argc, argv);
277   // { mpi_setup_end }
278
279      /////////////////////////////////
280      //   Generate the grid
281      /////////////////////////////////
282
283   // { create_grid_begin }
284      constexpr int dim = 2;
285      using Grid = UGGrid<dim>;
286      std::shared_ptr<Grid> grid = GmshReader<Grid>::read("l-shape.msh");
287
288      grid->globalRefine(2);
```

```
289
290    using GridView = Grid::LeafGridView;
291    GridView gridView = grid->leafGridView();
292  // { create_grid_end }
293
294  /////////////////////////////////////////////////////////
295  //   Stiffness matrix and right hand side vector
296  /////////////////////////////////////////////////////////
297
298  // { create_matrix_vector_begin }
299    using Matrix = BCRSMatrix<double>;
300    using Vector = BlockVector<double>;
301
302    Matrix stiffnessMatrix;
303    Vector b;
304  // { create_matrix_vector_end }
305
306  /////////////////////////////////////////////////////////
307  //   Assemble the system
308  /////////////////////////////////////////////////////////
309
310  // { setup_basis_begin }
311    Functions::LagrangeBasis<GridView,1> basis(gridView);
312
313    auto sourceTerm = [](const FieldVector<double,dim>& x){return -5.0;};
314  // { setup_basis_end }
315  // { call_assembler_begin }
316    assemblePoissonProblem(basis, stiffnessMatrix, b, sourceTerm);
317  // { call_assembler_end }
318
319    // Determine Dirichlet dofs by marking all degrees of freedom whose Lagrange nodes
320    // comply with a given predicate.
321  // { dirichlet_marking_begin }
322    auto predicate = [](auto x)
323    {
324      return x[0] < 1e-8
325          || x[1] < 1e-8
326          || (x[0] > 0.4999 && x[1] > 0.4999);
327    };
328
329    // Evaluating the predicate will mark all Dirichlet degrees of freedom
330    std::vector<bool> dirichletNodes;
331    Functions::interpolate(basis, dirichletNodes, predicate);
332  // { dirichlet_marking_end }
333
334  /////////////////////////////////////////
335  //   Modify Dirichlet rows
336  /////////////////////////////////////////
337  // { dirichlet_matrix_modification_begin }
338    // Loop over the matrix rows
339    for (size_t i=0; i<stiffnessMatrix.N(); i++)
340    {
341      if (dirichletNodes[i])
342      {
343        auto cIt    = stiffnessMatrix[i].begin();
344        auto cEndIt = stiffnessMatrix[i].end();
345        // Loop over nonzero matrix entries in current row
346        for (; cIt!=cEndIt; ++cIt)
347          *cIt = (cIt.index()==i) ? 1.0 : 0.0;
348      }
349    }
350  // { dirichlet_matrix_modification_end }
351
352    // Set Dirichlet values
353  // { dirichlet_rhs_modification_begin }
354    auto dirichletValues = [](auto x)
355    {
356      return (x[0]< 1e-8 || x[1] < 1e-8) ? 0 : 0.5;
357    };
358    Functions::interpolate(basis,b,dirichletValues, dirichletNodes);
359  // { dirichlet_rhs_modification_end }
360
361  /////////////////////////////////////////////////////////////////////////////
362  // Write matrix and load vector to files, to be used in later examples
363  /////////////////////////////////////////////////////////////////////////////
364  // { matrix_rhs_writing_begin }
365    std::string baseName = "getting-started-poisson-fem-"
366                         + std::to_string(grid->maxLevel()) + "-refinements";
367    storeMatrixMarket(stiffnessMatrix, baseName + "-matrix.mtx");
368    storeMatrixMarket(b, baseName + "-rhs.mtx");
369  // { matrix_rhs_writing_end }
370
371  /////////////////////////////
372  //   Compute solution
373  /////////////////////////////
374
375  // { algebraic_solving_begin }
```

```
376    // Choose an initial iterate that fulfills the Dirichlet conditions
377    Vector x(basis.size());
378    x = b;
379
380    // Turn the matrix into a linear operator
381    MatrixAdapter<Matrix,Vector,Vector> linearOperator(stiffnessMatrix);
382
383    // Sequential incomplete LU decomposition as the preconditioner
384    SeqILU<Matrix,Vector,Vector> preconditioner(stiffnessMatrix,
385                                    1.0);  // Relaxation factor
386
387    // Preconditioned conjugate gradient solver
388    CGSolver<Vector> cg(linearOperator,
389                        preconditioner,
390                        1e-5, // Desired residual reduction factor
391                        50,    // Maximum number of iterations
392                        2);    // Verbosity of the solver
393
394    // Object storing some statistics about the solving process
395    InverseOperatorResult statistics;
396
397    // Solve!
398    cg.apply(x, b, statistics);
399  // { algebraic_solving_end }
400
401    // Output result
402  // { vtk_output_begin }
403    VTKWriter<GridView> vtkWriter(gridView);
404    vtkWriter.addVertexData(x, "solution");
405    vtkWriter.write("getting-started-poisson-fem-result");
406  // { vtk_output_end }
407  }
```

B.3. Finite Volume Method for the Linear Transport Equation

File:	getting-started-transport-fv.cc
Equation:	Linear transport equation
Discretization:	Cell-centered finite volumes, with explicit time stepping
Grid:	Uniform quadrilateral grid, implemented with YaspGrid
Solver:	none
Distributed:	no
Discussed in:	Chapter 3.4

```
1   #include "config.h"
2
3   #include <iostream>
4   #include <vector>
5
6   #include <dune/common/parallel/mpihelper.hh>
7
8   #include <dune/grid/common/mcmgmapper.hh>
9   #include <dune/grid/yaspgrid.hh>
10  #include <dune/grid/io/file/vtk.hh>
11
12  // { using_namespace_dune_begin }
13  using namespace Dune;
14  // { using_namespace_dune_end }
15
16  // { evolve_signature_begin }
17  template<class GridView, class Mapper>
18  void evolve(const GridView& gridView,
19              const Mapper& mapper,
20              double dt,        // Time step size
21              std::vector<double>& c,
22              const std::function<FieldVector<double,GridView::dimension>
23                              (FieldVector<double,GridView::dimension>)> v,
24              const std::function<double
```

```
25                              (FieldVector<double,GridView::dimension>)> inflow)
26    // { evolve_signature_end }
27    {
28    // { evolve_init_begin }
29       // Grid dimension
30       constexpr int dim = GridView::dimension;
31
32       // Allocate a temporary vector for the update
33       std::vector<double> update(c.size());
34       std::fill(update.begin(), update.end(), 0.0);
35    // { evolve_init_end }
36
37       // Compute update vector
38    // { element_loop_begin }
39       for (const auto& element : elements(gridView))
40       {
41          // Element geometry
42          auto geometry = element.geometry();
43
44          // Element volume
45          double elementVolume = geometry.volume();
46
47          // Unique element number
48          typename Mapper::Index i = mapper.index(element);
49    // { element_loop_end }
50
51          // Loop over all intersections γ_ij with neighbors and boundary
52    // { intersection_loop_begin }
53          for (const auto& intersection : intersections(gridView,element))
54          {
55             // Geometry of the intersection
56             auto intersectionGeometry = intersection.geometry();
57
58             // Center of intersection in global coordinates
59             FieldVector<double,dim>
60                 intersectionCenter = intersectionGeometry.center();
61
62             // Velocity at intersection center v_ij
63             FieldVector<double,dim> velocity = v(intersectionCenter);
64
65             // Center of the intersection in local coordinates
66             const auto& intersectionReferenceElement
67                = ReferenceElements<double,dim-1>::general(intersection.type());
68             FieldVector<double,dim-1> intersectionLocalCenter
69                = intersectionReferenceElement.position(0,0);
70
71             // Normal vector scaled with intersection area: n_ij|γ_ij|
72             FieldVector<double,dim> integrationOuterNormal
73                = intersection.integrationOuterNormal(intersectionLocalCenter);
74
75             // Compute factor occuring in flux formula: ⟨v_ij, n_ij⟩|γ_ij|
76             double intersectionFlow = velocity*integrationOuterNormal;
77    // { intersection_loop_initend }
78
79    // { intersection_loop_mainbegin }
80             // Outflow contributions
81             update[i] -= c[i]*std::max(0.0,intersectionFlow)/elementVolume;
82
83             // Inflow contributions
84             if (intersectionFlow<=0)
85             {
86                // Handle interior intersection
87                if (intersection.neighbor())
88                {
89                   // Access neighbor
90                   auto j = mapper.index(intersection.outside());
91                   update[i] -= c[j]*intersectionFlow/elementVolume;
92                }
93
94                // Handle boundary intersection
95                if (intersection.boundary())
96                   update[i] -= inflow(intersectionCenter)
97                                         * intersectionFlow/elementVolume;
98             }
99    // { intersection_loopend }
100          } // End loop over all intersections
101       } // End loop over the grid elements
102    // { element_loop_end }
103
104    // { evolve_laststeps }
105       // Update the concentration vector
106       for (std::size_t i=0; i<c.size(); ++i)
107          c[i] += dt*update[i];
108    }
109    // { evolve_end }
```

```
110
111    // { main_begin }
112    int main(int argc, char *argv[])
113    {
114      // Set up MPI, if available
115      MPIHelper::instance(argc, argv);
116    // { main_signature_end }
117
118    // { create_grid_begin }
119      constexpr int dim = 2;
120      using Grid = YaspGrid<dim>;
121      Grid grid({1.0,1.0},      // Upper right corner, the lower left one is (0,0)
122               { 80, 80});     // Number of elements per direction
123
124      using GridView = Grid::LeafGridView;
125      GridView gridView = grid.leafGridView();
126    // { create_grid_end }
127
128      // Assigns a unique number to each element
129    // { create_concentration_begin }
130      MultipleCodimMultipleGeomTypeMapper<GridView>
131      mapper(gridView, mcmgElementLayout());
132
133      // Allocate a vector for the concentration
134      std::vector<double> c(mapper.size());
135    // { create_concentration_end }
136
137      // Initial concentration
138    // { lambda_initial_concentration_begin }
139      auto c0 = [](const FieldVector<double,dim>& x)
140      {
141        return (x.two_norm()>0.125 && x.two_norm()<0.5) ? 1.0 : 0.0;
142      };
143    // { lambda_initial_concentration_end }
144
145    // { sample_initial_concentration_begin }
146      // Iterate over grid elements and evaluate c0 at element centers
147      for (const auto& element : elements(gridView))
148      {
149        // Get element geometry
150        auto geometry = element.geometry();
151
152        // Get global coordinate of element center
153        auto global = geometry.center();
154
155        // Sample initial concentration c0 at the element center
156        c[mapper.index(element)] = c0(global);
157      }
158    // { sample_initial_concentration_end }
159
160      // Construct VTK writer
161    // { construct_vtk_writer_begin }
162      auto vtkWriter = std::make_shared<Dune::VTKWriter<GridView> >(gridView);
163      VTKSequenceWriter<GridView>
164          vtkSequenceWriter(vtkWriter,
165                          "getting-started-transport-fv-result");   // File name
166
167      // Write the initial values
168      vtkWriter->addCellData(c,"concentration");
169      vtkSequenceWriter.write(0.0);   // 0.0 is the current time
170    // { construct_vtk_writer_end }
171
172      // Now do the time steps
173    // { time_loop_begin }
174      double t=0;                      // Initial time
175      const double tend=0.6;           // Final time
176      const double dt=0.006;           // Time step size
177      int k=0;                         // Time step counter
178
179      // Inflow boundary values
180      auto inflow = [](const FieldVector<double,dim>& x)
181      {
182        return 0.0;
183      };
184
185      // Velocity field
186      auto v = [](const FieldVector<double,dim>& x)
187      {
188        return FieldVector<double,dim> (1.0);
189      };
190
191      while (t<tend)
192      {
193        // Apply finite volume scheme
194        evolve(gridView,mapper,dt,c,v,inflow);
195
196        // Augment time and time step counter
```

```
197        t += dt;
198        ++k;
199
200        // Write data. We do not have to call addCellData again!
201        vtkSequenceWriter.write(t);
202
203        // Print iteration number, time, and time step size
204        std::cout << "k=" << k << " t=" << t << std::endl;
205      }
206    // { time_loop_end }
207    }
208    // { main_end }
```

B.4. Local Grid Adaptivity Without Data Transfer

File:	grid-adaptivity.cc
Equation:	none
Discretization:	none
Grid:	Unstructured simplicial grid, implemented with UGGrid
Solver:	none
Distributed:	no
Discussed in:	Chapter 5.9.1

```
1    #include "config.h"
2
3    #include <iostream>
4
5    #include <dune/grid/io/file/vtk/vtkwriter.hh>
6
7    #include <dune/grid/uggrid.hh>
8    #include <dune/grid/utility/structuredgridfactory.hh>
9
10   #include <dune/common/parallel/mpihelper.hh>
11   #include <dune/common/exceptions.hh>
12
13   using namespace Dune;
14
15   // { sphere_begin }
16   template<int dim>
17   class Sphere
18   {
19     double radius_;
20     FieldVector<double, dim> center_;
21   public:
22     Sphere(const FieldVector<double, dim>& center, const double& radius)
23     : radius_(radius),
24       center_(center)
25     {}
26
27     double distanceTo(const FieldVector<double, dim>& point) const
28     {
29       return std::abs((center_ - point).two_norm() - radius_);
30     }
31
32     void displace(const FieldVector<double, dim>& increment)
33     {
34       center_ += increment;
35     }
36   };
37   // { sphere_end }
38
39
40   // { grid_setup_begin }
41   int main(int argc, char *argv[])
42   {
43     // Set up MPI if available
44     MPIHelper::instance(argc, argv);
45
46     constexpr int dim = 2;   // Grid and world dimension
47     using Grid = UGGrid<dim>;
```

```
48
49      // Start with a structured grid
50      const std::array<unsigned, dim> n = {8, 25};
51
52      const FieldVector<double, dim> lower = {0, 0};
53      const FieldVector<double, dim> upper = {6, 15};
54
55      std::shared_ptr<Grid> grid
56          = StructuredGridFactory<Grid>::createSimplexGrid(lower, upper, n);
57
58      using GridView = Grid::LeafGridView;
59      const GridView gridView = grid->leafGridView();
60      // { grid_setup_end }
61
62      // Create sphere
63      // { sphere_setup_begin }
64      Sphere<dim> sphere({3.0, 2.5}, 1.0);
65
66      // Set parameters
67      const int steps = 30;                // Total number of steps
68      const FieldVector<double, dim>
69          stepDisplacement = {0, 0.5};     // Sphere displacement per step
70
71      const double epsilon = 0.4;          // Thickness of the refined region
72                                           // around the sphere
73      const int levels = 3;                // Number of refinement levels
74      // { sphere_setup_end }
75
76      // { coarsening_refinement_begin }
77      for (int i = 0; i < steps; ++i)
78      {
79        std::cout << "Step " << i << std::endl;
80
81        // Coarsen everything
82        for (int k = 0; k < levels-1; ++k)
83        {
84          for (const auto& element : elements(gridView))
85            grid->mark(-1, element);
86
87          grid->preAdapt();
88          grid->adapt();
89          grid->postAdapt();
90        }
91
92        // Refine near the sphere
93        for (int k = 0; k < levels-1; ++k)
94        {
95          // Select elements that are close to the sphere for grid refinement
96          for (const auto& element : elements(gridView))
97            if (sphere.distanceTo(element.geometry().center()) < epsilon)
98              grid->mark(1, element);
99
100         grid->preAdapt();
101         grid->adapt();
102         grid->postAdapt();
103       }
104     // { coarsening_refinement_end }
105
106       // Write grid to file
107     // { writing_moving_begin }
108       VTKWriter<GridView> vtkWriter(gridView);
109       vtkWriter.write("refined_grid_"+std::to_string(i));
110
111       // Move sphere
112       sphere.displace(stepDisplacement);
113     }
114     // { writing_moving_end }
115     }
```

B.5. Local Grid Adaptivity with Data Transfer

File:	`grid-adaptivity-data-transfer.cc`
Equation:	none
Discretization:	First-order Lagrange finite elements
Grid:	Unstructured simplicial grid, implemented with `UGGrid`
Solver:	none
Distributed:	no
Discussed in:	Chapter 5.9.2

```
1    #include "config.h"
2
3    #include <iostream>
4    #include <map>
5
6    #include <dune/grid/io/file/vtk/vtkwriter.hh>
7
8    #include <dune/grid/uggrid.hh>
9    #include <dune/grid/utility/structuredgridfactory.hh>
10
11   #include <dune/common/parallel/mpihelper.hh>
12   #include <dune/common/exceptions.hh>
13
14   using namespace Dune;
15
16   template<int dim>
17   class Sphere
18   {
19     double radius_;
20     Dune::FieldVector<double, dim> center_;
21   public:
22     Sphere(const Dune::FieldVector<double, dim>& c, const double& r) : radius_(r), center_(c) {}
23
24     double distanceTo(const Dune::FieldVector<double, dim>& point) const
25     {
26       return std::abs((center_ - point).two_norm() - radius_);
27     }
28
29     void displace(const Dune::FieldVector<double, dim>& increment)
30     {
31       center_ += increment;
32     }
33   };
34
35   // Linear interpolation on a simplex
36   // { linear_interpolation_begin }
37   template<int dim>
38   double interpolate(const std::vector<double> values,
39                      FieldVector<double,dim> p)
40   {
41     assert(values.size() == dim+1);
42     double result = values[0];
43     for (std::size_t i=0; i<p.size(); i++)
44       result += p[i]*(values[i+1]-values[0]);
45     return result;
46   }
47   // { linear_interpolation_end }
48
49   // { grid_setup_begin }
50   int main(int argc, char *argv[])
51   {
52     // Set up MPI if available
53     MPIHelper::instance(argc, argv);
54
55     constexpr int dim = 2;    // Grid dimension
56     using Grid = UGGrid<dim>;
57
58     // Create UGGrid from structured triangle grid
59     const std::array<unsigned, dim> n = {8, 25};
60
61     const FieldVector<double, dim> lower = {0, 0};
62     const FieldVector<double, dim> upper = {6, 15};
63
64     std::shared_ptr<Grid> grid
65         = StructuredGridFactory<Grid>::createSimplexGrid(lower, upper, n);
```

```
66
67    using GridView = Grid::LeafGridView;
68    const GridView gridView = grid->leafGridView();
69    const auto& indexSet = gridView.indexSet();
70    const auto& idSet    = grid->localIdSet();
71  // { grid_setup_end }
72
73    // Create sphere
74  // { sphere_setup_begin }
75    Sphere<dim> sphere({3.0, 2.5}, 1.0);
76
77    // Set parameters
78    const int steps = 30;
79    const FieldVector<double, dim> stepDisplacement = {0, 0.5};
80
81    const double epsilon = 0.4;   // Thickness of the refined region
82                                  // around the sphere
83    const int levels = 2;
84  // { sphere_setup_end }
85
86    // Construct a piecewise linear function on the grid
87  // { function_setup_begin }
88    auto dataFunction = [](const FieldVector<double,dim>& x)
89    {
90      return std::sin(x[1]);
91    };
92
93    std::vector<double> data(gridView.size(dim));
94    for (auto&& v : vertices(gridView))
95      data[indexSet.index(v)] = dataFunction(v.geometry().corner(0));
96  // { function_setup_end }
97
98  // { coarsening_refinement_begin }
99    for (int i = 0; i < steps; ++i)
100   {
101     std::map<Grid::LocalIdSet::IdType, double> persistentContainer;
102
103     // Coarsen everything
104     for (int j = 0; j<levels; ++j)
105     {
106       for (const auto& element : elements(gridView))
107         grid->mark(-1, element);    // Mark element for coarsening
108
109       grid->preAdapt();
110
111       for (const auto& vertex : vertices(gridView))
112         persistentContainer[idSet.id(vertex)] = data[indexSet.index(vertex)];
113
114       grid->adapt();
115
116       data.resize(gridView.size(dim));
117
118       for (const auto& v : vertices(gridView))
119         data[indexSet.index(v)] = persistentContainer[idSet.id(v)];
120
121       grid->postAdapt();
122     }
123  // { coarsening_end }
124
125     // Refine near the sphere
126  // { refinement_begin }
127     for (int j = 0; j<levels; ++j)
128     {
129       // Select elements that are close to the sphere for grid refinement
130       for (const auto& element : elements(gridView))
131         if (sphere.distanceTo(element.geometry().center()) < epsilon)
132           grid->mark(1, element);
133
134       grid->preAdapt();
135
136       for (const auto& vertex : vertices(gridView))
137         persistentContainer[idSet.id(vertex)] = data[indexSet.index(vertex)];
138
139       grid->adapt();
140  // { refinement_after_modification }
141
142  // { refinement_data_interpolation_begin }
143       data.resize(gridView.size(dim));
144
145       for (const auto& element : elements(gridView))
146       {
147         if (element.isNew())
148         {
149           for (std::size_t k=0; k<element.subEntities(dim); k++)
150           {
151             auto father = element;
152             auto positionInFather
```

```
153                    = ReferenceElements<double,dim>::general(element.type())
154                                              .position(k,dim);
155
156            do
157            {
158              positionInFather
159                    = father.geometryInFather().global(positionInFather);
160              father = father.father();
161            } while (father.isNew());
162
163            // Extract corner values
164            std::vector<double> values(father.subEntities(dim));
165            for (std::size_t l=0; l<father.subEntities(dim); l++)
166              values[l] = persistentContainer[idSet.subId(father,l,dim)];
167
168            // Interpolate linearly on the ancestor simplex
169            data[indexSet.subIndex(element,k,dim)]
170                    = interpolate(values, positionInFather);
171          }
172        }
173        else
174          for (std::size_t k=0; k<element.subEntities(dim); k++)
175            data[indexSet.subIndex(element,k,dim)]
176                    = persistentContainer[idSet.subId(element,k,dim)];
177      }
178
179      grid->postAdapt();
180  // { refinement_data_interpolation_end }
181
182    }
183  // { coarsening_refinement_end }
184
185    // Write grid to file
186  // { writing_moving_begin }
187    VTKWriter<GridView> vtkWriter(gridView);
188    vtkWriter.addVertexData(data, "data");
189    vtkWriter.write("refined_grid_"+std::to_string(i));
190
191    // Move sphere
192    sphere.displace(stepDisplacement);
193  }
194  // { writing_moving_end }
195  }
```

B.6. The Poisson Equation on a Distributed Grid

File:	grid-distributed-poisson.cc
Equation:	Poisson equation
Discretization:	First-order Lagrange finite elements
Grid:	Unstructured simplicial grid, implemented with UGGrid
Solver:	CG solver with Jacobi preconditioner
Distributed:	yes
Discussed in:	Chapter 6.7

```
1   #include <config.h>
2
3   #include <vector>
4   #include <map>
5
6   #include <dune/geometry/quadraturerules.hh>
7
8   #include <dune/grid/uggrid.hh>
9   #include <dune/grid/io/file/gmshreader.hh>
10  #include <dune/grid/io/file/vtk/vtkwriter.hh>
11
12  #include <dune/istl/matrix.hh>
13  #include <dune/istl/bcrsmatrix.hh>
14  #include <dune/istl/bvector.hh>
15  #include <dune/istl/matrixindexset.hh>
16
17  #include <dune/functions/functionspacebases/lagrangebasis.hh>
18  #include <dune/functions/functionspacebases/interpolate.hh>
```

532

```
19
20
21     using namespace Dune;
22
23     // Compute the stiffness matrix for a single element
24     template<class LocalView, class Matrix>
25     void assembleElementStiffnessMatrix(const LocalView& localView,
26                                         Matrix& elementMatrix)
27     {
28       using Element = typename LocalView::Element;
29       constexpr int dim = Element::dimension;
30       auto element = localView.element();
31       auto geometry = element.geometry();
32
33       // Get set of shape functions for this element
34       const auto& localFiniteElement = localView.tree().finiteElement();
35
36       // Set all matrix entries to zero
37       elementMatrix.setSize(localView.size(),localView.size());
38       elementMatrix = 0;        // Fill the entire matrix with zeros
39
40       // Get a quadrature rule
41       int order = 2*(dim*localFiniteElement.localBasis().order()-1);
42       const auto& quadRule = QuadratureRules<double, dim>::rule(element.type(), order);
43
44       // Loop over all quadrature points
45       for (const auto& quadPoint : quadRule)
46       {
47         // Position of the current quadrature point in the reference element
48         const auto quadPos = quadPoint.position();
49
50         // The transposed inverse Jacobian of the map from the reference element
51         // to the grid element
52         const auto jacobian = geometry.jacobianInverseTransposed(quadPos);
53
54         // The multiplicative factor in the integral transformation formula
55         const auto integrationElement = geometry.integrationElement(quadPos);
56
57         // The gradients of the shape functions on the reference element
58         std::vector<FieldMatrix<double,1,dim> > referenceGradients;
59         localFiniteElement.localBasis().evaluateJacobian(quadPos,
60                                                          referenceGradients);
61
62         // Compute the shape function gradients on the grid element
63         std::vector<FieldVector<double,dim> > gradients(referenceGradients.size());
64         for (size_t i=0; i<gradients.size(); i++)
65           jacobian.mv(referenceGradients[i][0], gradients[i]);
66
67         // Compute the actual matrix entries
68         for (size_t p=0; p<elementMatrix.N(); p++)
69         {
70           auto localRow = localView.tree().localIndex(p);
71           for (size_t q=0; q<elementMatrix.M(); q++)
72           {
73             auto localCol = localView.tree().localIndex(q);
74             elementMatrix[localRow][localCol]
75               += ( gradients[p] * gradients[q] ) * quadPoint.weight() * integrationElement;
76           }
77         }
78       }
79     }
80
81
82     // Compute the source term for a single element
83     template<class LocalView>
84     void assembleElementVolumeTerm(
85             const LocalView& localView,
86             BlockVector<double>& localB,
87             const std::function<double(FieldVector<double,LocalView::Element::dimension>)> volumeTerm)
88     {
89       using Element = typename LocalView::Element;
90       auto element = localView.element();
91       constexpr int dim = Element::dimension;
92
93       // Set of shape functions for a single element
94       const auto& localFiniteElement = localView.tree().finiteElement();
95
96       // Set all entries to zero
97       localB.resize(localFiniteElement.size());
98       localB = 0;
99
100      // A quadrature rule
101      int order = dim;
102      const auto& quadRule = QuadratureRules<double, dim>::rule(element.type(), order);
103
104      // Loop over all quadrature points
105      for (const auto& quadPoint : quadRule)
```

```
106    {
107      // Position of the current quadrature point in the reference element
108      const FieldVector<double,dim>& quadPos = quadPoint.position();
109
110      // The multiplicative factor in the integral transformation formula
111      const double integrationElement = element.geometry().integrationElement(quadPos);
112
113      double functionValue = volumeTerm(element.geometry().global(quadPos));
114
115      // Evaluate all shape function values at this point
116      std::vector<FieldVector<double,1> > shapeFunctionValues;
117      localFiniteElement.localBasis().evaluateFunction(quadPos, shapeFunctionValues);
118
119      // Actually compute the vector entries
120      for (size_t p=0; p<localB.size(); p++)
121      {
122        auto localIndex = localView.tree().localIndex(p);
123        localB[localIndex] += shapeFunctionValues[p] * functionValue
124                            * quadPoint.weight() * integrationElement;
125      }
126    }
127  }
128
129  // Get the occupation pattern of the stiffness matrix
130  template<class Basis>
131  void getOccupationPattern(const Basis& basis, MatrixIndexSet& nb)
132  {
133    nb.resize(basis.size(), basis.size());
134
135    auto gridView = basis.gridView();
136
137    // A loop over all elements of the grid
138    auto localView = basis.localView();
139
140    for (const auto& element : elements(gridView))
141    {
142      localView.bind(element);
143
144      for (size_t i=0; i<localView.size(); i++)
145      {
146        // The global index of the i-th vertex of the element
147        auto row = localView.index(i);
148
149        for (size_t j=0; j<localView.size(); j++ )
150        {
151          // The global index of the j-th vertex of the element
152          auto col = localView.index(j);
153          nb.add(row,col);
154        }
155      }
156    }
157  }
158
159
160  // Assemble the Laplace stiffness matrix on the given grid view
161  template<class Basis>
162  void assemblePoissonProblem(const Basis& basis,
163                              BCRSMatrix<double>& matrix,
164                              BlockVector<double>& b,
165                              const std::function<
166                                  double(FieldVector<double,
167                                              Basis::GridView::dimension>)
168                                        > volumeTerm)
169  {
170    auto gridView = basis.gridView();
171
172    // MatrixIndexSets store the occupation pattern of a sparse matrix.
173    // They are not particularly efficient, but simple to use.
174    MatrixIndexSet occupationPattern;
175    getOccupationPattern(basis, occupationPattern);
176
177    // ... and give it the occupation pattern we want.
178    occupationPattern.exportIdx(matrix);
179
180    // Set all entries to zero
181    matrix = 0;
182
183    // Set b to correct length
184    b.resize(basis.dimension());
185
186    // Set all entries to zero
187    b = 0;
188
189    // A loop over all elements of the grid
190    auto localView = basis.localView();
191
192    for (const auto& element : elements(gridView,Partitions::interior))
```

```
193      {
194        // Now let's get the element stiffness matrix
195        // A dense matrix is used for the element stiffness matrix
196        localView.bind(element);
197
198        Matrix<double> elementMatrix;
199        assembleElementStiffnessMatrix(localView, elementMatrix);
200
201        for(size_t p=0; p<elementMatrix.N(); p++)
202        {
203          // The global index of the p-th degree of freedom of the element
204          auto row = localView.index(p);
205
206          for (size_t q=0; q<elementMatrix.M(); q++ )
207          {
208            // The global index of the q-th degree of freedom of the element
209            auto col = localView.index(q);
210            matrix[row][col] += elementMatrix[p][q];
211          }
212        }
213
214        // Now get the local contribution to the right-hand side vector
215        BlockVector<double> localB;
216        assembleElementVolumeTerm(localView, localB, volumeTerm);
217
218        for (size_t p=0; p<localB.size(); p++)
219        {
220          // The global index of the p-th vertex of the element
221          auto row = localView.index(p);
222          b[row] += localB[p];
223        }
224      }
225    }
226
227    // { lb_data_handle_begin }
228    template<class Grid, class AssociativeContainer>
229    struct LBVertexDataHandle
230      : public CommDataHandleIF<LBVertexDataHandle<Grid, AssociativeContainer>,
231                                typename AssociativeContainer::mapped_type>
232    {
233      LBVertexDataHandle(const std::shared_ptr<Grid>& grid,
234                         AssociativeContainer& dataContainer)
235        : idSet_(grid->localIdSet()), dataContainer_(dataContainer)
236      {}
237
238      bool contains(int dim, int codim) const
239      {
240        assert(dim == Grid::dimension);
241        return (codim == dim);   // Only vertices have data
242      }
243
244      bool fixedSize(int dim, int codim) const
245      {
246        return true;   // All vertices carry the same number of data items
247      }
248
249      template<class Entity>
250      size_t size(const Entity& entity) const
251      {
252        return 1;        // One data item per vertex
253      }
254
255      template<class MessageBuffer, class Entity>
256      void gather(MessageBuffer& buffer, const Entity& entity) const
257      {
258        auto id = idSet_.id(entity);
259        buffer.write(dataContainer_[id]);
260      }
261
262      template<class MessageBuffer, class Entity>
263      void scatter(MessageBuffer& buffer, const Entity& entity, size_t n)
264      {
265        assert(n==1);   // This data handle implementation
266                        // transfers only one data item.
267        auto id = idSet_.id(entity);
268        buffer.read(dataContainer_[id]);
269      }
270
271    private:
272      const typename Grid::LocalIdSet& idSet_;
273      AssociativeContainer& dataContainer_;
274    };
275    // { lb_data_handle_end }
276
277    // A DataHandle class to communicate and add vertex data
278    // { comm_data_handle_begin }
279    template<class GridView, class Vector>
```

```
280   struct VertexDataUpdate
281     : public Dune::CommDataHandleIF<VertexDataUpdate<GridView,Vector>,
282                                     typename Vector::value_type>
283   {
284     using DataType = typename Vector::value_type;
285
286     // Constructor
287     VertexDataUpdate(const GridView& gridView,
288                      const Vector& userDataSend,
289                      Vector& userDataReceive)
290       : gridView_(gridView),
291         userDataSend_(userDataSend),
292         userDataReceive_(userDataReceive)
293     {}
294
295     // True if data for this codim should be communicated
296     bool contains(int dim, int codim) const
297     {
298       return (codim == dim); // Only vertices have data
299     }
300
301     // True if data size per entity of given codim is constant
302     bool fixedSize(int dim, int codim) const
303     {
304       return true;            // All vertices carry the same number of data items
305     }
306
307     // How many objects of type DataType have to be sent for a given entity
308     template<class Entity>
309     size_t size(const Entity& e) const
310     {
311       return 1;                 // One data item per vertex
312     }
313
314     // Pack user data into message buffer
315     template<class MessageBuffer, class Entity>
316     void gather(MessageBuffer& buffer, const Entity& entity) const
317     {
318       auto index = gridView_.indexSet().index(entity);
319       buffer.write(userDataSend_[index]);
320     }
321
322     // Unpack user data from message buffer
323     template<class MessageBuffer, class Entity>
324     void scatter(MessageBuffer& buffer, const Entity& entity, size_t n)
325     {
326       assert(n==1);
327       DataType x;
328       buffer.read(x);
329
330       userDataReceive_[gridView_.indexSet().index(entity)] += x;
331     }
332
333   private:
334     const GridView gridView_;
335     const Vector& userDataSend_;
336     Vector& userDataReceive_;
337   };
338   // { comm_data_handle_end }
339
340   // { main_begin }
341   int main(int argc, char *argv[])
342   {
343     // Set up MPI
344     const MPIHelper& mpiHelper = MPIHelper::instance(argc, argv);
345   // { mpihelper_end }
346
347   ////////////////////////////////////
348   //   Generate the grid
349   ////////////////////////////////////
350
351   // { create_grid_begin }
352     constexpr int dim = 2;
353     using Grid = UGGrid<dim>;
354     using GridView = Grid::LeafGridView;
355
356     std::shared_ptr<Grid> grid = GmshReader<Grid>::read("l-shape-refined.msh");
357     auto gridView = grid->leafGridView();
358   // { create_grid_end }
359
360   // { sample_initial_iterate_begin }
361     std::vector<double> dataVector;
362
363     if (mpiHelper.rank()==0)
364     {
365       // The initial iterate as a function
366       auto initialIterate = [](auto x){return std::min(x[0],x[1]);};
```

```
367
368        // Sample on the grid vertices
369        dataVector.resize(gridView.size(dim));
370        for (const auto& vertex : vertices(gridView,
371                                  Dune::Partitions::interiorBorder))
372        {
373          auto index = gridView.indexSet().index(vertex);
374          dataVector[index] = initialIterate(vertex.geometry().corner(0));
375        }
376      }
377    // { sample_initial_iterate_end }
378
379    // { data_into_map_begin }
380      // Copy vertex data into associative container
381      using PersistentContainer = std::map<Grid::LocalIdSet::IdType, double>;
382      PersistentContainer persistentContainer;
383      const auto& idSet = grid->localIdSet();
384
385      for (const auto& vertex : vertices(gridView))
386        persistentContainer[idSet.id(vertex)]
387          = dataVector[gridView.indexSet().index(vertex)];
388    // { data_into_map_end }
389
390    // { load_balancing_begin }
391      // Distribute the grid and the data
392      LBVertexDataHandle<Grid, PersistentContainer>
393                              dataHandle(grid, persistentContainer);
394      grid->loadBalance(dataHandle);
395    // { load_balancing_end }
396
397    // { data_from_map_begin }
398      // Get gridView again after load-balancing, to make sure it is up-to-date
399      gridView = grid->leafGridView();
400
401      // Copy data back into the array
402      dataVector.resize(gridView.size(dim));
403
404      for (const auto& vertex : vertices(gridView))
405        dataVector[gridView.indexSet().index(vertex)]
406          = persistentContainer[idSet.id(vertex)];
407    // { data_from_map_end }
408
409
410    //////////////////////////////////////////////////////
411    //   Stiffness matrix and right hand side vector
412    //////////////////////////////////////////////////////
413
414    // { create_matrix_vector_begin }
415      using Matrix = BCRSMatrix<double>;
416      using Vector = BlockVector<double>;
417
418      Matrix stiffnessMatrix;
419      Vector b;
420
421      auto sourceTerm = [](const FieldVector<double,dim>& x){return -5.0;};
422
423      // Assemble the Poisson system in a first-order Lagrange space
424      Functions::LagrangeBasis<GridView,1> basis(gridView);
425      assemblePoissonProblem(basis, stiffnessMatrix, b, sourceTerm);
426    // { call_assembler_end }
427
428      // Obtain a consistent representation of the matrix diagonal
429    // { make_consistent_diagonal_begin }
430      Vector diagonal(basis.size());
431      for (std::size_t i=0; i<basis.size(); ++i)
432        diagonal[i] = stiffnessMatrix[i][i];
433
434      auto consistentDiagonal = diagonal;
435      VertexDataUpdate<GridView,Vector> matrixDataHandle(gridView,
436                                              diagonal,
437                                              consistentDiagonal);
438
439      gridView.communicate(matrixDataHandle,
440                            InteriorBorder_InteriorBorder_Interface,
441                            ForwardCommunication);
442    // { make_consistent_diagonal_end }
443
444      // Determine Dirichlet degrees of freedom by marking all degrees
445      // of freedom whose Lagrange nodes comply with a given predicate.
446    // { dirichlet_marking_begin }
447      auto dirichletPredicate = [](auto p)
448      {
449        return p[0]< 1e-8 || p[1] < 1e-8 || (p[0] > 0.4999 && p[1] > 0.4999);
450      };
451
452      // Interpolating the predicate will mark
453      // all desired Dirichlet degrees of freedom
```

```
454      std::vector<bool> dirichletNodes;
455      Functions::interpolate(basis, dirichletNodes, dirichletPredicate);
456      // { dirichlet_marking_end }
457
458      /////////////////////////////////////////////////////////////////
459      //     Modify Dirichlet matrix rows and load vector entries
460      /////////////////////////////////////////////////////////////////
461
462      // { dirichlet_modification_begin }
463      // Loop over the matrix rows
464      for (size_t i=0; i<stiffnessMatrix.N(); i++)
465      {
466        if (dirichletNodes[i])
467        {
468          auto cIt    = stiffnessMatrix[i].begin();
469          auto cEndIt = stiffnessMatrix[i].end();
470          // Loop over nonzero matrix entries in current row
471          for (; cIt!=cEndIt; ++cIt)
472            *cIt = (i==cIt.index()) ? 1.0 : 0.0;
473
474          // Modify corresponding load vector entry
475          b[i] = dataVector[i];
476        }
477      }
478      // { dirichlet_modification_end }
479
480      /////////////////////////////////
481      //   Compute solution
482      /////////////////////////////////
483
484      // { algebraic_solving_preprocess_begin }
485      // Set the initial iterate
486      Vector x(basis.size());
487      std::copy(dataVector.begin(), dataVector.end(), x.begin());
488
489      // Solver parameters
490      double reduction = 1e-3;  // Desired residual reduction factor
491      int maxIterations = 50;   // Maximum number of iterations
492      // { algebraic_solving_preprocess_end }
493
494      // Solve!
495      // { algebraic_solving_begin }
496      auto r = b;
497      stiffnessMatrix.mmv(x,r);               // r -= Ax
498      // { initial_residual_end }
499
500      // { make_residual_consistent_begin }
501      // Construct consistent representation of the data in r
502      auto rConsistent = r;
503
504      VertexDataUpdate<GridView,Vector> vertexUpdateHandle(gridView,
505                                                            r,
506                                                            rConsistent);
507
508      gridView.communicate(vertexUpdateHandle,
509                            InteriorBorder_InteriorBorder_Interface,
510                            ForwardCommunication);
511      // { make_residual_consistent_end }
512
513      // { global_initial_residual_begin }
514      double defect0 = r.dot(rConsistent);     // Norm on the local process
515      defect0 = grid->comm().sum(defect0);
516      defect0 = sqrt(defect0);
517      // { global_initial_residual_end }
518
519      // { output_header_begin }
520      if (mpiHelper.rank()==0)
521      {
522        std::cout << " Iteration          Defect       Rate" << std::endl;;
523        std::cout << "    0" << std::setw(16) << defect0 << std::endl;
524      }
525      // { output_header_end }
526
527      // { initial_direction_begin }
528      // Construct initial search direction in variable d by applying
529      // the Jacobi preconditioner to the residual in r.
530      Vector d(r.size());
531      for (std::size_t i=0; i<stiffnessMatrix.N(); ++i)
532      {
533        d[i] = 0;
534        if (std::abs(consistentDiagonal[i]) > 1e-5)   // Degree of freedom
535                                                      // is not on ghost vertex
536          d[i] = rConsistent[i] / consistentDiagonal[i];
537      }
538      // { initial_direction_end }
539
540      // { orthogonalization_begin }
```

```
541        double rho = d.dot(r);
542        rho = grid->comm().sum(rho);
543    // { orthogonalization_end }
544
545    // { loop_and_alpha_begin }
546        // Current residual norm
547        double defect=defect0;
548
549        for (int k=0; k<maxIterations; ++k)
550        {
551            // Step length in search direction d
552            Vector tmp(d.size());
553            stiffnessMatrix.mv(d,tmp);                // tmp=Ad^k
554            double alphaDenom = d.dot(tmp);           // Scalar product
555            alphaDenom = grid->comm().sum(alphaDenom);
556            double alpha = rho/alphaDenom;
557    // { loop_and_alpha_end }
558
559    // { update_iterate_begin }
560            x.axpy(alpha,d);            // Update iterate
561    // { update_iterate_end }
562    // { update_residual_begin }
563            r.axpy(-alpha,tmp);         // Update residual
564    // { update_residual_end }
565
566            // Convergence test
567    // { check_convergence_begin }
568            // Compute residual norm again
569            rConsistent = r;
570            gridView.communicate(vertexUpdateHandle,
571                                  InteriorBorder_InteriorBorder_Interface,
572                                  ForwardCommunication);
573
574            auto residualNorm = r.dot(rConsistent);
575            residualNorm = grid->comm().sum(residualNorm);
576            residualNorm = sqrt(residualNorm);
577
578            if (mpiHelper.rank()==0)
579            {
580                std::cout << std::setw(5)  << k+1 << " ";
581                std::cout << std::setw(16) << residualNorm << " ";
582                                          // Estimated convergence rate
583                std::cout << std::setw(16) << residualNorm/defect << std::endl;
584            }
585
586            defect = residualNorm;            // Update norm
587
588            if (defect<defect0*reduction)     // Convergence check
589                break;
590    // { check_convergence_end }
591
592            // Determine new search direction
593    // { compute_new_direction_begin }
594            // Precondition the residual
595            Vector preconditionedResidual(d.size());
596            for (std::size_t i=0; i<stiffnessMatrix.N(); i++)
597            {
598                preconditionedResidual[i] = 0;
599                if (std::abs(consistentDiagonal[i]) > 1e-5)    // Degree of freedom
600                                                               // not on ghost vertex
601                    preconditionedResidual[i] = rConsistent[i] / consistentDiagonal[i];
602            }
603
604            double rhoNext = preconditionedResidual.dot(r);
605            rhoNext = grid->comm().sum(rhoNext);
606            double beta = rhoNext/rho;
607
608            // Compute new search direction
609            d *= beta;
610            d += preconditionedResidual;
611            rho = rhoNext;              // Remember rho for the next iterate
612        }
613    // { algebraic_solving_end }
614
615        // Output result
616    // { vtk_output_begin }
617        // For visualization: Write the rank number for each element
618        MultipleCodimMultipleGeomTypeMapper<GridView>
619                                    elementMapper(gridView,mcmgElementLayout());
620        std::vector<int> ranks(elementMapper.size());
621        for (const auto& element : elements(gridView))
622            ranks[elementMapper.index(element)] = mpiHelper.rank();
623
624        VTKWriter<GridView> vtkWriter(gridView);
625        vtkWriter.addVertexData(x, "solution");
626        vtkWriter.addCellData(ranks, "ranks");
627        vtkWriter.write("grid-distributed-poisson-result");
```

```
628    // { vtk_output_end }
629    }
```

B.7. The Poisson Equation on a Distributed Grid Using ISTL Interfaces

File:	istl-distributed-poisson.cc
Equation:	Poisson equation
Discretization:	First-order Lagrange finite elements
Grid:	Unstructured simplicial grid, implemented with UGGrid
Solver:	CG solver with Jacobi preconditioner
Distributed:	yes
Discussed in:	Chapter 7.4.4

```
1    #include <config.h>
2
3    #include <vector>
4    #include <map>
5
6    #include <dune/geometry/quadraturerules.hh>
7
8    #include <dune/grid/uggrid.hh>
9    #include <dune/grid/io/file/gmshreader.hh>
10   #include <dune/grid/io/file/vtk/vtkwriter.hh>
11
12   #include <dune/istl/matrix.hh>
13   #include <dune/istl/bvector.hh>
14   #include <dune/istl/bcrsmatrix.hh>
15   #include <dune/istl/matrixindexset.hh>
16   #include <dune/istl/operators.hh>
17   #include <dune/istl/preconditioner.hh>
18   #include <dune/istl/scalarproducts.hh>
19   #include <dune/istl/solvers.hh>
20
21   #include <dune/functions/functionspacebases/lagrangebasis.hh>
22   #include <dune/functions/functionspacebases/interpolate.hh>
23
24
25   using namespace Dune;
26
27   // Compute the stiffness matrix for a single element
28   template<class LocalView, class Matrix>
29   void assembleElementStiffnessMatrix(const LocalView& localView,
30                                       Matrix& elementMatrix)
31   {
32     using Element = typename LocalView::Element;
33     constexpr int dim = Element::dimension;
34     auto element = localView.element();
35     auto geometry = element.geometry();
36
37     // Get set of shape functions for this element
38     const auto& localFiniteElement = localView.tree().finiteElement();
39
40     // Set all matrix entries to zero
41     elementMatrix.setSize(localView.size(),localView.size());
42     elementMatrix = 0;        // Fill the entire matrix with zeros
43
44     // Get a quadrature rule
45     int order = 2*(dim*localFiniteElement.localBasis().order()-1);
46     const auto& quadRule = QuadratureRules<double, dim>::rule(element.type(),
47                                                               order);
48
49     // Loop over all quadrature points
50     for (const auto& quadPoint : quadRule)
51     {
52       // Position of the current quadrature point in the reference element
53       const auto quadPos = quadPoint.position();
54
```

```
55      // The transposed inverse Jacobian of the map from the reference element
56      // to the grid element
57      const auto jacobian = geometry.jacobianInverseTransposed(quadPos);
58
59      // The multiplicative factor in the integral transformation formula
60      const auto integrationElement = geometry.integrationElement(quadPos);
61
62      // The gradients of the shape functions on the reference element
63      std::vector<FieldMatrix<double,1,dim> > referenceGradients;
64      localFiniteElement.localBasis().evaluateJacobian(quadPos,
65                                                       referenceGradients);
66
67      // Compute the shape function gradients on the real element
68      std::vector<FieldVector<double,dim> > gradients(referenceGradients.size());
69      for (size_t i=0; i<gradients.size(); i++)
70        jacobian.mv(referenceGradients[i][0], gradients[i]);
71
72      // Compute the actual matrix entries
73      for (size_t p=0; p<elementMatrix.N(); p++)
74      {
75        auto localRow = localView.tree().localIndex(p);
76        for (size_t q=0; q<elementMatrix.M(); q++)
77        {
78          auto localCol = localView.tree().localIndex(q);
79          elementMatrix[localRow][localCol]
80              += ( gradients[p] * gradients[q] ) * quadPoint.weight() * integrationElement;
81        }
82      }
83    }
84  }
85
86
87  // Compute the source term for a single element
88  template<class LocalView>
89  void assembleElementVolumeTerm(
90          const LocalView& localView,
91          BlockVector<double>& localB,
92          const std::function<double(FieldVector<double,LocalView::Element::dimension>)> volumeTerm)
93  {
94    using Element = typename LocalView::Element;
95    auto element = localView.element();
96    constexpr int dim = Element::dimension;
97
98    // Set of shape functions for a single element
99    const auto& localFiniteElement = localView.tree().finiteElement();
100
101    // Set all entries to zero
102    localB.resize(localFiniteElement.size());
103    localB = 0;
104
105    // A quadrature rule
106    int order = dim;
107    const auto& quadRule = QuadratureRules<double, dim>::rule(element.type(), order);
108
109    // Loop over all quadrature points
110    for (const auto& quadPoint ; quadRule)
111    {
112      // Position of the current quadrature point in the reference element
113      const FieldVector<double,dim>& quadPos = quadPoint.position();
114
115      // The multiplicative factor in the integral transformation formula
116      const double integrationElement = element.geometry().integrationElement(quadPos);
117
118      double functionValue = volumeTerm(element.geometry().global(quadPos));
119
120      // Evaluate all shape function values at this point
121      std::vector<FieldVector<double,1> > shapeFunctionValues;
122      localFiniteElement.localBasis().evaluateFunction(quadPos, shapeFunctionValues);
123
124      // Actually compute the vector entries
125      for (size_t p=0; p<localB.size(); p++)
126      {
127        auto localIndex = localView.tree().localIndex(p);
128        localB[localIndex] += shapeFunctionValues[p] * functionValue
129                            * quadPoint.weight() * integrationElement;
130      }
131    }
132  }
133
134  // Get the occupation pattern of the stiffness matrix
135  template<class Basis>
136  void getOccupationPattern(const Basis& basis, MatrixIndexSet& nb)
137  {
138    nb.resize(basis.size(), basis.size());
139
140    auto gridView = basis.gridView();
141
```

```
142      // A loop over all elements of the grid
143      auto localView = basis.localView();
144
145      for (const auto& element : elements(gridView))
146      {
147        localView.bind(element);
148
149        for (size_t i=0; i<localView.size(); i++)
150        {
151          // The global index of the i-th vertex of the element
152          auto row = localView.index(i);
153
154          for (size_t j=0; j<localView.size(); j++ )
155          {
156            // The global index of the j-th vertex of the element
157            auto col = localView.index(j);
158            nb.add(row,col);
159          }
160        }
161      }
162    }
163
164
165    // Assemble the Laplace stiffness matrix on the given grid view
166    template<class Basis>
167    void assemblePoissonProblem(const Basis& basis,
168                                BCRSMatrix<double>& matrix,
169                                BlockVector<double>& b,
170                                const std::function<double(FieldVector<double,Basis::GridView::dimension>)> volumeTerm)
171    {
172      auto gridView = basis.gridView();
173
174      // MatrixIndexSets store the occupation pattern of a sparse matrix.
175      // They are not particularly efficient, but simple to use.
176      MatrixIndexSet occupationPattern;
177      getOccupationPattern(basis, occupationPattern);
178
179      // ... and give it the occupation pattern we want.
180      occupationPattern.exportIdx(matrix);
181
182      // Set all entries to zero
183      matrix = 0;
184
185      // Set b to correct length
186      b.resize(basis.dimension());
187
188      // Set all entries to zero
189      b = 0;
190
191      // A loop over all elements of the grid
192      auto localView = basis.localView();
193
194      for (const auto& element : elements(gridView,Partitions::interior))
195      {
196        // Now let's get the element stiffness matrix
197        // A dense matrix is used for the element stiffness matrix
198        localView.bind(element);
199
200        Matrix<double> elementMatrix;
201        assembleElementStiffnessMatrix(localView, elementMatrix);
202
203        for(size_t p=0; p<elementMatrix.N(); p++)
204        {
205          // The global index of the p-th degree of freedom of the element
206          auto row = localView.index(p);
207
208          for (size_t q=0; q<elementMatrix.M(); q++ )
209          {
210            // The global index of the q-th degree of freedom of the element
211            auto col = localView.index(q);
212            matrix[row][col] += elementMatrix[p][q];
213          }
214        }
215
216        // Now get the local contribution to the right-hand side vector
217        BlockVector<double> localB;
218        assembleElementVolumeTerm(localView, localB, volumeTerm);
219
220        for (size_t p=0; p<localB.size(); p++)
221        {
222          // The global index of the p-th vertex of the element
223          auto row = localView.index(p);
224          b[row] += localB[p];
225        }
226      }
227    }
228
```

542

```
229   // { lb_data_handle_begin }
230   template<class Grid, class AssociativeContainer>
231   struct LBVertexDataHandle
232     : public CommDataHandleIF<LBVertexDataHandle<Grid, AssociativeContainer>,
233                                typename AssociativeContainer::mapped_type>
234   {
235     LBVertexDataHandle(const std::shared_ptr<Grid>& grid, AssociativeContainer& dataContainer)
236       : idSet_(grid->localIdSet()), dataContainer_(dataContainer)
237     {}
238
239     bool contains(int dim, int codim) const
240     {
241       assert(dim == Grid::dimension);
242       return (codim == dim);   // Only vertices have data
243     }
244
245     bool fixedSize(int dim, int codim) const
246     {
247       return true;       // All vertices carry the same number of data items
248     }
249
250     template<class Entity>
251     size_t size(const Entity& entity) const
252     {
253       return 1;          // One data item per vertex
254     }
255
256     template<class MessageBuffer, class Entity>
257     void gather(MessageBuffer& buffer, const Entity& entity) const
258     {
259       auto id = idSet_.id(entity);
260       buffer.write(dataContainer_[id]);
261     }
262
263     template<class MessageBuffer, class Entity>
264     void scatter(MessageBuffer& buffer, const Entity& entity, size_t n)
265     {
266       assert(n==1);   // This data handle implementations transfer only one data item.
267       auto id = idSet_.id(entity);
268       buffer.read(dataContainer_[id]);
269     }
270
271   private:
272     const typename Grid::LocalIdSet& idSet_;
273     AssociativeContainer& dataContainer_;
274   };
275   // { lb_data_handle_end }
276
277   // A DataHandle class to communicate and add vertex data
278   // { comm_data_handle_begin }
279   template<class GridView, class Vector>
280   struct VertexDataUpdate
281     : public Dune::CommDataHandleIF<VertexDataUpdate<GridView,Vector>,
282                                      typename Vector::value_type>
283   {
284     using DataType = typename Vector::value_type;
285
286     // Constructor
287     VertexDataUpdate(const GridView& gridView,
288                      const Vector& userDataSend,
289                      Vector& userDataReceive)
290       : gridView_(gridView),
291         userDataSend_(userDataSend),
292         userDataReceive_(userDataReceive)
293     {}
294
295     // True if data for this codim should be communicated
296     bool contains(int dim, int codim) const
297     {
298       return (codim == dim);      // Only vertices have data
299     }
300
301     // True if data size per entity of given codim is constant
302     bool fixedSize(int dim, int codim) const
303     {
304       return true;                // All vertices carry the same number of data items
305     }
306
307     // How many objects of type DataType have to be sent for a given entity
308     template<class Entity>
309     size_t size(const Entity& e) const
310     {
311       return 1;                   // One data item per vertex
312     }
313
314     // Pack user data into message buffer
315     template<class MessageBuffer, class Entity>
```

```
316      void gather(MessageBuffer& buffer, const Entity& entity) const
317      {
318        auto index = gridView_.indexSet().index(entity);
319        buffer.write(userDataSend_[index]);
320      }
321
322      // Unpack user data from message buffer
323      template<class MessageBuffer, class Entity>
324      void scatter(MessageBuffer& buffer, const Entity& entity, size_t n)
325      {
326        assert(n==1);
327        DataType x;
328        buffer.read(x);
329
330        userDataReceive_[gridView_.indexSet().index(entity)] += x;
331      }
332
333    private:
334      const GridView gridView_;
335      const Vector& userDataSend_;
336      Vector& userDataReceive_;
337    };
338    // { comm_data_handle_end }
339
340    // { preconditioner_begin }
341    template<class GridView, class Matrix, class Vector>
342    class JacobiPreconditioner : public Preconditioner<Vector,Vector>
343    {
344    public:
345      // Constructor
346      JacobiPreconditioner (const GridView& gridView, const Matrix& matrix)
347      : gridView_(gridView), matrix_(matrix)
348      {
349        Vector diagonal(matrix_.N());
350        for (std::size_t i=0; i<diagonal.size(); ++i)
351          diagonal[i] = matrix_[i][i];
352
353        consistentDiagonal_ = diagonal;
354        VertexDataUpdate<GridView,Vector> matrixDataHandle(gridView,
355                                                            diagonal,
356                                                            consistentDiagonal_);
357
358        gridView_.communicate(matrixDataHandle,
359                              All_All_Interface,
360                              ForwardCommunication);
361      }
362
363      // Prepare the preconditioner
364      virtual void pre(Vector& x, Vector& b) override
365      {}
366
367      // Apply the preconditioner
368      virtual void apply(Vector& v, const Vector& r) override
369      {
370        auto rConsistent = r;
371
372        VertexDataUpdate<GridView,Vector> vertexUpdateHandle(gridView_,
373                                                             r,
374                                                             rConsistent);
375
376        gridView_.communicate(vertexUpdateHandle,
377                              InteriorBorder_InteriorBorder_Interface,
378                              ForwardCommunication);
379
380        for (std::size_t i=0; i<matrix_.N(); i++)
381          v[i] = rConsistent[i] / consistentDiagonal_[i];
382      }
383
384      // Clean up
385      virtual void post(Vector& x) override
386      {}
387
388      // Category of the preconditioner
389      virtual SolverCategory::Category category() const override
390      {
391        return SolverCategory::sequential;
392      }
393
394    private:
395      const GridView gridView_;
396      const Matrix& matrix_;
397      Vector consistentDiagonal_;
398    };
399    // { preconditioner_end }
400
401    // Scalar product for pairs of additive and consistent vectors
402    template<class GridView, class Vector>
```

544

```
403    // { scalar_product_begin }
404    class AdditiveScalarProduct : public ScalarProduct<Vector>
405    {
406      using typename ScalarProduct<Vector>::field_type;
407      using typename ScalarProduct<Vector>::real_type;
408
409    public:
410      // Constructor
411      AdditiveScalarProduct (const GridView& gridView)
412        : gridView_(gridView)
413      {}
414
415      // Dot product of a consistent vector x and an additive vector y,
416      // or vice versa
417      virtual field_type dot (const Vector& x, const Vector& y) const override
418      {
419        return gridView_.comm().sum(x.dot(y));
420      }
421
422      // Norm of a vector given in additive representation
423      virtual real_type norm (const Vector& x) const override
424      {
425        // Construct consistent representation of x
426        auto xConsistent = x;
427
428        VertexDataUpdate<GridView,Vector> vertexUpdateHandle(gridView_,
429                                                             x,
430                                                             xConsistent);
431
432        gridView_.communicate(vertexUpdateHandle,
433                              InteriorBorder_InteriorBorder_Interface,
434                              ForwardCommunication);
435
436        // Local scalar product of x with itself
437        auto localNorm2 = x.dot(xConsistent);
438
439        // Sum over all subdomains
440        return std::sqrt(gridView_.comm().sum(localNorm2));
441      }
442
443      // Scalar product, linear operator, and preconditioner must be
444      // of the same category
445      virtual SolverCategory::Category category() const override
446      {
447        return SolverCategory::sequential;
448      }
449
450    private:
451      const GridView gridView_;
452    };
453    // { scalar_product_end }
454
455
456    // { main_begin }
457    int main(int argc, char *argv[])
458    {
459      // Set up MPI
460      const MPIHelper& mpiHelper = MPIHelper::instance(argc, argv);
461
462      // Set up the grid
463      constexpr int dim = 2;
464      using Grid = UGGrid<dim>;
465      using GridView = Grid::LeafGridView;
466
467      std::shared_ptr<Grid> grid = GmshReader<Grid>::read("l-shape-refined.msh");
468
469      std::vector<double> dataVector;
470      auto gridView = grid->leafGridView();
471
472      if (mpiHelper.rank()==0)
473      {
474        // The initial iterate as a function
475        auto initialIterate = [](auto p){return std::min(p[0],p[1]);};
476
477        // Sample on the grid vertices
478        dataVector.resize(gridView.size(dim));
479        for (const auto& vertex : vertices(gridView, Partitions::interiorBorder))
480        {
481          auto index = gridView.indexSet().index(vertex);
482          dataVector[index] = initialIterate(vertex.geometry().corner(0));
483        }
484      }
485
486      // Copy vertex data into associative container
487      std::map<Grid::LocalIdSet::IdType, double> persistentContainer;
488      const auto& idSet = grid->localIdSet();
489
```

```
490    for (const auto& vertex : vertices(gridView))
491      persistentContainer[idSet.id(vertex)]
492         = dataVector[gridView.indexSet().index(vertex)];
493
494    // Distribute the grid and the data
495    LBVertexDataHandle<Grid, std::map<Grid::LocalIdSet::IdType,double> >
496                              dataHandle(grid, persistentContainer);
497    grid->loadBalance(dataHandle);
498
499    // Get gridView object again after load-balancing,
500    // to make sure it is up-to-date
501    gridView = grid->leafGridView();
502
503    // Copy data back into the array
504    dataVector.resize(gridView.size(dim));
505
506    for (const auto& vertex : vertices(gridView))
507      dataVector[gridView.indexSet().index(vertex)]
508         = persistentContainer[idSet.id(vertex)];
509  // { grid_setup_end }
510
511  //////////////////////////////////////////////////////////
512  //   Stiffness matrix and right hand side vector
513  //////////////////////////////////////////////////////////
514
515  // { assembly_begin }
516    using Vector = BlockVector<double>;
517    using Matrix = BCRSMatrix<double>;
518
519    Vector rhs;
520    Matrix stiffnessMatrix;
521
522    // Assemble the Poisson system in a first-order Lagrange space
523    Functions::LagrangeBasis<GridView,1> basis(gridView);
524
525    auto sourceTerm = [](const FieldVector<double,dim>& x){return -5.0;};
526    assemblePoissonProblem(basis, stiffnessMatrix, rhs, sourceTerm);
527
528    // Determine Dirichlet degrees of freedom by marking all those
529    // whose Lagrange nodes comply with a given predicate.
530    auto dirichletPredicate = [](auto p)
531    {
532      return p[0]< 1e-8 || p[1] < 1e-8 || (p[0] > 0.4999 && p[1] > 0.4999);
533    };
534
535    // Interpolating the predicate will mark all Dirichlet degrees of freedom
536    std::vector<bool> dirichletNodes;
537    Functions::interpolate(basis, dirichletNodes, dirichletPredicate);
538
539  //////////////////////////////////////////////////
540  //   Modify Dirichlet matrix rows
541  //////////////////////////////////////////////////
542
543    // Loop over the matrix rows
544    for (size_t i=0; i<stiffnessMatrix.N(); i++)
545    {
546      if (dirichletNodes[i])
547      {
548        auto cIt    = stiffnessMatrix[i].begin();
549        auto cEndIt = stiffnessMatrix[i].end();
550        // Loop over nonzero matrix entries in current row
551        for (; cIt!=cEndIt; ++cIt)
552          *cIt = (i==cIt.index()) ? 1.0 : 0.0;
553      }
554    }
555
556    // Set Dirichlet values
557    for (std::size_t i=0; i<dirichletNodes.size(); i++)
558      if (dirichletNodes[i])
559        rhs[i] = dataVector[i];
560  // { assembly_end }
561
562  /////////////////////////////
563  //   Compute solution
564  /////////////////////////////
565
566  // { algebraic_solving_begin }
567    // Set the initial iterate
568    Vector x(basis.size());
569    std::copy(dataVector.begin(), dataVector.end(), x.begin());
570
571    // Set up the preconditioned conjugate-gradient solver
572    double reduction = 1e-3;  // Desired residual reduction factor
573    int maxIterations = 50;   // Maximum number of iterations
574
575    MatrixAdapter<Matrix,Vector,Vector> linearOperator(stiffnessMatrix);
576    JacobiPreconditioner<GridView,Matrix,Vector> preconditioner(gridView,
```

```
577                                                    stiffnessMatrix);
578    AdditiveScalarProduct<GridView,Vector> scalarProduct(gridView);
579
580    CGSolver<Vector> cg(linearOperator,
581                       scalarProduct,
582                       preconditioner,
583                       reduction,
584                       maxIterations,
585                       (mpiHelper.rank()==0) ? 2 : 0);  // Only rank 0
586                                                        // will print output
587
588    // Object storing some statistics about the solving process
589    InverseOperatorResult statistics;
590
591    // Solve!
592    cg.apply(x, rhs, statistics);
593  // { algebraic_solving_end }
594
595    // Output result
596    VTKWriter<GridView> vtkWriter(gridView);
597    vtkWriter.addVertexData(x, "solution");
598    vtkWriter.write("istl-distributed-poisson-result");
599  }
```

B.8. The Sequential AMG Preconditioner

File:	`istl-sequential-amg.cc`
Equation:	Poisson equation
Discretization:	First-order Lagrange finite elements
Grid:	Unstructured simplicial grid, implemented with `UGGrid`
Solver:	CG solver with AMG preconditioner
Distributed:	no
Discussed in:	Chapter 7.5.3

```
1   #include "config.h"
2
3   #include <iostream>
4
5   #include <dune/istl/bcrsmatrix.hh>
6   #include <dune/istl/bvector.hh>
7   #include <dune/istl/preconditioners.hh>
8   #include <dune/istl/paamg/amg.hh>
9   #include <dune/istl/solvers.hh>
10  #include <dune/istl/matrixmarket.hh>
11
12  using namespace Dune;
13
14  // { problem_setup_begin }
15  int main(int argc, char *argv[])
16  {
17    using Matrix = BCRSMatrix<double>;
18    using Vector = BlockVector<double>;
19    Matrix A;
20    Vector b;
21    loadMatrixMarket(A, "getting-started-poisson-fem-4-refinements-matrix.mtx");
22    loadMatrixMarket(b, "getting-started-poisson-fem-4-refinements-rhs.mtx");
23  // { problem_setup_end }
24
25    // Construct vector to hold the iterates
26  // { initial_iterate_begin }
27    Vector x(b.size());
28    x=b;
29  // { initial_iterate_end }
30
31    // Set up the smoother
32  // { smoother_setup_begin }
33    using Smoother = SeqSSOR<Matrix,Vector,Vector>;
34
35    Amg::SmootherTraits<Smoother>::Arguments smootherArgs;
36    smootherArgs.iterations = 3;
```

```
37      smootherArgs.relaxationFactor = 1;
38  // { smoother_setup_end }
39
40      // Set up the coarsening criterion
41  // { coarsening_setup_begin }
42      using Norm = Amg::FirstDiagonal;
43      using Criterion
44          = Amg::CoarsenCriterion<Amg::UnSymmetricCriterion<Matrix,Norm> >;
45
46      Criterion criterion(15,                    // Maximum number
47                                                 // of multigrid levels
48                          2000);                 // Create coarse levels until
49                                                 // problem size gets smaller
50                                                 // than this
51      criterion.setDefaultValuesIsotropic(2);    // Aggregate sizes and shapes
52      criterion.setAlpha(.67);                   // Connections above this value
53                                                 // are "strong"
54      criterion.setBeta(1.0e-4);                 // Connections below this value
55                                                 // are treated as zero
56      criterion.setGamma(1);                     // Number of
57                                                 // coarse-level iterations
58      criterion.setDebugLevel(2);                // Print some debugging output
59  // { coarsening_setup_end }
60
61  // { solver_setup_begin }
62      using LinearOperator = MatrixAdapter<Matrix,Vector,Vector>;
63      using AMG = Amg::AMG<LinearOperator,Vector,Smoother>;
64
65      LinearOperator linearOperator(A);
66      AMG amg(linearOperator, criterion, smootherArgs);
67
68      CGSolver<Vector> amgCG(linearOperator,
69                             amg,                // Preconditioner
70                             1e-5,               // Desired residual reduction
71                             50,                 // Maximum number of iterations
72                             2);                 // Verbosity
73  // { solver_setup_end }
74
75  // { solve_begin }
76      InverseOperatorResult r;
77      amgCG.apply(x,b,r);
78
79      std::cout << "CG+AMG did " << r.iterations << " iterations." << std::endl;
80  // { solve_end }
81  }
```

B.9. Dual Mortar Finite Elements

File:	localfunctions-use-dual-mortar-basis.cc
Equation:	none
Discretization:	Dual mortar finite elements
Grid:	Simplicial grid, implemented with UGGrid
Solver:	none
Distributed:	no
Discussed in:	Chapter 8.3.1

```
1   #include "config.h"
2
3   #include <iostream>
4
5   #include <dune/istl/io.hh>
6   #include <dune/istl/matrix.hh>
7
8   #include <dune/localfunctions/lagrange.hh>
9   #include <dune/localfunctions/dualmortarbasis.hh>
10
11  #include <dune/geometry/quadraturerules.hh>
12
13  #include <dune/grid/uggrid.hh>
14  #include <dune/grid/utility/structuredgridfactory.hh>
```

```
15
16    using namespace Dune;
17
18    // { use_dual_mortar_basis_begin }
19    int main(int argc, char *argv[])
20    {
21      // Set up MPI if available
22      MPIHelper::instance(argc, argv);
23
24      // Construct a 2x2 structured triangle grid
25      constexpr int dim = 2;
26      using Grid = UGGrid<dim>;
27
28      auto grid = StructuredGridFactory<Grid>::createSimplexGrid({0.0,0.0},
29                                                                 {1.0,1.0},
30                                                                 { 2,  2});
31      auto gridView = grid->leafGridView();
32    // { grid_setup_end }
33
34      // Set up Lagrange and dual finite elements for simplex elements
35    // { space_setup_begin }
36      LagrangeSimplexLocalFiniteElement<double,double,dim,1> lagrangeFE;
37      DualP1LocalFiniteElement<double,double,dim,true> dualFE;
38
39      // Matrix that will store the integrals
40      Matrix<double> facetMassMatrix;
41    // { space_setup_end }
42
43    // { intersection_loop_begin }
44      for (const auto& element : elements(gridView))
45      {
46        for (const auto& intersection : intersections(gridView, element))
47        {
48          if (intersection.boundary())
49          {
50            auto refElement = referenceElement<double,dim>(element.type());
51    // { intersection_loop_end }
52
53    // { zero_matrix_begin }
54            facetMassMatrix.setSize(lagrangeFE.size(), dualFE.size());
55            facetMassMatrix = 0;
56    // { zero_matrix_end }
57
58    // { quad_loop_begin }
59            constexpr auto intersectionDim
60              = std::decay_t<decltype(intersection)>::mydimension;
61            const auto& quad
62              = QuadratureRules<double,intersectionDim>::rule(intersection.type(),
63                                                              2);
64
65            for (const auto& quadPoint : quad)
66            {
67    // { quad_loop_end }
68              auto quadPosInElement
69                = intersection.geometryInInside().global(quadPoint.position());
70
71              std::vector<FieldVector<double,1> > lagrangeValues, dualValues;
72              lagrangeFE.localBasis().evaluateFunction(quadPosInElement,
73                                                       lagrangeValues);
74              dualFE.localBasis().evaluateFunction(quadPosInElement, dualValues);
75    // { evaluation_end }
76
77    // { integration_begin }
78              for (std::size_t i=0; i<lagrangeFE.size(); i++)
79              {
80                LocalKey lagrangeLocalKey
81                  = lagrangeFE.localCoefficients().localKey(i);
82
83                auto lagrangeSubEntities
84                  = refElement.subEntities(intersection.indexInInside(),
85                                           1,   // The codimension of facets
86                                           lagrangeLocalKey.codim());
87
88                if (lagrangeSubEntities.contains(lagrangeLocalKey.subEntity()))
89                {
90                  for (std::size_t j=0; j<dualFE.size(); j++)
91                  {
92                    LocalKey dualLocalKey = dualFE.localCoefficients().localKey(j);
93
94                    auto dualSubEntities
95                      = refElement.subEntities(intersection.indexInInside(),
96                                               1,   // The codimension of facets
97                                               dualLocalKey.codim());
98
99                    if (dualSubEntities.contains(dualLocalKey.subEntity()))
100                   {
101                     auto integrationElement
```

```
102                         = intersection.geometry()
103                             .integrationElement(quadPoint.position());
104                 facetMassMatrix[i][j] += quadPoint.weight()
105                             * integrationElement
106                             * lagrangeValues[i]
107                             * dualValues[j];
108              }
109            }
110          }
111        }
112      } // End of quadrature loop
113 // { quad_loop_final }
114
115      // Print the facet mass matrix, observe how it is diagonal!
116 // { print_matrix_begin }
117      printmatrix(std::cout, facetMassMatrix, "facetMassMatrix", "--");
118
119    } // End of 'if boundary'
120   } // End of intersection loop
121  } //End of element loop
122 // { intersection_loop_final }
123 }
```

B.10. The Stokes Equation Using Taylor–Hood Elements

File:	`functions-stokes.cc`
Equation:	Stokes equation
Discretization:	Taylor–Hood finite elements
Grid:	Uniform quadrilateral grid, implemented with `YaspGrid`
Solver:	GMRes
Distributed:	no
Discussed in:	Chapter 10.8

```
1  #include <config.h>
2
3  #include <array>
4  #include <vector>
5
6  #include <dune/common/indices.hh>
7
8  #include <dune/geometry/quadraturerules.hh>
9
10 #include <dune/grid/yaspgrid.hh>
11 #include <dune/grid/io/file/vtk/subsamplingvtkwriter.hh>
12
13 #include <dune/istl/matrix.hh>
14 #include <dune/istl/bcrsmatrix.hh>
15 #include <dune/istl/multitypeblockmatrix.hh>
16 #include <dune/istl/multitypeblockvector.hh>
17 #include <dune/istl/matrixindexset.hh>
18 #include <dune/istl/solvers.hh>
19 #include <dune/istl/preconditioners.hh>
20
21 #include <dune/functions/functionspacebases/interpolate.hh>
22 #include <dune/functions/functionspacebases/taylorhoodbasis.hh>
23 #include <dune/functions/backends/istlvectorbackend.hh>
24 #include <dune/functions/functionspacebases/powerbasis.hh>
25 #include <dune/functions/functionspacebases/compositebasis.hh>
26 #include <dune/functions/functionspacebases/lagrangebasis.hh>
27 #include <dune/functions/functionspacebases/subspacebasis.hh>
28 #include <dune/functions/functionspacebases/boundarydofs.hh>
29
30 #include <dune/functions/gridfunctions/discreteglobalbasisfunction.hh>
31 #include <dune/functions/gridfunctions/gridviewfunction.hh>
32
33 using namespace Dune;
34
```

```
35    // Compute the stiffness matrix for a single element
36    // { local_assembler_signature_begin }
37    template<class LocalView>
38    void assembleElementStiffnessMatrix(const LocalView& localView,
39                                         Matrix<double>& elementMatrix)
40    // { local_assembler_signature_end }
41    {
42      // Get the grid element from the local FE basis view
43      // { local_assembler_get_element_information_begin }
44      using Element = typename LocalView::Element;
45      const Element element = localView.element();
46
47      constexpr int dim = Element::dimension;
48      auto geometry = element.geometry();
49      // { local_assembler_get_element_information_end }
50
51      // Set all matrix entries to zero
52      // { initialize_element_matrix_begin }
53      elementMatrix.setSize(localView.size(), localView.size());
54      elementMatrix = 0;        // Fills the entire matrix with zeros
55      // { initialize_element_matrix_end }
56
57      // Get set of shape functions for this element
58      // { get_local_fe_begin }
59      using namespace Indices;
60      const auto& velocityLocalFiniteElement
61          = localView.tree().child(_0,0).finiteElement();
62      const auto& pressureLocalFiniteElement
63          = localView.tree().child(_1).finiteElement();
64      // { get_local_fe_end }
65
66      // Get a quadrature rule
67      // { begin_quad_loop_begin }
68      int order = 2*(dim*velocityLocalFiniteElement.localBasis().order()-1);
69      const auto& quad = QuadratureRules<double,dim>::rule(element.type(), order);
70
71      // Loop over all quadrature points
72      for (const auto& quadPoint : quad)
73      {
74      // { begin_quad_loop_end }
75      // { quad_loop_preamble_begin }
76        // The transposed inverse Jacobian of the map from the
77        // reference element to the element
78        const auto jacobianInverseTransposed
79            = geometry.jacobianInverseTransposed(quadPoint.position());
80
81        // The multiplicative factor in the integral transformation formula
82        const auto integrationElement
83            = geometry.integrationElement(quadPoint.position());
84      // { quad_loop_preamble_end }
85
86        //////////////////////////////////////////////////////////////////
87        //   Velocity--velocity coupling
88        //////////////////////////////////////////////////////////////////
89
90        // The gradients of the shape functions on the reference element
91      // { velocity_gradients_begin }
92        std::vector<FieldMatrix<double,1,dim> > referenceGradients;
93        velocityLocalFiniteElement.localBasis().evaluateJacobian(
94                                                   quadPoint.position(),
95                                                   referenceGradients);
96
97        // Compute the shape function gradients on the grid element
98        std::vector<FieldVector<double,dim> > gradients(referenceGradients.size());
99        for (size_t i=0; i<gradients.size(); i++)
100          jacobianInverseTransposed.mv(referenceGradients[i][0], gradients[i]);
101     // { velocity_gradients_end }
102
103       // Compute the actual matrix entries
104     // { velocity_velocity_coupling_begin }
105       for (size_t i=0; i<velocityLocalFiniteElement.size(); i++)
106         for (size_t j=0; j<velocityLocalFiniteElement.size(); j++ )
107           for (size_t k=0; k<dim; k++)
108           {
109             size_t row = localView.tree().child(_0,k).localIndex(i);
110             size_t col = localView.tree().child(_0,k).localIndex(j);
111             elementMatrix[row][col] += ( gradients[i] * gradients[j] )
112                               * quadPoint.weight() * integrationElement;
113           }
114     // { velocity_velocity_coupling_end }
115
116       //////////////////////////////////////////////////////////////////
117       //   Velocity--pressure coupling
118       //////////////////////////////////////////////////////////////////
119
120       // The values of the pressure shape functions
121     // { pressure_values_begin }
```

```
122         std::vector<FieldVector<double,1> > pressureValues;
123         pressureLocalFiniteElement
124             .localBasis().evaluateFunction(quadPoint.position(),pressureValues);
125  // { pressure_values_end }
126
127         // Compute the actual matrix entries
128  // { velocity_pressure_coupling_begin }
129      for (size_t i=0; i<velocityLocalFiniteElement.size(); i++)
130        for (size_t j=0; j<pressureLocalFiniteElement.size(); j++ )
131          for (size_t k=0; k<dim; k++)
132          {
133            size_t vIndex = localView.tree().child(_0,k).localIndex(i);
134            size_t pIndex = localView.tree().child(_1).localIndex(j);
135
136            auto value = gradients[i][k] * pressureValues[j]
137                       * quadPoint.weight() * integrationElement;
138            elementMatrix[vIndex][pIndex] += value;
139            elementMatrix[pIndex][vIndex] += value;
140          }
141  // { velocity_pressure_coupling_end }
142    }
143  }
144
145
146  // Set the occupation pattern of the stiffness matrix
147  template<class Basis, class Matrix>
148  void setOccupationPattern(const Basis& basis, Matrix& matrix)
149  {
150    enum {dim = Basis::GridView::dimension};
151
152    // MatrixIndexSets store the occupation pattern of a sparse matrix.
153    // They are not particularly efficient, but simple to use.
154    std::array<std::array<MatrixIndexSet, 2>, 2> nb;
155
156    // Set sizes of the 2x2 submatrices
157    for (size_t i=0; i<2; i++)
158      for (size_t j=0; j<2; j++)
159        nb[i][j].resize(basis.size({i}), basis.size({j}));
160
161    // A view on the FE basis on a single element
162    auto localView = basis.localView();
163
164    // Loop over all leaf elements
165    for(const auto& element : elements(basis.gridView()))
166    {
167      // Bind the local  view to the current element
168      localView.bind(element);
169
170      // Add element stiffness matrix onto the global stiffness matrix
171      for (size_t i=0; i<localView.size(); i++)
172      {
173        // Global index of the i-th local degree of freedom of the current element
174        auto row = localView.index(i);
175
176        for (size_t j=0; j<localView.size(); j++ )
177        {
178          // Global index of the j-th local degree of freedom of the current element
179          auto col = localView.index(j);
180
181          nb[row[0]][col[0]].add(row[1],col[1]);
182        }
183      }
184    }
185
186    // Give the matrix the occupation pattern we want.
187    using namespace Indices;
188    nb[0][0].exportIdx(matrix[_0][_0]);
189    nb[0][1].exportIdx(matrix[_0][_1]);
190    nb[1][0].exportIdx(matrix[_1][_0]);
191    nb[1][1].exportIdx(matrix[_1][_1]);
192  }
193
194
195  // { matrixentry_begin }
196  template<class Matrix, class MultiIndex>
197  decltype(auto) matrixEntry(
198          Matrix& matrix, const MultiIndex& row, const MultiIndex& col)
199  {
200    using namespace Indices;
201    if ((row[0]==0) && (col[0]==0))
202      return matrix[_0][_0][row[1]][col[1]][row[2]][col[2]];
203    if ((row[0]==0) && (col[0]==1))
204      return matrix[_0][_1][row[1]][col[1]][row[2]][0];
205    if ((row[0]==1) && (col[0]==0))
206      return matrix[_1][_0][row[1]][col[1]][0][col[2]];
207    return matrix[_1][_1][row[1]][col[1]];
208  }
```

```
209     // { matrixentry_end }
210
211
212     // Assemble the Laplace stiffness matrix on the given grid view
213     // { global_assembler_signature_begin }
214     template<class Basis, class Matrix>
215     void assembleStokesMatrix(const Basis& basis, Matrix& matrix)
216     // { global_assembler_signature_end }
217     {
218     // { setup_matrix_pattern_begin }
219        // Set matrix size and occupation pattern
220        setOccupationPattern(basis, matrix);
221
222        // Set all entries to zero
223        matrix = 0;
224     // { setup_matrix_pattern_end }
225
226        // A view on the FE basis on a single element
227     // { get_localview_begin }
228        auto localView = basis.localView();
229     // { get_localview_end }
230
231        // A loop over all elements of the grid
232     // { element_loop_and_bind_begin }
233        for (const auto& element : elements(basis.gridView()))
234        {
235           // Bind the local FE basis view to the current element
236           localView.bind(element);
237     // { element_loop_and_bind_end }
238
239           // Now let's get the element stiffness matrix
240           // A dense matrix is used for the element stiffness matrix
241     // { setup_element_stiffness_begin }
242           Dune::Matrix<double> elementMatrix;
243           assembleElementStiffnessMatrix(localView, elementMatrix);
244     // { setup_element_stiffness_end }
245
246           // Add element stiffness matrix onto the global stiffness matrix
247     // { accumulate_global_matrix_begin }
248           for (size_t i=0; i<elementMatrix.N(); i++)
249           {
250              // The global index of the i-th local degree of freedom
251              // of the current element
252              auto row = localView.index(i);
253
254              for (size_t j=0; j<elementMatrix.M(); j++ )
255              {
256                 // The global index of the j-th local degree of freedom
257                 // of the current element
258                 auto col = localView.index(j);
259                 matrixEntry(matrix, row, col) += elementMatrix[i][j];
260              }
261           }
262     // { accumulate_global_matrix_end }
263        }
264     }
265
266
267
268     // { main_begin }
269     int main(int argc, char *argv[])
270     {
271        // Set up MPI, if available
272        MPIHelper::instance(argc, argv);
273     // { mpi_setup_end }
274
275        //////////////////////////////////
276        //   Generate the grid
277        //////////////////////////////////
278
279     // { grid_setup_begin }
280        constexpr int dim = 2;
281        using Grid = YaspGrid<dim>;
282        FieldVector<double,dim> upperRight = {1, 1};
283        std::array<int,dim> nElements = {4, 4};
284        Grid grid(upperRight,nElements);
285
286        using GridView = typename Grid::LeafGridView;
287        GridView gridView = grid.leafGridView();
288     // { grid_setup_end }
289
290        ////////////////////////////////////////////////////////
291        //   Choose a finite element space
292        ////////////////////////////////////////////////////////
293
294     // { function_space_basis_begin }
295        using namespace Functions::BasisFactory;
```

```
296
297     constexpr std::size_t p = 1; // Pressure order for Taylor-Hood
298
299     auto taylorHoodBasis = makeBasis(
300             gridView,
301             composite(
302               power<dim>(
303                 lagrange<p+1>(),
304                 blockedInterleaved()),
305               lagrange<p>()
306             ));
307     // { function_space_basis_end }
308
309     //////////////////////////////////////////////////////
310     //   Stiffness matrix and right hand side vector
311     //////////////////////////////////////////////////////
312
313     // { linear_algebra_setup_begin }
314     using VelocityVector = BlockVector<FieldVector<double,dim>>;
315     using PressureVector = BlockVector<double>;
316     using Vector = MultiTypeBlockVector<VelocityVector, PressureVector>;
317
318     using Matrix00 = BCRSMatrix<FieldMatrix<double,dim,dim>>;
319     using Matrix01 = BCRSMatrix<FieldMatrix<double,dim,1>>;
320     using Matrix10 = BCRSMatrix<FieldMatrix<double,1,dim>>;
321     using Matrix11 = BCRSMatrix<double>;
322     using MatrixRow0 = MultiTypeBlockVector<Matrix00, Matrix01>;
323     using MatrixRow1 = MultiTypeBlockVector<Matrix10, Matrix11>;
324     using Matrix = MultiTypeBlockMatrix<MatrixRow0,MatrixRow1>;
325     // { linear_algebra_setup_end }
326
327     //////////////////////////////////////////////////////
328     //   Assemble the system
329     //////////////////////////////////////////////////////
330
331     // { rhs_assembly_begin }
332     Vector rhs;
333
334     auto rhsBackend = Functions::istlVectorBackend(rhs);
335
336     rhsBackend.resize(taylorHoodBasis);
337     rhs = 0;
338     // { rhs_assembly_end }
339
340     // { matrix_assembly_begin }
341     Matrix stiffnessMatrix;
342     assembleStokesMatrix(taylorHoodBasis, stiffnessMatrix);
343     // { matrix_assembly_end }
344
345     //////////////////////////////////////////////////////
346     // Set Dirichlet values.
347     // Only velocity components have Dirichlet boundary values
348     //////////////////////////////////////////////////////
349
350     // { initialize_boundary_dofs_vector_begin }
351     using VelocityBitVector = std::vector<std::array<char,dim> >;
352     using PressureBitVector = std::vector<char>;
353     using BitVector = TupleVector<VelocityBitVector, PressureBitVector>;
354
355     BitVector isBoundary;
356
357     auto isBoundaryBackend = Functions::istlVectorBackend(isBoundary);
358     isBoundaryBackend.resize(taylorHoodBasis);
359
360     using namespace Indices;
361     for (auto&& b0i : isBoundary[_0])
362       for (std::size_t j=0; j<b0i.size(); ++j)
363         b0i[j] = false;
364     std::fill(isBoundary[_1].begin(), isBoundary[_1].end(), false);
365     // { initialize_boundary_dofs_vector_end }
366
367     // { determine_boundary_dofs_begin }
368     Functions::forEachBoundaryDOF(
369             Functions::subspaceBasis(taylorHoodBasis, _0),
370             [&] (auto&& index) {
371               isBoundaryBackend[index] = true;
372             });
373     // { determine_boundary_dofs_end }
374
375     // { interpolate_dirichlet_values_begin }
376     using Coordinate = GridView::Codim<0> ::Geometry::GlobalCoordinate;
377     using VelocityRange = FieldVector<double,dim>;
378     auto&& g = [](Coordinate x)
379     {
380       return VelocityRange{0.0, (x[0] < 1e-8) ? 1.0 : 0.0};
381     };
382
```

```
383        Functions::interpolate(Functions::subspaceBasis(taylorHoodBasis, _0),
384                                rhs,
385                                g,
386                                isBoundary);
387    // { interpolate_dirichlet_values_end }
388
389    /////////////////////////////////////////
390    //   Modify Dirichlet rows
391    /////////////////////////////////////////
392
393    // Loop over the matrix rows
394    // { set_dirichlet_matrix_begin }
395      auto localView = taylorHoodBasis.localView();
396      for(const auto& element : elements(gridView))
397      {
398        localView.bind(element);
399        for (size_t i=0; i<localView.size(); ++i)
400        {
401          auto row = localView.index(i);
402          // If row corresponds to a boundary entry,
403          // modify it to be an identity matrix row.
404          if (isBoundaryBackend[row])
405            for (size_t j=0; j<localView.size(); ++j)
406            {
407              auto col = localView.index(j);
408              matrixEntry(stiffnessMatrix, row, col) = (i==j) ? 1 : 0;
409            }
410        }
411      }
412    // { set_dirichlet_matrix_end }
413
414    ////////////////////////////
415    //   Compute solution
416    ////////////////////////////
417    // { stokes_solve_begin }
418    // Initial iterate: Start from the rhs vector,
419    // that way the Dirichlet entries are already correct.
420    Vector x = rhs;
421
422    // Turn the matrix into a linear operator
423    MatrixAdapter<Matrix,Vector,Vector> stiffnessOperator(stiffnessMatrix);
424
425    // Fancy (but only) way to not have a preconditioner at all
426    Richardson<Vector,Vector> preconditioner(1.0);
427
428    // Construct the iterative solver
429    RestartedGMResSolver<Vector> solver(
430          stiffnessOperator,   // Operator to invert
431          preconditioner,      // Preconditioner
432          1e-10,               // Desired residual reduction factor
433          500,                 // Number of iterations between restarts,
434                               // here: no restarting
435          500,                 // Maximum number of iterations
436          2);                  // Verbosity of the solver
437
438    // Object storing some statistics about the solving process
439    InverseOperatorResult statistics;
440
441    // Solve!
442    solver.apply(x, rhs, statistics);
443    // { stokes_solve_end }
444
445    ////////////////////////////////////////////////////////////////////////////
446    //   Make a discrete function from the FE basis and the coefficient vector
447    ////////////////////////////////////////////////////////////////////////////
448
449    // { make_result_functions_begin }
450      using VelocityRange = FieldVector<double,dim>;
451      using PressureRange = double;
452
453      auto velocityFunction
454            = Functions::makeDiscreteGlobalBasisFunction<VelocityRange>(
455                  Functions::subspaceBasis(taylorHoodBasis, _0), x);
456      auto pressureFunction
457            = Functions::makeDiscreteGlobalBasisFunction<PressureRange>(
458                  Functions::subspaceBasis(taylorHoodBasis, _1), x);
459    // { make_result_functions_end }
460
461    ////////////////////////////////////////////////////////////////////////////
462    //   Write result to VTK file
463    //   We need to subsample, because the dune-grid VTKWriter cannot natively display
464    //   second-order functions
465    ////////////////////////////////////////////////////////////////////////////
466    // { vtk_output_begin }
467      SubsamplingVTKWriter<GridView> vtkWriter(
468            gridView,
469            refinementLevels(2));
```

```
470    vtkWriter.addVertexData(
471        velocityFunction,
472        VTK::FieldInfo("velocity", VTK::FieldInfo::Type::vector, dim));
473    vtkWriter.addVertexData(
474        pressureFunction,
475        VTK::FieldInfo("pressure", VTK::FieldInfo::Type::scalar, 1));
476    vtkWriter.write("stokes-taylorhood-result");
477    // { vtk_output_end }
478    }
```

B.11. The Linear Reaction–Diffusion Equation Using dune-pdelab

File:	`pdelab-linear-reaction-diffusion.cc`
Equation:	Linear reaction–diffusion equation
Discretization:	First and second-order Lagrange finite elements, Rannacher–Turek finite elements
Grid:	Uniform quadrilateral grid, implemented with `YaspGrid`
Solver:	CG solver with SSOR preconditioner
Distributed:	no
Discussed in:	Chapter 11.1

```
 1    #include "config.h"
 2
 3    #include <iostream>
 4    #include <vector>
 5
 6    #include <dune/grid/yaspgrid.hh>
 7    #include <dune/grid/io/file/vtk/subsamplingvtkwriter.hh>
 8
 9    #include <dune/functions/functionspacebases/lagrangebasis.hh>
10    #include <dune/functions/functionspacebases/rannacherturekbasis.hh>
11    #include <dune/functions/gridfunctions/discreteglobalbasisfunction.hh>
12
13    #include <dune/pdelab.hh>
14
15    using namespace Dune;
16
17    // { problem_description_begin }
18    template<class GridView>
19    struct ReactionDiffusionProblem
20    : public PDELab::ConvectionDiffusionModelProblem<GridView,double>
21    {
22      template<typename Element, typename Coord>
23      auto f(const Element& element, const Coord& xi) const
24      {
25        const Coord center = Coord(0.5);
26        auto distanceToCenter = (element.geometry().global(xi)-center).two_norm();
27        return (distanceToCenter <= 0.25) ? -10.0 : 10.0;
28      }
29
30      template<typename Element, typename Coord>
31      auto c(const Element& element, const Coord& xi) const
32      {
33        return 10.0;
34      }
35    };
36    // { problem_description_end }
37
38
39    // { solvereactiondiffusionproblem_begin }
40    template<class Basis>
41    void solveReactionDiffusionProblem(std::shared_ptr<Basis> basis,
42                                       std::string filename,
43                                       int averageNumberOfEntriesPerMatrixRow)
44    {
45    // { solvereactiondiffusionproblem_signature_end }
```

```
46      // Make grid function space
47      // { construct_function_space_begin }
48      using GridView = typename Basis::GridView;
49      using VectorBackend = PDELab::ISTL::VectorBackend<>;
50
51      using GridFunctionSpace = PDELab::Experimental::GridFunctionSpace<Basis,
52                                                                        VectorBackend,
53                                                                        PDELab::NoConstraints>;
54
55      GridFunctionSpace gridFunctionSpace(basis);
56      // { construct_function_space_end }
57
58      // Make grid operator
59      // { construct_element_assembler_begin }
60      using Problem = ReactionDiffusionProblem<GridView>;
61      Problem reactionDiffusionProblem;
62      using LocalOperator
63          = PDELab::ConvectionDiffusionFEM<Problem,
64                                           typename GridFunctionSpace::Traits::
65                                                         FiniteElementMap>;
66      LocalOperator localOperator(reactionDiffusionProblem);
67      // { construct_element_assembler_end }
68
69      // { construct_global_assembler_begin }
70      using MatrixBackend = PDELab::ISTL::BCRSMatrixBackend<>;
71      MatrixBackend matrixBackend(averageNumberOfEntriesPerMatrixRow);
72      using GridOperator
73          = PDELab::GridOperator<GridFunctionSpace,      // Trial function space
74                                 GridFunctionSpace,      // Test function space
75                                 LocalOperator,          // Element assembler
76                                 MatrixBackend,          // Data structure
77                                                         // for the stiffness matrix
78                                 double,                 // Number type for
79                                                         // solution vector entries
80                                 double,                 // Number type for
81                                                         // residual vector entries
82                                 double>;                // Number type for
83                                                         // stiffness matrix entries
84
85      GridOperator gridOperator(gridFunctionSpace,
86                                gridFunctionSpace,
87                                localOperator,
88                                matrixBackend);
89      // { construct_global_assembler_end }
90
91      // Select vector data type to hold the iterate
92      // { create_initial_iterate_begin }
93      using VectorContainer = PDELab::Backend::Vector<GridFunctionSpace,double>;
94      VectorContainer u(gridFunctionSpace,0.0); // Initial value
95      // { create_initial_iterate_end }
96
97      // Select a linear solver backend, and solve the linear problem
98      // { solving_begin }
99      using LinearSolverBackend = PDELab::ISTLBackend_SEQ_CG_SSOR;
100     LinearSolverBackend linearSolverBackend(5000,2);
101
102     using LinearProblemSolver
103         = PDELab::StationaryLinearProblemSolver<GridOperator,
104                                                 LinearSolverBackend,
105                                                 VectorContainer>;
106
107     LinearProblemSolver linearProblemSolver(gridOperator,
108                                             linearSolverBackend,
109                                             u,
110                                             1e-10);
111     linearProblemSolver.apply();
112     // { solving_end }
113
114     // Output as VTK file
115     // { vtk_writing_begin }
116     // Make a discrete function from the FE basis and the coefficient vector
117     auto uFunction
118         = Functions::makeDiscreteGlobalBasisFunction<double>(
119                                       *basis,
120                                       PDELab::Backend::native(u));
121
122     // We need to subsample, because the dune-grid VTKWriter
123     // cannot natively display second-order functions.
124     SubsamplingVTKWriter<GridView> vtkWriter(basis->gridView(),
125                                              refinementLevels(4));
126     vtkWriter.addVertexData(uFunction,
127                             VTK::FieldInfo("u",
128                                            VTK::FieldInfo::Type::scalar,
129                                            1));
130     vtkWriter.write(filename);
131     // { vtk_writing_end }
132 }
```

```
133
134
135
136    // { main_begin }
137    int main(int argc, char *argv[])
138    {
139      // Set up MPI, if available
140      MPIHelper::instance(argc,argv);
141
142      // Create a structured grid
143      constexpr int dim = 2;
144      using Grid = YaspGrid<dim>;
145      FieldVector<double,dim> bbox = {1.0, 1.0};
146      std::array<int,dim> elements = {8,8};
147      Grid grid(bbox,elements);
148
149      using GridView = Grid::LeafGridView;
150      GridView gridView = grid.leafGridView();
151    // { grid_setup_end }
152
153    // { simulation_begin }
154      // Solve boundary value problem using first-order Lagrange elements
155      using Q1Basis = Functions::LagrangeBasis<GridView,1>;
156      auto q1Basis = std::make_shared<Q1Basis>(gridView);
157
158      solveReactionDiffusionProblem(q1Basis,
159                                    "pdelab-linear-reaction-diffusion-result-q1",
160                                    9);    // Average number of matrix entries
161                                           // per row
162
163      // Solve boundary value problem using second-order Lagrange elements
164      using Q2Basis = Functions::LagrangeBasis<GridView,2>;
165      auto q2Basis = std::make_shared<Q2Basis>(gridView);
166
167      solveReactionDiffusionProblem(q2Basis,
168                                    "pdelab-linear-reaction-diffusion-result-q2",
169                                    25);   // Average number of matrix entries
170                                           // per row
171
172      // Solve boundary value problem using Rannacher-Turek elements
173      using RannacherTurekBasis = Functions::RannacherTurekBasis<GridView>;
174      auto rannacherTurekBasis = std::make_shared<RannacherTurekBasis>(gridView);
175
176      solveReactionDiffusionProblem(rannacherTurekBasis,
177                                    "pdelab-linear-reaction-diffusion-result-rt",
178                                    7);    // Average number of matrix entries
179                                           // per row
180    }
181    // { main_end }
```

B.12. p-Laplace Problem Using `dune-pdelab` and Finite Elements

File:	`pdelab-p-laplace.cc`
Equation:	p-Laplace equation with linear reaction term
Discretization:	Second-order Lagrange finite elements
Grid:	Uniform quadrilateral grid, implemented with `YaspGrid`
Solver:	Newton solver with line search
Distributed:	no
Discussed in:	Chapter 11.2.4

```
1    #include "config.h"
2
3    #include <iostream>
4    #include <vector>
5
6    #include <dune/common/parallel/mpihelper.hh>
7
```

```
 8    #include <dune/geometry/quadraturerules.hh>
 9
10    #include <dune/grid/io/file/vtk/subsamplingvtkwriter.hh>
11    #include <dune/grid/yaspgrid.hh>
12
13    #include <dune/functions/functionspacebases/lagrangebasis.hh>
14    #include <dune/functions/gridfunctions/discreteglobalbasisfunction.hh>
15
16    #include <dune/pdelab.hh>
17
18
19    using namespace Dune;
20
21    /** Local operator for solving the equation
22     *
23     *   - \Delta_p u + cu   = f   in \Omega
24     *   \nabla u \cdot n    = 0   on \partial\Omega
25     *
26     * with conforming finite elements on all types of grids in any dimension
27     *
28     */
29    // { localoperator_begin }
30    template<int dim>
31    class PLaplaceLocalOperator :
32      public PDELab::LocalOperatorDefaultFlags,
33      public PDELab::FullVolumePattern
34    {
35    public:
36      // Pattern assembly flags
37      static const bool doPatternVolume = true;
38
39      // Residual assembly flags
40      static const bool doAlphaVolume = true;
41    // { localoperator_header_end }
42
43    // { data_and_constructor_begin }
44      // The order of the p-Laplace term
45      const double p_;
46
47      // Source term
48      std::function<double(const FieldVector<double,dim>&)> f_;
49
50      // Reaction term
51      std::function<double(const FieldVector<double,dim>&)> c_;
52
53      PLaplaceLocalOperator(double p,
54                 const std::function<double(const FieldVector<double,dim>&)>& f,
55                 const std::function<double(const FieldVector<double,dim>&)>& c)
56      : p_(p),
57        f_(f),
58        c_(c)
59      {}
60    // { data_and_constructor_end }
61
62      // Volume contribution of the residual
63    // { alpha_volume_begin }
64      template<class EntityGeometry,
65               class LocalFunctionSpaceU, class Vector,
66               class LocalFunctionSpaceV, class Residual>
67      void alpha_volume(const EntityGeometry& entityGeometry,
68                        const LocalFunctionSpaceU& localFunctionSpaceU,
69                        const Vector& x,
70                        const LocalFunctionSpaceV& localFunctionSpaceV,
71                        Residual& residual) const
72      {
73        // Extract types of shape function values and gradients
74        using TrialFE  = typename LocalFunctionSpaceU::Traits::FiniteElementType;
75        using TestFE   = typename LocalFunctionSpaceV::Traits::FiniteElementType;
76        using LocalBasisU = typename TrialFE::Traits::LocalBasisType;
77        using LocalBasisV = typename TestFE::Traits::LocalBasisType;
78        using RangeU = typename LocalBasisU::Traits::RangeType;
79        using RangeV = typename LocalBasisV::Traits::RangeType;
80        using GradientU = typename LocalBasisU::Traits::JacobianType;
81        using GradientV = typename LocalBasisV::Traits::JacobianType;
82    // { alpha_volume_types_end }
83
84    // { alpha_volume_container_setup }
85        std::vector<RangeU> phi(localFunctionSpaceU.size());
86        std::vector<RangeV> theta(localFunctionSpaceV.size());
87
88        std::vector<GradientU> localGradPhi(localFunctionSpaceU.size());
89        std::vector<GradientV> localGradTheta(localFunctionSpaceV.size());
90
91        std::vector<GradientU> gradPhi(localFunctionSpaceU.size());
92        std::vector<GradientV> gradTheta(localFunctionSpaceV.size());
93    // { alpha_volume_setup_end }
94
```

```
95          // Loop over quadrature points
96   // { alpha_volume_quad_begin }
97        int intOrder
98            = (localFunctionSpaceU.finiteElement().localBasis().order()-1) * (p_-1)
99            + (localFunctionSpaceV.finiteElement().localBasis().order()-1);
100       const auto& quadRule
101           = QuadratureRules<double,dim>::rule(entityGeometry.entity().type(),
102                                               intOrder);
103
104       for (const auto& quadPoint : quadRule)
105       {
106   // { alpha_volume_quad_setup_end }
107
108   // { evaluate_shape_functions_begin }
109         // Evaluate basis functions on reference element
110         localFunctionSpaceU.finiteElement().localBasis()
111                   .evaluateFunction(quadPoint.position(),phi);
112         localFunctionSpaceV.finiteElement().localBasis()
113                   .evaluateFunction(quadPoint.position(),theta);
114
115         // Evaluate gradient of basis functions on reference element
116         localFunctionSpaceU.finiteElement().localBasis()
117                   .evaluateJacobian(quadPoint.position(),localGradPhi);
118         localFunctionSpaceV.finiteElement().localBasis()
119                   .evaluateJacobian(quadPoint.position(),localGradTheta);
120
121         // Transform gradients from reference element to grid element
122         const auto jacInvTransp
123             = entityGeometry.geometry()
124                   .jacobianInverseTransposed(quadPoint.position());
125
126         for (std::size_t i=0; i<localFunctionSpaceU.size(); i++)
127           jacInvTransp.mv(localGradPhi[i][0], gradPhi[i][0]);
128
129         for (std::size_t i=0; i<localFunctionSpaceV.size(); i++)
130           jacInvTransp.mv(localGradTheta[i][0], gradTheta[i][0]);
131   // { evaluate_shape_functions_end }
132
133   // { alpha_volume_compute_u_uh_begin }
134         // Compute u at integration point
135         RangeU u = 0;
136         for (std::size_t i=0; i<localFunctionSpaceU.size(); ++i)
137           u += x(localFunctionSpaceU,i)*phi[i];
138
139         // Compute gradient of u
140         GradientU gradU = 0;
141         for (std::size_t i=0; i<localFunctionSpaceU.size(); ++i)
142           gradU.axpy(x(localFunctionSpaceU,i),gradPhi[i]);
143   // { alpha_volume_compute_u_uh_end }
144
145   // { compute_alpha_volume_term_begin }
146         auto globalPos = entityGeometry.geometry().global(quadPoint.position());
147
148         // Integrate |∇u_h|^(p-2) ⟨∇u_h, ∇θ_i⟩ + cu_h θ_i − fθ_i
149         auto factor = quadPoint.weight()
150                     * entityGeometry.geometry()
151                             .integrationElement(quadPoint.position());
152
153         for (std::size_t i=0; i<localFunctionSpaceV.size(); ++i)
154         {
155           auto value = (std::pow(gradU[0].two_norm2(), 0.5*(p_-2))
156                                 * (gradU[0]*gradTheta[i][0])
157                 + c_(globalPos)*u*theta[i]
158                 - f_(globalPos)*theta[i]) * factor;
159           residual.accumulate(localFunctionSpaceV, i, value);
160         }
161   // { compute_alpha_volume_term_end }
162       }
163     }
164
165     // Jacobian of volume term
166   // { jacobian_volume_begin }
167     template<class EntityGeometry,
168              class LocalFunctionSpaceU, class Vector,
169              class LocalFunctionSpaceV, class Jacobian>
170     void jacobian_volume(const EntityGeometry& entityGeometry,
171                          const LocalFunctionSpaceU& localFunctionSpaceU,
172                          const Vector& x,
173                          const LocalFunctionSpaceV& localFunctionSpaceV,
174                          Jacobian& jacobian) const
175     {
176   // { jacobian_volume_signature_end }
177       using TrialFE = typename LocalFunctionSpaceU::Traits::FiniteElementType;
178       using TestFE  = typename LocalFunctionSpaceU::Traits::FiniteElementType;
179       using LocalBasisU = typename TrialFE::Traits::LocalBasisType;
180       using LocalBasisV = typename TestFE::Traits::LocalBasisType;
```

```
181    using RangeU = typename LocalBasisU::Traits::RangeType;
182    using RangeV = typename LocalBasisV::Traits::RangeType;
183    using GradientU = typename LocalBasisU::Traits::JacobianType;
184    using GradientV = typename LocalBasisV::Traits::JacobianType;
185
186    std::vector<RangeU> phi(localFunctionSpaceU.size());
187    std::vector<RangeV> theta(localFunctionSpaceV.size());
188
189    std::vector<GradientU> localGradPhi(localFunctionSpaceU.size());
190    std::vector<GradientV> localGradTheta(localFunctionSpaceV.size());
191
192    std::vector<GradientU> gradPhi(localFunctionSpaceU.size());
193    std::vector<GradientV> gradTheta(localFunctionSpaceV.size());
194  // { jacobian_volume_setup_end }
195
196    // Loop over quadrature points
197  // { jacobian_volume_quad_loop_begin }
198    int intOrder
199      = (localFunctionSpaceU.finiteElement().localBasis().order()-1)
200      * (p_-1)
201      * (localFunctionSpaceV.finiteElement().localBasis().order()-1);
202    const auto& quadRule
203      = QuadratureRules<double,dim>::rule(entityGeometry.entity().type(),
204                                          intOrder);
205
206    for (const auto& quadPoint : quadRule)
207    {
208      // Evaluate shape functions on reference element
209      localFunctionSpaceU.finiteElement().localBasis()
210                          .evaluateFunction(quadPoint.position(),phi);
211      localFunctionSpaceV.finiteElement().localBasis()
212                          .evaluateFunction(quadPoint.position(),theta);
213
214      // Evaluate gradients of shape functions on reference element
215      localFunctionSpaceU.finiteElement().localBasis()
216                          .evaluateJacobian(quadPoint.position(),localGradPhi);
217      localFunctionSpaceV.finiteElement().localBasis()
218                          .evaluateJacobian(quadPoint.position(),localGradTheta);
219
220      // Transform gradients from reference element to grid element
221      const auto jacInvTransp
222        = entityGeometry.geometry()
223                          .jacobianInverseTransposed(quadPoint.position());
224
225      for (std::size_t i=0; i<localFunctionSpaceU.size(); i++)
226        jacInvTransp.mv(localGradPhi[i][0],gradPhi[i][0]);
227
228      for (std::size_t i=0; i<localFunctionSpaceU.size(); i++)
229        jacInvTransp.mv(localGradTheta[i][0],gradTheta[i][0]);
230  // { jacobian_volume_evaluate_shape_functions_end }
231
232  // { jacobian_volume_u_uh_begin }
233      // Compute gradient of u
234      GradientU gradU(0.0);
235      for (std::size_t i=0; i<localFunctionSpaceU.size(); ++i)
236        gradU.axpy(x(localFunctionSpaceU,i),gradPhi[i]);
237  // { jacobian_volume_u_uh_end }
238
239  // { compute_jacobian_volume_term_begin }
240      auto factor = quadPoint.weight()
241                  * entityGeometry.geometry()
242                                  .integrationElement(quadPoint.position());
243      auto globalPos = entityGeometry.geometry().global(quadPoint.position());
244
245      // Integrate
246      for (std::size_t i=0; i<localFunctionSpaceV.size(); i++)
247        for (std::size_t j=0; j<localFunctionSpaceU.size(); j++)
248        {
249          auto value = (p_-2) * std::pow(gradU[0].two_norm2(), 0.5*p_-2)
250                     * (gradU[0] * gradPhi[j][0]) * (gradU[0] * gradTheta[i][0]);
251          value += std::pow(gradU[0].two_norm2(), 0.5*p_-1)
252                             * (gradTheta[i][0] * gradPhi[j][0]);
253          value += c_(globalPos) * theta[i] * phi[j];
254          jacobian.accumulate(localFunctionSpaceV, i,
255                              localFunctionSpaceU, j,
256                              value*factor);
257        }
258  // { compute_jacobian_volume_term_end }
259    }
260  }
261 };
262
263
264  // { main_begin }
265  int main(int argc, char *argv[])
266  {
267    // Initialize MPI
```

```
268    MPIHelper::instance(argc,argv);
269
270    constexpr int dim = 2;
271    using Grid = YaspGrid<dim>;
272    FieldVector<double,dim> upper = {1.0, 1.0};
273    std::array<int,dim> elements = {8, 8};
274    Grid grid(upper,elements);
275
276    using GridView = Grid::LeafGridView;
277    GridView gridView = grid.leafGridView();
278 // { grid_setup_end }
279
280    // Make grid function space
281 // { make_grid_function_space_begin }
282    using Basis = Functions::LagrangeBasis<GridView,2>;
283    auto basis = std::make_shared<Basis>(gridView);
284
285    using VectorBackend = PDELab::ISTL::VectorBackend<>;
286
287    using GridFunctionSpace
288        = PDELab::Experimental::GridFunctionSpace<Basis,
289                                                  VectorBackend,
290                                                  PDELab::NoConstraints>;
291
292    GridFunctionSpace gridFunctionSpace(basis);
293 // { make_grid_function_space_end }
294
295 // { make_coefficient_functions_begin }
296    auto f = [](const FieldVector<double,dim>& x)
297    {
298      FieldVector<double,dim> center(0.5);
299      auto distanceToCenter = (x-center).two_norm();
300      return (distanceToCenter <= 0.25) ? 100 : 0;
301    };
302
303    auto c = [](const FieldVector<double,dim>& x)
304    {
305      return 100.0;
306    };
307
308    double p = 5.0;
309 // { make_coefficient_functions_end }
310
311    // Make grid operator
312 // { make_local_operator_begin }
313    using LocalOperator = PLaplaceLocalOperator<dim>;
314    LocalOperator localOperator(p,f,c);
315 // { make_local_operator_end }
316
317 // { make_grid_operator_begin }
318    using MatrixBackend = PDELab::ISTL::BCRSMatrixBackend<>;
319    MatrixBackend matrixBackend(25);  // 25: Expected average number of entries
320                                      // per matrix row
321
322    using GridOperator
323        = PDELab::GridOperator<GridFunctionSpace,    // Trial function space
324                               GridFunctionSpace,    // Test function space
325                               LocalOperator,        // Element assembler
326                               MatrixBackend,        // Data structure
327                                                     // for the stiffness matrix
328                               double,               // Number type for
329                                                     // solution vector entries
330                               double,               // Number type for
331                                                     // residual vector entries
332                               double>;              // Number type for
333                                                     // stiffness matrix entries
334
335    GridOperator gridOperator(gridFunctionSpace,
336                              gridFunctionSpace,
337                              localOperator,
338                              matrixBackend);
339 // { make_grid_operator_end }
340
341    // Select a linear solver backend
342 // { solver_preparations_begin }
343    using LinearSolverBackend = PDELab::ISTLBackend_SEQ_CG_SSOR;
344    LinearSolverBackend linearSolverBackend(5000,  // Max. number of iterations
345                                            2);    // Verbosity level
346
347    // Make vector of coefficients
348    using U = PDELab::Backend::Vector<GridFunctionSpace,double>;
349    U u(gridFunctionSpace,0.0);   // Initial iterate
350 // { solver_preparations_end }
351
352    // Solve nonlinear problem
353 // { newton_begin }
354    PDELab::NewtonMethod<GridOperator,LinearSolverBackend>
```

```
355                                    newtonMethod(gridOperator,
356                                                 linearSolverBackend);
357
358     newtonMethod.setReduction(1e-10);           // Requested reduction
359                                                 // of the residual
360     newtonMethod.setMinLinearReduction(1e-4);   // Minimal reduction of the
361                                                 // inner linear solver
362     newtonMethod.setVerbosityLevel(2);          // Solver verbosity
363     newtonMethod.apply(u);
364     // { newton_end }
365
366     // Write VTK output file
367     // { vtk_writing_begin }
368     //  Make a discrete function from the finite element basis
369     // and the coefficient vector
370     auto uFunction
371         = Functions::makeDiscreteGlobalBasisFunction<double>(
372                                             *basis,
373                                             PDELab::Backend::native(u));
374
375     // We need to subsample, because the dune-grid VTKWriter
376     // cannot natively display second-order functions.
377     SubsamplingVTKWriter<GridView> vtkwriter(gridView, refinementLevels(2));
378     vtkwriter.addVertexData(uFunction,
379                             VTK::FieldInfo("u",
380                                            VTK::FieldInfo::Type::scalar,
381                                            1));
382     vtkwriter.write("pdelab-p-laplace-result");
383     // { vtk_writing_end }
384   }
```

B.13. The Linear Reaction-Diffusion Equation Using a DG Method

File: pdelab-dg-diffusion.cc
Equation: Linear reaction–diffusion equation
Discretization: Symmetric Interior Penalty Galerkin, with second-order discontinuous Lagrange finite elements
Grid: Nonconforming quadrilateral grid, implemented with UGGrid
Solver: CG solver with ILU preconditioner
Distributed: no
Discussed in: Chapter 11.2.6

```
1    #include "config.h"
2
3    #include <vector>
4
5    #include <dune/common/parallel/mpihelper.hh>
6    #include <dune/common/fvector.hh>
7
8    #include <dune/grid/uggrid.hh>
9    #include <dune/grid/utility/structuredgridfactory.hh>
10   #include <dune/grid/io/file/vtk/subsamplingvtkwriter.hh>
11
12   #include <dune/functions/functionspacebases/lagrangedgbasis.hh>
13   #include <dune/functions/gridfunctions/discreteglobalbasisfunction.hh>
14
15   #include <dune/pdelab.hh>
16
17   using namespace Dune;
18
19   // A local operator for solving the reaction-diffusion equation with SIPG
20   // { local_operator_begin }
21   template<class GridView>
22   class ReactionDiffusionDG
23   : public PDELab::LocalOperatorDefaultFlags,
24     public Dune::PDELab::FullVolumePattern,
```

```
25     public Dune::PDELab::FullSkeletonPattern,
26     public Dune::PDELab::FullBoundaryPattern
27  {
28  // { local_operator_signature_end }
29
30  // { diameter_begin }
31    template<class Geometry>
32    static auto diameter (const Geometry& geometry)
33    {
34      typename Geometry::ctype h = 0.0;
35      for (int i=0; i<geometry.corners(); i++)
36        for (int j=i+1; j<geometry.corners(); j++)
37          h = std::max(h, (geometry.corner(j)-geometry.corner(i)).two_norm());
38
39      return h;
40    }
41  // { diameter_end }
42
43  // { internal_members_begin }
44    double kappa_;
45
46    static constexpr int dim = GridView::dimension;
47
48    std::function<double(const FieldVector<double,dim>&)> c_;
49    std::function<double(const FieldVector<double,dim>&)> f_;
50    std::function<double(const FieldVector<double,dim>&)> neumann_;
51  // { internal_members_end }
52
53  public:
54  // { flags_begin }
55    // Residual assembly flags
56    enum { doAlphaVolume   = true };
57    enum { doAlphaSkeleton = true };
58    enum { doAlphaBoundary = true };
59
60    // Pattern assembly flags
61    enum { doPatternVolume = true };
62    enum { doPatternSkeleton = true };
63    enum { doPatternBoundary = true };
64  // { flags_end }
65
66  // { constructor_begin }
67    ReactionDiffusionDG(std::function<double(const FieldVector<double,dim>&)> c,
68                        std::function<double(const FieldVector<double,dim>&)> f,
69                        std::function<double(const FieldVector<double,dim>&)>
70                                                                     neumann,
71                        double kappa)
72    : kappa_(kappa),
73      c_(c),
74      f_(f),
75      neumann_(neumann)
76    {}
77  // { constructor_end }
78
79    // Volume integral depending on trial and test functions
80    template<class EntityGeometry,
81            class LocalFunctionSpaceU, class Vector,
82            class LocalFunctionSpaceV, class Residual>
83    void alpha_volume(const EntityGeometry& entityGeometry,
84                      const LocalFunctionSpaceU& localFunctionSpaceU,
85                      const Vector& x,
86                      const LocalFunctionSpaceV& localFunctionSpaceV,
87                      Residual& residual) const
88    {
89      using TrialFE  = typename LocalFunctionSpaceU::Traits::FiniteElementType;
90      using TestFE   = typename LocalFunctionSpaceV::Traits::FiniteElementType;
91      using RangeU   = typename TrialFE::Traits::LocalBasisType::Traits::RangeType;
92      using RangeV   = typename TestFE::Traits::LocalBasisType::Traits::RangeType;
93      using GradientU = typename TrialFE::Traits::LocalBasisType::Traits::JacobianType;
94      using GradientV = typename TestFE::Traits::LocalBasisType::Traits::JacobianType;
95      using size_type = typename LocalFunctionSpaceU::Traits::SizeType;
96
97      // Containers for shape function values and gradients
98      std::vector<RangeU> phi(localFunctionSpaceU.size());
99      std::vector<RangeV> theta(localFunctionSpaceV.size());
100
101      std::vector<GradientU> localGradPhi(localFunctionSpaceU.size());
102      std::vector<GradientV> localGradTheta(localFunctionSpaceV.size());
103
104      std::vector<GradientU> gradPhi(localFunctionSpaceU.size());
105      std::vector<GradientV> gradTheta(localFunctionSpaceV.size());
106
107      // Quadrature loop
108      const int quadOrder = 2 * localFunctionSpaceU.finiteElement().localBasis().order();
109      constexpr auto entityDim = EntityGeometry::Entity::mydimension;
110      const auto& quadRule
111        = QuadratureRules<double,entityDim>::rule(entityGeometry.entity().type(), quadOrder);
```

```
112
113    for (const auto& quadPoint : quadRule)
114    {
115      // Evaluate shape functions
116      localFunctionSpaceU.finiteElement().localBasis().evaluateFunction(quadPoint.position(),phi);
117      localFunctionSpaceV.finiteElement().localBasis().evaluateFunction(quadPoint.position(),theta);
118
119      // Compute value of u
120      RangeU u = 0.0;
121      for (size_type i=0; i<localFunctionSpaceU.size(); i++)
122        u += x(localFunctionSpaceU,i)*phi[i];  ·
123
124      // Evaluate gradients of shape functions
125      localFunctionSpaceU.finiteElement().localBasis().evaluateJacobian(quadPoint.position(), localGradPhi);
126      localFunctionSpaceV.finiteElement().localBasis().evaluateJacobian(quadPoint.position(), localGradTheta);
127
128      // Transform gradients of shape functions to element coordinates
129      auto jacInvTransp = entityGeometry.geometry().jacobianInverseTransposed(quadPoint.position());
130      for (size_type i=0; i<localFunctionSpaceU.size(); i++)
131        jacInvTransp.mv(localGradPhi[i][0],gradPhi[i][0]);
132
133      for (size_type i=0; i<localFunctionSpaceV.size(); i++)
134        jacInvTransp.mv(localGradTheta[i][0],gradTheta[i][0]);
135
136      // Compute gradient of u
137      GradientU gradU = 0.0;
138      for (size_type i=0; i<localFunctionSpaceU.size(); i++)
139        gradU.axpy(x(localFunctionSpaceU,i),gradPhi[i]);
140
141      // Evaluate reaction and volume terms
142      auto globalQuadPosition = entityGeometry.geometry().global(quadPoint.position());
143      auto c = c_(globalQuadPosition);
144      auto f = f_(globalQuadPosition);
145
146      // Integrate
147      auto factor = quadPoint.weight() * entityGeometry.geometry().integrationElement(quadPoint.position());
148      for (size_type i=0; i<localFunctionSpaceV.size(); i++)
149        residual.accumulate(localFunctionSpaceV,i,(gradU[0]*gradTheta[i][0] + c*u*theta[i] - f*theta[i]) * factor);
150    }
151  }
152
153  // Jacobian of volume contribution of the residual
154  template<class EntityGeometry,
155           class LocalFunctionSpaceU, class Vector,
156           class LocalFunctionSpaceV, class Jacobian>
157  void jacobian_volume(const EntityGeometry& entityGeometry,
158                       const LocalFunctionSpaceU& localFunctionSpaceU,
159                       const Vector& x,
160                       const LocalFunctionSpaceV& localFunctionSpaceV,
161                       Jacobian& jacobian) const
162  {
163    using TrialFE   = typename LocalFunctionSpaceU::Traits::FiniteElementType;
164    using TestFE    = typename LocalFunctionSpaceV::Traits::FiniteElementType;
165    using RangeU    = typename TrialFE::Traits::LocalBasisType::Traits::RangeType;
166    using RangeV    = typename TestFE::Traits::LocalBasisType::Traits::RangeType;
167    using GradientU = typename TrialFE::Traits::LocalBasisType::Traits::JacobianType;
168    using GradientV = typename TestFE::Traits::LocalBasisType::Traits::JacobianType;
169    using size_type = typename LocalFunctionSpaceU::Traits::SizeType;
170
171    // Containers for shape function values and gradients
172    std::vector<RangeU> phi(localFunctionSpaceU.size());
173    std::vector<RangeV> theta(localFunctionSpaceV.size());
174
175    std::vector<GradientU> localGradPhi(localFunctionSpaceU.size());
176    std::vector<GradientV> localGradTheta(localFunctionSpaceV.size());
177
178    std::vector<GradientU> gradPhi(localFunctionSpaceU.size());
179    std::vector<GradientV> gradTheta(localFunctionSpaceV.size());
180
181    // Loop over quadrature points
182    const int quadOrder = 2 * localFunctionSpaceU.finiteElement().localBasis().order();
183    constexpr auto entityDim = EntityGeometry::Entity::mydimension;
184    const auto& quadRule
185        = QuadratureRules<double,entityDim>::rule(entityGeometry.geometry().type(), quadOrder);
186
187    for (const auto& quadPoint : quadRule)
188    {
189      // Evaluate shape functions
190      localFunctionSpaceU.finiteElement().localBasis().evaluateFunction(quadPoint.position(),phi);
191      localFunctionSpaceV.finiteElement().localBasis().evaluateFunction(quadPoint.position(),theta);
192
193      // Evaluate gradients of shape functions
194      localFunctionSpaceU.finiteElement().localBasis().evaluateJacobian(quadPoint.position(),localGradPhi);
195      localFunctionSpaceV.finiteElement().localBasis().evaluateJacobian(quadPoint.position(),localGradTheta);
196
197      // Transform gradients of shape functions to real element
198      auto jacInvTransp = entityGeometry.geometry().jacobianInverseTransposed(quadPoint.position());
```

```
199          for (size_type i=0; i<localFunctionSpaceU.size(); i++)
200            jacInvTransp.mv(localGradPhi[i][0],gradPhi[i][0]);
201
202          for (size_type i=0; i<localFunctionSpaceV.size(); i++)
203            jacInvTransp.mv(localGradTheta[i][0],gradTheta[i][0]);
204
205          // Evaluate reaction term
206          auto c = c_(entityGeometry.geometry().global(quadPoint.position()));
207
208          // Integrate
209          auto factor = quadPoint.weight() * entityGeometry.geometry().integrationElement(quadPoint.position());
210          for (size_type i=0; i<localFunctionSpaceV.size(); i++)
211            for (size_type j=0; j<localFunctionSpaceU.size(); j++)
212              jacobian.accumulate(localFunctionSpaceV,
213                                  i,
214                                  localFunctionSpaceU,
215                                  j,
216                                  (gradPhi[j][0]*gradTheta[i][0] + c*phi[j]*theta[i]) * factor);
217        }
218      }
219
220      // Skeleton integral for a given configuration u_h
221      // Each intersection is only visited once
222      // { alpha_skeleton_interface_begin }
223      template<class IntersectionGeometry,
224               class LocalFunctionSpaceU, class Vector,
225               class LocalFunctionSpaceV, class Residual>
226      void alpha_skeleton(const IntersectionGeometry& intersectionGeometry,
227                          const LocalFunctionSpaceU& localFunctionSpaceUIn,
228                          const Vector& xIn,
229                          const LocalFunctionSpaceV& localFunctionSpaceVIn,
230                          const LocalFunctionSpaceU& localFunctionSpaceUOut,
231                          const Vector& xOut,
232                          const LocalFunctionSpaceV& localFunctionSpaceVOut,
233                          Residual& residualIn, Residual& residualOut) const
234      {
235      // { alpha_skeleton_interface_end }
236        using TrialFE = typename LocalFunctionSpaceU::Traits::FiniteElementType;
237        using TestFE = typename LocalFunctionSpaceV::Traits::FiniteElementType;
238        using LocalBasisU = typename TrialFE::Traits::LocalBasisType;
239        using LocalBasisV = typename TestFE::Traits::LocalBasisType;
240        using RangeU = typename LocalBasisU::Traits::RangeType;
241        using RangeV = typename LocalBasisV::Traits::RangeType;
242        using GradientU = typename LocalBasisU::Traits::JacobianType;
243        using GradientV = typename LocalBasisV::Traits::JacobianType;
244        using size_type = typename LocalFunctionSpaceV::Traits::SizeType;
245
246        // References to inside and outside elements
247        const auto& elementInside = intersectionGeometry.inside();
248        const auto& elementOutside = intersectionGeometry.outside();
249
250        // Intersection and element geometries
251        auto geo = intersectionGeometry.geometry();
252        auto geoIn  = elementInside.geometry();
253        auto geoOut = elementOutside.geometry();
254
255        // Geometries of intersection in local coordinates
256        // of elementInside and elementOutside
257        auto geoInInside  = intersectionGeometry.geometryInInside();
258        auto geoInOutside = intersectionGeometry.geometryInOutside();
259      // { alpha_skeleton_variable_setup_end }
260
261        // Intersection diameter
262      // { compute_diameter_begin }
263        auto h = diameter(geo);
264      // { compute_diameter_end }
265
266      // { alpha_skeleton_container_setup_begin }
267        // Shape function values
268        std::vector<RangeU> phiIn(localFunctionSpaceUIn.size());
269        std::vector<RangeV> thetaIn(localFunctionSpaceVIn.size());
270        std::vector<RangeU> phiOut(localFunctionSpaceUOut.size());
271        std::vector<RangeV> thetaOut(localFunctionSpaceVOut.size());
272
273        // Shape function gradients on the reference element
274        std::vector<GradientU> localGradPhiIn(localFunctionSpaceUIn.size());
275        std::vector<GradientV> localGradThetaIn(localFunctionSpaceVIn.size());
276        std::vector<GradientU> localGradPhiOut(localFunctionSpaceUOut.size());
277        std::vector<GradientV> localGradThetaOut(localFunctionSpaceVOut.size());
278
279        // Shape function gradients on the grid element
280        std::vector<GradientU> gradPhiIn(localFunctionSpaceUIn.size());
281        std::vector<GradientV> gradThetaIn(localFunctionSpaceVIn.size());
282        std::vector<GradientU> gradPhiOut(localFunctionSpaceUOut.size());
283        std::vector<GradientV> gradThetaOut(localFunctionSpaceVOut.size());
284      // { alpha_skeleton_container_setup_end }
285
```

```
286        // Loop over quadrature points
287   // { alpha_skeleton_quadrature_begin }
288        const auto quadOrder
289            = 2 * localFunctionSpaceUIn.finiteElement().localBasis().order();
290        constexpr auto intersectionDim = IntersectionGeometry::mydimension;
291        const auto& quadRule
292            = QuadratureRules<double,intersectionDim>::
293                             rule(intersectionGeometry.geometry().type(),
294                                  quadOrder);
295
296        for (const auto& quadPoint : quadRule)
297        {
298   // { alpha_skeleton_quadrature_end }
299        // Quadrature point position in local coordinates of adjacent elements
300        auto quadPointLocalIn  = geoInInside.global(quadPoint.position());
301        auto quadPointLocalOut = geoInOutside.global(quadPoint.position());
302
303        // Evaluate shape functions
304        localFunctionSpaceUIn.finiteElement().localBasis()
305                     .evaluateFunction(quadPointLocalIn,phiIn);
306        localFunctionSpaceUOut.finiteElement().localBasis()
307                      .evaluateFunction(quadPointLocalOut,phiOut);
308        localFunctionSpaceVIn.finiteElement().localBasis()
309                     .evaluateFunction(quadPointLocalIn,thetaIn);
310        localFunctionSpaceVOut.finiteElement().localBasis()
311                      .evaluateFunction(quadPointLocalOut,thetaOut);
312
313        // Evaluate gradients of shape functions
314        localFunctionSpaceUIn.finiteElement().localBasis()
315                     .evaluateJacobian(quadPointLocalIn,
316                                       localGradPhiIn);
317        localFunctionSpaceUOut.finiteElement().localBasis()
318                      .evaluateJacobian(quadPointLocalOut,
319                                        localGradPhiOut);
320        localFunctionSpaceVIn.finiteElement().localBasis()
321                     .evaluateJacobian(quadPointLocalIn,
322                                       localGradThetaIn);
323        localFunctionSpaceVOut.finiteElement().localBasis()
324                      .evaluateJacobian(quadPointLocalOut,
325                                        localGradThetaOut);
326   // { alpha_skeleton_shape_functions_end }
327
328        // Transform gradients of shape functions to grid element coordinates
329   // { alpha_skeleton_gradient_transform_begin }
330        auto jacInvTranspIn
331            = geoIn.jacobianInverseTransposed(quadPointLocalIn);
332        for (size_type i=0; i<localFunctionSpaceUIn.size(); i++)
333          jacInvTranspIn.mv(localGradPhiIn[i][0], gradPhiIn[i][0]);
334        for (size_type i=0; i<localFunctionSpaceVIn.size(); i++)
335          jacInvTranspIn.mv(localGradThetaIn[i][0], gradThetaIn[i][0]);
336
337        auto jacInvTranspOut
338            = geoOut.jacobianInverseTransposed(quadPointLocalOut);
339        for (size_type i=0; i<localFunctionSpaceUOut.size(); i++)
340          jacInvTranspOut.mv(localGradPhiOut[i][0], gradPhiOut[i][0]);
341        for (size_type i=0; i<localFunctionSpaceVOut.size(); i++)
342          jacInvTranspOut.mv(localGradThetaOut[i][0], gradThetaOut[i][0]);
343   // { alpha_skeleton_gradient_transform_end }
344
345   // { alpha_skeleton_u_u_h_begin }
346        // Compute values of u_h
347        RangeU uIn(0.0);
348        for (size_type i=0; i<localFunctionSpaceUIn.size(); i++)
349          uIn += xIn(localFunctionSpaceUIn,i)*phiIn[i];
350        RangeU uOut(0.0);
351        for (size_type i=0; i<localFunctionSpaceUOut.size(); i++)
352          uOut += xOut(localFunctionSpaceUOut,i)*phiOut[i];
353
354        // Compute gradients of u_h
355        GradientU gradUIn(0.0);
356        for (size_type i=0; i<localFunctionSpaceUIn.size(); i++)
357          gradUIn.axpy(xIn(localFunctionSpaceUIn,i), gradPhiIn[i]);
358        GradientU gradUOut(0.0);
359        for (size_type i=0; i<localFunctionSpaceUOut.size(); i++)
360          gradUOut.axpy(xOut(localFunctionSpaceUOut,i), gradPhiOut[i]);
361   // { alpha_skeleton_u_u_h_end }
362
363   // { alpha_skeleton_compute_values_begin }
364        // Unit normal from T_in to T_out
365        auto n_F = intersectionGeometry.unitOuterNormal(quadPoint.position());
366
367        // Integration factor
368        auto factor
369            = quadPoint.weight()*geo.integrationElement(quadPoint.position());
370
371        // Interior penalty term
372        auto interiorPenaltyTerm =  -0.5*(gradUIn[0]*n_F + gradUOut[0]*n_F);
```

```
373          for (size_type i=0; i<localFunctionSpaceVIn.size(); i++)
374            residualIn.accumulate(localFunctionSpaceVIn,
375                                  i,
376                                  interiorPenaltyTerm * thetaIn[i] * factor);
377          for (size_type i=0; i<localFunctionSpaceVOut.size(); i++)
378            residualOut.accumulate(localFunctionSpaceVOut,
379                                   i,
380                                   -interiorPenaltyTerm * thetaOut[i] * factor);
381
382          // Symmetric interior penalty term
383          for (size_type i=0; i<localFunctionSpaceVIn.size(); i++)
384            residualIn.accumulate(localFunctionSpaceVIn,
385                                  i,
386                                  -0.5 * (gradThetaIn[i][0]*n_F)
387                                       * (uIn-uOut) * factor);
388          for (size_type i=0; i<localFunctionSpaceVOut.size(); i++)
389            residualOut.accumulate(localFunctionSpaceVOut,
390                                   i,
391                                   -0.5 * (gradThetaOut[i][0]*n_F)
392                                        * (uIn-uOut) * factor);
393
394          // Coercivity term
395          auto coercivityTerm = (kappa_/h) * (uIn-uOut);
396          for (size_type i=0; i<localFunctionSpaceVIn.size(); i++)
397            residualIn.accumulate(localFunctionSpaceVIn,
398                                  i,
399                                  coercivityTerm * thetaIn[i] * factor);
400          for (size_type i=0; i<localFunctionSpaceVOut.size(); i++)
401            residualOut.accumulate(localFunctionSpaceVOut,
402                                   i,
403                                   -coercivityTerm * thetaOut[i] * factor);
404        }
405      }
406  // { alpha_skeleton_compute_values_end }
407
408  // { jacobian_skeleton_signature_begin }
409    template<class IntersectionGeometry,
410             class LocalFunctionSpaceU, class Vector,
411             class LocalFunctionSpaceV, class Jacobian>
412    void jacobian_skeleton(const IntersectionGeometry& intersectionGeometry,
413                           const LocalFunctionSpaceU& localFunctionSpaceUIn,
414                           const Vector& xIn,
415                           const LocalFunctionSpaceV& localFunctionSpaceVIn,
416                           const LocalFunctionSpaceU& localFunctionSpaceUOut,
417                           const Vector& xOut,
418                           const LocalFunctionSpaceV& localFunctionSpaceVOut,
419                           Jacobian& jacobianInIn, Jacobian& jacobianInOut,
420                           Jacobian& jacobianOutIn, Jacobian& jacobianOutOut)
421                                                                        const
422  // { jacobian_skeleton_signature_end }
423    {
424  // { jacobian_skeleton_setup_begin }
425      // Define types
426      using TrialFE = typename LocalFunctionSpaceU::Traits::FiniteElementType;
427      using TestFE = typename LocalFunctionSpaceV::Traits::FiniteElementType;
428      using LocalBasisU = typename TrialFE::Traits::LocalBasisType;
429      using LocalBasisV = typename TestFE::Traits::LocalBasisType;
430      using RangeU = typename LocalBasisU::Traits::RangeType;
431      using RangeV = typename LocalBasisV::Traits::RangeType;
432      using GradientU = typename LocalBasisU::Traits::JacobianType;
433      using GradientV = typename LocalBasisV::Traits::JacobianType;
434      using size_type = typename LocalFunctionSpaceV::Traits::SizeType;
435
436      // References to inside and outside elements
437      const auto& elementInside = intersectionGeometry.inside();
438      const auto& elementOutside = intersectionGeometry.outside();
439
440      // Get geometries
441      auto geo = intersectionGeometry.geometry();
442      auto geoIn = elementInside.geometry();
443      auto geoOut = elementOutside.geometry();
444
445      // Geometries of intersection in local coordinates
446      // of elementInside and elementOutside
447      auto geoInInside = intersectionGeometry.geometryInInside();
448      auto geoInOutside = intersectionGeometry.geometryInOutside();
449
450      // Intersection diameter
451      auto h = diameter(geo);
452
453      // Shape function values
454      std::vector<RangeU> phiIn(localFunctionSpaceUIn.size());
455      std::vector<RangeU> phiOut(localFunctionSpaceUOut.size());
456      std::vector<RangeV> thetaIn(localFunctionSpaceVIn.size());
457      std::vector<RangeV> thetaOut(localFunctionSpaceVOut.size());
458
459      // Shape function gradients on the reference element
```

```
460        std::vector<GradientU> localGradPhiIn(localFunctionSpaceUIn.size());
461        std::vector<GradientU> localGradPhiOut(localFunctionSpaceUOut.size());
462        std::vector<GradientU> localGradThetaIn(localFunctionSpaceVIn.size());
463        std::vector<GradientV> localGradThetaOut(localFunctionSpaceVOut.size());
464
465        // Shape function gradients on the grid element
466        std::vector<GradientU> gradPhiIn(localFunctionSpaceUIn.size());
467        std::vector<GradientU> gradPhiOut(localFunctionSpaceUOut.size());
468        std::vector<GradientV> gradThetaIn(localFunctionSpaceVIn.size());
469        std::vector<GradientV> gradThetaOut(localFunctionSpaceVOut.size());
470  // { jacobian_skeleton_setup_end }
471
472        // Loop over quadrature points
473  // { jacobian_skeleton_quad_setup_begin }
474        auto quadOrder
475            = 2*localFunctionSpaceUIn.finiteElement().localBasis().order();
476        constexpr auto intersectionDim = IntersectionGeometry::mydimension;
477        const auto& quadRule
478            = QuadratureRules<double,intersectionDim>::
479                              rule(intersectionGeometry.geometry().type(),
480                                   quadOrder);
481
482        for (const auto& quadPoint : quadRule)
483        {
484          // Position of quadrature point in local coordinates
485          // of inside and outside elements
486          auto quadPointLocalIn = geoInInside.global(quadPoint.position());
487          auto quadPointLocalOut = geoInOutside.global(quadPoint.position());
488
489          // Evaluate shape functions
490          localFunctionSpaceUIn.finiteElement().localBasis()
491                          .evaluateFunction(quadPointLocalIn,phiIn);
492          localFunctionSpaceUOut.finiteElement().localBasis()
493                          .evaluateFunction(quadPointLocalOut,phiOut);
494
495          localFunctionSpaceVIn.finiteElement().localBasis()
496                          .evaluateFunction(quadPointLocalIn,thetaIn);
497          localFunctionSpaceVOut.finiteElement().localBasis()
498                          .evaluateFunction(quadPointLocalOut,thetaOut);
499
500          // Evaluate gradients of shape functions
501          localFunctionSpaceUIn.finiteElement().localBasis()
502                          .evaluateJacobian(quadPointLocalIn,
503                                            localGradPhiIn);
504          localFunctionSpaceUOut.finiteElement().localBasis()
505                          .evaluateJacobian(quadPointLocalOut,
506                                            localGradPhiOut);
507
508          localFunctionSpaceVIn.finiteElement().localBasis()
509                          .evaluateJacobian(quadPointLocalIn,
510                                            localGradThetaIn);
511          localFunctionSpaceVOut.finiteElement().localBasis()
512                          .evaluateJacobian(quadPointLocalOut,
513                                            localGradThetaOut);
514
515          // Transform gradients of shape functions to element coordinates
516          auto jacInvTranspIn
517              = geoIn.jacobianInverseTransposed(quadPointLocalIn);
518          for (size_type i=0; i<localFunctionSpaceUIn.size(); i++)
519            jacInvTranspIn.mv(localGradPhiIn[i][0],gradPhiIn[i][0]);
520          for (size_type i=0; i<localFunctionSpaceVIn.size(); i++)
521            jacInvTranspIn.mv(localGradThetaIn[i][0],gradThetaIn[i][0]);
522
523          auto jacInvTranspOut
524              = geoOut.jacobianInverseTransposed(quadPointLocalOut);
525          for (size_type i=0; i<localFunctionSpaceUOut.size(); i++)
526            jacInvTranspOut.mv(localGradPhiOut[i][0],gradPhiOut[i][0]);
527          for (size_type i=0; i<localFunctionSpaceVOut.size(); i++)
528            jacInvTranspOut.mv(localGradThetaOut[i][0],gradThetaOut[i][0]);
529  // { jacobian_skeleton_quad_setup_end }
530
531  // { jacobian_skeleton_compute_matrices_begin }
532          // Unit normal from T_in to T_out
533          auto n_F = intersectionGeometry.unitOuterNormal(quadPoint.position());
534
535          // Integration factor
536          auto factor = quadPoint.weight()
537                      * geo.integrationElement(quadPoint.position());
538
539          // Fill jacobianInIn
540          for (size_type i=0; i<localFunctionSpaceVIn.size(); i++)
541          {
542            for (size_type j=0; j<localFunctionSpaceUIn.size(); j++)
543            {
544              jacobianInIn.accumulate(localFunctionSpaceVIn,
545                                      i,
546                                      localFunctionSpaceUIn,
```

```
547                                         j,
548                                         (-0.5 * (gradPhiIn[j][0]*n_F) * thetaIn[i]
549                                          - 0.5 * phiIn[j] * (gradThetaIn[i][0]*n_F)
550                                          + (kappa_ / h) * phiIn[j] * thetaIn[i] )
551                                                                          * factor);
552           }
553         }
554
555         // Fill jacobianInOut
556         for (size_type i=0; i<localFunctionSpaceVOut.size(); i++)
557         {
558           for (size_type j=0; j<localFunctionSpaceUIn.size(); j++)
559           {
560             jacobianInOut.accumulate(localFunctionSpaceVIn,
561                                      i,
562                                      localFunctionSpaceUOut,
563                                      j,
564                                      (-0.5 * (n_F*gradPhiOut[j][0]) * thetaIn[i]
565                                       + 0.5 * phiOut[j] * (n_F*gradThetaIn[i][0])
566                                       - (kappa_ / h) * phiOut[j] * thetaIn[i])
567                                                                          * factor);
568           }
569         }
570
571         // Fill jacobianOutIn
572         for (size_type i=0; i<localFunctionSpaceVIn.size(); i++)
573         {
574           for (size_type j=0; j<localFunctionSpaceUOut.size(); j++)
575           {
576             jacobianOutIn.accumulate(localFunctionSpaceVOut,
577                                      i,
578                                      localFunctionSpaceUIn,
579                                      j,
580                                      (0.5 * (n_F*gradPhiIn[j][0]) * thetaOut[i]
581                                       - 0.5 * phiIn[j] * (n_F*gradThetaOut[i][0])
582                                       - (kappa_ / h) * phiIn[j] * thetaOut[i])
583                                                                          * factor);
584           }
585         }
586
587         // Fill jacobianOutOut
588         for (size_type i=0; i<localFunctionSpaceVOut.size(); i++)
589         {
590           for (size_type j=0; j<localFunctionSpaceUOut.size(); j++)
591           {
592             jacobianOutOut.accumulate(localFunctionSpaceVOut,
593                                       i,
594                                       localFunctionSpaceUOut,
595                                       j,
596                                       (0.5 * (n_F*gradPhiOut[j][0]) * thetaOut[i]
597                                        + 0.5 * phiOut[j] * (n_F*gradThetaOut[i][0])
598                                        + (kappa_ / h) * phiOut[j] * thetaOut[i])
599                                                                          * factor);
600           }
601         }
602       }
603     }
604 // { jacobian_skeleton_compute_matrices_end }
605
606 // { alpha_boundary_begin }
607     // Boundary integral implementing the Neumann term
608     template<class IntersectionGeometry,
609              class LocalFunctionSpaceU, class Vector,
610              class LocalFunctionSpaceV, class Residual>
611     void alpha_boundary(const IntersectionGeometry& intersectionGeometry,
612                         const LocalFunctionSpaceU& localFunctionSpaceUIn,
613                         const Vector& xIn,
614                         const LocalFunctionSpaceV& localFunctionSpaceVIn,
615                         Residual& residualIn) const
616     {
617       using TestFE = typename LocalFunctionSpaceV::Traits::FiniteElementType;
618       using RangeV = typename TestFE::Traits::LocalBasisType::Traits::RangeType;
619       using size_type = typename LocalFunctionSpaceV::Traits::SizeType;
620
621       // Get geometry of intersection in local coordinates of element_inside
622       auto geoInInside = intersectionGeometry.geometryInInside();
623
624       std::vector<RangeV> thetaIn(localFunctionSpaceVIn.size());
625
626       // Loop over quadrature points
627       const int quadOrder
628           = 2*localFunctionSpaceVIn.finiteElement().localBasis().order();
629       constexpr auto intersectionDim = IntersectionGeometry::mydimension;
630       const auto& quadRule
631           = QuadratureRules<double,intersectionDim>::
632                                   rule(intersectionGeometry.geometry().type(),
633                                        quadOrder);
```

```
634
635        for (const auto& quadPoint : quadRule)
636        {
637          // Position of quadrature point in local coordinates of elements
638          auto quadPointLocalIn = geoInInside.global(quadPoint.position());
639
640          // Evaluate basis functions
641          localFunctionSpaceVIn.finiteElement().localBasis()
642                                 .evaluateFunction(quadPointLocalIn,thetaIn);
643
644          // Integration factor
645          auto factor
646             = quadPoint.weight()
647             * intersectionGeometry.geometry()
648                                  .integrationElement(quadPoint.position());
649
650          // Evaluate Neumann boundary condition
651          auto neumannValue
652             = neumann_(intersectionGeometry.geometry()
653                                  .global(quadPoint.position()));
654
655          // Integrate
656          for (size_type i=0; i<localFunctionSpaceVIn.size(); i++)
657            residualIn.accumulate(localFunctionSpaceVIn,
658                                  i,
659                                  -1 * neumannValue * thetaIn[i] * factor);
660        }
661      }
662    // { alpha_boundary_end }
663
664    // { jacobian_boundary_begin }
665      template<class IntersectionGeometry,
666               class LocalFunctionSpaceU, class Vector,
667               class LocalFunctionSpaceV, typename Jacobian>
668      void jacobian_boundary(const IntersectionGeometry& intersectionGeometry,
669                             const LocalFunctionSpaceU& localFunctionSpaceUIn,
670                             const Vector& xIn,
671                             const LocalFunctionSpaceV& localFunctionSpaceVIn,
672                             Jacobian& jacobianInIn) const
673      {
674        // Does not do anything---the Jacobian of the boundary term is zero.
675      }
676    // { jacobian_boundary_end }
677    };
678
679    // { main_begin }
680    int main(int argc, char *argv[])
681    {
682      // Initialize MPI, if present
683      MPIHelper::instance(argc, argv);
684
685      constexpr int dim = 2;
686      using Grid = UGGrid<dim>;
687      std::shared_ptr<Grid> grid
688         = StructuredGridFactory<Grid>::createCubeGrid({0.0,0.0},
689                                                       {1.0,1.0},
690                                                       { 4,  4});
691
692      // Nonconforming refinement near the domain center
693      grid->setClosureType(Grid::NONE);  // UGGrid-specific:
694                                         // turn off green closure
695
696      using Position = FieldVector<Grid::ctype,dim>;
697      for (const auto& element : elements(grid->leafGridView()))
698        if ( (element.geometry().center()-Position(0.5)).two_norm() < 0.25 )
699          grid->mark(1,element);
700
701      grid->preAdapt();
702      grid->adapt();
703      grid->postAdapt();
704
705      using GridView = Grid::LeafGridView;
706      const GridView gridView = grid->leafGridView();
707    // { grid_setup_end }
708
709    // { gfs_setup_begin }
710      // Make dune-functions basis
711      using DGBasis = Functions::LagrangeDGBasis<GridView,2>;
712      auto dgBasis = std::make_shared<DGBasis>(gridView);
713
714      // Make dune-pdelab grid function space
715      using VectorBackend = PDELab::ISTL::VectorBackend<>;
716      using GridFunctionSpace
717         = PDELab::Experimental::GridFunctionSpace<DGBasis,
718                                                   VectorBackend,
719                                                   PDELab::NoConstraints>;
720
```

```
721      GridFunctionSpace gridFunctionSpace(dgBasis);
722  // { gfs_setup_end }
723
724  // { make_parameter_functions_begin }
725      // Reaction strength
726      auto c = [](const FieldVector<double,dim>& x)
727      {
728        return 10.0;
729      };
730
731      // Source term
732      auto f = [](const FieldVector<double,dim>& x)
733      {
734        const auto center = FieldVector<double,dim>(0.5);
735        auto distanceToCenter = (x - center).two_norm();
736        return (distanceToCenter <= 0.25) ? -10.0 : 10.0;
737      };
738
739      // Neumann boundary data
740      auto neumann = [](const FieldVector<double,dim>& x)
741      {
742        return (x[0] >= 0.999 && x[1] >= 0.5) ? -2 : 0;
743      };
744  // { make_parameter_functions_end }
745
746  // { assembler_setup_begin }
747      // Make local operator
748      using LocalOperator = ReactionDiffusionDG<GridView>;
749      double kappa = 3;
750      LocalOperator localOperator(c,f,neumann,kappa);
751      using MatrixBackend = PDELab::ISTL::BCRSMatrixBackend<>;
752      MatrixBackend matrixBackend(45);    // 45: Expected approximate number
753                                          // of nonzeros per matrix row
754
755      using GridOperator
756          = PDELab::GridOperator<GridFunctionSpace,    // Trial function space
757                                 GridFunctionSpace,    // Test function space
758                                 LocalOperator,        // Element assembler
759                                 MatrixBackend,        // Data structure
760                                                       // for the stiffness matrix
761                                 double,               // Number type for
762                                                       // solution vector entries
763                                 double,               // Number type for
764                                                       // residual vector entries
765                                 double>;              // Number type for
766                                                       // stiffness matrix entries
767
768      GridOperator gridOperator(gridFunctionSpace,
769                                gridFunctionSpace,
770                                localOperator,
771                                matrixBackend);
772  // { assembler_setup_end }
773
774  // { construct_and_run_solver_begin }
775      // Make a vector of degrees of freedom, and initialize it with 0
776      using VectorContainer = typename GridOperator::Traits::Domain;
777      VectorContainer u(gridFunctionSpace,0.0);
778
779      // Make linear solver and solve problem
780      using LinearSolverBackend = PDELab::ISTLBackend_SEQ_CG_ILU0;
781      LinearSolverBackend linearSolverBackend(50,      // Maximum number
782                                                       // of iterations
783                                              2);      // Solver verbosity
784
785      using LinearProblemSolver
786          = PDELab::StationaryLinearProblemSolver<GridOperator,
787                                                  LinearSolverBackend,
788                                                  VectorContainer>;
789
790      LinearProblemSolver linearProblemSolver(gridOperator,
791                                              linearSolverBackend,
792                                              u,
793                                              1e-10);   // Desired relative
794                                                        // residual reduction
795      linearProblemSolver.apply();
796  // { construct_and_run_solver_end }
797
798  // { output_result_begin }
799      // Make a discrete function from the FE basis and the coefficient vector
800      auto uFunction
801          = Functions::makeDiscreteGlobalBasisFunction<double>(
802                                          *dgBasis,
803                                          PDELab::Backend::native(u));
804
805      // We need to subsample, because VTKWriter
806      // cannot natively display second-order functions.
807      SubsamplingVTKWriter<GridView> vtkWriter(gridView,
```

```
808                                  refinementLevels(5));
809    vtkWriter.addVertexData(uFunction,
810                           VTK::FieldInfo("u",
811                                          VTK::FieldInfo::Type::scalar,
812                                          1));
813    vtkWriter.write("pdelab-dg-diffusion-result");
814    // { output_result_end }
815    }
```

B.14. The Poisson Equation with Dirichlet Boundary Conditions, Using `dune-pdelab`

File:	`pdelab-poisson-dirichlet.cc`
Equation:	Poisson equation
Discretization:	First-order Lagrange finite elements
Grid:	Unstructured triangle grid, implemented with `UGGrid`
Solver:	CG solver with SSOR preconditioner
Distributed:	no
Discussed in:	Chapter 11.3.2

```
1    #include "config.h"
2
3    #include <memory>
4
5    #include <dune/grid/io/file/gmshreader.hh>
6    #include <dune/grid/io/file/vtk.hh>
7    #include <dune/grid/uggrid.hh>
8
9    #include <dune/functions/functionspacebases/lagrangebasis.hh>
10   #include <dune/functions/gridfunctions/discreteglobalbasisfunction.hh>
11   #include <dune/functions/functionspacebases/interpolate.hh>
12
13   #include <dune/pdelab.hh>
14
15
16   using namespace Dune;
17
18   // { parameterclass_begin }
19   template<class GridView, class Range>
20   class PoissonProblem
21   : public PDELab::ConvectionDiffusionModelProblem<GridView,Range>
22   {
23   public:
24     using Traits
25       = typename PDELab::ConvectionDiffusionModelProblem<GridView,Range>::
26                                                          Traits;
27     // Source term
28     auto f(const typename Traits::ElementType& element,
29          const typename Traits::DomainType& xi) const
30     {
31       return -5.0;
32     }
33
34     //! Boundary condition type function
35     auto bctype (const typename Traits::IntersectionType& intersection,
36              const typename Traits::IntersectionDomainType& xi) const
37     {
38       auto x = intersection.geometry().global(xi);
39       return ( x[0]< 1e-8 || x[1] < 1e-8 || (x[0] > 0.4999 && x[1] > 0.4999) )
40         ? PDELab::ConvectionDiffusionBoundaryConditions::Dirichlet
41         : PDELab::ConvectionDiffusionBoundaryConditions::Neumann;
42     }
43   };
44   // { parameterclass_end }
45
46   // { main_begin }
47   int main(int argc, char *argv[])
48   {
```

```
49     // Initialize MPI, if available
50     MPIHelper::instance(argc, argv);
51
52     constexpr int dim = 2;
53     using Grid = UGGrid<dim>;
54     std::shared_ptr<Grid> grid = GmshReader<Grid>::read("l-shape.msh");
55
56     grid->globalRefine(2);
57
58     using GridView = Grid::LeafGridView;
59     GridView gridView = grid->leafGridView();
60   // { grid_setup_end }
61
62   // { make_function_space_begin }
63     using Basis = Functions::LagrangeBasis<GridView,1>;
64     auto basis = std::make_shared<Basis>(gridView);
65
66     using VectorBackend = PDELab::ISTL::VectorBackend<>;
67     using Constraints = PDELab::ConformingDirichletConstraints;
68
69     using GridFunctionSpace
70         = PDELab::Experimental::GridFunctionSpace<Basis,
71                                                   VectorBackend,
72                                                   Constraints>;
73
74     GridFunctionSpace gridFunctionSpace(basis);
75   // { make_function_space_end }
76
77     // Assemble constraints on this space
78   // { make_bc_adapter_begin }
79     using Problem = PoissonProblem<GridView,double>;
80     Problem problem;
81     PDELab::ConvectionDiffusionBoundaryConditionAdapter<Problem> bctype(problem);
82   // { make_bc_adapter_end }
83
84   // { make_constraints_begin }
85     using ConstraintsContainer
86         = GridFunctionSpace::ConstraintsContainer<double>::Type;
87     ConstraintsContainer constraintsContainer;
88     PDELab::constraints(bctype, gridFunctionSpace, constraintsContainer);
89   // { make_constraints_end }
90
91     // Make grid operator
92   // { make_assembler_begin }
93     using LocalOperator
94         = PDELab::ConvectionDiffusionFEM<Problem,
95                                          GridFunctionSpace::Traits::
96                                                        FiniteElementMap>;
97     LocalOperator localOperator(problem);
98
99     using MatrixBackend = PDELab::ISTL::BCRSMatrixBackend<>;
100    MatrixBackend matrixBackend(7);
101
102    using GridOperator
103        = PDELab::GridOperator<GridFunctionSpace,
104                               GridFunctionSpace,
105                               LocalOperator,
106                               MatrixBackend,
107                               double,double,double,
108                               ConstraintsContainer,    // For trial space
109                               ConstraintsContainer>;   // For test space
110
111    GridOperator gridOperator(gridFunctionSpace,
112                              constraintsContainer,    // Trial space
113                              gridFunctionSpace,
114                              constraintsContainer,    // Test space
115                              localOperator,
116                              matrixBackend);
117  // { make_assembler_end }
118
119    // Make solution vector
120  // { make_solution_vector_begin }
121    using U = PDELab::Backend::Vector<GridFunctionSpace,double>;
122    U u(gridFunctionSpace,0.0);
123    auto g = [](auto x){return  (x[0]< 1e-8 || x[1] < 1e-8) ? 0 : 0.5;};
124    Functions::interpolate(*basis, PDELab::Backend::native(u), g);
125  // { make_solution_vector_end }
126
127  // { linear_solver_begin }
128    // Select a linear solver backend
129    using LinearSolverBackend = PDELab::ISTLBackend_SEQ_CG_SSOR;
130    LinearSolverBackend linearSolverBackend(50,2);
131
132    // Select linear problem solver
133    using LinearProblemSolver
134        = PDELab::StationaryLinearProblemSolver<GridOperator,
135                                                LinearSolverBackend,
```

574

```
136                                          U>;
137     LinearProblemSolver linearProblemSolver(gridOperator,
138                                             linearSolverBackend,
139                                             u,
140                                             1e-5);
141     // Solve linear problem.
142     linearProblemSolver.apply();
143   // { linear_solver_end }
144
145     // Graphical output
146   // { vtk_output_begin }
147     VTKWriter<GridView> vtkwriter(gridView);
148     auto uFunction
149         = Functions::makeDiscreteGlobalBasisFunction<double>(
150                                          *basis,
151                                          PDELab::Backend::native(u));
152     vtkwriter.addVertexData(uFunction,
153                             VTK::FieldInfo("u",
154                                     VTK::FieldInfo::Type::scalar,
155                                     1));
156     vtkwriter.write("pdelab-poisson-dirichlet-result");
157   // { vtk_output_end }
158   }
```

B.15. Demonstrating the Linear Algebra Backends

File:	pdelab-backends.cc
Equation:	Linear reaction–diffusion equation
Discretization:	Second-order Lagrange finite elements
Grid:	Uniform quadrilateral grid, implemented with YaspGrid
Solver:	CG solver with SSOR preconditioner, CG solver with Jacobi preconditioner, damped Richardson iteration
Distributed:	no
Discussed in:	Chapter 11.4

```
1    #include "config.h"
2
3    #include <iostream>
4    #include <vector>
5
6    #include <dune/grid/yaspgrid.hh>
7    #include <dune/grid/io/file/vtk/subsamplingvtkwriter.hh>
8
9    #include <dune/functions/functionspacebases/lagrangebasis.hh>
10   #include <dune/functions/gridfunctions/discreteglobalbasisfunction.hh>
11
12   #include <dune/pdelab.hh>
13
14   using namespace Dune;
15
16   template<class GridView, class Range>
17   class LinearReactionDiffusionProblem
18   : public PDELab::ConvectionDiffusionModelProblem<GridView,Range>
19   {
20   public:
21
22     template<typename Element, typename Coord>
23     auto f(const Element& element, const Coord& x) const
24     {
25       auto globalpos = element.geometry().global(x);
26       decltype(globalpos) midpoint(0.5);
27       globalpos -= midpoint;
28       return (globalpos.two_norm()<0.25) ? -10.0 : 10.0;
29     }
30
31     template<typename Element, typename Coord>
32     auto c(const Element& element, const Coord& x) const
33     {
34       return 10.0;
```

```
 35     }
 36  };
 37
 38  // { solve_istl_begin }
 39  template<class GridView>
 40  void solveReactionDiffusionProblemISTL(const GridView& gridView,
 41                                         std::string filename)
 42  {
 43     // Make grid function space
 44     using Basis = Functions::LagrangeBasis<GridView,2>;
 45     auto basis = std::make_shared<Basis>(gridView);
 46
 47     using VectorBackend = PDELab::ISTL::VectorBackend<>;
 48
 49     using GridFunctionSpace
 50        = PDELab::Experimental::GridFunctionSpace<Basis,
 51                                         VectorBackend,
 52                                         PDELab::NoConstraints>;
 53
 54     GridFunctionSpace gridFunctionSpace(basis);
 55  // { istl_gfs_end }
 56
 57  // { istl_gridoperator_begin }
 58     // Make grid operator
 59     LinearReactionDiffusionProblem<GridView,double> problem;
 60     using LocalOperator
 61        = PDELab::ConvectionDiffusionFEM<decltype(problem),
 62                                 typename GridFunctionSpace::Traits::
 63                                              FiniteElementMap>;
 64     LocalOperator localOperator(problem);
 65
 66     using MatrixBackend = PDELab::ISTL::BCRSMatrixBackend<>;
 67     using GridOperator
 68        = PDELab::GridOperator<GridFunctionSpace,    // Trial function space
 69                               GridFunctionSpace,    // Test function space
 70                               LocalOperator,        // Element assembler
 71                               MatrixBackend,        // Data structure
 72                                                     // for the stiffness matrix
 73                               double,               // Number type for
 74                                                     // solution vector entries
 75                               double,               // Number type for
 76                                                     // residual vector entries
 77                               double>;              // Number type for
 78                                                     // stiffness matrix entries
 79
 80     MatrixBackend matrixBackend(25);
 81     GridOperator gridOperator(gridFunctionSpace,
 82                               gridFunctionSpace,
 83                               localOperator,
 84                               matrixBackend);
 85  // { istl_gridoperator_end }
 86
 87  // { istl_solver_begin }
 88     // Select vector data type to hold the iterates
 89     using VectorContainer = PDELab::Backend::Vector<GridFunctionSpace,double>;
 90     VectorContainer u(gridFunctionSpace,0.0);     // Initial iterate
 91
 92     // Select a linear solver backend
 93     using LinearSolverBackend = PDELab::ISTLBackend_SEQ_CG_SSOR;
 94     LinearSolverBackend linearSolverBackend(5000,    // Maximal number
 95                                                      // of iterations
 96                                             2);      // Verbosity level
 97
 98     // Solve linear problem
 99     using LinearProblemSolver
100        = PDELab::StationaryLinearProblemSolver<GridOperator,
101                                      LinearSolverBackend,
102                                      VectorContainer>;
103     LinearProblemSolver linearProblemSolver(gridOperator,
104                                     linearSolverBackend,
105                                     u,
106                                     1e-10);
107     linearProblemSolver.apply();
108  // { istl_solver_end }
109
110  // { istl_vtk_begin }
111     // Output as VTK file
112     SubsamplingVTKWriter<GridView> vtkwriter(gridView,refinementLevels(2));
113     auto uFunction
114        = Functions::makeDiscreteGlobalBasisFunction<double>(
115                                     *basis,
116                                     PDELab::Backend::native(u));
117     vtkwriter.addVertexData(uFunction,
118                     VTK::FieldInfo("u",
119                              VTK::FieldInfo::Type::scalar,
120                              1));
121     vtkwriter.write(filename);
```

```
122    }
123    // { solve_istl_end }
124
125
126    #if HAVE_EIGEN
127    // { solve_eigen_begin }
128    template<class GridView>
129    void solveReactionDiffusionProblemEigen(const GridView& gridView,
130                                            std::string filename)
131    {
132      // Make grid function space
133      using Basis = Functions::LagrangeBasis<GridView,2>;
134      auto basis = std::make_shared<Basis>(gridView);
135
136      using VectorBackend = PDELab::Eigen::VectorBackend;
137
138      using GridFunctionSpace
139        = PDELab::Experimental::GridFunctionSpace<Basis,
140                                                  VectorBackend,
141                                                  PDELab::NoConstraints>;
142
143      GridFunctionSpace gridFunctionSpace(basis);
144
145      // Make grid operator
146      LinearReactionDiffusionProblem<GridView,double> problem;
147      using LocalOperator
148        = PDELab::ConvectionDiffusionFEM<decltype(problem),
149                                         typename GridFunctionSpace::Traits::
150                                                             FiniteElementMap>;
151      LocalOperator localOperator(problem);
152
153      using MatrixBackend = PDELab::Eigen::MatrixBackend<>;
154      using GridOperator
155        = PDELab::GridOperator<GridFunctionSpace,      // Trial function space
156                               GridFunctionSpace,      // Test function space
157                               LocalOperator,          // Element assembler
158                               MatrixBackend,          // Data structure
159                                                       // for the stiffness matrix
160                               double,                 // Number type for
161                                                       // solution vector entries
162                               double,                 // Number type for
163                                                       // residual vector entries
164                               double>;                // Number type for
165                                                       // stiffness matrix entries
166
167
168      MatrixBackend matrixBackend(25);
169      GridOperator gridOperator(gridFunctionSpace,
170                                gridFunctionSpace,
171                                localOperator,
172                                matrixBackend);
173
174      // Select vector data type to hold the iterate
175      using VectorContainer = PDELab::Backend::Vector<GridFunctionSpace,double>;
176      VectorContainer u(gridFunctionSpace,0.0);    // Initial iterate
177
178      // Select a linear solver backend
179      using LinearSolverBackend = PDELab::EigenBackend_CG_Diagonal_Up;
180      LinearSolverBackend linearSolverBackend(5000);
181
182      // Solve linear problem
183      using LinearProblemSolver
184        = PDELab::StationaryLinearProblemSolver<GridOperator,
185                                                LinearSolverBackend,
186                                                VectorContainer>;
187      LinearProblemSolver linearProblemSolver(gridOperator,
188                                              linearSolverBackend,
189                                              u,
190                                              1e-10);
191      linearProblemSolver.apply();
192
193      // Output as VTK file
194      SubsamplingVTKWriter<GridView> vtkwriter(gridView,refinementLevels(2));
195      auto uFunction
196        = Functions::makeDiscreteGlobalBasisFunction<double>(
197                                         *basis,
198                                         PDELab::Backend::native(u));
199      vtkwriter.addVertexData(uFunction,
200                              VTK::FieldInfo("u",
201                                             VTK::FieldInfo::Type::scalar,
202                                             1));
203      vtkwriter.write(filename);
204    }
205    // { solve_eigen_end }
206    #endif
207
208
```

```
209   template<class GridView>
210   void solveReactionDiffusionProblemISTLExplicit(const GridView& gridView,
211                                                 std::string filename)
212   {
213     // Make grid function space
214     using Basis = Functions::LagrangeBasis<GridView,2>;
215     auto basis = std::make_shared<Basis>(gridView);
216
217     using VectorBackend = PDELab::ISTL::VectorBackend<>;
218
219     using GridFunctionSpace = PDELab::Experimental::GridFunctionSpace<Basis,
220                                                       VectorBackend,
221                                                       PDELab::NoConstraints>;
222
223     GridFunctionSpace gridFunctionSpace(basis);
224
225     // Make grid operator
226     LinearReactionDiffusionProblem<GridView,double> problem;
227     using LocalOperator
228       = PDELab::ConvectionDiffusionFEM<decltype(problem),
229                             typename GridFunctionSpace::Traits::FiniteElementMap>;
230     LocalOperator localOperator(problem);
231
232     using MatrixBackend = PDELab::ISTL::BCRSMatrixBackend<>;
233     MatrixBackend matrixBackend(25);
234
235     using GridOperator
236       = PDELab::GridOperator<GridFunctionSpace,    // Trial function space
237                            GridFunctionSpace,    // Test function space
238                            LocalOperator,        // Element assembler
239                            MatrixBackend,        // Data structure
240                                                 // for the stiffness matrix
241                            double,               // Number type for
242                                                 // solution vector entries
243                            double,               // Number type for
244                                                 // residual vector entries
245                            double>;              // Number type for
246                                                 // stiffness matrix entries
247
248     GridOperator gridOperator(gridFunctionSpace,
249                             gridFunctionSpace,
250                             localOperator,
251                             matrixBackend);
252
253     // Vector holding the solver iterates
254   // { istl_explicit_make_iterate_begin }
255     using VectorContainer = typename GridOperator::Traits::Domain;
256     VectorContainer xContainer(gridFunctionSpace, 0.0);
257   // { istl_explicit_make_iterate_end }
258
259   // { construct_istl_explicit_begin }
260     // Evaluate residual at the zero configuration
261     typename GridOperator::Traits::Range rContainer(gridFunctionSpace, 0.0);
262
263     gridOperator.residual(xContainer,rContainer);
264
265     // Compute stiffness matrix
266     using MatrixContainer = typename GridOperator::Jacobian;
267     MatrixContainer matrixContainer(gridOperator, 0.0);
268
269     gridOperator.jacobian(xContainer,matrixContainer);
270   // { construct_istl_explicit_end }
271
272   // { get_native_data_begin }
273     auto& x = PDELab::Backend::native(xContainer);
274     auto& b = PDELab::Backend::native(rContainer);
275     b *= -1.0;
276     const auto& stiffnessMatrix = PDELab::Backend::native(matrixContainer);
277   // { get_native_data_end }
278
279   // { get_native_types_begin }
280     using DomainVector = std::decay_t<decltype(x)>;  // Decay from reference
281                                                     // to value type
282     using RangeVector  = std::decay_t<decltype(b)>;
283     using Matrix       = std::decay_t<decltype(stiffnessMatrix)>;
284   // { get_native_types_end }
285
286   // { solve_istl_explicit_begin }
287     // Turn the matrix into a linear operator
288     MatrixAdapter<Matrix,DomainVector,RangeVector>
289                                             linearOperator(stiffnessMatrix);
290
291     // SSOR as the preconditioner
292     SeqSSOR<Matrix,DomainVector,RangeVector> preconditioner(stiffnessMatrix,
293                                             1,       // Number of
294                                                     // iterations
295                                             1.0);   // Damping
```

578

```
296
297    // Preconditioned conjugate-gradient solver
298    CGSolver<DomainVector> cg(linearOperator,
299                             preconditioner,
300                             1e-4,    // Desired residual reduction factor
301                             50,      // Maximum number of iterations
302                             2);      // Verbosity level
303
304    // Object storing some statistics about the solving process
305    InverseOperatorResult statistics;
306
307    // Solve!
308    cg.apply(x, b, statistics);
309  // { solve_istl_explicit_end }
310
311    // Output as VTK file
312    SubsamplingVTKWriter<GridView> vtkwriter(gridView,Dune::refinementLevels(2));
313    auto uFunction = Functions::makeDiscreteGlobalBasisFunction<double>(*basis, x);
314    vtkwriter.addVertexData(uFunction, VTK::FieldInfo("u", VTK::FieldInfo::Type::scalar, 1));
315    vtkwriter.write(filename);
316  }
317
318
319  // { solve_simple_begin }
320  template<class GridView>
321  void solveReactionDiffusionProblemSimple(const GridView& gridView,
322                                           std::string filename)
323  {
324    // Make grid function space
325    using Basis = Functions::LagrangeBasis<GridView,2>;
326    auto basis = std::make_shared<Basis>(gridView);
327
328    using VectorBackend = PDELab::Simple::VectorBackend<>;
329
330    using GridFunctionSpace
331      = PDELab::Experimental::GridFunctionSpace<Basis,
332                                                VectorBackend,
333                                                PDELab::NoConstraints>;
334
335    GridFunctionSpace gridFunctionSpace(basis);
336  // { simple_gfs_end }
337
338    // Make grid operator
339  // { simple_gridoperator_begin }
340    LinearReactionDiffusionProblem<typename Basis::GridView,double> problem;
341    using LocalOperator
342      = PDELab::ConvectionDiffusionFEM<decltype(problem),
343                                       typename GridFunctionSpace::Traits::
344                                                             FiniteElementMap>;
345    LocalOperator localOperator(problem);
346
347    using MatrixBackend = PDELab::Simple::SparseMatrixBackend<>;
348    using GridOperator
349      = PDELab::GridOperator<GridFunctionSpace,    // Trial function space
350                             GridFunctionSpace,    // Test function space
351                             LocalOperator,        // Element assembler
352                             MatrixBackend,        // Data structure
353                                                   // for the stiffness matrix
354                             double,               // Number type for
355                                                   // solution vector entries
356                             double,               // Number type for
357                                                   // residual vector entries
358                             double>;              // Number type for
359                                                   // stiffness matrix entries
360
361
362    GridOperator gridOperator(gridFunctionSpace,
363                              gridFunctionSpace,
364                              localOperator);
365  // { simple_gridoperator_end }
366
367  // { simple_algebraic_problem_begin }
368    // Vector for the iterates
369    typename GridOperator::Traits::Domain xContainer(gridFunctionSpace,0.0);
370
371    // Evaluate residual at the zero vector
372    typename GridOperator::Traits::Range rContainer(gridFunctionSpace, 0.0);
373
374    gridOperator.residual(xContainer,rContainer);
375    rContainer *= -1.0;
376
377    // Compute stiffness matrix
378    using MatrixContainer = typename GridOperator::Traits::Jacobian;
379    MatrixContainer matrixContainer(gridOperator, 0.0);
380
381    gridOperator.jacobian(xContainer,matrixContainer);
382  // { simple_algebraic_problem_end }
```

579

```
383
384   // { simple_solution_begin }
385     double omega = 0.2;
386     for (int i=0; i<200; i++)
387     {
388       // Damped Richardson iteration
389       auto correction = rContainer;
390       matrixContainer.usmv(-1.0,xContainer,correction);
391       xContainer.axpy(omega,correction);
392     }
393   // { simple_solution_end }
394
395       // Output as VTK file
396       SubsamplingVTKWriter<GridView> vtkwriter(gridView,Dune::refinementLevels(2));
397       auto uFunction = Functions::makeDiscreteGlobalBasisFunction<double>(*basis, PDELab::Backend::native(xContainer));
398       vtkwriter.addVertexData(uFunction, VTK::FieldInfo("u", VTK::FieldInfo::Type::scalar, 1));
399       vtkwriter.write(filename);
400   }
401
402
403   int main(int argc, char *argv[])
404   {
405     // Set up MPI, if available
406     MPIHelper::instance(argc,argv);
407
408     // Create a structured 8x8 grid
409     constexpr int dim = 2;
410     FieldVector<double,dim> l = {1.0, 1.0};
411     std::array<int,dim> n = {4,4};
412     YaspGrid<2> grid(l,n);
413     auto gridView=grid.leafGridView();
414
415     // ISTL backend
416     solveReactionDiffusionProblemISTL(gridView,
417                                       "pdelab-backends-result-istl");
418
419   #if HAVE_EIGEN
420     // Eigen backend
421     solveReactionDiffusionProblemEigen(gridView,
422                                        "pdelab-backends-result-eigen");
423   #else
424   #warning Skipping the Eigen example, because the Eigen library was not found.
425     std::cerr << "Skipping the Eigen example, because the Eigen library was not found." << std::endl;
426   #endif
427
428     // ISTL backend, but giving the solver explicitly
429     solveReactionDiffusionProblemISTLExplicit(gridView,
430                                               "pdelab-backends-result-istl-explicit");
431
432     // 'simple' backend
433     solveReactionDiffusionProblemSimple(gridView,
434                                         "pdelab-backends-result-simple");
435   }
```

B.16. Error Estimation and Adaptive Grid Refinement Using `dune-pdelab`

File:	`pdelab-adaptivity.cc`
Equation:	Poisson equation
Discretization:	First-order Lagrange finite elements
Grid:	Unstructured triangle grid, implemented with UGGrid
Solver:	CG solver with SSOR preconditioner
Distributed:	no
Discussed in:	Chapter 11.5.2

```cpp
1   #include "config.h"
2
3   #include <iostream>
4   #include <vector>
5
6   #include <dune/grid/io/file/gmshreader.hh>
7   #include <dune/grid/io/file/vtk/subsamplingvtkwriter.hh>
8   #include <dune/grid/uggrid.hh>
9
10  #include <dune/functions/functionspacebases/lagrangebasis.hh>
11  #include <dune/functions/functionspacebases/interpolate.hh>
12  #include <dune/functions/gridfunctions/discreteglobalbasisfunction.hh>
13
14  #include <dune/pdelab.hh>
15
16
17  using namespace Dune;
18
19  template<class GridView, class Range>
20  class PoissonProblem
21  : public PDELab::ConvectionDiffusionModelProblem<GridView,Range>
22  {
23  public:
24    using Traits = typename PDELab::ConvectionDiffusionModelProblem<GridView,Range>::Traits;
25
26    // Source term
27    auto f(const typename Traits::ElementType& element,
28           const typename Traits::DomainType& xi) const
29    {
30      return -5.0;
31    }
32
33    //! boundary condition type function
34    auto bctype (const typename Traits::IntersectionType& intersection,
35                 const typename Traits::IntersectionDomainType& xi) const
36    {
37      auto x = intersection.geometry().global(xi);
38      return ( x[0]< 1e-8 || x[1] < 1e-8 || (x[0] > 0.4999 && x[1] > 0.4999) )
39        ? PDELab::ConvectionDiffusionBoundaryConditions::Dirichlet
40        : PDELab::ConvectionDiffusionBoundaryConditions::Neumann;
41    }
42  };
43
44  // { local_estimator_begin }
45  template<class Problem>
46  class ResidualErrorEstimator
47    : public PDELab::LocalOperatorDefaultFlags
48  {
49    Problem problem_;
50
51  public:
52    ResidualErrorEstimator(const Problem& problem)
53    : problem_(problem)
54    {}
55  // { local_estimator_setup_end }
56
57    // Residual assembly flags
58  // { local_estimator_flags_begin }
59    static const bool doAlphaVolume   = true;
60    static const bool doAlphaSkeleton = true;
61    static const bool doAlphaBoundary = true;
62  // { local_estimator_flags_end }
63
64  // { alpha_volume_begin }
65    template<class EntityGeometry,
66             class LocalFunctionSpaceU, class Vector,
67             class LocalFunctionSpaceV, class Residual>
68    void alpha_volume(const EntityGeometry& entityGeometry,
69                      const LocalFunctionSpaceU& localFunctionSpaceU,
70                      const Vector& x,
71                      const LocalFunctionSpaceV& localFunctionSpaceV,
72                      Residual& residual) const
73    {
74      using RangeField
75        = typename LocalFunctionSpaceU::Traits::
76                   FiniteElement::Traits::LocalBasisType::Traits::RangeType;
77
78      // Element diameter
79      auto h_T = diameter(entityGeometry.geometry());
80
81      // Loop over quadrature points
82      int intOrder = localFunctionSpaceU.finiteElement().localBasis().order();
83      constexpr auto entityDim = EntityGeometry::Entity::mydimension;
84      const auto& quadRule
85        = QuadratureRules<double,entityDim>::rule(entityGeometry.geometry()
86                                                    .type(),
87                                                  intOrder);
```

```
88
89      for (const auto& quadPoint : quadRule)
90      {
91        // Laplace of u_h is always zero,
92        // because we are using first-order elements on simplices
93        RangeField laplaceU = 0.0;
94
95        // Evaluate source term function f
96        auto f = problem_.f(entityGeometry.entity(),quadPoint.position());
97
98        // Integrate h_T^2 (f + Δu_h)^2
99        auto factor = quadPoint.weight()
100                    * entityGeometry.geometry()
101                          .integrationElement(quadPoint.position());
102        auto value = std::pow(h_T * (f + laplaceU), 2);
103        residual.accumulate(localFunctionSpaceV, 0, value * factor);
104      }
105    }
106  // { alpha_volume_end }
107
108  // { alpha_skeleton_begin }
109    template<class IntersectionGeometry,
110             class LocalFunctionSpaceU, class Vector,
111             class LocalFunctionSpaceV, class Residual>
112    void alpha_skeleton(const IntersectionGeometry& intersectionGeometry,
113                        const LocalFunctionSpaceU& localFunctionSpaceUIn,
114                        const Vector& xIn,
115                        const LocalFunctionSpaceV& localFunctionSpaceVIn,
116                        const LocalFunctionSpaceU& localFunctionSpaceUOut,
117                        const Vector& xOut,
118                        const LocalFunctionSpaceV& localFunctionSpaceVOut,
119                        Residual& residualIn, Residual& residualOut) const
120    {
121      // Extract type of shape function gradients
122      using TrialFE = typename LocalFunctionSpaceU::Traits::FiniteElementType;
123      using LocalBasisU = typename TrialFE::Traits::LocalBasisType;
124      using GradientU = typename LocalBasisU::Traits::JacobianType;
125      using size_type = typename LocalFunctionSpaceU::Traits::SizeType;
126
127      auto insideGeometry = intersectionGeometry.inside().geometry();
128      auto outsideGeometry = intersectionGeometry.outside().geometry();
129
130      auto h_F = diameter(intersectionGeometry.geometry());
131
132      auto geometryInInside = intersectionGeometry.geometryInInside();
133      auto geometryInOutside = intersectionGeometry.geometryInOutside();
134
135      std::vector<GradientU> localGradPhiIn(localFunctionSpaceUIn.size());
136      std::vector<GradientU> localGradPhiOut(localFunctionSpaceUOut.size());
137
138      // Loop over quadrature points and integrate the jump term
139      const int intOrder
140          = 2*localFunctionSpaceUIn.finiteElement().localBasis().order();
141      constexpr auto intersectionDim = IntersectionGeometry::mydimension;
142      const auto& quadRule
143          = QuadratureRules<double,intersectionDim>::rule(
144                                 intersectionGeometry.geometry().type(),
145                                 intOrder);
146
147      for (const auto& quadPoint : quadRule)
148      {
149        // Position of quadrature point in local coordinates of elements
150        auto quadPointLocalIn = geometryInInside.global(quadPoint.position());
151        auto quadPointLocalOut = geometryInOutside.global(quadPoint.position());
152
153        // Evaluate gradient of basis functions
154        localFunctionSpaceUIn.finiteElement().localBasis()
155                            .evaluateJacobian(quadPointLocalIn,localGradPhiIn);
156        localFunctionSpaceUOut.finiteElement().localBasis()
157                            .evaluateJacobian(quadPointLocalOut,localGradPhiOut);
158
159        // Compute gradient of u
160        GradientU localGradUIn(0.0);
161        for (size_type i=0; i<localFunctionSpaceUIn.size(); i++)
162          localGradUIn.axpy(xIn(localFunctionSpaceUIn,i),localGradPhiIn[i]);
163        GradientU localGradUOut(0.0);
164        for (size_type i=0; i<localFunctionSpaceUOut.size(); i++)
165          localGradUOut.axpy(xOut(localFunctionSpaceUOut,i),localGradPhiOut[i]);
166
167        GradientU gradUIn;
168        auto jacIn = insideGeometry.jacobianInverseTransposed(quadPointLocalIn);
169        jacIn.mv(localGradUIn[0],gradUIn[0]);
170
171        GradientU gradUOut;
172        auto jacOut
173            = outsideGeometry.jacobianInverseTransposed(quadPointLocalOut);
```

```
174            jacOut.mv(localGradUOut[0],gradUOut[0]);
175
176            // Integrate
177            const auto n_F
178                = intersectionGeometry.unitOuterNormal(quadPoint.position());
179            auto jumpSquared = std::pow((n_F*gradUIn[0])-(n_F*gradUOut[0]), 2);
180            auto factor = quadPoint.weight()
181                          * intersectionGeometry.geometry()
182                                      .integrationElement(quadPoint.position());
183
184            // Accumulate indicator
185            residualIn.accumulate(localFunctionSpaceVIn,
186                                  0,
187                                  0.5 * h_F * jumpSquared * factor);
188            residualOut.accumulate(localFunctionSpaceVOut,
189                                   0,
190                                   0.5 * h_F * jumpSquared * factor);
191        }
192    }
193 // { alpha_skeleton_end }
194
195 // { alpha_boundary_begin }
196    template<class IntersectionGeometry,
197             class LocalFunctionSpaceU, class Vector,
198             class LocalFunctionSpaceV, class Residual>
199    void alpha_boundary(const IntersectionGeometry& intersectionGeometry,
200                        const LocalFunctionSpaceU& localFunctionSpaceUIn,
201                        const Vector& xIn,
202                        const LocalFunctionSpaceV& localFunctionSpaceVIn,
203                        Residual& residualIn) const
204    {
205        using TrialFE = typename LocalFunctionSpaceU::Traits::FiniteElementType;
206        using LocalBasisU = typename TrialFE::Traits::LocalBasisType;
207        using GradientU = typename LocalBasisU::Traits::JacobianType;
208        using size_type = typename LocalFunctionSpaceU::Traits::SizeType;
209
210        auto insideGeometry = intersectionGeometry.inside().geometry();
211
212        auto geometryInInside = intersectionGeometry.geometryInInside();
213
214        // Intersection diameter
215        auto h_F = diameter(intersectionGeometry.geometry());
216
217        std::vector<GradientU> localGradPhi(localFunctionSpaceUIn.size());
218
219        // Loop over quadrature points and integrate the jump term
220        const int intOrder
221            = localFunctionSpaceUIn.finiteElement().localBasis().order();
222        constexpr auto intersectionDim = IntersectionGeometry::mydimension;
223        const auto& quadRule
224            = QuadratureRules<double,intersectionDim>::rule(
225                                      intersectionGeometry.geometry().type(),
226                                      intOrder);
227
228        for (const auto& quadPoint : quadRule)
229        {
230            // Skip if quadrature point is not on Neumann boundary
231            auto bcType = problem_.bctype(intersectionGeometry.intersection(),
232                                          quadPoint.position());
233            if (bcType != PDELab::ConvectionDiffusionBoundaryConditions::Neumann)
234                return;
235
236            // Position of quadrature point in local coordinates of element
237            auto quadPointLocalIn = geometryInInside.global(quadPoint.position());
238
239            // Evaluate gradient of trial shape functions
240            localFunctionSpaceUIn.finiteElement().localBasis()
241                                      .evaluateJacobian(quadPointLocalIn,
242                                                        localGradPhi);
243
244            // Evaluate gradient of u_h
245            GradientU localGradU(0.0);
246            for (size_type i=0; i<localFunctionSpaceUIn.size(); i++)
247                localGradU.axpy(xIn(localFunctionSpaceUIn,i), localGradPhi[i]);
248
249            GradientU gradU;
250            auto jac = insideGeometry.jacobianInverseTransposed(quadPointLocalIn);
251            jac.mv(localGradU[0], gradU[0]);
252
253            // Evaluate Neumann boundary value function
254            auto neumann = problem_.j(intersectionGeometry.intersection(),
255                                      quadPoint.position());
256
257            // Integrate
258            auto factor = quadPoint.weight()
259                          * intersectionGeometry.geometry()
260                                      .integrationElement(quadPoint.position());
```

```
261
262        const auto n
263            = intersectionGeometry.unitOuterNormal(quadPoint.position());
264
265        residualIn.accumulate(localFunctionSpaceVIn,
266                              0,
267                              h_F * std::pow((neumann - n*gradU[0]), 2)
268                                                              * factor);
269      }
270    }
271  // { alpha_boundary_end }
272
273  private:
274    template<class Geometry>
275    static auto diameter (const Geometry& geometry)
276    {
277      typename Geometry::ctype h = 0.0;
278      for (int i=0; i<geometry.corners(); i++)
279        for (int j=i+1; j<geometry.corners(); j++)
280          h = std::max(h, (geometry.corner(j)-geometry.corner(i)).two_norm());
281
282      return h;
283    }
284  };
285
286
287  // { main_begin }
288  int main(int argc, char *argv[])
289  {
290    // Initialize MPI, if available
291    MPIHelper::instance(argc, argv);
292
293    constexpr int dim = 2;
294    using Grid = UGGrid<dim>;
295    std::shared_ptr<Grid> grid = GmshReader<Grid>::read("l-shape.msh");
296    using GridView = Grid::LeafGridView;
297    auto gridView = grid->leafGridView();
298  // { grid_setup_end }
299
300  // { define_threshold_begin }
301    const double estimatedErrorThreshold = 0.1;
302  // { define_threshold_end }
303
304  // { make_function_space_begin }
305    // Make grid function space
306    using Basis = Functions::LagrangeBasis<GridView,1>;
307    auto basis = std::make_shared<Basis>(gridView);
308
309    using Constraints = PDELab::ConformingDirichletConstraints;
310    using VectorBackend = PDELab::ISTL::VectorBackend<>;
311
312    using GridFunctionSpace
313        = PDELab::Experimental::GridFunctionSpace<Basis,
314                                                  VectorBackend,
315                                                  Constraints>;
316
317    GridFunctionSpace gridFunctionSpace(basis);
318  // { make_function_space_end }
319
320    // Assemble Dirichlet constraints on this space
321  // { make_constraints_begin }
322    PoissonProblem<GridView,double> poissonProblem;
323    PDELab::
324      ConvectionDiffusionBoundaryConditionAdapter<decltype(poissonProblem)>
325                                                  bctype(poissonProblem);
326
327    using ConstraintsContainer
328        = GridFunctionSpace::ConstraintsContainer<double>::Type;
329    ConstraintsContainer constraintsContainer;
330    PDELab::constraints(bctype, gridFunctionSpace, constraintsContainer);
331  // { make_constraints_end }
332
333    // Make solution vector
334  // { make_solution_vector_begin }
335    using U = PDELab::Backend::Vector<GridFunctionSpace,double>;
336    U u(gridFunctionSpace,0.0);
337    auto g = [](auto x){return  (x[0]< 1e-8 || x[1] < 1e-8) ? 0 : 0.5;};
338    Functions::interpolate(*basis, PDELab::Backend::native(u), g);
339  // { make_solution_vector_end }
340
341  // { refinement_loop_begin }
342    std::size_t i=0;     // Loop variable, for data output
343    while (true)         // Loop termination condition
344                         // is near the middle of the loop body
345    {
346      std::cout << "Iteration: " << i
347              << ", highest level in grid: " << grid->maxLevel()
```

584

```
348                      << ", degrees of freedom: " << gridFunctionSpace.globalSize()
349                          << std::endl;
350     // { refinement_loop_begin_end }
351
352         // Make grid operator
353     // { make_assembler_and_solver_begin }
354         using LocalOperator
355             = PDELab::ConvectionDiffusionFEM<decltype(poissonProblem),
356                                     GridFunctionSpace::Traits::
357                                                 FiniteElementMap>;
358         LocalOperator localOperator(poissonProblem);
359
360         using MatrixBackend = PDELab::ISTL::BCRSMatrixBackend<>;
361         MatrixBackend matrixBackend(7);      // Expected average number
362                                     // of matrix entries per row
363         using GridOperator = PDELab::GridOperator<GridFunctionSpace,
364                                     GridFunctionSpace,
365                                     LocalOperator,
366                                     MatrixBackend,
367                                     double,double,double,
368                                     ConstraintsContainer,
369                                     ConstraintsContainer>;
370         GridOperator gridOperator(gridFunctionSpace, constraintsContainer,
371                             gridFunctionSpace, constraintsContainer,
372                             localOperator,
373                             matrixBackend);
374
375         // Select a linear solver backend
376         using LinearSolverBackend = PDELab::ISTLBackend_SEQ_CG_SSOR;
377         LinearSolverBackend linearSolverBackend(5000,    // Maximum number
378                                     // of iterations
379                                 1);     // Verbosity level
380
381         // Select linear problem solver
382         using LinearProblemSolver
383             = PDELab::StationaryLinearProblemSolver<GridOperator,
384                                     LinearSolverBackend,
385                                     U>;
386         LinearProblemSolver linearProblemSolver(gridOperator,
387                                     linearSolverBackend,
388                                     u,
389                                     1e-10);
390
391         // Solve linear problem
392         linearProblemSolver.apply();
393     // { make_assembler_and_solver_end }
394
395         // File output
396     // { vtk_output_begin }
397         VTKWriter<GridView> vtkwriter(gridView);
398         auto uFunction
399             = Functions::makeDiscreteGlobalBasisFunction<double>(
400                                     *basis,
401                                     PDELab::Backend::native(u));
402         vtkwriter.addVertexData(uFunction,
403                         VTK::FieldInfo("u",
404                                 VTK::FieldInfo::Type::scalar,
405                                 1));
406         vtkwriter.write("pdelab-adaptivity-result-" + std::to_string(i));
407     // { vtk_output_end }
408
409         // Preparation: Define types for the computation of the error estimate eta
410     // { estimation_preparation_begin }
411         using P0Basis = Functions::LagrangeBasis<GridView,0>;
412         auto p0Basis = std::make_shared<P0Basis>(gridView);
413
414         using P0GridFunctionSpace
415             = PDELab::Experimental::GridFunctionSpace<P0Basis,
416                                     VectorBackend,
417                                     PDELab::NoConstraints>;
418
419         P0GridFunctionSpace p0GridFunctionSpace(p0Basis);
420         using U0 = PDELab::Backend::Vector<P0GridFunctionSpace,double>;
421         U0 etaSquared(p0GridFunctionSpace,0.0);
422     // { estimation_preparation_end }
423
424         // Compute estimated error eta
425     // { estimator_gfs_begin }
426         using EstimatorLocalOperator
427             = ResidualErrorEstimator<decltype(poissonProblem)>;
428         EstimatorLocalOperator estimatorLocalOperator(poissonProblem);
429         using EstimatorGridOperator = PDELab::GridOperator<GridFunctionSpace,
430                                     P0GridFunctionSpace,
431                                     EstimatorLocalOperator,
432                                     MatrixBackend,
433                                     double,double,double>;
434         EstimatorGridOperator estimatorGridOperator(gridFunctionSpace,
```

```
435                                                     p0GridFunctionSpace,
436                                                     estimatorLocalOperator,
437                                                     matrixBackend);
438    // { estimator_gfs_end }
439
440    // { estimator_call_begin }
441        estimatorGridOperator.residual(u,etaSquared);
442    // { estimator_call_end }
443
444    // { termination_begin }
445        // Terminate if desired accuracy has been reached
446        double totalEstimatedError = std::sqrt(std::accumulate(etaSquared.begin(),
447                                                               etaSquared.end(),
448                                                               0.0));
449        std::cout << "Total estimated error: " << totalEstimatedError << std::endl;
450
451        if (totalEstimatedError < estimatedErrorThreshold)
452          break;
453    // { termination_end }
454
455        // Adapt the grid locally...
456    // { error_fraction_begin }
457        double alpha = 0.4;      // Refinement fraction
458        double refineThreshold;  // Refinement threshold
459        double beta = 0.0;       // Coarsening fraction
460        double coarsenThreshold; // Coarsening threshold
461        int verbose = 0;         // No screen output
462
463        PDELab::error_fraction(etaSquared,
464                               alpha, beta,
465                               refineThreshold, coarsenThreshold,
466                               verbose);
467    // { error_fraction_end }
468
469    // { mark_begin }
470        PDELab::mark_grid(*grid, etaSquared, refineThreshold, coarsenThreshold);
471    // { mark_end }
472    // { adapt_begin }
473        PDELab::adapt_grid(*grid, gridFunctionSpace, u, 2 );
474    // { adapt_end }
475
476    // { dirichlet_reinterpolation_begin }
477        // Reassemble the Dirichlet constraints
478        PDELab::constraints(bctype,gridFunctionSpace,constraintsContainer);
479
480        // Interpolate the Dirichlet values function on the entire domain!
481        PDELab::Backend::Vector<GridFunctionSpace,double>
482                                    dirichletValues(gridFunctionSpace);
483        Functions::interpolate(*basis,
484                               PDELab::Backend::native(dirichletValues),
485                               g);
486
487        // Copy all Dirichlet degrees of freedom into the actual solution vector
488        PDELab::copy_constrained_dofs(constraintsContainer,dirichletValues,u);
489
490        // Increment loop counter
491        i++;
492      }
493    // { dirichlet_reinterpolation_end }
494    }
```

Bibliography

[1]   D. Abrahams and A. Gurtovoy. *C++ Template Metaprogramming*. Addison-Wesley, 2005.

[2]   M. Alkämper, A. Dedner, R. Klöfkorn, and M. Nolte. "The DUNE-ALUGrid Module". In: *Archive of Numerical Software* 4.1 (2016), pp. 1–28.

[3]   E. L. Allgower and K. Georg. *Introduction to Numerical Continuation Methods*. Vol. 45. Classics in Applied Mathematics. SIAM, 2003.

[4]   D. N. Arnold, F. Brezzi, B. Cockburn, and L. D. Marini. "Unified Analysis of Discontinuous Galerkin Methods for Elliptic Problems". In: *SIAM J. Numer. Anal.* 39.5 (2002), pp. 1749–1779.

[5]   M. Bader. *Space-Filling Curves – An Introduction with Applications in Scientific Computing*. Springer, 2013.

[6]   S. Balay, S. Abhyankar, M. F. Adams, J. Brown, P. Brune, K. Buschelman, V. Eijkhout, W. D. Gropp, D. Kaushik, M. G. Knepley, L. C. McInnes, K. Rupp, B. F. Smith, and H. Zhang. *PETSc Users Manual*. Tech. rep. ANL-95/11 - Revision 3.5. Argonne National Laboratory, 2014. URL: http://www.mcs.anl.gov/petsc.

[7]   S. Balay, W. D. Gropp, L. C. McInnes, and B. F. Smith. "Efficient Management of Parallelism in Object Oriented Numerical Software Libraries". In: *Modern Software Tools in Scientific Computing*. Ed. by E. Arge, A. M. Bruaset, and H. P. Langtangen. Birkhäuser Press, 1997, pp. 163–202.

[8]   W. Bangerth, R. Hartmann, and G. Kanschat. "deal.II – a General Purpose Object Oriented Finite Element Library". In: *ACM Trans. Math. Softw.* 33.4 (2007), pp. 24/1–24/27.

[9]   R. E. Bank. "Hierarchical bases and the finite element method". In: *Acta Numerica* 5 (Jan. 1996), pp. 1–43. DOI: 10.1017/S0962492900002610.

[10] R. E. Bank, A. H. Sherman, and A. Weiser. "Some Refinement Algorithms and Data Structures for Regular Local Mesh Refinement". In: *Scientific Computing. Applications of Mathematics and Computing to the Physical Sciences*. Ed. by R. Stepleman. North-Holland, 1983, pp. 3–17.

[11] J. W. Barrett and C. M. Elliott. "Finite Element Approximation of the Dirichlet Problem Using the Boundary Penalty Method". In: *Numerische Mathematik* 49 (1986), pp. 343–366.

© Springer Nature Switzerland AG 2020
O. Sander, *DUNE — The Distributed and Unified Numerics Environment*, Lecture Notes in Computational Science and Engineering 140, https://doi.org/10.1007/978-3-030-59702-3

[12]   T. Barth, R. Herbin, and M. Ohlberger. "Finite Volume Methods: Foundation and Analysis". In: *Encyclopedia of Computational Mechanics. 2nd edition.* Wiley, 2017. DOI: 10.1002/9781119176817.ecm2010.

[13]   P. Bastian, K. Birken, K. Johannsen, S. Lang, N. Neuß, H. Rentz–Reichert, and C. Wieners. "UG – a flexible Software toolbox for solving partial differential equations". In: *Comp. Vis. Sci* 1 (1997), pp. 27–40.

[14]   P. Bastian and M. Blatt. "On the Generic Parallelisation of Iterative Solvers for the Finite Element Method". In: *Int. J. Computational Science and Engineering* 4.1 (2008), pp. 56–69. DOI: 10.1504/IJCSE.2008.021112.

[15]   P. Bastian, M. Blatt, A. Dedner, C. Engwer, R. Klöfkorn, R. Kornhuber, M. Ohlberger, and O. Sander. "A Generic Interface for Parallel and Adaptive Scientific Computing. Part II: Implementation and Tests in DUNE". In: *Computing* 82.2-3 (2008), pp. 121–138.

[16]   P. Bastian, M. Blatt, A. Dedner, C. Engwer, R. Klöfkorn, M. Ohlberger, and O. Sander. "A Generic Interface for Parallel and Adaptive Scientific Computing. Part I: Abstract Framework". In: *Computing* 82.2-3 (2008), pp. 103–119.

[17]   P. Bastian, M. Blatt, A. Dedner, N.-A. Dreier, C. Engwer, R. Fritze, C. Gräser, C. Grüninger, D. Kempf, R. Klöfkorn, M. Ohlberger, and O. Sander. "The DUNE Framework: Basic Concepts and Recent Developments". In: *Computers and Mathematics with Applications* (2020). DOI: 10.1016/j.camwa.2020.06.007.

[18]   P. Bastian, C. Engwer, J. Fahlke, M. Geveler, D. Göddeke, O. Iliev, O. Ippisch, R. Milk, J. Mohring, S. Müthing, M. Ohlberger, D. Ribbrock, and S. Turek. "Advances Concerning Multiscale Methods and Uncertainty Quantification in EXA-DUNE". In: *Software for Exascale Computing – SPPEXA 2013-2015.* Ed. by H.-J. Bungartz, P. Neumann, and W. Nagel. Vol. 113. Lecture Notes in Computational Science and Engineering. Springer, 2016, pp. 25–43. DOI: 10.1007/978-3-319-40528-5_2.

[19]   P. Bastian, C. Engwer, J. Fahlke, M. Geveler, D. Göddeke, O. Iliev, O. Ippisch, R. Milk, J. Mohring, S. Müthing, M. Ohlberger, D. Ribbrock, and S. Turek. "Hardware-Based Efficiency Advances in the EXA-DUNE Project". In: *Software for Exascale Computing – SPPEXA 2013-2015.* Ed. by H.-J. Bungartz, P. Neumann, and W. Nagel. Vol. 113. Lecture Notes in Computational Science and Engineering. Springer, 2016, pp. 3–23. DOI: 10.1007/978-3-319-40528-5_1.

[20]   P. Bastian, F. Heimann, and S. Marnach. "Generic implementation of finite element methods in the distributed and unified numerics environment (DUNE)". In: *Kybernetika* 46.2 (2010), pp. 294–315.

[21] L. Beirão da Veiga, F. Brezzi, A. Cangiani, G. Manzini, L. D. Marini, and A. Russo. "Basic Principles of Virtual Element Methods". In: *Mathematical Models and Methods in Applied Sciences* 23.01 (2013), pp. 199–214. DOI: 10. 1142/S0218202512500492.

[22] W. Benger. "Visualization of General Relativistic Tensor Fields via a Fiber Bundle Data Model". PhD thesis. Freie Universität Berlin, 2005.

[23] C. Bernardi, Y. Maday, and A. Patera. "Domain Decomposition by the Mortar Element Method". In: *Asymptotic and Numerical Methods for Partial Differential Equations with Critical Parameters*. Ed. by H. Kaper, M. Garbey, and G. Pieper. Vol. 384. NATO ASI Series (Series C: Mathematical and Physical Sciences). Springer, 1993, pp. 269–286. DOI: 10.1007/978-94-011-1810-1_17.

[24] G. Berti. "Generic Software Components for Scientific Computing". PhD thesis. Technische Universität Cottbus, 2000.

[25] J. Bey. "Tetrahedral Grid Refinement". In: *Computing* 55.4 (1995), pp. 355–378.

[26] P. Binev, W. Dahmen, and R. DeVore. "Adaptive finite element methods with convergence rates". In: *Numerische Mathematik* 97.2 (2004), pp. 219–268.

[27] A. Björner, M. L. Vergnas, B. Sturmfels, N. White, and G. M. Ziegler. *Oriented Matroids*. 2nd edition. Vol. 46. Encyclopedia of mathematics and its applications. Cambridge University Press, 1999.

[28] BLAST Forum. "Basic Linear Algebra Subprograms Technical (BLAST) Forum Standard". In: *International Journal of High Performance Applications and Supercomputing* 16.1 (2001). URL: http://www.netlib.org/blas/blast-forum/.

[29] M. Blatt and P. Bastian. "The Iterative Solver Template Library". In: *Applied Parallel Computing. State of the Art in Scientific Computing* Ed. by B. Kagström, E. Elmroth, J. Dongarra, and J. Wasniewski. Lecture Notes in Scientific Computing 4699. 2007, pp. 666–675. URL: 10.1007/978-3-540-75755-9%5C_82.

[30] M. Blatt. "A Parallel Algebraic Multigrid Method for Elliptic Problems with Highly Discontinuous Coefficients". PhD thesis. Universität Heidelberg, 2010.

[31] M. Blatt, O. Ippisch, and P. Bastian. "A massively parallel algebraic multigrid preconditioner based on aggregation for elliptic problems with heterogeneous coefficients". In: *ArXiv e-prints* (Sept. 2012). arXiv: 1209.0960 [math.NA].

[32] D. Boffi, F. Brezzi, and M. Fortin. *Mixed Finite Element Methods and Applications*. Springer, 2013.

[33] R. F. Boisvert, R. Pozo, and K. A. Remington. *The Matrix Market Exchange Formats: Initial Design*. NISTIR 5935. National Institute of Standards and Technology, 1996.

[34] M. Bolten and H. Rittich. "Fourier Analysis of Periodic Stencils in Multigrid Methods". In: *SIAM Journal on Scientific Computing* 40.3 (2018), A1642–A1668. DOI: 10.1137/16M1073959.

[35] E. G. Boman, Ü. V. Çatalyurek, C. Chevalier, and K. D. Devine. "The Zoltan and Isorropia parallel toolkits for combinatorial scientific computing: Partitioning, ordering, and coloring". In: *Scientific Programming* 20.2 (2012), pp. 129–150.

[36] N. Botta, C. Ionescu, C. Linstead, and R. Klein. *Structuring distributed relation-based computations with SCDRC*. Tech. rep. PIK Report No. 103, Potsdam Institute for Climate Impact Research, 2006.

[37] D. Braess. *Finite Elemente*. Springer, 2013.

[38] A. Bressan, S. Čanić, M. Garavello, M. Herty, and B. Piccoli. "Flows on networks: recent results and perspectives". In: *EMS Surv. Math. Sci.* 1.1 (2014), pp. 47–111.

[39] F. Brooks. *The Mythical Man–Month*. Addison–Wesly, 1975.

[40] A. Burri, A. Dedner, R. Klöfkorn, and M. Ohlberger. "An efficient implementation of an adaptive and parallel grid in DUNE". In: *Proc. of the 2nd Russian–German Advanced Research Workshop on Computational Science and High Performance Computing*. 2005.

[41] Y. Chen, T. A. Davis, W. W. Hager, and S. Rajamanickam. "Algorithm 887: CHOLMOD, Supernodal Sparse Cholesky Factorization and Update/Downdate". In: *ACM Trans. Math. Softw.* 35.3 (2008), 22:1–22:14. DOI: 10.1145/1391989.1391995.

[42] P. Ciarlet. *The finite element method for elliptic problems*. SIAM, 2002.

[43] T. H. Cormen, C. E. Leiserson, and R. L. Rivest. *Introduction to Algorithms*. MIT Press, 1990.

[44] J. A. Cottrell, T. J. R. Hughes, and Y. Bazilevs. *Isogeometric Analysis*. Wiley, 2009.

[45] M. Crouzeix and P.-A. Raviart. "Conforming and nonconforming finite element methods for solving the stationary Stokes equations I". In: *Revue française d'automatique informatique recherche opérationnelle* 7.R3 (1973), pp. 33–75.

[46] T. A. Davis. "Algorithm 832: UMFPACK V4.3—an unsymmetric-pattern multifrontal method". In: *ACM Transactions on Mathematical Software (TOMS)* 30.2 (2004), pp. 196–199.

[47] T. A. Davis. *Direct Methods for Sparse Linear Systems*. SIAM, 2006. DOI: 10.1137/1.9780898718881.

[48] A. Dedner, R. Klöfkorn, M. Nolte, and M. Ohlberger. "A generic interface for parallel and adaptive scientific computing: Abstraction principles and the DUNE-FEM module". In: *Computing* 90.3 (2011), pp. 165–196.

[49] A. Dedner and M. Nolte. "Construction of Local Finite Element Spaces Using the Generic Reference Elements". In: *Advances in DUNE*. Ed. by A. Dedner, B. Flemisch, and R. Klöfkorn. Springer, 2012, pp. 3–16.

[50] A. Dedner and M. Nolte. "The Dune Python Module". In: *arXiv e-prints* (2018). arXiv: 1807.05252 [cs.MS].

[51] P. Deuflhard. *Newton Methods for Nonlinear Problems*. Springer, 2006.

[52] P. Deuflhard and A. Hohmann. *Numerische Mathematik I*. de Gruyter, 1991.

[53] P. Deuflhard and M. Weiser. *Adaptive Numerical Solution of PDEs*. de Gruyter, 2012.

[54] R. Diestel. *Graph Theory*. 5th edition. Springer, 2016.

[55] V. Dolejší and M. Feistauer. *Discontinuous Galerkin Method*. Springer Series in Computational Mathematics 48. Springer, 2015.

[56] J. Dongarra. *Freely Available Software for Linear Algebra*. online. last checked on Jul. 21. 2020. URL: http://www.netlib.org/utk/people/JackDongarra/la-sw.html.

[57] W. Dörfler. "A Convergent Adaptive Algorithm for Poisson's Equation". In: *SIAM J. Numer. Anal.* 33.3 (1996), pp. 1106–1124.

[58] J. Douglas and T. Dupont. "Interior Penalty Procedures for Elliptic and Parabolic Galerkin Methods". In: *Computing Methods in Applied Sciences*. Ed. by R. Glowinski and L. J.L. Vol. 58. Lecture Notes in Physics. Springer, 1976, pp. 207–216.

[59] K. Driesen and U. Hölzle. "The Direct Cost of Virtual Function Calls in C++". In: *Proceedings of the 11th ACM SIGPLAN Conference on Object-oriented Programming, Systems, Languages, and Applications*. OOPSLA '96. ACM, 1996, pp. 306–323.

[60] G. Dziuk and C. Elliott. "Surface Finite Elements for Parabolic Equations". In: *Journal of Computational Mathematics* 25.4 (2007), pp. 385–407.

[61] C. Engwer, C. Gräser, S. Müthing, and O. Sander. "Function space bases in the dune-functions module". In: *ArXiv e-prints* (2018). eprint: 1806.09545 (cs.MS).

[62] C. Engwer, C. Gräser, S. Müthing, and O. Sander. "The interface for functions in the dune-functions module". In: *Archive of Numerical Software* 5.1 (2017), pp. 95–105. DOI: 10.11588/ans.2017.1.27683.

[63] A. Ern and J.-L. Guermond. *Theory and Practice of Finite Elements*. Springer, 2004.

[64] L. C. Evans. *Partial Differential Equations*. 2nd edition. American Mathematical Society, 2010.

[65] R. Eymard, G. Henry, R. Herbin, F. Hubert, R. Klöfkorn, and G. Manzini. "3D Benchmark on Discretization Schemes for Anisotropic Diffusion Problems on General Grids". In: *Finite Volumes for Complex Applications VI Problems & Perspectives*. Ed. by J. Fort, J. Fürst, J. Halama, R. Herbin, and F. Hubert. Vol. 4. Springer Proceedings in Mathematics. Springer, 2011, pp. 895–930. DOI: 10.1007/978-3-642-20671-9_89.

[66] R. Eymard, T. Gallouët, and R. Herbin. "Finite Volume Methods". In: *Handbook of Numerical Analysis*. Ed. by P. Ciarlet and J. Lions. Vol. 7. Elsevier, 2000, pp. 713–1018.

[67] H. Fahs. "Discontinuous Galerkin Method for Time-Domain Electromagnetics on Curvilinear Domains". In: *Applied Mathematical Sciences* 4.19 (2010), pp. 943–958.

[68] R. Falgout, J. Jones, and U. Yang. "The Design and Implementation of hypre, a Library of Parallel High Performance Preconditioners". In: *Numerical Solution of Partial Differential Equations on Parallel Computers*. Ed. by A. M. Bruaset and A. Tveito. Vol. 51. Lecture Notes in Computational Science and Engineering. Springer, 2006, pp. 267–294. DOI: 10.1007/3-540-31619-1_8.

[69] A. Fomins and B. Oswald. "Dune-CurvilinearGrid: Parallel Dune Grid Manager for Unstructured Tetrahedral Curvilinear Meshes". In: *arXiv e-prints* (2016). arXiv: 1612.02967 [cs.CG].

[70] H. Freudenthal. "Simplizialzerlegungen von beschränkter Flachheit". In: *Ann. Math.* 43.3 (1942), pp. 580–582.

[71] T.-P. Fries and T. Belytschko. "The extended/generalized finite element method: An overview of the method and its applications". In: *International Journal for Numerical Methods in Engineering* 84.3 (2010), pp. 253–304. DOI: 10.1002/nme.2914.

[72] E. Gamma, R. Helm, R. Johnson, and J. Vlissides. *Design Patterns: Elements of Reusable Object-Oriented Software*. Addison–Wesley, 1994.

[73] C. Gersbacher. "The Dune-PrismGrid Module". In: *Advances in DUNE*. Ed. by A. Dedner, B. Flemisch, and R. Klöfkorn. Springer, 2012, pp. 33–44.

[74] C. Geuzaine and F. Remacle. *Gmsh Reference Manual*. 2015. URL: http://www.geuz.org/gmsh/doc/texinfo/gmsh.pdf.

[75] G. H. Golub and C. F. V. Loan. *Matrix Computations*. 3rd edition. The John Hopkins University Press, 1996.

[76] S. Götschel, M. Weiser, and A. Schiela. "Solving Optimal Control Problems with the Kaskade7 Finite Element Toolbox," in: *Advances in DUNE*. Ed. by A. Dedner, B. Flemisch, and R. Klöfkorn. 2012, pp. 101–112.

[77] P. Gottschling, T. Witkowski, and A. Voigt. "Integrating object-oriented and generic programming paradigms in real-world software environments: Experiences with AMDiS and MTL4". In: *POOSC 2008 at ECOOP08*. Paphros, Cyprus, 2008.

[78] J. Grande. *Red–Green Refinement of Simplicial Meshes in D Dimensions.* Tech. rep. 436. IGPM, RWTH Aachen, 2015. URL: https://www.igpm.rwth-aachen.de/Download/reports/pdf/IGPM436.pdf.

[79] T. Granlund and the GMP development team. *GNU MP – The GNU multiple precision arithmetic library.* Version 6.1.0. Free Software Foundation. 2015. URL: https://gmplib.org/gmp-man-6.1.0.pdf.

[80] C. Gräser, M. Kahnt, and R. Kornhuber. "Numerical approximation of multi-phase Penrose–Fife systems". In: *Comput. Methods Appl. Math.* 16.4 (2016), pp. 523–542. DOI: 10.1515/cmam-2016-0020.

[81] C. Gräser, R. Kornhuber, and U. Sack. "On hierarchical error estimators for time-discretized phase field models". In: *Numerical Mathematics and Advanced Applications 2009.* Ed. by G. Kreiss, P. Lötstedt, A. Malqvist, and M. Neytcheva. 2010, pp. 397–405. DOI: 10.1007/978-3-642-11795-4_42.

[82] C. Gräser and O. Sander. "The dune-subgrid Module and Some Applications". In: *Computing* 8.4 (2009), pp. 269–290.

[83] C. Gräser and O. Sander. "Truncated Nonsmooth Newton Multigrid Methods for Block-Separable Minimization Problems". In: *IMA Journal of Numerical Analysis* 39.1 (2018), pp. 454–481.

[84] A. Greenbaum. *Iterative Methods for Solving Linear Systems.* SIAM, 1997.

[85] A. Griewank and A. Walther. *Evaluating derivatives: principles and techniques of algorithmic differentiation.* 2nd edition. SIAM, 2008.

[86] S. Gross and A. Reusken. *Numerical Methods for Two-phase Incompressible Flows.* Vol. 40. Series in Computational Mathematics. Springer, 2011.

[87] G. Guennebaud, B. Jacob, et al. *Eigen v3.* 2010. URL: http://eigen.tuxfamily.org.

[88] W. Hackbusch. *Multi-Grid Methods and Applications.* Springer, 1985. DOI: 10.1007/978-3-662-02427-0.

[89] W. Hackbusch and A. Reusken. "Analysis of a damped nonlinear multilevel method". In: *Numerische Mathematik* 55.2 (1989), pp. 225–246. DOI: 10.1007/BF01406516.

[90] A. Hatcher. *Algebraic Topology.* Cambridge University Press, 2002.

[91] J. Heinonen, T. Kilpeläinen, and O. Martio. *Nonlinear potential theory of degenerate elliptic equations.* Dover Publications, 2006.

[92] J. Hennessy and D. Patterson. *Computer Architecture – A Quantitative Approach.* 5th edition. Morgan Kaufmann, 2011.

[93] M. A. Heroux, R. A. Bartlett, V. E. Howle, R. J. Hoekstra, J. J. Hu, T. G. Kolda, R. B. Lehoucq, K. R. Long, R. P. Pawlowski, E. T. Phipps, A. G. Salinger, H. K. Thornquist, R. S. Tuminaro, J. M. Willenbring, A. Williams, and K. S. Stanley. "An overview of the Trilinos project". In: *ACM Trans. Math. Softw.* 31.3 (2005), pp. 397–423. DOI: 10.1145/1089014.1089021.

[94] J. S. Hesthaven and T. Warburton. *Nodal Discontinuous Galerkin Methods.* Springer, 2008.

[95] W. Huang and R. D. Russell. *Adaptive Moving Mesh Methods.* Springer, 2011.

[96] J. Hubička. *Devirtualization in C++.* online blog. (at least) seven parts, last checked on Jul. 21. 2020. 2014. URL: http://hubicka.blogspot.de/2014/01/devirtualization-in-c-part-1.html.

[97] International Organization for Standardization. *C++ extensions for concepts.* ISO/IEC. 2015. URL: https://www.iso.org/standard/64031.html.

[98] International Organization for Standardization. *ISO/IEC 14882:2011 Programming Language C++.* 2011.

[99] International Organization for Standardization. *ISO/IEC 14882:2017 Programming Language C++.* 2017.

[100] O. Ippisch and M. Blatt. "Scalability of $\mu\varphi$ and the Parallel Algebraic Multigrid Solver in DUNE-ISTL". In: *Hierarchical Methods for Dynamics in Complex Molecular Systems.* Vol. 10. Forschungszentrum Jülich GmbH, 2012, pp. 527–532.

[101] V. John. *Finite Element Methods for Incompressible Flow Problems.* Springer, 2016.

[102] S. Josefsson. *The Base16, Base32, and Base64 Data Encodings.* RFC 4648. Network Working Group, 2006. URL: https://tools.ietf.org/html/rfc4648.

[103] N. M. Josuttis. *The C++ Standard Library – A Tutorial and Reference.* 2nd edition. Addison Wesley Longman, 2012.

[104] JVAPen. *Small Object Optimization.* last accessed 11-March-2019. URL: https://riptutorial.com/cplusplus/example/31654/small-object-optimization.

[105] G. Karypis and K. Schloegel. *PARMETIS – Parallel Graph Partitioning and Sparse Matrix Ordering Library, Version 4.0.* University of Minnesota, Department of Computer Science and Engineering. Minneapolis, MN 55455, Mar. 2013.

[106] R. C. Kirby, A. Logg, M. E. Rognes, and A. R. Terrel. "Common and unusual finite elements". In: *Automated Solution of Differential Equations by the Finite Element Method.* Springer, 2012. Chap. 3, pp. 95–119.

[107] Kitware. *VTK File Formats (for VTK Version 4.2).* 2003. URL: www.vtk.org/img/file-formats.pdf.

[108] R. Klöfkorn and M. Nolte. "Performance Pitfalls in the Dune Grid Interface". In: *Advances in DUNE.* Ed. by A. Dedner, B. Flemisch, and R. Klöfkorn. Springer, 2012, pp. 45–58.

[109] T. Koch, D. Gläser, K. Weishaupt, S. Ackermann, M. Beck, B. Becker, S. Burbulla, H. Class, E. Coltman, S. Emmert, T. Fetzer, C. Grüninger, K. Heck, J. Hommel, T. Kurz, M. Lipp, F. Mohammadi, S. Scherrer, M. Schneider, G. Seitz, L. Stadler, M. Utz, F. Weinhardt, and B. Flemisch. "DuMu^x 3 – an open-source simulator for solving flow and transport problems in porous media with a focus on model coupling". In: *Computers and Mathematics with Applications* (2020). DOI: 10.1016/j.camwa.2020.02.012.

[110] A. Koenig and B. E. Moo. "Templates and Duck Typing". In: *Dr. Dobb's* (2005). online. URL: www.drdobbs.com/templates-and-duck-typing/184401971.

[111] R. Kornhuber and R. Krause. "Robust multigrid methods for vector-valued Allen–Cahn equations with logarithmic free energy". In: *Comp. Vis. Sci* 9.2 (2006), pp. 103–116.

[112] I. Kossaczký. "A recursive approach to local mesh refinement in two and three dimensions". In: *Journal of Computational and Applied Mathematics* 55.3 (1994), pp. 275–288.

[113] T. Kröger and T. Preusser. "Stability of the 8-tetrahedra shortest-interior-edge partitioning method". In: *Numerische Mathematik* 109 (2008), pp. 435–457. DOI: 10.1007/s00211-008-0148-8.

[114] M. Kronbichler and K. Kormann. "A generic interface for parallel cell-based finite element operator application". In: *Computers & Fluids* 63.Supplement C (2012), pp. 135–147. DOI: 10.1016/j.compfluid.2012.04.012.

[115] R. J. LeVeque. *Finite Volume Methods for Hyperbolic Problems*. Cambridge University Press, 2002.

[116] X. S. Li. "An Overview of SuperLU: Algorithms, Implementation, and User Interface". In: *Transactions on Mathematical Software (TOMS)* 31.3 (Sept. 2005), pp. 302–325.

[117] K.-A. Lie, *An Introduction to Reservoir Simulation Using MATLAB/GNU Octave: User Guide for the MATLAB Reservoir Simulation Toolbox (MRST)*. Cambridge University Press, 2019. DOI: 10.1017/9781108591416.

[118] P. Lindqvist. *Notes on the p-Laplace equation*. Report. University of Jyväskylä, Department of Mathematics and Statistics, 2006. URL: http://urn.fi/URN:ISBN:951-39-2586-2.

[119] K. Lipnikov, G. Manzini, and M. Shashkov. "Mimetic finite difference method". In: *Journal of Computational Physics* 257, Part B (2014), pp. 1163–1227. DOI: 10.1016/j.jcp.2013.07.031.

[120] A. Logg, K.-A. Mardal, and G. N. Wells, eds. *Automated Solution of Differential Equations by the Finite Element Method*. Springer, 2012. DOI: 10.1007/978-3-642-23099-8.

[121] F. Luporini, D. A. Ham, and P. H. J. Kelly. "An Algorithm for the Optimization of Finite Element Integration Loops". In: *ACM Trans. Math. Softw. (TOMS)* 44.1 (2017). DOI: 10.1145/3054944.

[122]   N. Moës, J. Dolbow, and T. Belytschko. "A finite element method for crack growth without remeshing". In: *Int. J. Numer. Methods Eng.* 46.1 (1999), pp. 131–150.

[123]   S. Müthing. "A Flexible Framework for Multi Physics and Multi Domain PDE Simulations". PhD thesis. Universität Stuttgart, 2015.

[124]   S. Müthing and P. Bastian. "Dune-MultidomainGrid: A Metagrid Approach to Subdomain Modeling". In: *Advances in DUNE*. Ed. by A. Dedner, B. Flemisch, and R. Klöfkorn. Springer, 2012, pp. 59–73.

[125]   T. D. Ngo, A. Fourno, and B. Noetinger. "Modeling of transport processes through large-scale discrete fracture networks using conforming meshes and open-source software". In: *Journal of Hydrology* 554 (2017), pp. 66–79. DOI: 10.1016/j.jhydrol.2017.08.052.

[126]   E. Niebler. *Concept checking in C++11*. online blog, http://ericniebler.com/2013/11/23/concept-checking-in-c11. last checked on Dec. 8. 2015. 2013.

[127]   J. Nitsche. "Über ein Variationsprinzip zur Lösung von Dirichlet-Problemen bei Verwendung von Teilräumen, die keinen Randbedingungen unterworfen sind". In: *Abh. Math. Sem. Univ. Hamburg* 36 (1971), pp. 9–15.

[128]   R. H. Nochetto, K. G. Siebert, and A. Veeser. "Theory of adaptive finite element methods: An introduction". In: *Multiscale, Nonlinear and Adaptive Approximation*. Ed. by R. DeVore and A. Kunoth. Springer, 2009, pp. 409–542. DOI: 10.1007/978-3-642-03413-8_12.

[129]   K. B. Ølgaard and G. N. Wells. "Optimizations for quadrature representations of finite element tensors through automated code generation". In: *ACM Trans. Math. Softw. (TOMS)* 37.1 (2010). DOI: 10.1145/1644001.1644009.

[130]   J. M. Ortega and W. C. Rheinboldt. *Iterative Solution of Nonlinear Equations in Several Variables*. Classics in Applied Mathematics. SIAM, 1970.

[131]   F. Pellegrini. *PT-Scotch and libPTScotch 6.0 User's Guide*. Tech. rep. Université Bordeaux 1 & LaBRI, UMR CNRS 5800, Bacchus team, INRIA Bordeaux Sud-Ouest, 2012.

[132]   F. Pellegrini. *Scotch and libScotch 6.0 User's Guide*. Tech. rep. Université Bordeaux 1 & LaBRI, UMR CNRS 5800, Bacchus team, INRIA Bordeaux Sud-Ouest, 2012.

[133]   S. Pissanetzky. *Sparse Matrix Technology*. Academic Press, 1984.

[134]   T. Plewa, T. Linde, and V. G. Weirs, eds. *Adaptive Mesh Refinement – Theory and Applications*. Vol. 41. Lecture Notes in Computational Science and Engineering. Springer, 2004.

[135]   D. Ponting. "Corner point geometry in reservoir simulation". In: *Proceedings of the 1st European Conference on Mathematics of Oil Recovery*. Ed. by P. King. Clarendon Press, 1989, pp. 45–65. DOI: 10.3997/2214-4609.201411305.

[136]   R. Rannacher and S. Turek. "Simple Nonconforming Quadrilateral Stokes Element". In: *Numerical Methods for Partial Differential Equations* 8 (1992), pp. 97–111.

[137]   M. C. Rivara. "Mesh Refinement Processes Based on the Generalized Bisection of Simplices". In: *SIAM J. Numer. Anal.* 21.3 (1984), pp. 604–613.

[138]   Y. Saad. *Iterative methods for sparse linear systems.* 2nd edition. SIAM, 2003.

[139]   Y. Saad and M. H. Schultz. "GMRES: A generalized minimal residual algorithm for solving nonsymmetric linear systems". In: *SIAM Journal on Scientific and Statistical Computing* 7 (1986), pp. 856–869.

[140]   O. Sander, T. Koch, N. Schröder, and B. Flemisch. "The Dune FoamGrid implementation for surface and network grids". In: *Archive of Numerical Software* 5.1 (2017), pp. 217–244. DOI: 10.11588/ans.2017.1.28490.

[141]   A. Schmidt and K. G. Siebert. *Design of Adaptive Finite Element Software – The Finite Element Toolbox ALBERTA.* Vol. 42. Lecture Notes in Computer Science and Engineering. Springer, 2005.

[142]   R. Schreiber and H. B. Keller. "Driven cavity flows by efficient numerical techniques". In: *Journal of Computational Physics* 49.2 (1983), pp. 310–333.

[143]   W. Schroeder, K. Martin, and B. Lorensen. *The Visualization Toolkit.* Kitware, 2006.

[144]   B. Schupp. "Entwicklung eines effizienten Verfahrens zur Simulation kompressibler Strömungen in 3D auf Parallelrechnern". PhD thesis. Albert-Ludwigs-Universität Freiburg, Mathematische Fakultät, 1999.

[145]   C. Schwab. *p and hp-Finite Element Methods.* Oxford Science Publications, 1998.

[146]   J. R. Shewchuk. *An Introduction to the Conjugate Gradient Method Without the Agonizing Pain.* Tech. rep. School of Computer Science, Carnegie Mellon University, 1994.

[147]   J. R. Shewchuk. *What is a Good Linear Finite Element? Interpolation, Conditioning, Anisotropy, and Quality Measures.* Tech. rep. Department of Electrical Engineering and Computer Science, University of California at Berkeley, 2002.

[148]   B. Smith, P. Bjørstad, and W. Gropp. *Domain Decomposition – Parallel Multilevel Methods for Elliptic Partial Differential Equations.* Cambridge University Press, 1996.

[149]   A. H. Stroud. *Approximate Calculation of Multiple Integrals.* Prentice–Hall, 1971.

[150]   B. Stroustrup. *The C++ Programming Language.* 3rd edition. Addison–Wesley, 2007.

[151] A. Sutton, B. Stroustrup, and G. Dos Reis. *Concepts Lite: Constraining Templates with Predicates*. Tech. rep. N3580. JTC1/SC22/WG21 – The C++ Standards Committee, 2013. URL: http://www.open-std.org/jtc1/sc22/wg21/docs/papers/2013/n3580.pdf.

[152] R. L. Taylor. *FEAP — A Finite Element Analysis Program, Version 8.4 User Manual*. Tech. rep. University of California at Berkely, 2013. URL: http://www.ce.berkeley.edu/projects/feap/manual84.pdf.

[153] J. F. Thompson, B. K. Soni, and N. P. Weatherill, eds. *Handbook of Grid Generation*. CRC Press, 1999.

[154] A. Toselli and O. Widlund. *Domain Decomposition Methods – Algorithms and Theory*. Springer, 2004.

[155] U. Trottenberg, C. Oosterlee, and A. Schüller. *Multigrid*. Elsevier, 2001.

[156] T. Veldhuizen. "Expression Templates". In: *C++ Report* 7.5 (1995), pp. 26–31.

[157] R. Verfürth. *A Posteriori Error Estimation Techniques for Finite Element Methods*. Oxford University Press, 2013.

[158] A. Wächter and L. T. Biegler. "On the Implementation of a Primal–Dual Interior Point Filter Line Search Algorithm for Large-Scale Nonlinear Programming". In: *Mathematical Programming* 106.1 (2006), pp. 25–57.

[159] Wikipedia. *Camel case*. last accessed 9-February-2020. 2020. URL: https://en.wikipedia.org/wiki/Camel_case.

[160] Wikipedia. *Endianness*. last accessed 9-February-2020. 2020. URL: https://en.wikipedia.org/wiki/Endianness.

[161] Wikipedia. *Spaghetti code*. last accessed 9-February-2020. 2020. URL: https://en.wikipedia.org/wiki/Spaghetti_code.

[162] J. Wloka. *Partial differential equations*. Cambridge University Press, 1987.

[163] B. Wohlmuth and R. Krause. "Monotone Methods on Nonmatching Grids for Nonlinear Contact Problems". In: *SIAM Journal on Scientific Computing* 25.1 (2003), pp. 324–347.

[164] B. I. Wohlmuth. *Discretization Methods and Iterative Solvers Based on Domain Decomposition*. Vol. 17. Lecture Notes in Computer Science and Engineering. Springer, 2001.

[165] S. Zhang. "Successive Subdivisions of Tetrahedra and Multigrid Methods on Tetrahedral Meshes". In: *Houston Journal of Mathematics* 21.3 (1995), pp. 541–556.

[166] M. Zlámal. "Curved Elements in the Finite Element Method. I". In: *SIAM Journal on Numerical Analysis* 10.1 (1973), pp. 229–240.

Index

© Springer Nature Switzerland AG 2020
O. Sander, *DUNE — The Distributed and Unified Numerics Environment*, Lecture Notes
in Computational Science and Engineering 140, https://doi.org/10.1007/978-3-030-59702-3

Editorial Policy

1. Volumes in the following three categories will be published in LNCSE:

i)   Research monographs
ii) Tutorials
iii) Conference proceedings

Those considering a book which might be suitable for the series are strongly advised to contact the publisher or the series editors at an early stage.

2. Categories i) and ii). Tutorials are lecture notes typically arising via summer schools or similar events, which are used to teach graduate students. These categories will be emphasized by Lecture Notes in Computational Science and Engineering. **Submissions by interdisciplinary teams of authors are encouraged.** The goal is to report new developments – quickly, informally, and in a way that will make them accessible to non-specialists. In the evaluation of submissions timeliness of the work is an important criterion. Texts should be well-rounded, well-written and reasonably self-contained. In most cases the work will contain results of others as well as those of the author(s). In each case the author(s) should provide sufficient motivation, examples, and applications. In this respect, Ph.D. theses will usually be deemed unsuitable for the Lecture Notes series. Proposals for volumes in these categories should be submitted either to one of the series editors or to Springer-Verlag, Heidelberg, and will be refereed. A provisional judgement on the acceptability of a project can be based on partial information about the work: a detailed outline describing the contents of each chapter, the estimated length, a bibliography, and one or two sample chapters – or a first draft. A final decision whether to accept will rest on an evaluation of the completed work which should include

– at least 100 pages of text;
– a table of contents;
– an informative introduction perhaps with some historical remarks which should be accessible to readers unfamiliar with the topic treated;
– a subject index.

3. Category iii). Conference proceedings will be considered for publication provided that they are both of exceptional interest and devoted to a single topic. One (or more) expert participants will act as the scientific editor(s) of the volume. They select the papers which are suitable for inclusion and have them individually refereed as for a journal. Papers not closely related to the central topic are to be excluded. Organizers should contact the Editor for CSE at Springer at the planning stage, see *Addresses* below.

In exceptional cases some other multi-author-volumes may be considered in this category.

4. Only works in English will be considered. For evaluation purposes, manuscripts may be submitted in print or electronic form, in the latter case, preferably as pdf- or zipped ps-files. Authors are requested to use the LaTeX style files available from Springer at http:// www.springer.com/gp/authors-editors/book-authors-editors/manuscript-preparation/5636 (Click on LaTeX Template → monographs or contributed books).

For categories ii) and iii) we strongly recommend that all contributions in a volume be written in the same LaTeX version, preferably LaTeX2e. Electronic material can be included if appropriate. Please contact the publisher.

Careful preparation of the manuscripts will help keep production time short besides ensuring satisfactory appearance of the finished book in print and online.

5. The following terms and conditions hold. Categories i), ii) and iii):

Authors receive 50 free copies of their book. No royalty is paid.
Volume editors receive a total of 50 free copies of their volume to be shared with authors, but no royalties.

Authors and volume editors are entitled to a discount of 40 % on the price of Springer books purchased for their personal use, if ordering directly from Springer.

6. Springer secures the copyright for each volume.

Addresses:

Timothy J. Barth
NASA Ames Research Center
NAS Division
Moffett Field, CA 94035, USA
barth@nas.nasa.gov

Michael Griebel
Institut für Numerische Simulation
der Universität Bonn
Wegelerstr. 6
53115 Bonn, Germany
griebel@ins.uni-bonn.de

David E. Keyes
Mathematical and Computer Sciences
and Engineering
King Abdullah University of Science
and Technology
P.O. Box 55455
Jeddah 21534, Saudi Arabia
david.keyes@kaust.edu.sa

and

Department of Applied Physics
and Applied Mathematics
Columbia University
500 W. 120 th Street
New York, NY 10027, USA
kd2112@columbia.edu

Risto M. Nieminen
Department of Applied Physics
Aalto University School of Science
and Technology
00076 Aalto, Finland
risto.nieminen@aalto.fi

Dirk Roose
Department of Computer Science
Katholieke Universiteit Leuven
Celestijnenlaan 200A
3001 Leuven-Heverlee, Belgium
dirk.roose@cs.kuleuven.be

Tamar Schlick
Department of Chemistry
and Courant Institute
of Mathematical Sciences
New York University
251 Mercer Street
New York, NY 10012, USA
schlick@nyu.edu

Editor for Computational Science
and Engineering at Springer:

Martin Peters
Springer-Verlag
Mathematics Editorial IV
Tiergartenstrasse 17
69121 Heidelberg, Germany
martin.peters@springer.com

Lecture Notes
in Computational Science
and Engineering

24. T. Schlick, H.H. Gan (eds.), *Computational Methods for Macromolecules: Challenges and Applications.*

25. T.J. Barth, H. Deconinck (eds.), *Error Estimation and Adaptive Discretization Methods in Computational Fluid Dynamics.*

26. M. Griebel, M.A. Schweitzer (eds.), *Meshfree Methods for Partial Differential Equations.*

27. S. Müller, *Adaptive Multiscale Schemes for Conservation Laws.*

28. C. Carstensen, S. Funken, W. Hackbusch, R.H.W. Hoppe, P. Monk (eds.), *Computational Electromagnetics.*

29. M.A. Schweitzer, *A Parallel Multilevel Partition of Unity Method for Elliptic Partial Differential Equations.*

30. T. Biegler, O. Ghattas, M. Heinkenschloss, B. van Bloemen Waanders (eds.), *Large-Scale PDE-Constrained Optimization.*

31. M. Ainsworth, P. Davies, D. Duncan, P. Martin, B. Rynne (eds.), *Topics in Computational Wave Propagation*. Direct and Inverse Problems.

32. H. Emmerich, B. Nestler, M. Schreckenberg (eds.), *Interface and Transport Dynamics*. Computational Modelling.

33. H.P. Langtangen, A. Tveito (eds.), *Advanced Topics in Computational Partial Differential Equations*. Numerical Methods and Diffpack Programming.

34. V. John, *Large Eddy Simulation of Turbulent Incompressible Flows*. Analytical and Numerical Results for a Class of LES Models.

35. E. Bänsch (ed.), *Challenges in Scientific Computing - CISC 2002.*

36. B.N. Khoromskij, G. Wittum, *Numerical Solution of Elliptic Differential Equations by Reduction to the Interface.*

37. A. Iske, *Multiresolution Methods in Scattered Data Modelling.*

38. S.-I. Niculescu, K. Gu (eds.), *Advances in Time-Delay Systems.*

39. S. Attinger, P. Koumoutsakos (eds.), *Multiscale Modelling and Simulation.*

40. R. Kornhuber, R. Hoppe, J. Périaux, O. Pironneau, O. Wildlund, J. Xu (eds.), *Domain Decomposition Methods in Science and Engineering.*

41. T. Plewa, T. Linde, V.G. Weirs (eds.), *Adaptive Mesh Refinement – Theory and Applications.*

42. A. Schmidt, K.G. Siebert, *Design of Adaptive Finite Element Software*. The Finite Element Toolbox ALBERTA.

43. M. Griebel, M.A. Schweitzer (eds.), *Meshfree Methods for Partial Differential Equations II.*

44. B. Engquist, P. Lötstedt, O. Runborg (eds.), *Multiscale Methods in Science and Engineering.*

45. P. Benner, V. Mehrmann, D.C. Sorensen (eds.), *Dimension Reduction of Large-Scale Systems.*

46. D. Kressner, *Numerical Methods for General and Structured Eigenvalue Problems.*

47. A. Boriçi, A. Frommer, B. Joó, A. Kennedy, B. Pendleton (eds.), *QCD and Numerical Analysis III.*

48. F. Graziani (ed.), *Computational Methods in Transport.*

49. B. Leimkuhler, C. Chipot, R. Elber, A. Laaksonen, A. Mark, T. Schlick, C. Schütte, R. Skeel (eds.), *New Algorithms for Macromolecular Simulation.*

50. M. Bücker, G. Corliss, P. Hovland, U. Naumann, B. Norris (eds.), *Automatic Differentiation: Applications, Theory, and Implementations.*

51. A.M. Bruaset, A. Tveito (eds.), *Numerical Solution of Partial Differential Equations on Parallel Computers.*

52. K.H. Hoffmann, A. Meyer (eds.), *Parallel Algorithms and Cluster Computing.*

53. H.-J. Bungartz, M. Schäfer (eds.), *Fluid-Structure Interaction.*

54. J. Behrens, *Adaptive Atmospheric Modeling.*

55. O. Widlund, D. Keyes (eds.), *Domain Decomposition Methods in Science and Engineering XVI.*

56. S. Kassinos, C. Langer, G. Iaccarino, P. Moin (eds.), *Complex Effects in Large Eddy Simulations.*

57. M. Griebel, M.A Schweitzer (eds.), *Meshfree Methods for Partial Differential Equations III.*

58. A.N. Gorban, B. Kégl, D.C. Wunsch, A. Zinovyev (eds.), *Principal Manifolds for Data Visualization and Dimension Reduction.*

59. H. Ammari (ed.), *Modeling and Computations in Electromagnetics: A Volume Dedicated to Jean-Claude Nédélec.*

60. U. Langer, M. Discacciati, D. Keyes, O. Widlund, W. Zulehner (eds.), *Domain Decomposition Methods in Science and Engineering XVII.*

61. T. Mathew, *Domain Decomposition Methods for the Numerical Solution of Partial Differential Equations.*

62. F. Graziani (ed.), *Computational Methods in Transport: Verification and Validation.*

63. M. Bebendorf, *Hierarchical Matrices.* A Means to Efficiently Solve Elliptic Boundary Value Problems.

64. C.H. Bischof, H.M. Bücker, P. Hovland, U. Naumann, J. Utke (eds.), *Advances in Automatic Differentiation.*

65. M. Griebel, M.A. Schweitzer (eds.), *Meshfree Methods for Partial Differential Equations IV.*

66. B. Engquist, P. Lötstedt, O. Runborg (eds.), *Multiscale Modeling and Simulation in Science.*

67. I.H. Tuncer, Ü. Gülcat, D.R. Emerson, K. Matsuno (eds.), *Parallel Computational Fluid Dynamics 2007.*

68. S. Yip, T. Diaz de la Rubia (eds.), *Scientific Modeling and Simulations.*

69. A. Hegarty, N. Kopteva, E. O'Riordan, M. Stynes (eds.), *BAIL* 2008 – *Boundary and Interior Layers.*

70. M. Bercovier, M.J. Gander, R. Kornhuber, O. Widlund (eds.), *Domain Decomposition Methods in Science and Engineering XVIII.*

71. B. Koren, C. Vuik (eds.), *Advanced Computational Methods in Science and Engineering.*

72. M. Peters (ed.), *Computational Fluid Dynamics for Sport Simulation.*

73. H.-J. Bungartz, M. Mehl, M. Schäfer (eds.), *Fluid Structure Interaction II - Modelling, Simulation, Optimization.*

74. D. Tromeur-Dervout, G. Brenner, D.R. Emerson, J. Erhel (eds.), *Parallel Computational Fluid Dynamics 2008.*

75. A.N. Gorban, D. Roose (eds.), *Coping with Complexity: Model Reduction and Data Analysis.*

76. J.S. Hesthaven, E.M. Rønquist (eds.), *Spectral and High Order Methods for Partial Differential Equations.*

77. M. Holtz, *Sparse Grid Quadrature in High Dimensions with Applications in Finance and Insurance.*

78. Y. Huang, R. Kornhuber, O.Widlund, J. Xu (eds.), *Domain Decomposition Methods in Science and Engineering XIX.*

79. M. Griebel, M.A. Schweitzer (eds.), *Meshfree Methods for Partial Differential Equations V.*

80. P.H. Lauritzen, C. Jablonowski, M.A. Taylor, R.D. Nair (eds.), *Numerical Techniques for Global Atmospheric Models.*

81. C. Clavero, J.L. Gracia, F.J. Lisbona (eds.), *BAIL 2010 – Boundary and Interior Layers, Computational and Asymptotic Methods.*

82. B. Engquist, O. Runborg, Y.R. Tsai (eds.), *Numerical Analysis and Multiscale Computations.*

83. I.G. Graham, T.Y. Hou, O. Lakkis, R. Scheichl (eds.), *Numerical Analysis of Multiscale Problems.*

84. A. Logg, K.-A. Mardal, G. Wells (eds.), *Automated Solution of Differential Equations by the Finite Element Method.*

85. J. Blowey, M. Jensen (eds.), *Frontiers in Numerical Analysis - Durham 2010.*

86. O. Kolditz, U.-J. Gorke, H. Shao, W. Wang (eds.), *Thermo-Hydro-Mechanical-Chemical Processes in Fractured Porous Media - Benchmarks and Examples.*

87. S. Forth, P. Hovland, E. Phipps, J. Utke, A. Walther (eds.), *Recent Advances in Algorithmic Differentiation.*

88. J. Garcke, M. Griebel (eds.), *Sparse Grids and Applications.*

89. M. Griebel, M.A. Schweitzer (eds.), *Meshfree Methods for Partial Differential Equations VI.*

90. C. Pechstein, *Finite and Boundary Element Tearing and Interconnecting Solvers for Multiscale Problems.*

91. R. Bank, M. Holst, O. Widlund, J. Xu (eds.), *Domain Decomposition Methods in Science and Engineering XX.*

92. H. Bijl, D. Lucor, S. Mishra, C. Schwab (eds.), *Uncertainty Quantification in Computational Fluid Dynamics.*

93. M. Bader, H.-J. Bungartz, T. Weinzierl (eds.), *Advanced Computing.*

94. M. Ehrhardt, T. Koprucki (eds.), *Advanced Mathematical Models and Numerical Techniques for Multi-Band Effective Mass Approximations.*

95. M. Azaïez, H. El Fekih, J.S. Hesthaven (eds.), *Spectral and High Order Methods for Partial Differential Equations ICOSAHOM 2012.*

96. F. Graziani, M.P. Desjarlais, R. Redmer, S.B. Trickey (eds.), *Frontiers and Challenges in Warm Dense Matter.*

97. J. Garcke, D. Pflüger (eds.), *Sparse Grids and Applications – Munich 2012.*

98. J. Erhel, M. Gander, L. Halpern, G. Pichot, T. Sassi, O. Widlund (eds.), *Domain Decomposition Methods in Science and Engineering XXI.*

99. R. Abgrall, H. Beaugendre, P.M. Congedo, C. Dobrzynski, V. Perrier, M. Ricchiuto (eds.), *High Order Nonlinear Numerical Methods for Evolutionary PDEs - HONOM 2013.*

100. M. Griebel, M.A. Schweitzer (eds.), *Meshfree Methods for Partial Differential Equations VII.*

101. R. Hoppe (ed.), *Optimization with PDE Constraints - OPTPDE 2014.*

102. S. Dahlke, W. Dahmen, M. Griebel, W. Hackbusch, K. Ritter, R. Schneider, C. Schwab, H. Yserentant (eds.), *Extraction of Quantifiable Information from Complex Systems.*

103. A. Abdulle, S. Deparis, D. Kressner, F. Nobile, M. Picasso (eds.), *Numerical Mathematics and Advanced Applications - ENUMATH 2013.*

104. T. Dickopf, M.J. Gander, L. Halpern, R. Krause, L.F. Pavarino (eds.), *Domain Decomposition Methods in Science and Engineering XXII.*

105. M. Mehl, M. Bischoff, M. Schäfer (eds.), *Recent Trends in Computational Engineering - CE2014.* Optimization, Uncertainty, Parallel Algorithms, Coupled and Complex Problems.

106. R.M. Kirby, M. Berzins, J.S. Hesthaven (eds.), *Spectral and High Order Methods for Partial Differential Equations - ICOSAHOM'14.*

107. B. Jüttler, B. Simeon (eds.), *Isogeometric Analysis and Applications 2014.*

108. P. Knobloch (ed.), *Boundary and Interior Layers, Computational and Asymptotic Methods – BAIL 2014.*

109. J. Garcke, D. Pflüger (eds.), *Sparse Grids and Applications – Stuttgart 2014.*

110. H. P. Langtangen, *Finite Difference Computing with Exponential Decay Models.*

111. A. Tveito, G.T. Lines, *Computing Characterizations of Drugs for Ion Channels and Receptors Using Markov Models.*

112. B. Karazösen, M. Manguoğlu, M. Tezer-Sezgin, S. Göktepe, Ö. Uğur (eds.), *Numerical Mathematics and Advanced Applications - ENUMATH 2015.*

113. H.-J. Bungartz, P. Neumann, W.E. Nagel (eds.), *Software for Exascale Computing - SPPEXA 2013-2015.*

114. G.R. Barrenechea, F. Brezzi, A. Cangiani, E.H. Georgoulis (eds.), *Building Bridges: Connections and Challenges in Modern Approaches to Numerical Partial Differential Equations.*

115. M. Griebel, M.A. Schweitzer (eds.), *Meshfree Methods for Partial Differential Equations VIII.*

116. C.-O. Lee, X.-C. Cai, D.E. Keyes, H.H. Kim, A. Klawonn, E.-J. Park, O.B. Widlund (eds.), *Domain Decomposition Methods in Science and Engineering XXIII.*

117. T. Sakurai, S.-L. Zhang, T. Imamura, Y. Yamamoto, Y. Kuramashi, T. Hoshi (eds.), *Eigenvalue Problems: Algorithms, Software and Applications in Petascale Computing.* EPASA 2015, Tsukuba, Japan, September 2015.

118. T. Richter (ed.), *Fluid-structure Interactions.* Models, Analysis and Finite Elements.

119. M.L. Bittencourt, N.A. Dumont, J.S. Hesthaven (eds.), *Spectral and High Order Methods for Partial Differential Equations ICOSAHOM 2016.* Selected Papers from the ICOSAHOM Conference, June 27-July 1, 2016, Rio de Janeiro, Brazil.

120. Z. Huang, M. Stynes, Z. Zhang (eds.), *Boundary and Interior Layers, Computational and Asymptotic Methods BAIL 2016.*

121. S.P.A. Bordas, E.N. Burman, M.G. Larson, M.A. Olshanskii (eds.), *Geometrically Unfitted Finite Element Methods and Applications.* Proceedings of the UCL Workshop 2016.

122. A. Gerisch, R. Penta, J. Lang (eds.), *Multiscale Models in Mechano and Tumor Biology*. Modeling, Homogenization, and Applications.

123. J. Garcke, D. Pflüger, C.G. Webster, G. Zhang (eds.), *Sparse Grids and Applications - Miami 2016*.

124. M. Schäfer, M. Behr, M. Mehl, B. Wohlmuth (eds.), *Recent Advances in Computational Engineering*. Proceedings of the 4th International Conference on Computational Engineering (ICCE 2017) in Darmstadt.

125. P.E. Bjørstad, S.C. Brenner, L. Halpern, R. Kornhuber, H.H. Kim, T. Rahman, O.B. Widlund (eds.), *Domain Decomposition Methods in Science and Engineering XXIV*. 24th International Conference on Domain Decomposition Methods, Svalbard, Norway, February 6–10, 2017.

126. F.A. Radu, K. Kumar, I. Berre, J.M. Nordbotten, I.S. Pop (eds.), *Numerical Mathematics and Advanced Applications – ENUMATH 2017*.

127. X. Roca, A. Loseille (eds.), *27th International Meshing Roundtable*.

128. Th. Apel, U. Langer, A. Meyer, O. Steinbach (eds.), *Advanced Finite Element Methods with Applications*. Selected Papers from the 30th Chemnitz Finite Element Symposium 2017.

129. M. Griebel, M.A. Schweitzer (eds.), *Meshfree Methods for Partial Differential Equations IX*.

130. S. Weißer, BEM-based Finite Element *Approaches on Polytopal Meshes*.

131. V.A. Garanzha, L. Kamenski, H. Si (eds.), *Numerical Geometry, Grid Generation and Scientific Computing*. Proceedings of the 9th International Conference, NUMGRID2018/Voronoi 150, Celebrating the 150th Anniversary of G. F. Voronoi, Moscow, Russia, December 2018.

132. H. van Brummelen, A. Corsini, S. Perotto, G. Rozza (eds.), *Numerical Methods for Flows*.

133. H. van Brummelen, C. Vuik, M. Möller, C. Verhoosel, B. Simeon, B. Jüttler (eds.), *Isogeometric Analysis and Applications 2018*.

134. S.J. Sherwin, D. Moxey, J. Peiro, P.E. Vincent, C. Schwab (eds.), *Spectral and High Order Methods for Partial Differential Equations ICOSAHOM 2018*.

135. G.R. Barrenechea, J. Mackenzie (eds.), *Boundary and Interior Layers, Computational and Asymptotic Methods BAIL 2018*.

136. H.-J. Bungartz, S. Reiz, B. Uekermann, P. Neumann, W.E. Nagel (eds.), *Software for Exascale Computing - SPPEXA 2016–2019*.

137. M. D'Elia, M. Gunzburger, G. Rozza (eds.), *Quantification of Uncertainty: Improving Efficiency and Technology*.

138. R. Haynes, S. MacLachlan, X.-C. Cai, L. Halpern, H.H. Kim, A. Klawonn, O. Widlund (eds.), *Domain Decomposition Methods in Science and Engineering XXV*.

139. F.J. Vermolen, C. Vuik (eds.), *Numerical Mathematics and Advanced Applications ENUMATH 2019*.

For further information on these books please have a look at our mathematics catalogue at the following URL: www.springer.com/series/3527

Monographs in Computational Science and Engineering

For further information on this book, please have a look at our mathematics catalogue at the following URL: www.springer.com/series/7417

Texts in Computational Science and Engineering

19. J. A. Trangenstein, *Scientific Computing*. Volume II - Eigenvalues and Optimization.

20. J. A. Trangenstein, *Scientific Computing*. Volume III - Approximation and Integration.

21. H. P. Langtangen, K.-A. Mardal, *Introduction to Numerical Methods for Variational Problems*.

22. T. Lyche, *Numerical Linear Algebra and Matrix Factorizations*.

For further information on these books please have a look at our mathematics catalogue at the following URL: www.springer.com/series/5151

Printed in the United States
by Baker & Taylor Publisher Services